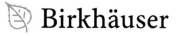

Matthias Dehmer
Editor

Structural Analysis of Complex Networks

Editor
Ao. Prof. Dr. habil. Matthias Dehmer
Institute for Bioinformatics and Translational Research
The Health and Life Sciences University
UMIT-Private Universität für Gesundheitswissenschaften
Eduard Wallnöfer-Zentrum 1
A-6060 Hall in Tirol, Austria

and

Institute of Discrete Mathematics and Geometry
Vienna University of Technology
Wiedner Hauptstrasse 8-10
A-1040 Vienna, Austria

ISBN 978-0-8176-4788-9   e-ISBN 978-0-8176-4789-6
DOI 10.1007/978-0-8176-4789-6
Springer Dordrecht Heidelberg London New York

Library of Congress Control Number: 2010938359

Mathematics Subject Classification: 05C05, 05C12, 05C75, 05C80, 05C85, 05D40, 68R10, 90B10, 92E10, 94C15

© Springer Science+Business Media, LLC 2011
All rights reserved. This work may not be translated or copied in whole or in part without the written permission of the publisher (Springer Science+Business Media, LLC, 233 Spring Street, New York, NY 10013, USA), except for brief excerpts in connection with reviews or scholarly analysis. Use in connection with any form of information storage and retrieval, electronic adaptation, computer software, or by similar or dissimilar methodology now known or hereafter developed is forbidden.
The use in this publication of trade names, trademarks, service marks, and similar terms, even if they are not identified as such, is not to be taken as an expression of opinion as to whether or not they are subject to proprietary rights.

Printed on acid-free paper

www.birkhauser-science.com

Preface

Because of the increasing complexity and growth of real-world networks, their analysis by using classical graph-theoretic methods is oftentimes a difficult procedure. Thus, there is a strong need to combine graph-theoretic methods with mathematical techniques from other scientific disciplines, such as machine learning, statistics, and information theory, for analyzing complex networks more adequately.

The book *Structural Analysis of Complex Networks* presents theoretical as well as practice-oriented results for structurally exploring complex networks. Hence, the book does not only focus on classical graph-theoretical methods, it also shows the usefulness and potential of structural graph theory as a tool for solving interdisciplinary problems. Special emphasis is given to methods and areas which can be roughly summarized as follows:

- Graph-theoretical applications in, e.g., structural biology, computational biology, mathematical chemistry, and computational linguistics;
- Graph classes;
- General structural properties of networks;
- Graph colorings;
- Graph polynomials;
- Information measures for graphs, e.g., graph entropies;
- Metrical properties of graphs;
- Partitions and decompositions;
- Quantitative graph measures.

This book is intended for an interdisciplinary audience, covering topics from artificial intelligence, computer science, computational and systems biology, cognitive science, computational linguistics, discrete mathematics, machine learning, mathematical chemistry, and statistics, and it contains nineteen chapters that have been peer-reviewed according to the standards of international journals in applied mathematics. The chapters and some of their interrelations can be briefly described as follows.

Emmert-Streib starts the volume by surveying basic structural properties of complex networks, important graph classes, and graph measures used when performing network analysis quantitatively. The latter relates to determining the structural similarity between graphs and their structural complexity using entropic measures.

v

Further concepts used to explore networks structurally are provided by the next chapters authored by *Borowiecki*, *Goddard* et al., and *Ananchuen* et al. In particular, these chapters present techniques of graph partitioning, distances in graphs, and domination in graphs, respectively. The chapter written by *Fujii* also discusses entropy measures, but for infinite directed graphs. However, these measures are obtained by using operator theory and, hence, are differently defined than the ones presented in the chapter by Emmert-Streib; those are derived based on Shannon's entropy and can be interpreted as the structural information content of a graph. Then, the chapters authored by *Matsumoto*, *Kovář*, and *Brešar* et al. investigate multifaceted problems, like exploring infinite labeled graphs to study presentations of symbolic dynamical systems, special graph decompositions, and the examination of geodetic sets in graphs, which represents an important problem using metrical properties of graphs. *Ellis-Monaghan* et al. provide two chapters in this volume on graph polynomials: The first one emphasizes the Tutte polynomial and some closely related graph polynomials. The second chapter by *Ellis-Monaghan* et al. sheds light on interpretations of concrete polynomials and on interrelations between other graph polynomials and the Tutte polynomial. The problem of reconstructing graphs by examining specific properties of polynomials, here, the zeros of Krawtchouk polynomials, is tackled in the next chapter by *Stoll*. Quantitative methods to calculate the structural similarity or distance between two graphs have already been mentioned in Emmert-Streib's chapter. In particular, classical measures based on determining isomorphic subgraphs have already been mentioned there. The chapter written by *Lauri* treats the graph similarity problem in a similar manner, namely based on the number of common vertex-deleted subgraphs, and examines aspects of the computational complexity for calculating the mentioned graph similarity measure. The idea of defining structural distances between graphs is tackled in the next chapter, written by *Benadé*. More precisely, a chromatic metric is defined, and, remarkably, by applying this metric, the maximum distance between any two graphs is at most three.

The last six chapters of this volume use graph-theoretic techniques to solve challenging problems in, e.g., applied mathematics, computer science, quantum chemistry, electrical engineering, computational linguistics, structural biology and RNA structure analysis, computational biology, and mathematical chemistry. The chapter authored by *Cioabă* gives a broad overview on results for relating important structural properties of a graph to its eigenvalues. Also, *Cioabă* surveys important applications of graph spectra in subfields of the just-mentioned disciplines. To demonstrate the great potential of novel graph classes within computational linguistics, *Mehler* introduces a graph class consisting of hierarchical graphs called Minimum Spanning Markovian Trees and shows its usefulness by outlining concrete applications within semiotic network analysis. The chapter contributed by *Scripps* et al. starts by reviewing techniques to mine general complex networks, but mainly focuses on link-based classification, which often appears as an important problem in Web mining. The next two chapters, authored by *Washietl* et al. and *Mason* et al., explore graph-based problems in structural and computational biology, respectively. In particular, *Washietl* et al. investigate RNA structures represented by

Preface vii

graphs and review graph-theoretical methods for describing and comparing such structures. A problem that is currently of considerable interest in biological network analysis is addressed by *Mason* et al. and deals with surveying methods for predicting protein function based on complex interaction networks. An area in which graph-theoretical models and techniques have been intensely applied so far is mathematical chemistry. The volume concludes by presenting a chapter about a graph class that is meaningful in mathematical chemistry: *Vukičević* presents techniques for determining the existence and enumeration of what are called perfect matchings that correspond to Kekulé structures, which are well known in mathematical chemistry.

Many colleagues, whether consciously or unconsciously, provided input, help, and support before and during the formation of this book. In particular, I would like to thank Hamid Arabnia, Alireza Ashrafi, Alexandru T. Balaban, Subhash Basak, Igor Bass, Agnieszka Bergel, David Bialy, Danail Bonchev, Stefan Borgert, Mieczysław Borowiecki, Monique Borusiak, Ulrike Brandt, Mathieu Dutour, Michael Drmota, Abdol-Hossein Esfahanian, Maria Fonoberova, Bernhard Gittenberger, Arno Homburg, Jürgen Kilian, Elena Konstantinova, Reinhard Kutzelnigg, Dmitrii Lozovanu, Alexander Mehler, Tomás Madaras, Abbe Mowshowitz, Marina Popovscaia, Fred Sobik, Stefan Shetschew, Doru Stefanescu, Thomas Stoll, Kurt Varmuza, Ilona Wesarg, Bohdan Zelinka, Dongxiao Zhu, and all authors and co-authors of this book. I apologize to any who inadvertently have not been named.

I am deeply grateful to Armin Graber from UMIT for his strong support and for providing such a stimulating working atmosphere. Many thanks to Isabella Fritz, Bernd Haas, Gerd Lorünser, Brigitte Senn-Kircher, and Klaus Weinberger for their help and fruitful discussions. Moreover, I thank Frank Emmert-Streib for the extremely fruitful collaboration and many stimulating discussions we had over several years. Frank also provided the figures used to design the front cover of this book.

In addition, I would like to thank editors Tom Grasso, Rebecca Biega, and Regina Gorenshteyn from Birkhäuser Publishing (Boston), who have always been available and helpful. Last but not least, I would like to thank my wife Jana and my family — Marion Dehmer-Sehn and Werner Dehmer — for their unfailing support and encouragement.

Finally, I hope that this book will help to extend the enthusiasm and joy that I feel for this field to others, and that it will inspire people to apply graph theory to different scientific areas for the solution of challenging and interdisciplinary problems.

Hall in Tirol, April 2010 Matthias Dehmer

Contents

Preface v

Contributors xi

1 **A Brief Introduction to Complex Networks and Their Analysis** 1
Frank Emmert-Streib

2 **Partitions of Graphs** 27
Mieczysław Borowiecki

3 **Distance in Graphs** 49
Wayne Goddard and Ortrud R. Oellermann

4 **Domination in Graphs** 73
Nawarat Ananchuen, Watcharaphong Ananchuen, and Michael D. Plummer

5 **Spectrum and Entropy for Infinite Directed Graphs** 105
Jun Ichi Fujii

6 **Application of Infinite Labeled Graphs to Symbolic Dynamical Systems** 137
Kengo Matsumoto

7 **Decompositions and Factorizations of Complete Graphs** 169
Petr Kovář

8 **Geodetic Sets in Graphs** 197
Boštjan Brešar, Matjaž Kovše, and Aleksandra Tepeh

ix

9 Graph Polynomials and Their Applications I:
The Tutte Polynomial 219
Joanna A. Ellis-Monaghan and Criel Merino

10 Graph Polynomials and Their Applications II:
Interrelations and Interpretations 257
Joanna A. Ellis-Monaghan and Criel Merino

11 Reconstruction Problems for Graphs, Krawtchouk
Polynomials, and Diophantine Equations 293
Thomas Stoll

12 Subgraphs as a Measure of Similarity 319
Josef Lauri

13 A Chromatic Metric on Graphs 335
Gerhard Benadé

14 Some Applications of Eigenvalues of Graphs 357
Sebastian M. Cioabă

15 Minimum Spanning Markovian Trees: Introducing
Context-Sensitivity into the Generation of Spanning Trees 381
Alexander Mehler

16 Link-Based Network Mining 403
Jerry Scripps, Ronald Nussbaum, Pang-Ning Tan,
and Abdol-Hossein Esfahanian

17 Graph Representations and Algorithms in Computational
Biology of RNA Secondary Structure 421
Stefan Washietl and Tanja Gesell

18 Inference of Protein Function from the Structure
of Interaction Networks 439
Oliver Mason, Mark Verwoerd, and Peter Clifford

19 Applications of Perfect Matchings in Chemistry 463
Damir Vukičević

Index 483

Contributors

Nawarat Ananchuen Department of Mathematics, Faculty of Science, Silpakorn University, Nakorn Pathom 73000, Thailand, nawarat@su.ac.th

Watcharaphong Ananchuen School of Liberal Arts, Sukhothai Thammathirat Open University, Nonthaburi 11120, Thailand, laasawat@stou.ac.th

Gerhard Benadé School of Computer Science, Statistics and Mathematics, North-West University, Potchefstroom, South Africa, gerhard.benade@nwu.ac.za

Mieczysław Borowiecki Faculty of Mathematics, Computer Science and Econometrics, University of Zielona Góra, Podgórna 50, 65-246 Zielona Góra, Poland, M.Borowiecki@wmie.uz.zgora.pl

Boštjan Brešar Faculty of Natural Sciences and Mathematics, University of Maribor, Koroška 160, 2000 Maribor, Slovenia, bostjan.bresar@uni-mb.si

Sebastian M. Cioabă Department of Mathematical Sciences, University of Delaware, 501 Ewing Hall, Newark, DE 19716-2553, USA, cioaba@math.udel.edu

Peter Clifford Hamilton Institute, NUI Maynooth, Maynooth, Ireland, p@pclifford.net

Joanna A. Ellis-Monaghan Department of Mathematics, Saint Michael's College, One Winooski Park, Colchester, VT 05439, USA
and
Department of Mathematics and Statistics, University of Vermont, 16 Colchester Avenue, Burlington, VT 05405, USA, jellis-monaghan@smcvt.edu

Frank Emmert-Streib Computational Biology and Machine Learning, Center for Cancer Research and Cell Biology, School of Medicine, Dentistry and Biomedical Sciences, Queen's University Belfast, 97 Lisburn Road, Belfast BT9 7BL, UK, v@bio-complexity.com

Abdol-Hossein Esfahanian Computer Science and Engineering Department, 3115 Engineering Building, Michigan State University, East Lansing, MI 48824-1226, USA, esfahanian@cse.msu.edu

Jun Ichi Fujii Department of Arts and Sciences (Information Science), Osaka Kyoiku University, Asahigaoka, Kashiwara, Osaka 582-8582, Japan, fujii@cc.osaka-kyoiku.ac.jp

Tanja Gesell Center for Integrative Bioinformatics Vienna, Max F. Perutz Laboratories, Dr. Bohr-Gasse 9, 1030 Vienna, Austria
and
University of Vienna, Medical University of Vienna, and University of Veterinary Medicine, Vienna, Austria, tanja.gesell@univie.ac.at

Wayne Goddard School of Computing and Department of Mathematical Sciences, Clemson University, Clemson, SC 29634-1906, USA, goddard@clemson.edu

Petr Kovář Department of Applied Mathematics, Technical University Ostrava, 17. listopadu, 708 33 Ostrava–Poruba, Czech Republic, petr.kovar@vsb.cz

Matjaž Kovše Faculty of Natural Sciences and Mathematics, University of Maribor, Koroška 160 2000 Maribor, Slovenia matjaz.kovse@uni-mb.si

Josef Lauri Department of Mathematics, University of Malta, Tal-Qroqq, Malta, josef.lauri@um.edu.mt

Oliver Mason Hamilton Institute, NUI Maynooth, Maynooth, Ireland, oliver.mason@nuim.ie

Kengo Matsumoto Department of Mathematics, Joetsu University of Education, Joetsu 943-8512, Japan, kengo@juen.ac.jp

Alexander Mehler Goethe-University Frankfurt am Main, Senckenberganlage 31, 60325 Frankfurt am Main, Germany, Mehler@em.uni-frankfurt.de

Criel Merino Instituto de Matemáticas, Universidad Nacional Autónoma de México, Area de la Investigación Científica, Circuito Exterior, C.U., Coyoacán, 04510 México D.F., México, merino@matem.unam.mx

Ronald Nussbaum Computer Science and Engineering Department, 3115 Engineering Building, Michigan State University, East Lansing, MI 48824-1226, USA, ronald@cse.msu.edu

Ortrud R. Oellermann Department of Mathematics and Statistics, University of Winnipeg, Winnipeg, MB R3B 2E9, Canada o.oellermann@uwinnipeg.ca

Michael D. Plummer Department of Mathematics, Vanderbilt University, Nashville, TN 37240, USA, michael.d.plummer@vanderbilt.edu

Jerry Scripps School of Computing and Information Systems, 1 Campus Drive, Grand Valley State University, Allendale, MI 49401, USA, scrippsj@gvsu.edu

Thomas Stoll Faculty of Mathematics, School of Computer Science, University of Waterloo, Waterloo, ON, Canada, tstoll@cs.uwaterloo.ca

Contributors

Pang-Ning Tan Computer Science and Engineering Department, 3115 Engineering Building, Michigan State University, East Lansing, MI 48824-1226, USA, ptan@cse.msu.edu

Aleksandra Tepeh University of Maribor, FEECS, Smetanova 17, 2000 Maribor, Slovenia, aleksandra.tepeh@uni-mb.si

Mark Verwoerd Hamilton Institute, NUI Maynooth, Maynooth, Ireland, mark.verwoerd@nuim.ie

Damir Vukičević Faculty of Mathematics and Natural Sciences, University of Split, Nikole Tesle 12, HR-21000 Split, Croatia, vukicevi@pmfst.hr

Stefan Washietl EMBL-European Bioinformatics Institute, Wellcome Trust Genome Campus, Hinxton, Cambridge CB10 1SD, UK
and
Institute for Theoretical Chemistry, University of Vienna, Währingerstraße 17, A-1090 Vienna, Austria, washietl@ebi.ac.uk

Chapter 1
A Brief Introduction to Complex Networks and Their Analysis

Frank Emmert-Streib

Abstract In this chapter we present a brief introduction to complex networks and their analysis. We review important network classes and properties thereof as well as general analysis methods. The focus of this chapter is on the structural analysis of networks, however, information-theoretic methods are also discussed.

Keywords Complex networks · Centrality measures · Comparative network analysis · Module detection · Information-theoretic measures

MSC2000: Primary 05C90; Secondary 65C60, 46N60, 94A17

1.1 Introduction

Discrete objects representing graphs have been studied for a long time. Among the first who studied graphs are Euler [56] and Cayley [30]. Interestingly, the origin of the term *graph* dates back to König in the 1930s [81], less than 100 years ago. The interest in graphs and their analysis is manifold. From a theoretical point of view the categorization and the analysis of properties of graphs [20, 44, 54, 70] as well as the development of graph algorithms [37, 57] are important problems that have been studied extensively. From an applied point of view it has been realized that graphs can represent physical [70], biological [51, 78, 101], or sociological objects [68, 104], e.g., a crystal or protein structure or the acquaintance network among a group of people. Recently, networks have been also employed in data analysis and machine learning [3, 51, 99]. In the following we use the terms *graph* and *network* interchangeably although they do not mean precisely the same thing. Usually,

---

F. Emmert-Streib (✉)
Computational Biology and Machine Learning, Center for Cancer Research
and Cell Biology, School of Medicine, Dentistry and Biomedical Sciences,
Queen's University Belfast, 97 Lisburn Road, Belfast BT9 7BL, UK
e-mail: v@bio-complexity.com

M. Dehmer (ed.), *Structural Analysis of Complex Networks*,
DOI 10.1007/978-0-8176-4789-6_1, © Springer Science+Business Media, LLC 2011

a *graph* refers first of all to a mathematical object regardless of its realization in, e.g., nature, whereas a *network* represents a "real-world" object rather than a pure mathematical one. Because we want to focus on applied aspects of graphs most of the time we prefer the expression *network*.

In this chapter we provide a brief introduction to complex networks and their analysis. In Sect. 1.2 we review some important network classes. In Sect. 1.3 we present some methods for the structural analysis of networks that help, e.g., to characterize them as a whole or allow us to identify specific nodes in the network with certain properties. In Sect. 1.3.5 we present important methods to analyze networks comparatively. This means that means these measures always compare two graphs with each other and provide, hence, similarity or dissimilarity measures for this comparison. Such methods are especially useful for data analysis or machine learning because they allow a combination with, e.g., clustering methods to extract regularities from the obtained, e.g., similarity values for corpora of networks. Section 1.3.6 presents a method for the identification of community or module structure of networks as important, e.g., for the analysis of communication networks. In Sect. 1.4 we discuss information-theoretic measures and this chapter finishes in Sect. 1.5 with a short summary and conclusions.

1.2 Important Network Classes

We begin this chapter by reviewing well-known network classes. In the following we mainly restrict ourselves to undirected unweighted networks, however, most of the presented networks can be generalized easily.

1.2.1 Simple Networks

A simple network consists of *regular* connections among the nodes. One of the most prominent examples therefore is the two-dimensional lattice as shown in Fig. 1.1. Here each node is connected to its nearest neighbors. Despite its simplicity, such networks have been used extensively, e.g., in physics to study phenomena like ferromagnetism with the Ising model [32]. Other examples of this class are linear chains or nonrectangular lattices as used, e.g., in the context of protein structure prediction to model protein folding [74].

1.2.2 Random Networks

Random networks have been extensively studied by Erdös and Rényi [54, 55]. A random graph with N nodes is obtained by connecting every pair of nodes with

1 A Brief Introduction to Complex Networks and Their Analysis

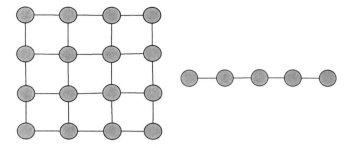

Fig. 1.1 *Left*: Regular two-dimensional lattice. *Right*: Linear, regular chain

probability p. The expected number of edges for a (undirected) network constructed this way is

$$E(N) = p\frac{N(N-1)}{2}. \tag{1.1}$$

1.2.3 Degree Distribution

The degree distribution of a node i in a random network is binomial

$$P(k_i = k) = \binom{N-1}{k} p^k (1-p)^{N-1-k}, \tag{1.2}$$

because the maximal degree of node i is $N-1$, the probability that the vertex has k links is $p^k(1-p)^{N-1-k}$ and there are $\binom{N-1}{k}$ possibilities to choose k links from $N-1$ nodes. In the limit $N \to \infty$ (1.2) becomes

$$P(k_i = k) = \frac{z^k \exp(-z)}{k!}. \tag{1.3}$$

Here $z = p(N-1)$ is the expected number of links for a node. This means that the degree distribution of a node in a random network can be approximated by the Poisson distribution for large N. For this reason random networks are also called Poisson random networks [96].

Furthermore one can show that the degree distribution of a random network (instead of just a node) also approximately follows a Poisson distribution

$$P(X_k = r) = \frac{z^r \exp(-z)}{r!}, \tag{1.4}$$

meaning that there are $X_k = r$ nodes in the network having degree k [2].

1.2.4 Clustering Coefficient

In general the clustering coefficient of a node i is defined as the fraction E_i of existing connections among its k_i nearest neighbors divided by the total number of possible connections,

$$C_i = \frac{2E_i}{k_i(k_i - 1)}. \tag{1.5}$$

This corresponds to the probability that two nearest neighbors of i are connected with each other. However, the probability in a random graph that two nodes are connected with each other is $C_i = p$. This can be approximated by

$$C_i \sim \frac{z}{N}, \tag{1.6}$$

because the mean degree of a node is $z = p(N - 1) \sim pN$.

1.2.5 Small-World Networks

Small-world networks were introduced by Watts and Strogatz [115]. They can be obtained via the following algorithm. First, arrange all nodes on a ring and connect each node with its $k/2$ nearest neighbors. Second, start with an arbitrary node i and rewire its connection to its nearest neighbor on, e.g., the left side with probability p_{rw} to any other node j in the network. If node i and j are already connected reject this selection and change nothing. Then choose the next node in the ring in a, e.g., clockwise direction and repeat this procedure. Third, after all next neighbor connections have been checked repeat this procedure for the second and all higher next neighbors successively. This algorithm guarantees that each connection occurring in the network is chosen exactly once to test for a rewiring with probability p_{rw}. The rewiring probability p_{rw} controls the disorder of the resulting topology. For $p_{rw} = 0$ the regular topology is conserved, whereas $p_{rw} = 1.0$ leads to a random network. Intermediate values $0 < p_{rw} < 1$ give a topological structure that is between regular and random.

1.2.6 Scale-Free Networks

Neither random nor small-world networks have a property frequently observed in real-world networks, namely a scale-free behavior of the degrees

$$P(k) \sim k^{-\gamma}. \tag{1.7}$$

1 A Brief Introduction to Complex Networks and Their Analysis

---

which means that there is no "top" or "bottom"
1: $t = 0$
2: Start with N_0 unconnected nodes
3: **repeat**
4:   Add one new node to the existing network consisting thus far of N_t nodes.
5:   Connect the new node to e ($\leq N_0$) nodes from the existing network. A node is chosen based on the degree distribution of the node,

$$p_i = \frac{k_i}{\sum_j k_j},$$

6:   $t = t + 1$
7:   $N_t = N_{t-1} + 1$
8: **until** $N_t = N$

Algorithm 1.1: Generation of a scale-free network (*preferential attachment*)

To explain this feature Barabasi and Albert introduced a model [4] now known as the Barabasi–Albert (BA) or *preferential attachment* model [96] that results in so-called scale-free networks which have a degree distribution following a power law [4]. The major difference between the *preferential attachment* model and the other algorithms described above to generate random or small-world networks is that the *preferential attachment* model does not assume a fixed number of nodes N and then start to rewire them with fixed probability with other nodes but N grows and is connected with a certain probability (is not constant) to other nodes depending on their degree. In Algorithm 1.1 we provide a principle algorithm to generate a scale-free network.

1.2.7 Trees

A graph G is a *tree* if it has no loops (cycles) in G. This means that a tree is an acyclic graph. Alternatively, upon removal of an edge a connected tree becomes unconnected. Trees were first studied by Cayley [30, 31] and are, in addition to their importance for graph theory, an important data structure in computer science which appears in many different algorithms. Figure 1.2 shows two trees. We want to emphasize that a tree does not represent a hierarchy, in the graph of a tree. This is in contrast to rooted trees. A rooted tree is obtained, e.g., from a tree by the identification of a so-called root node which forms the start of a hierarchy. In Fig. 1.3 we show two rooted trees obtained from the tree on the right-hand side of Fig. 1.2. The important difference is that rooted trees are ordered; they have a "top" corresponding to the root node and a "bottom" corresponding to the leaf nodes having no children. It is interesting to note that there is no restriction to the degree a node can have. Apparently, the minimal number is one because otherwise

Fig. 1.2 Trees

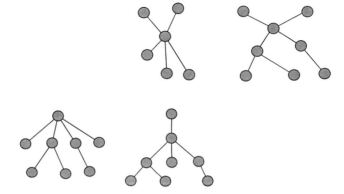

Fig. 1.3 Rooted trees obtained from the tree on the right-hand side in Fig. 1.2. *Left*: Root node is the third node from the top. *Right*: Root node is the leftmost node at the *top*

Fig. 1.4 Rooted binary trees

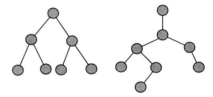

the tree would be unconnected, however, other than that it is arbitrary. This brings us to the next subclass of trees, rooted binary trees.

A special case of a rooted tree is a rooted binary tree. A node in a binary tree has at most two children. Figure 1.4 shows two examples of rooted binary trees, frequently just called binary trees. In Fig. 1.4 one can see that a node has at most two children (maximal degree of a node is three). Intermediate nodes can have only one child. This holds also for the root nodes as the right figure shows. Finally, we want to mention that a disjoint union of trees is called a forest.

1.2.8 Generalized Trees

The graph class of directed *generalized trees* was introduced in [39, 90]. Generalized trees are an important extension to trees maintaining their characteristic of having a hierarchy but in addition allowing a richer connectivity among nodes. Before we provide a formal definition we give a motivation for their introduction visualized by Fig. 1.5. On the left side in Fig. 1.5, a (normal) tree is shown. The dotted horizontal lines should remind the reader that a tree is a hierarchical graph and the dotted lines explicitly represent the hierarchy levels. These lines are included for didactic reasons only and are not actually part of the tree. On the right-hand side of Fig. 1.5 a generalized tree is shown obtained from the tree on the left side by including two additional edges, labeled E_2 and E_3. In general, edges that connect nodes on the

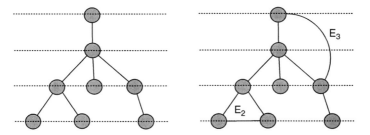

Fig. 1.5 *Left*: Tree. *Right*: Generalized tree

same hierarchical level are of type E_2 and edges that connect nodes on different hierarchical levels that are farther apart than one level are of type E_3 (formally defined below). From this, we note that every generalized tree will result in a tree after deleting *all* edges of type E_2 and E_3 from the generalized tree. Vice versa, starting from a tree and including edges of type E_2 and/or type E_3 results in a generalized tree. A formal definition is given as follows [50].

Definition 1 (Generalized Tree). A generalized tree GT_i is defined by a vertex set V, an edge set E, a level set L, and a multilevel function $\mathcal{L}_i$. The vertex and edge set define the connectivity and the level set and the multilevel function induce a hierarchy between the nodes of GT_i. The index $i \in V$ indicates the root.

The multilevel function is defined as follows.

Definition 2 (Multilevel Function). The function $\mathcal{L}_i : V \setminus \{i\} \to L$ is called a multilevel function.

The multilevel function $\mathcal{L}_i$ assigns to all nodes except i an element $l \in L$ that corresponds to the level it will be assigned. From these definitions it is immediately clear that a generalized tree is similar to a graph but additionally equipped with a level set L and a multilevel function $\mathcal{L}_i$ introducing a node grouping corresponding to the introduction of a hierarchy between nodes and sets thereof.

Definition 3 (Edge Types). A generalized tree GT_i has three edge types:

- Edges with $|\mathcal{L}_i(m) - \mathcal{L}_i(n)| = 1$ are called kernel edges (E_1).
- Edges with $|\mathcal{L}_i(m) - \mathcal{L}_i(n)| = 0$ are called cross edges (E_2).
- Edges with $|\mathcal{L}_i(m) - \mathcal{L}_i(n)| > 1$ are called up edges (E_3).

We want to remark that for directed generalized trees edge type E_3 will be split into two edge types: one for up and another for down links. Using Definition 1 a tree is characterized by $|\mathcal{L}_i(m) - \mathcal{L}_i(n)| = 1$ for all node pairs (m, n).

From the given definitions and the visualization in Fig. 1.5 it is apparent that a generalized tree is between a tree and a graph. It is hierarchical like a tree, but can contain cycles like a graph which is not hierarchical.

1.3 Structural Network Analysis

In this section, we summarize measures to characterize structural properties of networks.

1.3.1 Degree Distribution

Degree distributions [16, 88] can be calculated by

$$P(k) := \frac{|\delta_k(v)|}{N},\qquad(1.8)$$

where $|\delta_k(v)|$ denotes the number of vertices in the network G of degree k and N denotes the size of G (number of nodes). Equation (1.8) is just the proportion of vertices in G having degree k. The degree k_i of node i is the number of links connected with node i. From this, it follows that (1.8) also has the meaning that a randomly chosen node in the network has, with probability $P(k)$, k links to other nodes.

It was an interesting and important finding that many real world networks like the World Wide Web (WWW), the Internet, social networks, citation networks, or food webs [1, 16, 19, 22, 29, 42] are not Poisson distributed but follow a power law

$$P(k) \sim k^{-\gamma}, \quad \gamma > 1.\qquad(1.9)$$

1.3.2 Clustering Coefficient

The clustering coefficient C_i is a local measure defined for every node i. It is defined as the fraction of connections (E_i) between nearest neighbors of i among each other divided by the maximum number of such connections

$$C_i = \frac{2E_i}{k_i(k_i - 1)}.\qquad(1.10)$$

The clustering coefficient can be interpreted as the probability that two nearest neighbors of i are connected with each other.

1.3.3 Path-Based Measures

Graph-theoretical quantities or properties are often used to characterize special types or classes of complex networks. For example, it turned out that average path lengths

1 A Brief Introduction to Complex Networks and Their Analysis

and diameters of certain biological networks are rather small compared to the size of a network [86, 88, 98]. Related to this is the so-called "small-world" property [115] that has been observed in a number of network types, e.g., social, metabolic, and protein interaction networks in molecular biology [86, 88, 98]. We give a brief overview of path-based network measures [17, 21, 23, 67, 71, 107].

Distance Matrix

$(d(v_i, v_j))_{v_i, v_j \in V}$. $d(v_i, v_j)$ denotes the shortest distance (path) between v_i and v_j measured in the number of edges or nodes that are between start node v_i and end node v_j.

Mean or Characteristic Distance

$$\bar{d}(G) := \frac{1}{\binom{N}{2}} \sum_{v_i \neq v_j \in V} d(v_i, v_j). \tag{1.11}$$

E is the total number of edges in the network.

j-Sphere

The set
$$S_j(v_i, G) := \{v \in V \mid d(v_i, v) = j, \ j \geq 1\}, \tag{1.12}$$

is called the j-sphere of v_i regarding G. Starting from v_i, the cardinality $|S_j(v_i, G)|$ denotes the number of vertices that have a shortest distance equal to j.

Eccentricity, Diameter, and Radius

Let $G = (V, E)$ be a graph. Then,

$$\sigma(v) = \max_{u \in V} d(u, v), \tag{1.13}$$

is called the eccentricity of $v \in V$.

$$\rho(G) = \max_{v \in V} \sigma(v), \tag{1.14}$$

and

$$r(G) = \min_{v \in V} \sigma(v), \tag{1.15}$$

are called the diameter and radius of G, respectively.

Degree, Degree Statistics, and Edge Density

For undirected graphs $G = (V, E)$, $k_v = \sum_i A_{v,v_i}$ equals the number of edges which are adjacent to $v \in V$. k_v is called the degree of node v. From this, one obtains straightforwardly the following degree measures for the whole network:

$$k = k(G) := \sum_{v \in V} \frac{k_v}{N}, \tag{1.16}$$

$$\sigma_k(G) := \frac{1}{N-1} \sum_{v \in V} (k_v - k)^2, \tag{1.17}$$

and

$$\tau_k(G) := \frac{1}{N} \sum_{v \in V} |k_v - k|. \tag{1.18}$$

Equation (1.16) is the mean degree of the network, (1.17) is the variance of the degree, and (1.18) is the mean of absolute distances between k_v and k. Finally, the edge density of G is defined as

$$\beta(G) := \frac{E}{\binom{N}{2}}. \tag{1.19}$$

Further network statistics and advanced aspects can be found in, e.g., [21, 67, 71, 107].

1.3.4 Centrality Measures

Identifying *important* vertices in networks is an interesting problem that has gained much attention especially in the context of communication networks. For example, the communication among a group of humans forms a communication network. Social scientists in the late 1940s developed graph-theoretical measures to detect *important* vertices in networks. An important class of such measures is based on the centrality concept [66, 69, 114] which intuitively tries to identify nodes that are *central* to the communication within the network among all nodes. There are two fundamentally different types of centrality measures [58, 59]. The first type of measures evaluates the centrality of each node in a network and is called *point centrality* measures where the word "point" refers to a node or vertex. The second type is called *graph centrality* measures because it assigns a centrality value to the *whole* network.

Point Centrality

In the following we provide some examples of point centrality measures.

Degree Centrality

For an undirected graph $G = (V, E)$, the degree centrality of a vertex $v \in V$ is simply defined as its degree; i.e.,

$$C_D(v) = k_v. \tag{1.20}$$

For a directed network, the degree centrality can be analogously defined by using the definition of in-degree and out-degree.

Betweenness Centrality

The centrality index *betweenness* is based on shortest paths found in the network [5, 6, 18, 58, 66, 83, 88, 103, 104, 114] and is defined by

$$C_B(v_k) = \sum_{v_i, v_j \in V, v_i \neq v_j} \frac{\sigma_{v_i v_j}(v_k)}{\sigma_{v_i v_j}}. \tag{1.21}$$

Here $\sigma_{v_i v_j}$ denotes the number of shortest paths from v_i to v_j and $\sigma_{v_i v_j}(v_k)$ the number of shortest paths from v_i to v_j that include node v_k. That means

$$\frac{\sigma_{v_i v_j}(v_k)}{\sigma_{v_i v_j}}, \tag{1.22}$$

is the probability that node v_k lies on a shortest path connecting v_i with v_j. Hence, $C_B(v_k)$ evaluates the appearance of node v_k on all shortest paths in a network.

Closeness Centrality

The centrality index *closeness* tries to measure how close a node is to other nodes in the network. This is done in terms of communication distance as measured by the number of edges between two nodes if connected via the shortest path.

$$C_C(v_k) = \frac{1}{\sum_{i=1}^{N} d(v_k, v_i)}. \tag{1.23}$$

Here $d(v_k, v_i)$ is the number of edges on a shortest path between node v_k and v_i. In the case where there are multiple shortest paths connecting v_k with v_i, $d(v_k, v_i)$ is unchanged.

Graph Centrality

To evaluate the centrality of whole networks instead of single nodes in the network *graph centrality* measures have been suggested which form extensions to the three measures discussed above [59]. The basic idea is to use these individual measures to obtain an average characteristic for the whole network. It has been suggested to calculate

$$C_x = \frac{\sum_{i=1}^{N} C_x(v^m) - C_x(v_i)}{C_x^{max}}. \tag{1.24}$$

Here x stands for any of the three point centrality measures,

$$C_x(v^m) = \max_i \{C_x(v_i)\}, \tag{1.25}$$

for the maximal value of $C_x(v_i)$ that can be found in the network and C_x^{max} for the maximal value possible for a network with N nodes,

$$C_x^{max} = \max_{G \in \mathcal{G}(N)} \sum_{i=1}^{N} C_x(v^m) - C_x(v_i). \tag{1.26}$$

Degree Centrality

For the degree centrality of a network one obtains

$$C_d = \frac{\sum_{i=1}^{N} C_d(v^m) - C_d(v_i)}{N^2 - 3N + 2}. \tag{1.27}$$

Intuitively, the denominator is obtained by remembering that the maximal degree of a node is $N - 1$, hence, $C_x(v^m) - C_x(v_i) = N - 2$ because the minimal degree is one. This number multiplied by $N - 1$ (one node needs to have degree one) gives the denominator in (1.27).

Betweenness Centrality

For the betweenness centrality of a network [58] one obtains

$$C_b = \frac{\sum_{i=1}^{N} C_d(v^m) - C_d(v_i)}{N^3 - 4N^2 + 5N - 2}. \tag{1.28}$$

1 A Brief Introduction to Complex Networks and Their Analysis

Closeness Centrality

For the closeness centrality of a network one obtains

$$C_c = \frac{2N-3}{N^3 - 4N^2 + 5N - 2} \sum_{i=1}^{N} C_d(v^m) - C_d(v_i). \tag{1.29}$$

Extended Centrality Measures

In addition to the classical centrality measures described above there are many extensions. Here we present some of these.

Eigenvector Centrality

The eigenvector centrality, a point centrality measure, was introduced by Bonacich [15]. The key idea of eigenvector centrality is to express that an important vertex is connected to important neighbors. To define the eigenvector centrality measures one needs to find the eigenvector of the adjacency matrix A of graph G with the largest eigenvalue. Then eigenvector centrality is given by

$$C_e = x^m = \frac{1}{\lambda^m} A x^m. \tag{1.30}$$

Here λ^m is the largest eigenvalue and x^m the corresponding eigenvector solving the equation

$$\lambda^m x^m = A x^m. \tag{1.31}$$

Hence C_e is the principle eigenvector of A.

We want to emphasize that eigenvector centrality is a point centrality measure because each vertex in the network obtains a value corresponding to the component of C_e. In [83], advanced properties and further possibilities to compute eigenvector centrality measures are presented.

Joint Betweenness Centrality

The centrality measures discussed above, including *betweenness*, form a family of measures [58] that have been introduced with the purpose of analyzing communication networks. It is interesting to note that all point centrality measures focus solely on one vertex in the network. In the context of gene networks, which also form communication networks, the identification of a function of genes is an outstanding problem. It has been suggested to identify the unknown function of a gene by associating it with the known function of another gene. Because all measures from the centrality family are point measures they cannot be used for such studies. For this reason it has been suggested that an extension involving more than one node be called *joint betweenness* (JB) [47]. JB is a natural extension of the betweenness centrality evaluating the joint occurrence of two nodes on shortest communication paths in the network. Formally, it is defined as

$$C_{jb}(v_m, v_n) = \sum_{v_i, v_j \in V, v_i \neq v_j} \frac{\sigma_{v_i, v_j}(v_m, v_n)}{\sigma_{v_i, v_j}}. \tag{1.32}$$

Here σ_{v_i, v_j} is the number of shortest paths connecting node v_i with node v_j and $\sigma_{v_i, v_j}(v_m, v_n)$ is the number of shortest paths connecting v_i with v_j that contain the nodes v_m and v_n. Similar to other measures of the centrality family, it is sometimes more useful to use a different normalization. In fact, for the analysis conducted in [47] the following modification has been used,

$$C_{jb}(v_m, v_n) = \sum_{v_i, v_j \in V, v_i \neq v_j} \frac{\sigma_{v_i, v_j}(v_m, v_n)}{\sigma_{max}}. \tag{1.33}$$

Here σ_{max} is defined as

$$\sigma_{max} = \max_{v_i, v_j} \{\sigma_{v_i, v_j}\}. \tag{1.34}$$

1.3.5 Comparative Network Analysis

In this section measures structurally comparing whole networks are reviewed.

Measures Based on Isomorphic Relations

Classical graph similarity or distance methods deal with finding appropriate measures which are based on isomorphic and subgraph relations [75–77, 108, 109, 117]. A prominent example of such a measure is the Zelinka-distance [117], where this graph distance is based on the principle that two graphs are more similar, the bigger the common induced isomorphic subgraph is. First, Zelinka introduced this measure for unlabeled graphs with the same number of vertices. Later, Sobik [108, 109] and Kaden [75–77] generalized this measure for arbitrary graphs allowing them to have even different order. It is known that the subgraph isomorphism problem is NP-complete [113]. This implies for large graphs that these methods can be computationally demanding. A key result for exact graph matching was found by Zelinka [117].

1 A Brief Introduction to Complex Networks and Their Analysis

Theorem 1. *Let G, $\tilde{G}$ be unlabeled graphs without loops and multiple edges. Further, let $|V| = |\tilde{V}| = n$. $\overline{SUB}_m(G)$ denotes the set of induced subgraphs of order m. $G^{\star}$ denotes the isomorphism classes of such graphs in which G lies and let*

$$SUB_m(G) := \{G^{\star}| G \in \overline{SUB}_m(G)\}. \tag{1.35}$$

$SUB_m(G)$ is just the set of isomorphism classes in which the induced subgraphs of G with order m lie. Then,

$$d_Z(G, \tilde{G}) := n - SIM(G, \tilde{G}), \tag{1.36}$$

is a graph metric, where

$$SIM(G, \tilde{G}) := \max\{m|SUB_m(G) \cap SUB_m(\tilde{G}) \neq \emptyset\}, \tag{1.37}$$

holds.

Sobik [108,109] and Kaden [75–77] generalized this theorem by considering labeled graphs with a different number of vertices.

Theorem 2. *Let $G := (V, E, f_V, f_E, A_V, A_E)$ be a finite, labeled, and directed graph. A_V, A_E denote finite, nonempty vertex and edge alphabets and $f_V : V \to A_V$, $f_E : E \to A_E$ the associated vertex and edge labeling functions. Now, let G and $\tilde{G}$ be finite labeled graphs of arbitrary orders. Then,*

$$d_S(G, \tilde{G}) := \max\{|G|, |\tilde{G}|\} - SIM(G, \tilde{G}), \tag{1.38}$$

is a graph metric.

Another classical graph distance measure based on the maximum common subgraph of two graphs has been found by Bunke et al. [25,26,28].

Theorem 3. *Let G and $\tilde{G}$ be graphs and let G_{MCS} be their maximum common subgraph. Then, the distance measure*

$$d_{MCS}(G, \tilde{G}) := 1 - \frac{|V|_{MCS}}{\max(|V|, |\tilde{V}|)}, \tag{1.39}$$

is a graph metric.

Measures Based on Graph Transformations

In contrast to graph similarity measures from the exact graph matching paradigm, i.e., those based on isomorphic relations, a well-known class of graph similarity measures from inexact graph matching is based on graph transformations. The main idea behind this concept is not to match graphs exactly because one often wants to take structural errors of the underlying graphs into account. Therefore, this concept is often referred to as error-tolerant graph matching [25, 26, 28, 91]. For example, the so-called graph edit distance (GED) [24, 27, 28, 91] is a prominent example of such a graph similarity measure. The definition of GED is based on weighted transformation steps, e.g., deletions, substitutions, and insertions of vertices and edges, and, hence, the distance of two graphs is defined as the minimum cost of graph transformations that transform (map) one graph into another graph. The key result of error-tolerant graph matching, originally stated by Bunke [24], can be expressed as follows.

Theorem 4. *Let $d(G, \tilde{G})$ be the costs for determining the optimal inexact match between G and $\tilde{G}$ where an optimal inexact match is a sequence of graph transformations that transforms a graph G to $\tilde{G}$ by producing minimal edit costs. Then, it holds that $d(G, \tilde{G})$ is a graph metric.*

Regarding the computational complexity of GED, we want to remark that for unlabeled graphs there is no algorithm to compute GED efficiently [28, 91, 118]. For uniquely labeled graphs, it has been proven [43] that the computational complexity to compute GED is $O(|V|^2)$.

Measures Based on Graph Grammars

Methods to determine the similarity or distance between graphs based on graph grammars also belong to the paradigm of inexact graph matching. A classical contribution in this field has been made by Gernert [63, 64]. We want to note that the application of grammar-based measures is complex because the underlying graph grammar is quite often difficult to obtain.

Methods Based on Machine Learning Techniques

Machine learning techniques can be divided into two major categories: supervised and unsupervised learning methods [38, 72]. A newly developed supervised learning method, based on support vector machines [38], to measure the structural similarity of graphs is based on using so-called graph kernels [62, 73]. A graph kernel is a function $K : \mathcal{G} \times \mathcal{G} \longrightarrow \mathbb{R}$, for $G \in \mathcal{G}$, that maps the data implicitly into a high-dimensional feature space. For example, some graph kernels are based on the principle to determine the frequency of certain subgraph patterns of the given graph set and then to apply a proper kernel function to the obtained subgraphs. Following this principle Horváth et al. [73] proposed a graph kernel that is based on mapping graphs into cyclic graph patterns. Besides cycle-based graph kernels, so-called random-walk-based kernels [62, 80] are also often used to define graph kernels [36].

Another method to detect the structural similarity of graphs is based on dynamic programming [7]. In the following we just give an outline of the main construction steps of this similarity measuring for directed generalized trees (the generalization to networks can be found in [49]).

1. Transform the generalized trees in linear structures, called property strings
2. Derive similarity scores from the alignments of the property strings in order to measure the structural similarity of generalized trees

This means that we transform a graph similarity problem into a string similarity problem to develop an efficient graph similarity measure. More precisely, the main idea of our similarity measure is based on the derivation of property strings for each generalized tree and then to align the property strings by a dynamic programming technique [7]. We call these strings property strings because their components represent structural properties of the generalized trees. From the resulting alignment we obtain a value of the scoring function which is minimized during the alignment process. The similarity of two generalized trees is then given as the cumulation of local similarity functions which weight two alignment types: out-degree and in-degree alignments on a generalized tree level. Let $\hat{\mathcal{H}}^1$ and $\hat{\mathcal{H}}^2$ be generalized trees. Then the problem of determining the structural similarity between $\hat{\mathcal{H}}^1$ and $\hat{\mathcal{H}}^2$ is equivalent to finding the optimal alignment of the property strings. The key result is given in the following theorem by Dehmer [39].

Theorem 5. *Let $\hat{\mathcal{H}}^1, \hat{\mathcal{H}}^2$ be generalized trees, $0 \leq i \leq \rho$, $\rho := \max(h_1, h_2)$.*

$$d_1(\hat{\mathcal{H}}^1, \hat{\mathcal{H}}^2) := \frac{\sum_{i=0}^{\rho} \lambda_i \cdot \gamma^{fin}(i)}{\sum_{i=0}^{\rho} \lambda_i}, \tag{1.40}$$

$$d_2(\hat{\mathcal{H}}^1, \hat{\mathcal{H}}^2) := \frac{\sum_{i=0}^{\rho} \gamma^{fin}(i)}{\rho + 1}, \tag{1.41}$$

$$d_3(\hat{\mathcal{H}}^1, \hat{\mathcal{H}}^2) := \frac{\prod_{i=0}^{\rho} \gamma^{fin}(i)}{d_2(\hat{\mathcal{H}}^1, \hat{\mathcal{H}}^2)}, \tag{1.42}$$

is a family $(d_i(\hat{\mathcal{H}}^1, \hat{\mathcal{H}}^2))_{1 \leq i \leq 3}$ of backward similarity measures, where $\gamma^{fin}(i)$ is the weighted sum of the in- and out-degree alignments. Further it holds that $(d_i(\hat{\mathcal{H}}^1, \hat{\mathcal{H}}^2))_{1 \leq i \leq 3} \in [0, 1]$.

Here $d_i(\hat{\mathcal{H}}^1, \hat{\mathcal{H}}^2)$ are different similarity measures that allow us to emphasize different structural aspects of the networks. We just note that this method can be generalized to labeled generalized trees [39] and even networks [49].

1.3.6 Community or Module Detection

A community or a module of a network corresponds to a subgraph. Examples of modules are social groups [65, 116], pathways in molecular biological processes [46], or domains in proteins [53]. In the following we present results based on

Newman and Girvan [97] who introduced not only a measure (Q) to quantify modules in a network but also suggested an algorithm to find them.

Let A be the adjacency matrix of network G and

$$A_{ij} = \begin{cases} 1 & \text{if } i \text{ is directly connected to } j \\ 0 & \text{otherwise.} \end{cases} \tag{1.43}$$

We suppose that the set of vertices is partitioned in S communities and that vertex i belongs to community c_k if $\delta(c_k, i) = 1$, with $\delta(i, j)$ being the Kronecker delta which is one if $i = j$ and zero otherwise. The degree of a node is given by

$$k_i = \sum_j A_{ij}. \tag{1.44}$$

The *modularity* Q, introduced by Newman and Girvan [97], of a network is now defined by

$$Q = \frac{1}{2E} \sum_{i,j} \left(A_{ij} - \frac{k_i k_j}{2E} \delta(c_i, c_j) \right). \tag{1.45}$$

Q evaluates the difference between the fraction of edges found within communities minus the fraction of edges found by random connections (see also (1.49)). The modularity (1.45) can be written in an alternative form by introducing

$$e_{ij} = \frac{1}{2E} \sum_{mn} A_{mn} \delta(c_m, i) \delta(c_n, j) \tag{1.46}$$

$$a_i = \frac{1}{2E} \sum_m k_m \delta(c_m, i). \tag{1.47}$$

Equation (1.46) corresponds to the fraction of edges that connect community i with community j. For an undirected network G, $e_{ij} = e_{ji}$ holds. Equation (1.47) is the fraction of edges that is connected to nodes in community i. Utilizing (1.46) and (1.47) and

$$\delta(c_m, c_n) = \sum_i \delta(c_m, i) \delta(c_n, i), \tag{1.48}$$

gives

$$Q = \sum_i \left(e_{ii} - a_i^2 \right). \tag{1.49}$$

Because $a_i = \sum_j e_{ij}$ (e_{ii} counts the connections within community i) [97] we can also write

$$Q = \sum_i e_{ii} - \sum_{imn} e_{mi} e_{in} = \text{Tr } \mathbf{e} - ||\mathbf{e}^2||. \tag{1.50}$$

Here the norm corresponds to the sum of the matrix elements.

1 A Brief Introduction to Complex Networks and Their Analysis

For a given network G the task of finding communities or modules corresponds to finding a partitioning of the set of vertices of the network that maximizes the modularity Q. Exhaustive enumeration is prohibitive for most networks because the number of partitions N nodes can be distributed among M nonempty communities is given by the Stirling number of the second kind S_N^M [97]. For example, for $N = 100$ and $M = 3$ there are already $S_{1,000}^{20} = 8.5 \times 10^{46}$ possibilities to partition the network.

Because testing all partitions exhaustively is not possible we need to find the communities via an optimization algorithm. It is clear that there are various ways to define optimization algorithms for this task. In this chapter we present just one simple algorithm suggested by Newman and Girvan [97] that can be applied even for large networks consisting of tens of thousands of nodes. This algorithm performs a greedy optimization at each step and is shown in Algorithm 1.2. Starting from a configuration where each node forms a community the algorithm proceeds by merging at each step two communities that result in the largest change of ΔQ. Because ΔQ can also be negative this results not only in an increase of Q_{t+1}. The optimal partition is now obtained by finding the maximum of all Q_t. This also gives the level at which the dendrogram should be cut. This cut will then give the optimal partitions of the vertices of the network. The advantage of the algorithm is that it is not necessary to calculate Q_{t+1} anew at each step. It is possible to calculate ΔQ instead by

$$\Delta Q = 2(e_{mn} - a_m a_n). \tag{1.51}$$

Equation (1.51) can be obtained by the identities

$$e'_{ii} = e_{mn} + e_{nn} + e_{mn} + e_{nm} \tag{1.52}$$

$$a'_i = a_m + a_n. \tag{1.53}$$

```
 1: t = 0
 2: Start with M = N communities {c₁, ..., c_N} (each node is a community)
 3: Calculate Q
 4: Q_t = Q
 5: repeat
 6:     Calculate ΔQ(c_i, c_j) for all communities
 7:     Merge community c_m with c_n if ΔQ(c_m, c_n) = max_{i,j}{ΔQ(c_i, c_j)}
 8:     Q_{t+1} = Q_t + ΔQ
 9:     t = t + 1
10:     M = N - t
11: until M = 1
12: Q* = max_t Q_t
```

Algorithm 1.2: Agglomerative clustering of communities [97]

We use the prime ($'$) as an abbreviation for the values of e at step $t + 1$ and i symbolizes the newly formed community merging communities m and n. The update of the new $\mathbf{e}'$ matrix is completed by

$$e'_{ik} = e_{mk} + e_{nk} \tag{1.54}$$

$$e'_{ki} = e_{ik}, \tag{1.55}$$

for $k \neq i$. We want to emphasize that ΔQ can be calculated by just evaluating (1.52) and (1.53). The new complete matrix $\mathbf{e}'$ will only be calculated once after the decision to merge community m with n has been made. The overall time complexity of this algorithm is $O(N^2)$ [97] and, hence, applicable to large networks.

Finding communities or modules in complex networks has attracted much attention because of the importance of the problem for many scientific fields. For this reason the algorithm presented above is just one approach to this problem. The interested reader is referred to [33, 45, 97, 98] for more algorithms.

1.4 Information-Theoretic Methods

In many scientific areas, e.g., biology, chemistry, linguistics, and physics, it is known that the underlying system can be described or represented as an interaction network among its components [12, 89]. Such networks, e.g., protein–protein, signaling, reaction, or metabolic networks, have been intensely investigated, especially in computer science, computational biology, and computational chemistry [12, 35, 46, 61, 111]. Once a network is theoretically or experimentally inferred, existing methods from quantitative network analysis are basically applicable for investigating such networks structurally as described in the proceeding sections.

A different approach to analyzing complex networks can be obtained by combining methods from information theory and statistics. Particularly, it turns out that information-theoretic methods, e.g., entropy-based methods, are powerful tools to investigate complex networks [9, 12, 34, 84, 85, 105, 110]. Here we give just a brief summary of information-theoretic and statistical methods to analyze or compare networks:

- Classical information measures, e.g., entropy, conditional entropy, and mutual information applied to complex networks [8, 9, 11–14, 60, 79, 105, 110]
- Entropic measures for characterizing graph classes, e.g., perfect graphs [82, 106]
- Compression-based module identification [102]
- Information-theoretic measures to determine the structural information content of a network [9, 40, 41, 92–95, 100, 112]
- Complexity measures for networks based on Kolmogorov complexity [10, 11, 87]
- Information-theoretic measures of robustness of complex (gene) networks [48, 52]

1 A Brief Introduction to Complex Networks and Their Analysis

If we assume that N denotes the number of vertices of a graph G, n denotes the number of different (obtained) sets of vertices, $|N_i|$ is the number of elements in the ith set of vertices, and we set $P_i = \frac{N_i}{N}$, then Shannon's entropy formula can be stated as [11]

$$I_t(G) = N \log(N) - \sum_{i=1}^{n} N_i \log(N_i), \tag{1.56}$$

or

$$I_m(G) = -\sum_{i=1}^{n} P_i \log(P_i). \tag{1.57}$$

Equation (1.56) represents the total information and (1.57) the mean information content of G. From (1.56) and (1.57) we see that there are no free parameters because the quantities P_i are completely determined by the chosen partitioning [41]. This can be a problem if such an entropy measure should be used to analyze networks obtained, e.g., from an experiment for which expert knowledge is available regarding possible outcomes. For this reason, Dehmer [40, 41] introduces a parametric entropy measure

$$I_{f^V}(G) := -\sum_{i=1}^{N} \frac{f^V(v_i)}{\sum_{j=1}^{N} f^V(v_j)} \log\left(\frac{f^V(v_i)}{\sum_{j=1}^{N} f^V(v_j)}\right), \tag{1.58}$$

where

$$f^V(v_i) := \alpha^{c_1|S_1(v_i,G)|+c_2|S_2(v_i,G)|+\cdots+c_{\rho(G)}|S_{\rho(G)}(v_i,G)|},$$
$$c_k > 0, 1 \leq k \leq \rho(G), \alpha > 0. \tag{1.59}$$

Here, $|S_j(v_i, G)|$ denotes the cardinality of a j-sphere (see Sect. 1.3.3) of v_i regarding an undirected and connected graph G. $f^V(v_i)$ represents a so-called information functional that is based on metrical graph properties (see [107] for details). Equation (1.58) represents a family of parametric entropy measures. We generalized the classical entropy measure [41], i.e., (1.56) and (1.57), because the new measure allows us to weight structural characteristics of a graph by adapting the free parameters α and c_k correspondingly. As a corollary we note that it is now possible to analyze the spread of information within a network.

1.5 Conclusions

We finish this chapter by pointing out that there are many more approaches for the analysis of complex networks. Depending on the point of view, especially in the context of machine learning problems, only some methods give meaningful results with respect to certain applications. For this reason it is of the utmost importance to demonstrate that a chosen approach is appropriate to solve a task by application to a test dataset for which the performance can be evaluated objectively. That means, it is important to bear in mind that the characteristics of the investigated networks have a profound impact on the obtained results. Due to the fact that it is often difficult or even impossible to imagine all possibilities of network characteristics to properly take them into account when deriving a method, such an application is not of mere practical interest but also provides valuable feedback for the design of methods because the analysis of the data relentlessly reveals shortcomings of the method itself. Hence, the design and application of such methods are connected imperatively. For this reason it is no surprise that fields like computational biology, web mining, computational linguistics, quantitative finance, or sociology have triggered a wealth of novel methods for the analysis of complex networks, some of them presented in this chapter, that possibly would otherwise not have been invented. We are excited to see the developments over the next years and expect an ever-increasing interest in complex networks and their analysis due to their omnipresence in the real world.

Acknowledgments I would like to thank Matthias Dehmer for fruitful discussions.

References

1. Adamic L, Huberman B (2000) Power-law distribution of the world wide web. Science 287:2115
2. Albert R, Barabasi AL (2002) Statistical mechanics of complex networks. Rev Mod Phys 74:47
3. Bakir GH, Hofmann T, Schölkopf B, Smola AJ, Taskar B, Vishwanathan SVN (eds) (2007) Predicting structured data. MIT Press, Cambridge, MA
4. Barabasi AL, Albert R (1999) Emergence of scaling in random networks. Science 206: 509–512
5. Bavelas A (1948) A mathematical model for group structure. Hum Organ 7:16–30
6. Bavelas A (1950) Communication patterns in task-oriented groups. J Acoust Soc Am 22: 725–730
7. Bellman R (1957) Dynamic programming. International Series. Princeton University Press, Princeton, NJ
8. Bonchev D (1979) Information indices for atoms and molecules. Match 7:65–113
9. Bonchev D (1983) Information theoretic indices for characterization of chemical structures. Research Studies Press, Chichester
10. Bonchev D (1995) Kolmogorov's information, shannon's entropy, and topological complexity of molecules. Bulg Chem Commun 28:567–582
11. Bonchev D (2003) Complexity in chemistry. Introduction and fundamentals. Taylor & Francis, London (Philadelphia, PA)
12. Bonchev D, Rouvray DH (2005) Complexity in chemistry, biology, and ecology. Mathematical and computational chemistry. Springer, Berlin
13. Bonchev D, Trinajstić N (1977) Information theory, distance matrix and molecular branching. J Chem Phys 67:4517–4533
14. Bonchev D, Balaban AT, Mekenyan OG (1980) Generalization of the graph center concept, and derived topological centric indexes. J Chem Inf Comput Sci 20(2):106–113
15. Bonacich P (1972) Factoring and weighting approaches to status scores and clique identification. J Math Sociol 2:113–120

1 A Brief Introduction to Complex Networks and Their Analysis

16. Bornholdt S, Schuster HG (2003) Handbook of graphs and networks: from the genome to the internet. Wiley, New York, NY
17. Bornholdt S, Schuster HG (eds) (2003) Handbook of graphs and networks: from the genome to the internet. Wiley, New York, NY
18. Brandes U (2001) A faster algorithm for betweenness centrality. J Math Sociol 25(2):163–177
19. Brandes U, Erlebach T (2005) Network analysis. Lecture notes in computer science. Springer, Berlin
20. Brandstädt A, Le VB, Sprinrand JP (1999) Graph classes. A survey. SIAM Monographs on Discrete Mathematics and Applications
21. Brinkmeier M, Schank T (2005) Network statistics. In Brandes U, Erlebach T (eds) Network analysis. Lecture notes in computer science. Springer, Berlin, pp 293–317
22. Broder A, Kumar R, Maghoul F, Raghavan P, Rajagopalan S, Stata R, Tomkins A, Wiener J (2000) Graph structure in the web: experiments and models. In: Proceedings of the 9th WWW conference, Amsterdam
23. Buckley F, Harary F (1990) Distance in graphs. Addison-Wesley, Reading, MA
24. Bunke H (1983) What is the distance between graphs? Bull EATCS 20:35–39
25. Bunke H (1997) On a relation between graph edit distance and maximum common subgraph. Pattern Recognit Lett 18(9):689–694
26. Bunke H (1998) A graph distance metric based on the maximum common subgraph. Pattern Recognit Lett 19(3):255–259
27. Bunke H, Allermann G (1983) A metric on graphs for structural pattern recognition. In: Schussler HW (ed) Proceedings of 2nd European signal processing conference EUSIPCO, pp 257–260
28. Bunke H, Neuhaus M (2007) Graph matching. Exact and error-tolerant methods and the automatic learning of edit costs. In: Cook D, Holder LB (eds) Mining graph data. Wiley, New York, NY, pp 17–32
29. Carrière SJ, Kazman R (1997) Webquery: searching and visualizing the web through connectivity. Computer Networks and ISDN Systems 29(8–13):1257–1267
30. Cayley A (1857) On the theory of analytic forms called trees. Philos Mag 13:19–30
31. Cayley A (1875) On the analytical forms called trees, with application to the theory of chemical combinatorics. Report of the British Association for the Advancement of Science, pp 257–305
32. Chowdhury D, Stauffer D (2000) Principles of equilibrium statistical mechanics. Wiley-VCH, Weinheim
33. Clauset A, Newman MEJ, Moore C (2004) Finding community structure in very large networks. Phys Rev E 70:066111
34. Claussen JC (2007) Characterization of networks by the offdiagonal complexity. Physica A 365–373:321–354
35. Claussen JC (2007) Offdiagonal complexity: a computationally quick network complexity measure – application to protein networks and cell division. In: Deutsch A, Bravo de la Parra R et al (eds) Mathematical modeling of biological systems, vol II. Birkhäuser, Boston, MA, pp 303–311
36. Cook D, Holder LB (2007) Mining graph data. Wiley, New York, NY
37. Cormen T, Leiserson CE, Rivest RL, Leiserson C, Rivest R (2001) Introduction to algorithms. MIT Press, Cambridge, MA
38. Cristianini N, Shawe-Taylor J (2000) An introduction to support vector machines. Cambridge University Press, Cambridge
39. Dehmer M (2006) Strukturelle Analyse web-basierter Dokumente. Multimedia und Telekooperation. Deutscher Universitäts Verlag, Wiesbaden
40. Dehmer M (2008) A novel method for measuring the structural information content of networks. Cybern Syst 39:825–842
41. Dehmer M, Emmert-Streib F (2008) Structural information content of chemical networks. Zeitschrift für Naturforschung, Part A 63a:155–159
42. Deo N, Gupta P (2001) World wide web: a graph-theoretic perspective. Technical report, Department of Computer Science, University of Central Florida

43. Dickinson PJ, Bunke H, Dadej A, Kraetzl M (2004) Matching graphs with unique node labels. Pattern Anal Appl 7:243–266
44. Diestel R (2000) Graph theory. Springer, Berlin
45. Duch J, Arenas A (2005) Community detection in complex networks using extremal optimization. Phys Rev E, 72:027104
46. Emmert-Streib F (2007) The chronic fatigue syndrome: a comparative pathway analysis. J Comput Biol 14(7):961–972
47. Emmert-Streib F, Chen L, Storey J (2007) Functional annotation of genes in *Saccharomyces cerevisiae* based on joint betweenness. arXiv:0709.3291
48. Emmert-Streib F, Dehmer M (2007) Global information processing in gene networks: fault tolerance. In: Proceedings of the bio-inspired models of network, information, and computing systems, Bionetics 2007, art. no. 4610138, pp 326–329
49. Emmert-Streib F, Dehmer M, Kilian J (2005) Classification of large graphs by a local tree decomposition. In: Arabnia HR, Scime A (eds) Proceedings of DMIN'05, international conference on data mining, Las Vegas, June 20–23, pp 200–207
50. Emmert-Streib F, Dehmer M (2007) Topolocial mappings between graphs, trees and generalized trees. Appl Math Comput 186(2):1326–1333
51. Emmert-Streib F, Dehmer M (eds) (2008) Analysis of microarray data: a network based approach. Wiley-VCH, Weinheim
52. Emmert-Streib F, Dehmer M (2005) Robustness in scale-free networks: comparing directed and undirected networks. Int J Mod Phys C 19(5):717–726
53. Emmert-Streib F, Mushegian A (2007) A topological algorithm for identification of structural domains of proteins. BMC Bioinformatics 8:237
54. Erdös P, Rényi A (1959) On random graphs. Publicationes Mathematicae 6:290–297
55. Erdös P, Rényi A (1960) On the evolution of random graphs. Publications of Mathematical Institute of the Hungarian Academy of Sciences 5:17–61
56. Euler L (1736) Solutio problematis ad geometriam situs pertinentis. Comentarii Academiae Scientiarum Imperialis Petropolitanae 8:128–140
57. Even S (1979) Algorithms. Computer Science Press, Potomac, MD
58. Freeman LC (1977) A set of measures of centrality based on betweenness. Sociometry 40:35–41
59. Freeman LC (1979) Centrality in social networks: conceptual clarification. Soc Networks 1:215–239
60. Fujii JI, Yuki S (1997) Entropy and coding for graphs. Int J Math Stat Sci 6(1):63–77
61. Gagneur J, Krause R, Bouwmeester T, Casari G (2004) Modular decomposition of protein–protein interaction networks. Genome Biol 5:R57
62. Gärtner T, Flach PA, Wrobel S (2003) On graph kernels: hardness results and efficient alternatives. In: COLT, pp 129–143
63. Gernert D (1979) Measuring the similarity of complex structures by means of graph grammars. Bull EATCS 7:3–9
64. Gernert D (1981) Graph grammars which generate graphs with specified properties. Bull EATCS 13:13–20
65. Gleiser PM, Danon L (2003) Community structure in jazz. Advances in complex systems 6(4):565–574
66. Hage P, Harary F (1995) Eccentricity and centrality in networks. Soc Networks 17:57–63
67. Halin R (1989) Graphentheorie. Akademie Verlag, Berlin
68. Harary F (1959) Status and contrastatus. Sociometry 22:23–43
69. Harary F (1965) Structural models. An introduction to the theory of directed graphs. Wiley, NY
70. Harary F (1967) Graph theory and theoretical physics. Academic, New York, NY
71. Harary F (1969) Graph theory. Addison-Wesley, Reading, MA
72. Hastie T, Tibshirani R, Friedman JH (2001) The elements of statistical learning. Springer, Berlin

73. Horváth T, Gärtner T, Wrobel S (2004) Cyclic pattern kernels for predictive graph mining. In: Proceedings of the 2004 ACM SIGKDD international conference on knowledge discovery and data mining, pp 158–167
74. Hsu H-P, Mehra V, Grassberger P (2003) Structure optimization in an off-lattice protein model. Phys Rev E 68(3):037703
75. Kaden F (1982) Graphmetriken und Distanzgraphen. ZKI-Informationen, Akademie der Wissenschaften DDR 2(82):1–63
76. Kaden F (1983) Halbgeordnete Graphmengen und Graphmetriken. In: Proceedings of the conference graphs, hypergraphs, and applications DDR, pp 92–95
77. Kaden F (1986) Graphmetriken und Isometrieprobleme zugehöriger Distanzgraphen. ZKI-Informationen, Akademie der Wissenschaften DDR, pp 1–100
78. Kauffman SA (1969) Metabolic stability and epigenesis in randomly constructed genetic nets. J Theor Biol 22:437–467
79. Kieffer J, Yang E (1997) Ergodic behavior of graph entropy. Electronic Research Announcements of the American Mathematical Society 3:11–16
80. Kondor RI, Lafferty J (2002) Diffusion kernels on graphs and other discrete input spaces. In: Machine learning: Proceedings of the 19th international conference, Morgan Kaufmann, San Mateo, CA
81. König D (1936) Theorie der endlichen und unendlichen Graphen. Chelsea, New York, NY
82. Körner J (1973) Coding of an information source having ambiguous alphabet and the entropy of graphs. Transactions of the 6th Prague conference on information theory, pp 411–425
83. Koschützki D, Lehmann KA, Peters L, Richter S, Tenfelde-Podehl D, Zlotkowski O (2005) Clustering. In: Brandes U, Erlebach T (eds) Centrality indices. Lecture notes in computer science. Springer, Berlin, pp 16–61
84. Kullback S (1959) Information theory and statistics. Wiley, New York, NY
85. Kullback S, Leibler RA (1951) On information and sufficiency. Ann Math Stat 22(1):79–86
86. Laubenbacher RC (2007) Modeling and simulation of biological networks. In: Proceedings of symposia in applied mathematics. American Mathematical Society, Providence, RI
87. Li M, Vitányi P (1997) An introduction to Kolmogorov complexity and its applications. Springer, Berlin
88. Mason O, Verwoerd M (2007) Graph theory and networks in biology. IET Syst Biol 1(2): 89–119
89. Mehler A (2006) In search of a bridge between network analysis in computational linguistics and computational biology – a conceptual note. In: Proceedings of the 2006 international conference on bioinformatics & computational biology (BIOCOMP'06), 2006, Las Vegas, Nevada, USA, pp 496–500
90. Mehler A, Dehmer M, Gleim R (2005) Towards logical hypertext structure. a graph-theoretic perspective. In: Proceedings of I2CS'04. Lecture notes. Springer, Berlin, pp 136–150
91. Messmer BT, Bunke H (1998) A new algorithm for error-tolerant subgraph isomorphism detection. IEEE Trans Pattern Anal Mach Intell 20(5):493–504
92. Mowshowitz A (1968) Entropy and the complexity of the graphs I: an index of the relative complexity of a graph. Bull Math Biophys 30:175–204
93. Mowshowitz A (1968) Entropy and the complexity of graphs II: the information content of digraphs and infinite graphs. Bull Math Biophys 30:225–240
94. Mowshowitz A (1968) Entropy and the complexity of graphs III: graphs with prescribed information content. Bull Math Biophys 30:387–414
95. Mowshowitz A (1968) Entropy and the complexity of graphs IV: entropy measures and graphical structure. Bull Math Biophys 30:533–546
96. Newman MEJ (2003) The structure and function of complex networks. SIAM Rev 45: 167–256
97. Newman MEJ, Girvan M (2004) Finding and evaluating community structures in networks. Phys Rev E 69:026113
98. Newman MEJ (2006) Modularity and community structure in networks. Proc Natl Acad Sci USA 103:8577–8582

99. Pearl J (1998) Probabilistic reasoning in intelligent systems. Morgan Kaufmann, Los Altos, CA
100. Rashewsky N (1955) Life, information theory, and topology. Bull Math Biophys 17:229–235
101. Roberts F (1989) Applications of combinatorics and graph theory to the biological and social sciences series. IMA volumes in mathematics and its applications. Springer, Berlin
102. Rosvall M, Bergstrom CT (2007) An information-theoretic framework for resolving community structure in complex networks. In: Proc Natl Acad Sci USA 104(18):7327–31
103. Sabidussi G (1966) The centrality index of a graph. Psychometrika 31:581–603
104. Scott F (2001) Social network analysis. Sage, Beverly Hills, CA
105. Shannon CE, Weaver W (1997) The mathematical theory of communication. University of Illinois Press, Champaign, IL
106. Simonyi G (2001) Perfect graphs and graph entropy. An updated survey. In: Ramirez-Alfonsin J, Reed B (eds) Perfect graphs. Wiley, New York, NY, pp 293–328
107. Skorobogatov VA, Dobrynin AA (1988) Metrical analysis of graphs. MATCH 23:105–155
108. Sobik F (1982) Graphmetriken und Klassifikation strukturierter Objekte. ZKI-Informationen, Akademie der Wissenschaften DDR 2(82):63–122
109. Sobik F (1986) Modellierung von Vergleichsprozessen auf der Grundlage von Ähnlichkeitsmaßen für Graphen. ZKI-Informationen, Akademie der Wissenschaften DDR 4:104–144
110. Solé RV, Valverde S (2004) Information theory of complex networks: on evolution and architectural constraints. In: Lecture notes in physics, vol 650, pp 189–207
111. Temkin O, Zeigarnik AV, Bonchev D (1996) Chemical reaction networks. A graph-theoretical approach. CRC Press, West Palm Beach, FL
112. Trucco E (1956) A note on the information content of graphs. Bull Math Biol 18(2):129–135
113. Ullmann JR (1976) An algorithm for subgraph isomorphism. J ACM 23(1):31–42
114. Wasserman S, Faust K (1994) Social network analysis: methods and applications. Structural analysis in the social sciences. Cambridge University Press, Cambridge
115. Watts DJ, Strogatz SH (1998) Collective dynamics of 'small-world' networks. Nature 393:440–442
116. Zachary W (1977) An information flow model for conflict and fission in small groups. J Anthropol Res 33:452–473
117. Zelinka B (1975) On a certain distance between isomorphism classes of graphs. Časopis pro p̌est. Mathematiky 100:371–373
118. Zhang K, Statman R, Shasha D (1992) On the editing distance between unordered labeled trees. Inform Process Lett 42(3):133–139

Chapter 2
Partitions of Graphs

Mieczysław Borowiecki

Abstract Many difficult optimization problems on graphs become tractable when restricted to some classes of graphs, usually to hereditary classes. A large part of these problems can be expressed in the vertex partitioning formalism, i.e., by partitioning of the vertex set of a given graph into subsets $V_1, \ldots, V_k$ called colour classes, satisfying certain constraints either internally or externally, or both internally and externally. These requirements may be conveniently captured by the symmetric k-by-k matrix M. Concepts which are modeled by M-partitions fall naturally into the three types; each is represented in this work by some problem.

Any minimal reducible bound for a hereditary property is in some sense the best possible partition. A number of such partitions are given.

Clustering is a central optimization problem (among many others) with applications in various disciplines, e.g., computational biology, communications networks, image processing, pattern analysis [41,53,57,60], and numerous other fields. Some new results on k-clustering of graphs are proved.

Another type of M-partition is a matching cutset. The main known results on this subject are collected.

The last part of this work is devoted to acyclic partitions of graphs where we consider important classes of graphs and their acyclic reducible bounds.

For each partition type the complexity of considered problems is given. Also a number of open problems are presented.

Keywords Colourings · Partitions · Hereditary properties · k-clustering · Domination · Cut matching · Complexity

MSC2000: Primary 05C15, 05C70; Secondary 05C12, 05C69, 05C85, 05C90

M. Borowiecki (✉)
Faculty of Mathematics, Computer Science and Econometrics,
University of Zielona Góra, Podgórna 50, 65-246 Zielona Góra, Poland
e-mail: M.Borowiecki@wmie.uz.zgora.pl

M. Dehmer (ed.), *Structural Analysis of Complex Networks*,
DOI 10.1007/978-0-8176-4789-6_2, © Springer Science+Business Media, LLC 2011

2.1 Introduction and Notation

We present some of the basic definitions, notation, and terminology used in this chapter. Other terminology will be introduced as it naturally occurs in the text and those concepts not defined can be found in [4, 30, 59].

We consider finite undirected graphs without loops or multiple edges. The vertex set and the edge set of graph G are denoted by $V(G)$ and $E(G)$, respectively, and $\mathcal{I}$ is used to denote the class of these graphs.

Many difficult (NP-hard) optimization problems on graphs become tractable when restricted to some classes of graphs, usually to hereditary classes. A large part of these problems can be expressed in the vertex partitioning formalism, i.e., by partitioning of the vertices of a given graph into subsets $V_1, \ldots, V_k$ called *colour classes*, satisfying certain constraints either internally or externally, or both internally and externally. These requirements may be conveniently captured by the symmetric k-by-k matrix M in which the diagonal entries $m_{ii} = \mathcal{P}_i$ encode the internal restrictions on the sets V_i and the off-diagonal entries $m_{ij} = \mathcal{P}_{ij}$ $(i \neq j)$ encode the restriction on the edges between V_i and V_j. Formally, it can be defined as follows.

Definition 2.1.1. Let M be a fixed symmetric $k \times k$ matrix with entries $m_{ii} = \mathcal{P}_i$ and $m_{ij} = \mathcal{P}_{ij} \subseteq \mathcal{B}$ for $i \neq j$, where $\mathcal{B}$ is the class of all bipartite graphs.

An M-*colouring* (*partition*) of a graph G is a partition of vertices of G into k subsets $V_1, \ldots, V_k$ corresponding to the rows (and columns) of the matrix M such that the subgraph of G induced by V_i has the property $\mathcal{P}_i$ for $i = 1, \ldots, k$. Vertices of the set V_i are said to be i-*coloured*. For every two distinct colours i and j, the subgraph induced by all the edges linking an i-coloured vertex and a j-coloured vertex has the property $\mathcal{P}_{ij}$, $1 \leq i, j \leq k$.

Notice that properties $\mathcal{P}_{ij}$ always form some classes of bipartite graphs. A graph G is *bipartite* if it admits a vertex partition $V(G) = V_1 \cup V_2$ such that every edge of G joins two different V_i's.

Graph-theoretical concepts which are modeled by M-partitions fall naturally into three types:

(T1) $m_{ii} = \mathcal{P}_i$, $m_{ij} = \mathcal{P}_{ij} = \mathcal{B}$ $(i \neq j)$, $i, j = 1, \ldots, k$; i.e., no restrictions between colour classes.

(T2) $m_{ii} = \mathcal{I}$, $m_{ij} = \mathcal{P}_{ij} \subseteq \mathcal{B}$ $(i \neq j)$, $i, j = 1, \ldots, k$; i.e., no internal restrictions.

(T3) $m_{ii} = \mathcal{P}_i$, $m_{ij} = \mathcal{P}_{ij} \subseteq \mathcal{B}$ $(i \neq j)$, $i, j = 1, \ldots, k$.

2.1.1 Examples of M-Partitions

A $(\mathcal{P}_1, \ldots, \mathcal{P}_k)$-*partition* of G is defined as an M-partition of type (T1).

In the case when all $\mathcal{P}_i = \mathcal{O}$, where $\mathcal{O}$ is the class of edgeless graphs, we have a k-*colouring* of G. Thus a k-*colouring* of G is an M-partition of G, where the matrix M has $\mathcal{O}$ on the main diagonal and $\mathcal{B}$ everywhere else.

The smallest k for which there is a k-colouring of G is called the *chromatic number* of G and is denoted by $\chi(G)$.

Karp [40] has shown that the k-colouring problem ($k \geq 3$) is NP-complete for graphs and up to this time no polynomial-time algorithm is known for this problem. However, the k-colouring problem becomes much easier when we restrict the inputs to special classes of graphs.

An $\boldsymbol{M}$-partition of type (T1) of G is called k-*clustering*, if

$$m_{ii} = \mathcal{P} = \{G \in \mathcal{I} : \mathrm{diam}(G) \leq k\}.$$

Let (V_1, V_2) be an $\boldsymbol{M}$-partition of G. Denote by $\mathcal{M}$ the property of graphs for which the edges between the set V_1 and V_2 induce a matching, i.e., any two of them have no common endvertex.

An $\boldsymbol{M}$-partition of G such that $m_{ii} = \mathcal{I}$, $m_{12} = m_{21} = \mathcal{M} \subseteq \mathcal{B}$ is called a *matching cutset* of G.

A *path* in G is an alternating sequence of distinct vertices and edges beginning and ending with vertices in which each edge joins the vertex before it to the one following it. The first and the least vertex of this sequence is called the *endvertex* of the path. The path with n vertices is denoted by P_n. A *cycle* in G is a path with at least three vertices together with an edge joining its endvertices. A cycle with n vertices is denoted by C_n. The *length* of a path (cycle) is the number of edges in it. A graph G is called *acyclic* if G does not contain a cycle. Let us denote by $\mathcal{D}_1$ the class of all acyclic graphs.

An $\boldsymbol{M}$-partition of a graph G such that $m_{ii} = \mathcal{O}$ and $m_{ij} = \mathcal{D}_1$ for $i \neq j$, $1 \leq i, j \leq k$ is called an *acyclic colouring* of G.

The minimum k such that G has an acyclic k-colouring is the *acyclic chromatic number* of G, denoted by $\chi_a(G)$.

Similarly, for a class $\mathcal{P}$ of graphs, the *acyclic chromatic number of* $\mathcal{P}$, denoted by $\chi_a(\mathcal{P})$, is defined as the maximum $\chi_a(G)$ over all graphs $G \in \mathcal{P}$, assuming that $\chi_a(G)$ is finite for all $G \in \mathcal{P}$.

2.1.2 Hereditary Properties

Two graphs G and H are *isomorphic* if there is a bijection $f : V(G) \to V(H)$ such that $uv \in E(G)$ if and only if $f(u)f(v) \in E(H)$.

A *property of graphs* is any nonempty class of graphs from $\mathcal{I}$ which is closed under isomorphisms. We use the terms *class of graphs* and *property of graphs* interchangeably.

Graph H is a *subgraph* of graph G if $V(H) \subseteq V(G)$ and $E(H) \subseteq E(G)$ and is denoted by $H \subseteq G$. A subgraph H of G is *induced* if every pair of vertices in H which are adjacent in G are also adjacent in H. This fact is denoted by $H \leq G$.

Definition 2.1.2. A property $\mathcal{P}$ of graphs is said to be *induced hereditary* (*hereditary*, also called *monotone*) if whenever $G \in \mathcal{P}$ and $H \le G$ ($H \subseteq G$), then also $H \in \mathcal{P}$ and *additive* if it is closed under disjoint union, i.e., if every component of G has property $\mathcal{P}$, then $G \in \mathcal{P}$.

Obviously any hereditary property is induced hereditary, too.

Following [11] we list some examples and notations of hereditary properties of graphs.

$$
\begin{aligned}
\mathcal{O} &= \{G \in \mathcal{I} : E(G) = \emptyset\}, \\
\mathcal{O}^k &= \{G \in \mathcal{I} : \chi(G) \le k\}, \\
\mathcal{S}_k &= \{G \in \mathcal{I} : \text{ the maximum degree } \Delta(G) \le k\}, \\
\mathcal{O}_k &= \{G \in \mathcal{I} : \text{ each component of } G \text{ has order at most } k + 1\}, \\
\mathcal{D}_k &= \{G \in \mathcal{I} : G \text{ is } k\text{-degenerate;}\\
&\qquad \text{i.e., the minimum degree } \delta(H) \le k \text{ for each } H \subseteq G\}, \\
\mathcal{T}_k &= \{G \in \mathcal{I} : G \text{ contains no subgraph homeomorphic to } K_{k+2} \text{ or} \\
&\qquad K_{\lfloor \frac{k+3}{2} \rfloor, \lceil \frac{k+3}{2} \rceil}\}, \\
\mathcal{I}_k &= \{G \in \mathcal{I} : G \text{ does not contain } K_{k+2}\}.
\end{aligned}
$$

Observe that:

$\mathcal{O}$ is the class of edgeless graphs,
$\mathcal{O}^k$ is the class of *k-colourable* graphs,
$\mathcal{S}_k$ is the class of graphs with degree bounded by k,
$\mathcal{D}_1$ is the class of *acyclic* graphs,
$\mathcal{T}_2$ is the class of *outerplanar* graphs,
$\mathcal{T}_3$ is the class of *planar* graphs.

Additionally, let us denote by $\mathcal{LF} = \mathcal{D}_1 \cap \mathcal{S}_2$, the class of *linear forests*.

A hereditary property $\mathcal{P}$ can be uniquely determined by the set of *minimal forbidden subgraphs* which can be defined as follows.

Definition 2.1.3. $F_{\subseteq}(\mathcal{P}) = \{G \in \mathcal{I} : G \notin \mathcal{P}, \text{ but each proper subgraph } H \text{ of } G \text{ belongs to } \mathcal{P}\}.$

Let $\mathcal{F}$ be a family of graphs, $Forb_{\subseteq}(\mathcal{F})$ is defined to be the property of all graphs having no subgraph isomorphic to any graph from $\mathcal{F}$. Thus $\mathcal{P} = Forb_{\subseteq}(F_{\subseteq}(\mathcal{P}))$. In such a manner is defined, for example, the property $\mathcal{I}_k = Forb_{\subseteq}(\{K_{k+2}\})$, also called the class of K_{k+2}-*free* graphs.

2 Partitions of Graphs 31

Similarly, the set of minimal forbidden subgraphs $F_{\leq}(\mathcal{P})$ for an induced hereditary property $\mathcal{P}$ is defined.

Let us denote by $\mathbf{L}$ the set of all hereditary and by $\mathbf{L}_{\leq}$ induced hereditary properties of graphs, and let the corresponding sets of additive properties be denoted by $\mathbf{L}^a$ and $\mathbf{L}^a_{\leq}$, respectively. The sets $\mathbf{L}$, $\mathbf{L}^a$, $\mathbf{L}_{\leq}$ and $\mathbf{L}^a_{\leq}$ partially ordered by the set inclusion, form complete distributive lattices with the set intersection as the meet operation [17]. Obviously $(\mathbf{L}, \subseteq)$ is a proper sublattice of $(\mathbf{L}_{\leq}, \subseteq)$.

We now consider some examples of partitions of outerplanar and planar graphs.

In [21] it has been proved that each outerplanar graph has an $(\mathcal{LF}, \mathcal{LF})$-partition. An algorithmic proof of that fact is given in [18]. Each outerplanar graph also has an $(\mathcal{O}, \mathcal{D}_1)$-partition. A natural question arises: does a property $\mathcal{P} \subset \mathcal{LF}$ exist such that each outerplanar graph has a $(\mathcal{P}, \mathcal{LF})$-partition? In other words: is the $(\mathcal{LF}, \mathcal{LF})$-partition the best possible for the class of outerplanar graphs? An answer is given later on.

It is well known [26] that every planar graph has vertex arboricity at most 3; i.e., it has a $(\mathcal{D}_1, \mathcal{D}_1, \mathcal{D}_1)$-partition.

Stein [55] (see also [8, 37]) strengthened this result by proving that every planar graph can be vertex partitioned into two forests and one edgeless graph; i.e., an planar graph has an $(\mathcal{O}, \mathcal{D}_1, \mathcal{D}_1)$-partition.

Another strengthening was obtained by Poh [51] and independently by Goddard [36]. They proved that every planar graph has an $(\mathcal{LF}, \mathcal{LF}, \mathcal{LF})$-partition.

2.1.3 Reducibility

To more precisely analyse different partitions and compare them, we need some new notation.

A property $\mathcal{R} = \mathcal{P}_1 \circ \mathcal{P}_2 \circ \cdots \circ \mathcal{P}_k$ is defined as the set of all graphs having a $(\mathcal{P}_1, \mathcal{P}_2, \ldots, \mathcal{P}_k)$-partition. If $\mathcal{P}_1 = \cdots = \mathcal{P}_k = \mathcal{P}$, then we write $\mathcal{P}_1 \circ \cdots \circ \mathcal{P}_k = \mathcal{P}^k$.

It is easy to see that $\mathcal{P}_1 \circ \cdots \circ \mathcal{P}_k$ is (induced) hereditary and additive whenever $\mathcal{P}_1, \mathcal{P}_2, \ldots, \mathcal{P}_k$ are (induced) hereditary and additive, respectively.

An (induced) hereditary property $\mathcal{R}$ is said to be *reducible* if there exist two (induced) hereditary properties $\mathcal{P}_1$ and $\mathcal{P}_2$ such that $\mathcal{R} = \mathcal{P}_1 \circ \mathcal{P}_2$ and *irreducible*, otherwise.

Definition 2.1.4. (Mihók and Toft, see [39]) For a given irreducible property $\mathcal{P} \in \mathbf{L}$, a reducible property $\mathcal{R} \in \mathbf{L}$ is called a *minimal reducible bound* for $\mathcal{P}$ if $\mathcal{P} \subset \mathcal{R}$ and for each reducible property $\mathcal{R}' \subset \mathcal{R}$, $\mathcal{P} \nsubseteq \mathcal{R}'$.

We consider reducibility and minimal reducible bounds only in the lattice $\mathbf{L}^a$. The family of all minimal reducible bounds for $\mathcal{P}$ in this lattice is denoted by $B(\mathcal{P})$.

Example 1. The class $\mathcal{O}^2$ of bipartite graphs is the smallest reducible property in the lattice $\mathbf{L}^a$.

Obviously it is the unique minimal reducible bound for $\mathcal{P}$ if and only if $\mathcal{P} \subset \mathcal{O}^2$. In general, finding the set of minimal reducible bounds for a given irreducible property is a difficult problem. The existence of the set $\boldsymbol{B}(\mathcal{P})$ for any property $\mathcal{P}$ is proved in [5].

2.1.4 Examples of Some Reducible Bounds

From the previously mentioned results for outerplanar and planar graphs we can write:

1. $\mathcal{T}_2 \subset \mathcal{O} \circ \mathcal{D}_1$,
2. $\mathcal{T}_2 \subset \mathcal{LF}^2$,
3. Four Colour Theorem implies: $\mathcal{T}_3 \subset (\mathcal{O}^2 \cap \mathcal{T}_3)^2$,
4. [55]: $\mathcal{T}_3 \subset \mathcal{O} \circ \mathcal{D}_1^2$ was improved by Thomassen [56]: $\mathcal{T}_3 \subset \mathcal{D}_1 \circ (\mathcal{D}_2 \cap \mathcal{T}_3)$,
5. [36, 51]: $\mathcal{T}_3 \subset \mathcal{LF}^3$.

2.1.5 Minimal Reducible Bounds for Outerplanar and Planar Graphs

Theorem 2.1.5 ([47]). $\boldsymbol{B}(\mathcal{T}_2) = \{\mathcal{O} \circ \mathcal{D}_1, \mathcal{LF} \circ \mathcal{LF}\}$.

From this theorem it follows that for each reducible property $\mathcal{R} = \mathcal{P} \circ \mathcal{Q}$ such that $\mathcal{T}_2 \subset \mathcal{R} \subset \mathcal{LF}^2$ there is an outerplanar graph G which is not $(\mathcal{P}, \mathcal{Q})$-partitionable. The same holds for $\mathcal{O} \circ \mathcal{D}_1$. In this sense, the bounds given above are best possible.

For a class of planar graphs the problem of minimal reducible bounds is much harder. Until now any minimal reducible bound for $\mathcal{T}^3$ is not known. In [12] a few minimal reducible bounds for some interesting subclasses of planar graphs are given.

Theorem 2.1.6 ([12]).

1. $\boldsymbol{B}(\mathcal{T}_3 \cap \mathcal{D}_2) = \{\mathcal{O} \circ \mathcal{D}_1\}$,
2. $\boldsymbol{B}(\mathcal{T}_3 \cap \mathcal{D}_3) \ni \mathcal{D}_1 \circ \mathcal{D}_1$,
3. $\boldsymbol{B}(\mathcal{T}_3 \cap \mathcal{O}^3) \ni \mathcal{O} \circ (\mathcal{O}^2 \cap \mathcal{T}_3)$.

We are not able to find other minimal reducible bounds for 3-colourable planar graphs. We can only show:

Theorem 2.1.7 ([12]). *Barnette's Conjecture, if true, gives a minimal reducible bound $\mathcal{D}_1^2$ for $\mathcal{T}_3 \cap \mathcal{O}^3$.*

Theorem 2.1.7 gives an unexpected relation to a well-known conjecture of Barnette which says that whenever G is a 3-connected bipartite 3-regular planar graph, then G has a Hamiltonian cycle. This conjecture is true if and only if each 3-colourable planar graph has a vertex partition into two subsets such that each of them induces a forest.

2.1.6 Minimal Reducible Bounds for Some Other Classes of Graphs

In [15] a subclass of planar graphs (slightly wider than the class of outerplanar graphs) called 1-*nonouterplanar* was considered. For this class of graphs, in contrast with the class of outerplanar graphs (see Theorem 2.1.5), there is an infinite number of minimal reducible bounds.

Let us define a few properties:

$\mathcal{UC}_i = \{G \in I$: each component of G contains at most one cycle of length i and no cycle of any other length$\}$,

$\mathcal{UC}_i^k = \{G \in \mathcal{UC}_i$: if G contains a cycle (of length i), then the minimum degree in G of the vertices of this cycle is at most $k + 2\}$.

$\nabla_k = \{G \in I$: each component of G belongs to $\mathcal{UC}_3^k \cup \bigcup_{i \geq k+2} \mathcal{UC}_{2i+1}\}$.

For convenience, let $\nabla_\infty = \mathcal{UC}_3$.

For a plane graph G let $\mathrm{Int}(G)$ denote the set of vertices not belonging to the external face. If G is a connected planar graph, we define $\mathrm{int}(G)$ to be the minimum value of $|\mathrm{Int}(G)|$ over all plane embeddings of G.

If G is a planar graph with r components $H_1, \ldots, H_r$ then we define

$$\mathrm{int}(G) = \max\{\mathrm{int}(H_i) : 1 \leq i \leq r\}.$$

If $\mathrm{int}(G) \leq k$ then G is said to be k-*nonouterplanar* and we denote this property by $\mathcal{NOP}_k$, i.e., $\mathcal{NOP}_k = \{G \in \mathcal{T}_3 : \mathrm{int}(G) \leq k\}$.

It easy to see that $\mathcal{NOP}_0 = \mathcal{T}_2$ and $\mathcal{NOP}_\infty = \mathcal{T}_3$.

Theorem 2.1.8 ([15]).

$$\boldsymbol{B}(\mathcal{NOP}_1) = \{\mathcal{LF} \circ \mathcal{LF}, \ \mathcal{O}_1 \circ \mathcal{D}_1\} \cup \{\mathcal{O} \circ \nabla_k : \ k = 0, 1, \ldots, \infty\}.$$

The minimal reducible bound for the class of triangle free graphs is trivial and it follows from the theorem given by Nešetřil and Rödl.

Theorem 2.1.9 ([50]). *Let $\boldsymbol{F}(\mathcal{P})$ be a finite set of 2-connected graphs. Then for every graph G of property $\mathcal{P}$ there is a graph H of property $\mathcal{P}$ such that for any partition (V_1, V_2) of $V(H)$ there is an i, $i = 1$ or $i = 2$, for which the subgraph $H[V_i]$ contains G.*

From this theorem we immediately have the following.

Corollary 2.1.10 ([48]). *Let $\boldsymbol{F}(\mathcal{P})$ be a finite set of 2-connected graphs. Then the property $\mathcal{P}$ has exactly one minimal reducible bound. $\mathcal{O} \circ \mathcal{P}$.*

Corollary 2.1.10 implies that the class $\mathcal{I}_k$ of K_{k+2}-free graphs has only one trivial minimal reducible bound $\mathcal{O} \circ \mathcal{I}_k$.

For the class of graphs with a bounded order of components and the class of k-degenerate graphs we have the following sets of minimal reducible bounds.

Theorem 2.1.11 ([48]). *For any positive integer k,*

$$B(\mathcal{O}_k) = \{\mathcal{O}_p \circ \mathcal{O}_q : p + q + 1 = k\},$$
$$B(\mathcal{D}_k) = \{\mathcal{D}_p \circ \mathcal{D}_q : p + q + 1 = k\}.$$

A *k-tree* is a graph defined inductively as follows. A clique of order k is a k-tree. If G is a k-tree of order n, $n \geq k$, and K is a clique of G of order k, then the graph obtained from G by adding a new vertex and joining it by new edges to all vertices of K is a k-tree of order $n + 1$. Any subgraph of a k-tree is a *partial k-tree*. Let us denote the class of all partial k-trees by $\mathcal{P}T_k$. Obviously this class is hereditary.

Similar relations as above hold for the class of partial k-trees.

Theorem 2.1.12 ([52]). *For any positive integer k,*

$$B(\mathcal{P}T_k) = \{\mathcal{P}T_p \circ \mathcal{P}T_q : p + q + 1 = k\}.$$

Partitions and minimal reducible bounds for minor hereditary properties were considered in [19].

2.1.7 Complexity of Some Selected Graph Partition Problems

We begin with complexity of partition problems for the class of planar graphs.

Theorem 2.1.13. *The following partition problems are NP-complete:*

1. *[38]: $(\mathcal{O}, \mathcal{D}_1)$-partition for planar graphs G with $\Delta(G) \leq 4$,*
2. *[38]: $(\mathcal{D}_1, \mathcal{D}_1)$-partition even for maximal planar graphs,*
3. *[35]: $(\mathcal{O}, \mathcal{O}, \mathcal{O})$-partition for planar graphs G with $\Delta(G) \leq 4$,*
4. *[38]: $(\mathcal{O}, \mathcal{O}, \mathcal{D}_1)$-partition for planar graphs in which each face has size 3 or 4.*

Conjecture 2.1.14 ([38]). *$(\mathcal{O}, \mathcal{O}, \mathcal{D}_1)$-partition is NP-complete for maximal planar graphs.*

Below are presented partition problems for some interesting classes of graphs. General discussions on partitions with respect to hereditary or induced hereditary properties can be found in [2, 33].

Theorem 2.1.15. *The following partition problems for graphs (in general) are NP-complete:*

1. *[35]: $(\mathcal{O}, \mathcal{O}^2)$-partition,*
2. *[35]: $(\mathcal{O}, \mathcal{D}_1)$-partition,*
3. *[46]: $(\mathcal{O}, \mathcal{S}_1)$-partition,*
4. *[20]: $(\mathcal{O}, Forb_{\leq}(\{C_4, 2K_2, C_5\}))$-partition,*

2 Partitions of Graphs   35

5. [20]: $(\mathcal{O}, Forb_{\leq}(\{P_4\}))$-partition,
6. [20]: $(\mathcal{O}, Forb_{\leq}(\{P_4, C_4\}))$-partition,
7. [20]: $(\mathcal{O}, \mathcal{I}_1)$-partition.

2.2 k-Clustering

Another type of (T1) partition of graphs with properties $\mathcal{P}_i$, $i = 1, \ldots, k$, which are not hereditary is k-clustering. Clustering is a central optimization problem (among many others) with applications in various disciplines, e.g., computational biology, communications networks, image processing, pattern analysis [41, 53, 57, 60], and numerous other fields.

From a general point of view the goal of clustering is to find the groups (*clusters*) that are both homogeneous and well separated; i.e., elements within the same cluster should be similar and elements in different clusters dissimilar. The process of generating the clusters is called *clustering*. In the graph-theoretic approach to clustering, one builds from the data a *similarity graph* in which vertices correspond to objects and edges connect two vertices with similarity values above some predefined threshold. Typical objectives include: minimizing the maximum diameter of a cluster (*k-clustering*), minimizing the average distance between pairs of clustered points (*k-clustered sum*), and many others.

2.2.1 Minimum k-Clustering

Problem 2.2.1. k-CLUSTERING
Given a graph $G = (V, E)$ and a positive integer l, determine whether there is a partition of V into at most l subsets such that each of these subsets induces a subgraph of G with diameter at most k.

Problem 2.2.2. MINIMUM k-CLUSTERING
Given a graph $G = (V, E)$, find the smallest integer l such that there is a partition of V into l subsets each inducing a subgraph of G with diameter at most k.

The number of clusters in a minimum k-clustering of a graph G is denoted by $cl_k(G)$.

Theorem 2.2.3 ([23]). *Problem k-CLUSTERING is NP-complete for $l \geq 3$.*

Linear time algorithms for k-clustering on trees and some classes of perfect graphs were given in [1, 32].

The 1-CLUSTERING problem is equivalent to the problem PARTITION INTO CLIQUES which is NP-complete for any fixed $l \geq 3$. A graph G has 1-CLUSTERING into at most l clusters if and only if the complement of G has an l-colouring.

The graph 2-CLUSTERING problem is related to the NP-complete problem DOMINATING SET. A subset $D \subseteq V$ is said to be a *dominating set* of the graph $G = (V, E)$ if for every vertex $u \in V - D$ there exists a vertex $v \in D$ such that $uv \in E$. The minimum cardinality of a dominating set of G is denoted by $\gamma(G)$.

Problem 2.2.4. DOMINATING SET

Given a graph $G = (V, E)$ and a positive integer k, decide whether G has a dominating set S with $|S| \leq k$.

Theorem 2.2.5 ([29]). *For any graph G*

$$cl_2(G) \leq \gamma(G).$$

Proof. Let $D = \{w_1, \ldots, w_l\}$ be a dominating set of G. 2-clustering $\{Q_1, \ldots, Q_l\}$ of G with $w_i \in Q_i$ for all $i = 1, \ldots, l$ can be generated by assigning remaining vertices v to Q_i for which $vw_i \in E$. Thus a dominating set of cardinality $\gamma(G)$ induces a clustering of G into the smallest number of subsets for which each subgraph induced by Q_i has a dominating vertex. $\square$

2.2.2 Strongly Chordal Graphs

A class of *chordal graphs*, denoted by $\mathcal{C}$, is defined by the set of minimal forbidden induced subgraphs as follows:

$$\boldsymbol{F}_{\leq}(\mathcal{C}) = \{C_n : n \geq 4\}.$$

A graph $G = (V, E)$ is said to be *strongly chordal* if every induced subgraph contains a simple vertex, where a vertex v of a graph G is *simple* if the set $\{N[u] : u \in N[v]\}$ can be linearly ordered by inclusion; i.e., the closed neighbourhoods form a chain under inclusion; see Fig. 2.1. A class of strongly chordal graphs is denoted by $\mathcal{SC}$.

A *Sun* S_r is a chordal graph on $2r$ vertices, for some $r \geq 3$, whose vertex set can be partitioned into two sets: $U = \{u_1, \ldots, u_r\}$ and $W = \{w_1, \ldots, w_r\}$ such that:

(a) $S_r[U]$ is a complete graph (r-clique) and W is an independent set in S_r
(b) for each i and j, w_i is adjacent to u_j if and only if $i = j$ or $i \equiv j + 1 \pmod{r}$.

Theorem 2.2.6 ([31]). *A graph is strongly chordal if and only if it does not contain as an induced subgraph a cycle of length greater than three or Sun S_r, $r \geq 3$.*

Theorem 2.2.7 ([29]). *Let G be a strongly chordal graph. Then*

$$cl_2(G) = \gamma(G).$$

2 Partitions of Graphs

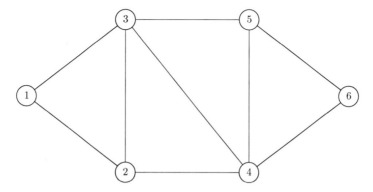

Fig. 2.1 A strongly chordal graph with a strong ordering (1, 2, 3, 4, 5, 6). The vertex 1 is simple: $N[1] = \{1, 2, 3\} \subseteq N[2] = \{1, 2, 3, 4\} \subseteq N[3] = \{1, 2, 3, 4, 5\}$

Consequently, there is a linear time algorithm to compute an optimal 2-clustering of a strongly chordal graph that is given with a strong elimination ordering. This algorithm applies a linear time algorithm to compute a minimum dominating set for strongly chordal graphs [31].

2.2.3 Hereditary Clique-Helly Graphs

A set family $\mathcal{F}$ has the *Helly property* if every collection of pairwise-intersecting sets from $\mathcal{F}$ has a common element.

In other words, if $\{S_1, \ldots, S_m\} \subseteq \mathcal{F}$, then

$$(\forall i, j \ S_i \cap S_j \neq \emptyset) \Rightarrow \left(\bigcap_{k=1}^{m} S_k \neq \emptyset \right).$$

A *clique* of a graph G is a maximal complete subgraph of G. We also use *"clique"* to mean *"vertex set of a clique"*.

A graph G is *clique-Helly* if the cliques of G have the Helly property, and G is *hereditary clique-Helly* if every induced subgraph of G is clique-Helly. Let us denote the class of hereditary clique-Helly graphs by $\mathcal{HCH}$. The set of minimal forbidden subgraphs of $\mathcal{HCH}$ is given by $\mathbf{F}_{\leq}(\mathcal{HCH}) = \{H_0 = S_3, H_1, H_2, H_3\}$, where H_1 is obtained from S_r by adding the edge w_1w_2, H_2 from H_1 by adding the edge w_2w_3, and H_3 from H_2 by adding the edge w_1w_3.

Let us denote the class of hereditary clique-Helly chordal graphs by $\mathcal{HCHC}$; i.e., let

$$\mathcal{HCHC} = \mathcal{HCH} \cap \mathcal{C}.$$

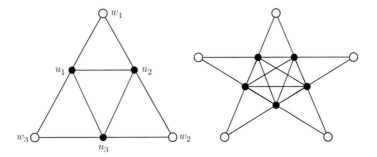

Fig. 2.2 Hajós graph: S_3 and Sun: S_5

Since each graph H_i, $i = 1, 2, 3$, from $F_{\leq}(\mathcal{HCH})$ contains C_4 as an induced subgraph, then the minimal forbidden subgraphs of $\mathcal{HCHC}$ are the Hajós graph S_3 and cycles of length greater than 3.

Consequently, it follows that

$$\mathcal{SC} \subset \mathcal{HCHC} \subset \mathcal{C}.$$

Since $S_5 \in \mathcal{HCHC} - \mathcal{SC}$ and $S_3 \in \mathcal{C} - \mathcal{HCHC}$, the above inclusions are proper.

Theorem 2.2.8. *There is a polynomial-time recognition algorithm for $\mathcal{HCHC}$ graphs.*

2.2.4 Minimum k-Clustering in $\mathcal{HCHC}$

Theorem 2.2.9. *Let $G \in \mathcal{HCHC}$. Then*

$$cl_2(G) = \gamma(G).$$

Proof. By Theorem 2.2.5 we have $cl_2(G) \leq \gamma(G)$. We prove that for $G \in \mathcal{HCHC}$, $cl_2(G) \geq \gamma(G)$.

Let $G \in \mathcal{HCHC}$ and $\{Q_1, \ldots, Q_l\}$ be 2-clusters of G with $l = cl_2(G)$. By heredity, $Q_i \in \mathcal{HCHC}$ and diam $(Q_i) \leq 2$ for $i = 1, \ldots, l$. Now, it is enough to prove that each Q_i has a universal vertex. Consider two cases.

Case 1. Q_i does not contain the Sun S_r, $r \geq 4$, $i = 1, \ldots, l$.
Hence each cluster Q_i is strongly chordal and Q_i has a universal vertex x_i, $i = 1, \ldots, l$. Thus the set $\{x_1, \ldots, x_l\}$ is dominating in G; i.e., $\gamma(G) \leq l = cl_2(G)$.

Case 2. For some i, $1 \leq i \leq l$, Q_i contains a Sun S_r, $r \geq 4$.
Let us denote Q_i and S_r briefly by Q and S, respectively, and let $V(S) = \{u_1, \ldots, u_r, w_1, \ldots, w_r\}$ with $G[\{u_1, \ldots, u_r\}] = K_r$. Consider three vertices in W with consecutive labels: w_t, w_{t+1}, w_{t+2}. It is easy to see that $d_S(w_t, w_{t+2}) = 3$

2 Partitions of Graphs   39

and therefore in S there is an induced path $w_t u_{t+1} u_{t+2} w_{t+2}$ of length three. Since $d_Q(w_t, w_{t+2}) = 2$, then there is $x \in V(Q) - V(S)$, a common neighbour of both w_t and w_{t+2}. Since Q is chordal, the vertex x has to be adjacent to u_{t+1} and u_{t+2}. If $w_{t+1} x \notin E$, then the set $\{w_t, u_{t+1}, w_{t+1}, u_{t+2}, w_{t+2}, x\}$ induces the Hajós graph, a contradiction. Thus $w_{t+1} x \in E$. It holds for any vertices with consecutive labels in the independent set W, thus x is adjacent to all vertices of S. Hence for each pair of vertices w_i and w_{i+2} there is a common neighbour $y \in V(Q) - V(S)$ with the same properties as above, possible $x = y$.

Let $X = \{x_1, \ldots, x_t\}$ be a set of vertices of $V(Q) - V(S)$ each of which dominates $V(S)$. If $x_i x_j \notin E$, then any two vertices of W with consecutive labels, say w_1, w_2 together with x_i, x_j, induce in Q a cycle C_4, which contradicts the chordality. It follows that the subgraph induced in Q by the set $\{x_1, \ldots, x_t\}$ is a clique.

Let $Y = V(Q) - (V(S) \cup X)$. If $|Y| \leq 1$, then any vertex of X is a universal vertex of Q. Assume that $|Y| \geq 2$. Let $y_1, y_2 \in Y$ and suppose that y_1, y_2 do not have a common neighbour in Y. Then either $d(y_1, y_2) \geq 3$ or $y_1 y_2 \in E$. In the first case, we have a contradiction; in the second, there is in Q an induced cycle C_4, which contradicts chordality of Q. Hence any two vertices of Y have a common neighbour in X. Since $Q[X \cup Y] \in \mathcal{HCHC}$, the Helly property implies that there is a vertex $x \in X$ such that $xy \in E$ for all $y \in Y$. Thus the vertex x is a universal vertex of Q, which completes the proof. $\qquad\square$

Problem 2.2.10. VERTEX COVER

Given a graph $G = (V, E)$ and an integer k, $1 \leq k \leq |V|$. Is there a *vertex cover* of cardinality $\leq k$ for G, i.e., a subset V' of V with $|V'| \leq k$ such that V' contains at least one vertex from every edge in E?

Theorem 2.2.11. *Problem* DOMINATING SET *is* NP-*complete for* $\mathcal{HCHC}$ *graphs.*

Proof. We transform the VERTEX COVER problem which remains NP-complete even when restricted to planar triangle free graphs [35] to DOMINATION SET in $\mathcal{HCHC}$, in the following way: Let $G = (V, E)$ be a planar triangle-free graph. Construct the graph $H = (U, F)$ with the vertex set $U = V \cup E$ and the edge set

$$F = \{vv' : v, v' \in V, v \neq v'\} \cup \{ve : v \in V, e \in E \text{ and } v \in e\}.$$

Claim 1. $H \in \mathcal{HCHC}$.

Proof. It is easy to see that H is chordal. Now it is enough to prove that H does not contain the Sun S_3 as an induced subgraph. On the contrary, suppose that $S_3 \leq H$. It implies that w_1, w_2, w_3 and u_1, u_2, u_3 of S_3 correspond to some vertices, say e_1, e_2, e_3 and v_1, v_2, v_3, respectively, in H. By the above and the construction of H it follows that the graph G contains a triangle induced by the set $\{v_1, v_2, v_3\}$, a contradiction. $\qquad\square$

Claim 2. G has a vertex cover of cardinality d if and only if H has a dominating set of cardinality d.

Fig. 2.3 Graph classes and domination set complexity

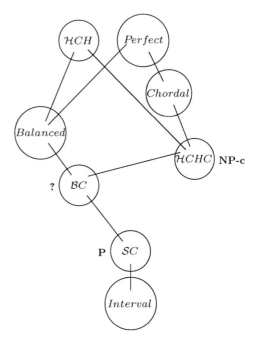

Proof. Let C be a vertex cover of G with $|C| = d$. Since $H[V]$ is a clique, every vertex of V is dominated by each vertex of C. Every $e \in E$ is covered by some v in C, thus e is dominated by v in H. Therefore C is a dominating set of cardinality d in H.

Suppose now that H has a dominating set D of cardinality d. If $D \subseteq V$, then D is also a vertex cover of G. (Each vertex $e \in E$ in H is dominated by some element of D, thus each $e \in E(G)$ has at least one vertex in D.) If $D \cap E \neq \emptyset$, i.e., say $e = vv' \in E$ is in D we may replace e by v or v' and get a new dominating set D' of the same cardinality. Applying this procedure to all $e \in D \cap E$ we will finally get a dominating set $D^* \subseteq V$ with $|D^*| = d$. □

Hence the NP-completeness of the DOMINATING SET problem in $\mathcal{HCHC}$ follows from the VERTEX COVER problem for planar triangle-free graphs. □

2.2.5 Balanced Graphs

A hypergraph H is said to be *balanced* if every odd cycle has an edge containing three vertices of the cycle; see [4]. In other words, H is balanced if and only if its incidence matrix contains no square submatrix of an odd cycle.

A graph G is called *balanced* if its clique hypergraph is balanced.

2 Partitions of Graphs   41

Theorem 2.2.12 ([28]). *There is a polynomial-time recognition algorithm for balanced graphs.*

Let us denote the class of balanced chordal graphs by $\mathcal{BC}$.

Theorem 2.2.13 ([45]). *A graph $G \in \mathcal{BC}$ if and only if it is odd Sun-free chordal.*

From Theorem 2.2.9 we have

Corollary 2.2.14. *Let $G \in \mathcal{BC}$. Then*

$$cl_2(G) = \gamma(G).$$

Open Problem 2.2.15. *Problem* DOMINATING SET *remains open for the class $\mathcal{BC}$.*

2.3 Matching Cutset

A set M of independent edges in a graph G is called a *matching*. A set $F \subseteq E(G)$ is a cutset in G if $G - F$ has more components than G. If a cutset is a matching of G then it is called a *matching cutset*. The problem of recognizing graphs with a matching cutset is well studied in the literature. For a survey of what is known to date, see [20, 27, 44, 58]. Historically, the first theorem for matching cutset complexity was given by Chvatál [25].

We list some important results concerning this problem. Some of them deal with well-known classes of graphs.

Theorem 2.3.1 ([25]). *It is* NP-*complete to recognize graphs with a matching cutset even if the input is restricted to graphs with $\Delta = 4$.*

The next two results yield conditions on vertex degrees of bipartite graphs sufficient to guarantee NP-completeness of matching cutsets.

Theorem 2.3.2 ([49]). MATCHING CUTSET *is* NP-*complete, even if the input is restricted to bipartite graphs of minimum degree* 2.

Theorem 2.3.3 ([44]). MATCHING CUTSET *is* NP-*complete, even if the input is restricted to bipartite graphs with one colour class consisting only of vertices of degree* 3 *and the other colour class consisting only of vertices of degree* 4.

Below are presented a few graph classes for which the MATCHING CUTSET problem is polynomial. The *line graph of G*, denoted by $L(G)$, is the graph the vertex set of which is the edge set of G and two vertices of $L(G)$ are adjacent if and only if, as edges in G, they are adjacent. A graph H is called a *line graph* if there is a graph G such that H is isomorphic to $L(G)$.

Theorem 2.3.4 ([49]). *Let $G = (V, E)$ be a line graph. Then we can determine in $O(|E|)$ time whether G has a matching cutset.*

Theorem 2.3.5 ([49]). *Let $G = (V, E)$ be a graph without induced cycles of length ≤ 4. Then we can determine in $O(|V|^3|E|)$ time whether G has a matching cutset.*

Theorem 2.3.6 ([16]). *Let G be a graph with $\mathrm{diam}(G) = 2$. Then the* MATCHING CUTSET *problem for G can be solved in polynomial time.*

Open Problem 2.3.7. *Find k which is a boundary that separates* NP-*complete instances of diameter k of the* MATCHING CUTSET *problem from polynomially solvable ones.*

2.4 Acyclic Colourings

Acyclic colourings have been studied extensively over the past 30 years. Several authors have been able to determine $\chi_a(\mathcal{P})$ for some classes $\mathcal{P}$ of graphs such as graphs of maximum degree 3, considered by Grűnbaum in [37] and of maximum degree 4, studied by Burstein in [24]. The acyclic chromatic number of planar graphs was determined by Borodin in 1979; see [8] for details. Planar graphs with "large" girth, outerplanar, and 1-planar graphs were also considered; see, for instance, [9, 10].

2.4.1 Selected Results

Theorem 2.4.1 ([42]). *It is an* NP-*complete problem to decide for a given graph G and $k \geq 3$ if the acyclic chromatic number of G is at most k.*

In 2004 Skulrattanakulchai [54] proved that there is a linear time algorithm that acyclically colours any graph of maximum degree 3 in four colours.

Theorem 2.4.2 ([34]). *For any graph G of maximum degree 5, $\chi_a(G) \leq 9$ and there exists a linear time algorithm to acyclically colour G in at most nine colours.*

Authors suspect that the upper bound of nine colours in the case $\Delta(G) = 5$ is not tight.

Theorem 2.4.3 ([37]).
$$\chi_a(\mathcal{S}_3) \leq 4.$$

Theorem 2.4.4 ([8]).
$$\chi_a(\mathcal{T}_3) \leq 5.$$

The bound $\chi_a(\mathcal{S}_3) \leq 4$ proved by Grűnbaum in Theorem 2.4.3 is the best possible. Moreover, Kostochka and Mel'nikov [43] proved that there are bipartite 2-degenerate planar graphs which are not acyclically 4-colourable. Acyclic colourings turned out to be useful for obtaining results about other types of colourings, see [39].

2.4.2 Brooks-Type Results

The well-known theorem of Brooks [22] relates the chromatic number of a graph to its maximum degree.

Theorem 2.4.5. *For any connected graph G,*

$$\chi(G) \leq \Delta(G) + 1$$

with equality if and only if either $\Delta(G) = 2$ and G is an odd cycle or $\Delta(G) \neq 2$ and G is a complete graph.

There are many generalisations of the Brooks theorem. These theorems are called Brooks-type results. For the acyclic chromatic number finding a sharp upper bound as a function of maximum degree seems to be an extremely hard problem.

Theorem 2.4.6 ([3]).
$$\chi_a(\mathcal{S}_\Delta) \leq C\Delta^{4/3}.$$

Theorem 2.4.7 ([34]). *For a graph G of maximum degree $\Delta \geq 5$ we have*

$$\chi_a(G) \leq \frac{\Delta(\Delta - 1)}{2}.$$

Open Problem 2.4.8. *Find a sharp upper bound for $\chi_a(G)$ as a function of $\Delta(G)$.*

2.4.3 Improper Acyclic Colouring

Studies have begun in [7] of acyclic colourings of graphs with respect to hereditary properties of graphs. Namely, they have considered outerplanar, planar graphs, and graphs with bounded degree; see [6,7]. They call such acyclic colouring improper.

Formally, an *improper acyclic colouring* of graphs is an M-partition with $m_{ii} = \mathcal{P}_i$ and $m_{ij} = \mathcal{D}_1$ for $i \neq j$, $1 \leq i, j \leq k$. We denote it briefly by $((\mathcal{P}_1, \ldots, \mathcal{P}_k), \mathcal{D}_1)$.

The class of graphs having $((\mathcal{P}_1, \ldots, \mathcal{P}_k), \mathcal{D}_1)$-partition is denoted by $\mathcal{P}_1 \odot \cdots \odot \mathcal{P}_k$.

An additive hereditary property $\mathcal{R}$ is said to be *acyclic reducible* in $\mathbf{L}^a$ if there are nontrivial additive hereditary properties $\mathcal{P}_1, \mathcal{P}_2$ such that $\mathcal{R} = \mathcal{P}_1 \odot \mathcal{P}_2$ and *acyclic irreducible* in $\mathbf{L}^a$, otherwise.

It is easy to see that the smallest acyclic reducible property in $\mathbf{L}^a$ is the property $\mathcal{O} \odot \mathcal{O} = \mathcal{D}_1$. Obviously, $\mathcal{D}_1$ is the unique minimal acyclic reducible bound for $\mathcal{P}$ if and only if $\mathcal{P} \subset \mathcal{D}_1$.

A maximal outerplanar graph G with at least three vertices is called a *2-path of order* $n = 2p$, if G consists of two paths $P_1 = (x_1, x_2, \ldots, x_p)$, $P_2 = (y_1, y_2, \ldots, y_p)$ and additional edges: $x_i y_i$, $i = 1, \ldots, p$ and $x_j y_{j+1}$ for $j = 1, \ldots, p-1$. For an odd $n = 2p-1$ a 2-path H is defined as $H = G - x_p$, where G is a 2-path of even order. A 2-path of order n is denoted by P_n^2.

A maximal outerplanar graph G with at least three vertices is called a *fan of order* n, if G is obtained from a star $K_{1,n-1}$ by joining all vertices of degree one by a path. A fan of order n is denoted by F_n.

Additionally we assume that the graph K_1 and K_2 is a *trivial 2-path* and a *trivial fan*. For each $n \leq 5$ there is exactly one (up to isomorphism) maximal outerplanar graph which is a 2-path and a fan.

Some acyclic reducible bounds for the class of outerplanar graphs can be found in [13].

Theorem 2.4.9 ([13]).

$$T_2 \subseteq \mathcal{O} \odot Forb(S_3, P_6^2),$$

$$T_2 \subseteq \mathcal{O} \odot Forb(S_3, F_6).$$

We have an interesting proposition which characterises fans and 2-paths in the class of maximal outerplanar graphs. It was used in the proof of the next corollary and theorem and gives some light on an open problem formulated at the end of this section.

Proposition 2.4.10 ([13]). *Let G be a maximal outerplanar graph of order $n \geq 3$. Then*

(a) G is a fan if and only if neither $S_3 \subseteq G$ nor $P_6^2 \subseteq G$.
(b) G is a 2-path if and only if neither $S_3 \subseteq G$ nor $F_6 \subseteq G$.

Let us recall that a *block* of a given graph G is defined to be a maximal connected subgraph of G without a cutvertex.

A *fan* (*2-path*) *tree* is a connected graph G every block of which is a fan (2-path).

Let us define the property $\mathcal{FT}$ ($\mathcal{PT}$) as the class of all fan (2-path) trees and their subgraphs. Both classes are hereditary and form a proper subclass of all outerplanar graphs.

From the definition of $\mathcal{FT}$ it follows that S_3 and P_6^2 do not belong to $\mathcal{FT}$. Similarly, S_3 and F_6 do not belong to $\mathcal{PT}$. It implies the following corollary.

2 Partitions of Graphs

Corollary 2.4.11.

$$\mathcal{FT} \subset Forb(S_3, P_6^2),$$

$$\mathcal{PT} \subset Forb(S_3, F_6).$$

Because of the above corollary, the next theorem gives two acyclic reducible bounds for outerplanar graphs which are better than those in Theorem 2.4.9.

Theorem 2.4.12.

$$\mathcal{T}_2 \subset \mathcal{O} \odot \mathcal{FT},$$

$$\mathcal{T}_2 \subset \mathcal{O} \odot \mathcal{PT}.$$

Open Problem 2.4.13. *Find at least one minimal acyclic reducible bound for the class of outerplanar graphs.*

Theorem 2.4.14 ([7]).

$$S_3 \subseteq S_1 \odot S_1 \odot S_1.$$

Conjecture 2.4.15 ([7]).

$$S_3 \subseteq S_2 \odot S_2.$$

Theorem 2.4.16 ([14]).

1. *Boiron, Sopena, and Vignal's conjecture is true; i.e., $S_3 \subseteq S_2 \odot S_2$ holds.*
2. *There is a polynomial time algorithm to acyclically (S_2, S_2)-colour each graph $G \in S_2$.*
3. *It is NP-complete to determine whether a graph $G \in S_4$ has an acyclic (S_2, S_2)-colouring.*

The first statement of Theorem 2.4.16 cannot be generalized for $k = 4$. It was observed by Hałuszczak [Hałuszczak M, 2007, private communication]. But this observation can be extended for all $k \geq 4$.

Theorem 2.4.17. $S_k \nsubseteq S_{k-1} \odot S_{k-1}$, *for $k \geq 4$.*

Proof. Let $G = C_k + D_{k-2}$, where by D_n the edgeless graph of order n is denoted; i.e., the graph G is obtained by joining each vertex of cycle of order k with each vertex of edgeless graph of order $k - 2$. Obviously, the graph $G \in S_k$ but $G \notin S_{k-1} \odot S_{k-1}$. We left an easy proof of that fact to a reader. □

Some other type results on acyclic colouring of graphs are presented in [6].

References

1. Abbas N (1995) Graph clustering: complexity, sequential and parallel algorithms. Ph.D. thesis, University of Alberta, Edmonton, Canada
2. Alekseev VE, Farrugia A, Lozin VV (2004) New results on generalized graph coloring. Discrete Math Theor Comput Sci 6:215–222

3. Alon N, McDiarmid C, Reed B (1991) Acyclic coloring of graphs. Random Struct Algorithm 2:277–288
4. Berge C (1987) Hypergraphs. Gauthier-Villars, Paris
5. Berger A (2001) Minimal forbidden subgraphs of reducible graph properties. Discuss Math Graph Theory 21:111–117
6. Boiron P, Sopena E, Vignal L (1999) Acyclic improper colorings of graphs. J Graph Theory 32:97–107
7. Boiron P, Sopena E, Vignal L (1999) Acyclic improper colourings of graphs with bounded degree. DIMACS Series Discrete Math Theor Comput Sci 49:1–9
8. Borodin OV (1979) On acyclic colorings of planar graphs. Discrete Math 25:211–236
9. Borodin OV, Kostochka AV, Raspaud A, Sopena E (2001) Acyclic colourings of 1-planar graphs. Discrete Appl Math 114:29–41
10. Borodin OV, Kostochka AV, Woodall DR (1999) Acyclic colorings of planar graphs with large girth. J Lond Math Soc 60:344–352
11. Borowiecki M, Broere I, Frick M, Mihók P, Semanišin G (1997) A survey of hereditary properties of graphs. Discuss Math Graph Theory 17:5–50
12. Borowiecki M, Broere I, Mihók P (2000) Minimal reducible bounds for planar graphs. Discrete Math 212:19–27
13. Borowiecki M, Fiedorowicz A (2006) On partitions of hereditary properties of graphs. Discuss Math Graph Theor 26:377–387
14. Borowiecki M, Fiedorowicz A, Jesse-Józefczyk K, Sidorowicz E (2010) Acyclic colourings of graphs with bounded degree. Discrete Math Theor Comput Sci 12:59–74
15. Borowiecki M, Hałuszczak M, Skowrońska M (2000) Minimal reducible bounds for 1-nonouterplanar graphs. Discrete Math 219:9–15
16. Borowiecki M, Jesse-Józefczyk K (2008) Matching cutsets in graphs of diameter 2. Theor Comput Sci 407:574–582
17. Borowiecki M, Mihók P (1991) Hereditary properties of graphs. In: Kulli VR (ed) Advances in graph theory. Vishwa International Publication, Gulbarga, pp 41–68
18. Borowiecki P (1993) $\mathcal{P}$-bipartitions of graphs. Vishwa Int J Graph Theory 2:109–116
19. Borowiecki P, Ivančo J (1997) $\mathcal{P}$-bipartitions of minor hereditary properties. Discuss Math Graph Theory 17:89–93
20. Brandstädt A, Le VB, Szymczak T (1998) The complexity of some problems related to GRAPH 3-COLORABILITY. Discrete Appl Math 89:59–73
21. Broere I, Mynhardt CM (1984) Generalized colorings of outerplanar and planar graphs. In: Proceedings of the fifth international conference on the theory and applications of graphs with special emphasis on algorithms and computer science. Wiley-Interscience, New York, pp 151–161
22. Brooks RL (1941) On colouring the nodes of a network. Proc Camb Philos Soc 37:194–197
23. Brucker P (1978) On the complexity of clustering problems. In: Henn R, Korte B, Oettli W (eds) Optimization and operations research. Lecture notes in economics and mathematical systems, vol 157. Springer, New York, pp 45–54
24. Burstein MI (1979) Every 4-valent graph has an acyclic 5-coloring. Soobšč. Akad Gruzin SSR 93:21–24 (in Russian)
25. Chvátal V (1984) Recognizing decomposable graphs. J Graph Theory 8:51–53
26. Chartrand G, Kronk HV (1969) The point arboricity of planar graphs. J Lond Math Soc 44:612–616
27. Coreli DG, Fonlupt J (1993) Stable set bonding in perfect graphs and parity graphs. J Combin Theory B 59:1–14
28. Dahlhaus E, Manuel P, Miller M (1998) Maximum h-colourable subgraph problem in balanced graphs. Inform Process Lett 65:301–303
29. Deogun JS, Kratsch D, Steiner G (1997) An approximation algorithm for clustering graphs with dominating diametral path. Inform Process Lett 61:121–127
30. Diestel R (1997) Graph theory. Springer, Berlin
31. Faber M (1983) Characterizations of strongly chordal graphs. Discrete Math 43:173–189

2 Partitions of Graphs

32. Farley A, Hedetniemi S, Proskurowski A (1981) Partitioning trees: matching, domination and maximum diameter. Int J Comput Inform Sci 10:55–61
33. Farrugia A (2004) Vertex-partitioning into fixed additive induced-hereditary properties is NP-hard. Electron J Combinator 11, #R46
34. Fertin G, Raspaud A (2008) Acyclic coloring of graphs of maximum degree five: nine colors are enough. Inform Process Lett 105:65–72
35. Garey MR, Johnson DS, Stockmeyer L (1976) Some simplified NP-complete graph problems. Theor Comput Sci 1:237–267
36. Goddard W (1991) Acyclic colorings of planar graphs. Discrete Math 91:91–94
37. Grűnbaum B (1973) Acyclic coloring of planar graphs. Israel J Math 14:390–412
38. Hakimi SL, Schmeichel E, Weinstein J (1990) Partitioning planar graphs into independent sets and forests. Congr Numer 78:109–118
39. Jensen TR, Toft B (1995) Graph coloring problems. Wiley-Interscience, New York
40. Karp RM (1972) Reducibility among combinatorial problems. In: Miller RE, Thatcher JW (eds) Complexity of computer computations. Plenum, New York, pp 85–104
41. Kershenbaum A (1993) Telecommunications network design algorithms. McGraw-Hill, New York
42. Kostochka AV (1978) Ph.D. thesis, Novosibirsk
43. Kostochka AV, Mel'nikov LS (1976) Note to the paper of Grűnbaum on acyclic colorings. Discrete Math 14:403–406
44. Lee VB, Randerath B (2003) On stable cutsets in line graphs. Theor Comput Sci 301:463–475
45. Lehel J, Tuza Zs (1986) Neighborhood perfect graphs. Discrete Math 61:93–101
46. Mahadev NVR, Peled VN (1995) Threshold graphs and related topics. Ann Discrete Math 56:170
47. Mihók P (1996) On the minimal reducible bound for outerplanar and planar graphs. Discrete Math 150:431–435
48. Mihók P (2001) Minimal reducible bounds for the class of k-degenerate graphs. Discrete Math 236:273–279
49. Moshi AM (1989) Matching cutsets in graphs. J Graph Theory 13:527–536
50. Nešetřil J, Rődl V (1976) Partitions of vertices. Comment Math Univ Carolinae 17:85–95
51. Poh KS (1990) On the linear vertex-arboricity of a planar graph. J Graph Theory 14:73–75
52. Semanišin G (2004) Minimal reducible bounds for induced-hereditary properties. Discrete Math 286:163–170
53. Sharan R, Maron-Katz A, Shamir R (2003) CLICK and EXPANDER: a system for clustering and visualizing gene expression data. Bioinformatics 19:1787–1799
54. Skulrattanakulchai S (2004) Acyclic colorings of subcubic graphs. Inform Process Lett 92:161–167
55. Stein SK (1971) B-sets and planar graphs. Pac J Math 37:217–224
56. Thomassen C (1995) Decomposing a planar graph into degenerate graphs. J Combin Theory B 65:305–314
57. Tou JT, Gonzalez RC (1974) Pattern recognition principles. Addison Wesley, Reading, MA
58. Tucker A (1983) Coloring graphs with stable cutsets. J Combin Theory B 34:258–267
59. West DB (2001) Introduction to graph theory, 2nd edn. Prentice Hall, Upper Saddle River
60. Wu Z, Leahy R (1993) An optimal graph theoretic approach to data clustering: theory and its application to image segmentation. IEEE Trans Pattern Anal Mach Intell 15:1101–1113

Chapter 3
Distance in Graphs

Wayne Goddard and Ortrud R. Oellermann

Abstract The distance between two vertices is the basis of the definition of several graph parameters including diameter, radius, average distance and metric dimension. These invariants are examined, especially how they relate to one another and to other graph invariants and their behaviour in certain graph classes. We also discuss characterizations of graph classes described in terms of distance or shortest paths. Finally, generalizations are considered.

Keywords Graph · Distance · Diameter · Radius · Steiner distance

MSC2000: Primary 05C12; Secondary 05C20, 05C38

3.1 Overview of Chapter

The distance between two vertices in a graph is a simple but surprisingly useful notion. It has led to the definition of several graph parameters such as the diameter, the radius, the average distance and the metric dimension. In this chapter we examine these invariants; how they relate to one another and other graph invariants and their behaviour in certain graph classes. We also discuss characterizations of graph classes that have properties that are described in terms of distance or shortest paths. We later consider generalizations of shortest paths connecting pairs of vertices to shortest trees, called Steiner trees, that connect three or more vertices.

W. Goddard (✉)
School of Computing and Department of Mathematical Sciences,
Clemson University, Clemson, SC 29634-1906, USA
e-mail: goddard@clemson.edu

M. Dehmer (ed.), *Structural Analysis of Complex Networks*,
DOI 10.1007/978-0-8176-4789-6_3, © Springer Science+Business Media, LLC 2011

3.2 Distance, Diameter and Radius

A *path* in a graph is a sequence of distinct vertices, such that adjacent vertices in the sequence are adjacent in the graph. For an unweighted graph, the *length* of a path is the number of edges on the path. For an (edge) weighted graph, the length of a path is the sum of the weights of the edges on the path. We assume that all weights are nonnegative and that all graphs are connected. We start with undirected graphs.

The *distance* between two vertices u and v, denoted $d(u, v)$, is the length of a shortest $u - v$ path, also called a $u - v$ *geodesic*. The distance function is a metric on the vertex set of a (weighted) graph G. In particular, it satisfies the *triangle inequality*:

$$d(a, b) \leq d(a, c) + d(c, b)$$

for all vertices a, b, c of G. This follows from the fact that, if you want to get from a to b, then one possibility is to go via vertex c.

3.2.1 Diameter and Radius

Two of the most commonly observed parameters of a graph are its radius and diameter. The *diameter* of a connected graph G, denoted $\text{diam}(G)$, is the maximum distance between two vertices. The *eccentricity* of a vertex is the maximum distance from it to any other vertex. The *radius*, denoted $\text{rad}(G)$, is the minimum eccentricity among all vertices of G. Of course the diameter is the maximum eccentricity among all vertices.

For a (weighted) undirected graph G:

$$\text{rad}(G) \leq \text{diam}(G) \leq 2\,\text{rad}(G).$$

The upper bound follows from the triangle inequality, where c is a vertex of minimum eccentricity.

The radius and diameter are easily computed for simple graphs.

Fact 1 *1. Complete graphs:* $\text{diam}(K_n) = \text{rad}(K_n) = 1$ *(for $n \geq 2$).*
2. Complete bipartite graphs: $\text{diam}(K_{m,n}) = \text{rad}(K_{m,n}) = 2$ *(if n or m is at least 2).*
3. Path on n vertices: $\text{diam}(P_n) = n - 1$; $\text{rad}(P_n) = \lceil (n-1)/2 \rceil$.
4. Cycle on n vertices: $\text{diam}(C_n) = \text{rad}(C_n) = \lfloor n/2 \rfloor$.

Note that cycles and complete graphs are vertex-transitive, so the radius and diameter are automatically the same (every vertex has the same eccentricity).

The *centre* $C(G)$ is the subgraph induced by the set of vertices of minimum eccentricity. Graphs G where $\text{rad}(G) = \text{diam}(G)$ are called *self-centred*.

3 Distance in Graphs 51

A famous result, due originally to Jordan [44], is that:

Fact 2 *For trees T, the diameter equals either $2\operatorname{rad}(T)$ or $2\operatorname{rad}(T) - 1$. In the first case the center is a single vertex, and in the second the centre is a pair of adjacent vertices.*

The *derivative T'* of a tree T is the tree obtained by deleting the leaves of T. Suppose the kth derivative $T^{(k)}$ of T has been defined for some $k \geq 1$. Then the $(k + 1)$st derivative of T is defined by $T^{(k+1)} = (T^{(k)})'$. It can be shown that the centre of a tree is $T^{(\lfloor diam(T)/2 \rfloor)}$.

For general graphs, Harary and Norman [40] showed the following.

Fact 3 *The centre of a graph forms a connected subgraph, and is contained inside a block of the graph.*

In general, there are no structural restrictions on the centre of a graph. Indeed, Hedetniemi (see [9]) showed that every graph is the centre of some graph.

3.2.2 Bounds for the Radius and Diameter

We look next at upper bounds involving the minimum degree δ or maximum degree Δ.

Theorem 1. *For a graph G of order n:*

1. $\operatorname{diam}(G) \leq n - \Delta(G) + 1$.
2. *[71]* $\operatorname{rad}(G) \leq (n - \Delta(G))/2 + 1$.
3. *If $\delta(G) \geq n/2$, then $\operatorname{diam}(G) \leq 2$.*

And these bounds are sharp.

One can also consider the problem of maximizing or minimizing these parameters given the number of vertices and edges. Both the radius and diameter are minimized by the star (and any supergraph thereof), so that case is not interesting.

For a given number n of vertices and $m \geq n - 1$ of edges, the diameter and radius are maximized by the "path-complete" and the "cycle-complete" graphs, respectively. These are defined as follows. A *cycle-complete* graph is obtained by taking disjoint copies of a cycle of even length and a complete graph, and joining three consecutive vertices of the cycle to all vertices in the complete graph. The radius is half the length of the cycle. This graph was introduced by Vizing [71]. An example is given in Fig. 3.1.

A *path-complete* graph is obtained by taking disjoint copies of a path and complete graph, and joining an end-vertex of the path to *one or more* vertices of the complete graph. It is not hard to show that there is a unique path-complete graph

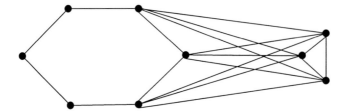

Fig. 3.1 A cycle-complete graph

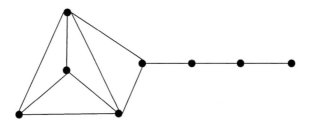

Fig. 3.2 The path-complete graph $PK_{8,11}$

with n vertices and m edges; this is denoted as $PK_{n,m}$. This graph was introduced by Harary [38]. An example is given in Fig. 3.2.

Theorem 2. 1. *[71] For a given number n of vertices and $m \geq n - 1$ of edges, the radius is maximized by the cycle-complete graphs.*
2. *[38] For a given number n of vertices and $m \geq n - 1$ of edges, the diameter is maximized by the path-complete graph $PK_{n,m}$.*

It is not hard to show that if a graph G has large diameter, then its complement $\bar{G}$ has small diameter:

Fact 4 1. *If $\mathrm{diam}(G) > 3$, then $\mathrm{diam}(\bar{G}) \leq 2$.*
2. *If $\mathrm{diam}(G) = 3$, then $\mathrm{diam}(\bar{G}) \leq 3$.*

3.2.3 Changes in Diameter and Radius with Edge and Vertex Removal

Removing an edge can never decrease the radius or the diameter of a graph. Indeed, removing a bridge disconnects the graph. So we consider cyclic edges.

Fact 5 *If e is a cyclic edge of graph G, then $\mathrm{rad}(G) \leq \mathrm{rad}(G - e) \leq 2\,\mathrm{rad}(G)$ and $\mathrm{diam}(G) \leq \mathrm{diam}(G - e) \leq 2\,\mathrm{diam}(G)$.*

Both upper bounds are attainable. Removing a cyclic edge can easily double the diameter; for example, it does so in the odd cycles. Plesník [61] showed that if every

3 Distance in Graphs

Fig. 3.3 Large increase in the diameter/radius of $G - v$

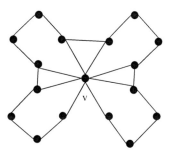

edge's removal doubles the diameter, then the graph is a Moore graph, i.e., a graph of diameter d and girth $2d + 1$ for some $d \geq 1$. For example, Moore graphs include the complete graphs, the odd cycles, the Petersen graph and the Hoffman–Singleton graph (see [7]). Removing a cyclic edge can also double the radius. Such a graph can be constructed by taking two equal-sized cycles, and sticking them together along one edge e.

There is a natural spanning tree with the same radius as the original graph, sometimes called a *breadth-first search* tree. This tree has diameter at most double its radius, and hence at most double the original radius. Buckley and Lewinter [8] determined which graphs have a diameter-preserving spanning tree.

Removing a vertex can both increase and decrease these parameters. Removing a cut-vertex from a disconnected graph results in a disconnected graph, so we do not consider such vertices.

Fact 6 *If v is a non-cut-vertex of graph G, then* $\mathrm{rad}(G - v) \geq \mathrm{rad}(G) - 1$ *and* $\mathrm{diam}(G - v) \geq \mathrm{diam}(G) - 1$.

The above bound is attained, for example, by an even-order path where v is an end-vertex of the path. There is no upper bound on $\mathrm{rad}(G - v)$ or $\mathrm{diam}(G - v)$ if G is 2-connected. To see this, let G be obtained from m cycles $C_i : v_{i_1} v_{i_2} \ldots v_{i_{2d+1}} v_{i_1}$ for $1 \leq i \leq m$ by identifying the vertices v_{i_1} for $1 \leq i \leq m$ into a vertex v and then adding the edges $v_{i_{2d+1}} v_{(i+1)_2}$ for $i = 1, 2, \ldots, m - 1$. Then $\mathrm{diam}(G) = 2d$ and $\mathrm{diam}(G - v) = 2dm - 1$. Since m can be made as large as we wish, there is no upper bound on $\mathrm{diam}(G - v)$ in terms of $\mathrm{diam}(G)$. Figure 3.3 shows this construction with $d = 2$ and $m = 4$.

3.2.4 Matrices and Walks

A *walk* is a sequence of not necessarily distinct vertices such that each vertex in the sequence except the first one is adjacent to the previous one. Suppose the vertices of a graph G are labeled $v_1, \ldots, v_n$. Then the *adjacency matrix* A of G is defined as the $n \times n$ matrix whose (i, j)-th entry is 1 if there is an edge joining vertices v_i and v_j, and 0 otherwise. The *Laplacian matrix* is defined as $A - D$ where D is the diagonal matrix whose (i, i)-entry is the degree of the vertex v_i. The eigenvalues of A are referred to as the *eigenvalues* of G, and the eigenvalues of L as the *Laplacian eigenvalues* of G. The following fact is well-known:

Fact 7 *The (i, j)-th entry of the power A^k gives the number of walks of length k from v_i to v_j.*

Bounds on the diameter of a graph in terms of the number of eigenvalues of these matrices follow from Fact 7 (see, for example, [17]):

Theorem 3. *Let G be a connected graph and b the number of distinct eigenvalues of G. Then*

$$\text{diam}(G) \leq b - 1.$$

The same result holds if b is the number of distinct Laplacian eigenvalues of G.

3.3 Other Measures of Centrality

Apart from the centre of a graph, there are several other centrality measures. These have many applications. The centre is for emergency facility location: the response time must be minimized in the worst case. For biological graphs, the centre vertices might be the most important (see, for example, [42]).

Another measure of centrality is the "median" of a graph. The *status* $\sigma(v)$ of a vertex v is the sum of the distances from v to all other vertices. The vertices having minimum status $\sigma(v)$ form the *median* of the graph, which is denoted by $M(G)$. For example, the median might be a good place to locate a mall: the average driving distance is minimized.

There is no intrinsic connection between the centre and the median. Indeed, they can be arbitrarily far apart. Slater [67] considered whether there are other measures of centrality that "connect" the centre and the median of a graph. He defined for a graph G, integer $k \geq 2$ and vertex u of G,

$$r_k(u) = \max \left\{ \sum_{s \in S} d(u, s) \mid S \subseteq V(G), \ |S| = k \right\}.$$

Thus, $r_k(u)$ is the sum of the k largest vertex distances to u. In particular, $r_1(u)$ is its eccentricity, and $r_{n-1}(u)$ is its status.

The *k-centrum*, $C(G; k)$, of G is the subgraph induced by those vertices u for which $r_k(u)$ is a minimum. Thus, $C(G; 2) = C(G)$ and $C(G; n - 1) = M(G)$. It is shown in [67] that the k-centrum of every tree consists of a single vertex or a pair of adjacent vertices. Further, every vertex on the shortest path from the centre to the median of a tree belongs to the k-centrum for some k between 1 and n.

In the case of trees, another centrality measure is well known. Suppose T is a tree and v a vertex of T. Then a *branch* at v is a maximal subtree containing v as a leaf.

The *weight* at a vertex v of T is the maximum number of edges in any branch at v. A vertex of T is a *centroid vertex* of T if it has minimum weight among all vertices of T and the subgraph induced by the centroid vertices is called the *centroid* of T. That is, a centroid vertex is one whose removal leaves the smallest maximum component order. However, Zelinka [77] observed that the centroid and the median coincide.

Another centrality idea is a "central path". This concept can be defined in one of two ways that parallel the centre and median, respectively. If H is a subgraph of a graph G and v is any vertex of G, then the distance from v to H is the minimum distance from v to a vertex of H. The *eccentricity* of H is the distance of a vertex v farthest from H, and the *status* of H is the sum of the distances of every vertex of G to H. A path P is a *path centre* of G if P has minimum eccentricity among all paths of G and has minimum length among all such paths. Similarly the *path median* of G is a path of minimum status in G. The path median of a network may indicate a good choice for a subway line, for example; in that case many individuals will use this line, so it is desirable that the average distance from the users' homes to the line is minimized. On the other hand the problem of finding the path centre of a network has application to the problem of building a highway between two major cities in such a way that the furthest distance from the highway to any town in a collection of other "important" towns is minimized. The problem of finding the path centre of a tree was solved independently in [16, 68]. Further types of centres are discussed in the book [7]. See [32] for more of the early history of centrality.

3.4 Special Graph Classes

3.4.1 Chordal Graphs

A *chordal graph* is one where every cycle of length greater than 3 has a chord. For example, trees and maximal outerplanar graphs are chordal. A *simplicial vertex* is one whose neighbourhood is complete. It was first shown by Dirac [21] and later by Lekkerkerker and Boland [49] that every chordal graph has a simplicial vertex (indeed at least two such vertices). Since induced subgraphs of chordal graphs are still chordal, chordal graphs have a *simplicial elimination ordering*; that is, an ordering $v_1, v_2, \ldots, v_n$ of the vertices such that the neighbourhood of v_i is complete in the induced graph $\langle \{v_i, v_{i+1}, \ldots, v_n\} \rangle$. Using induction one can show that every graph that has a simplicial elimination ordering is chordal.

The centre of chordal graphs was investigated by Laskar and Shier [48]. For example, they showed that the centre of a chordal graph is connected, and provided the following generalization of Fact 2.

Theorem 4. *For a chordal graph G, $\operatorname{diam}(G) \geq 2 \operatorname{rad}(G) - 1$.*

3.4.2 Cartesian Products

Distance behaves nicely in Cartesian products of graphs. The Cartesian product of two graphs $F = (V_1, E_1)$ and $H = (V_2, E_2)$, written $F \times H$, has vertex set $V_1 \times V_2$, and two vertices (u_1, v_1) and (u_2, v_2) are adjacent if either $u_1 = u_2$ and $v_1 v_2 \in E_2$ or $v_1 = v_2$ and $u_1 u_2 \in E_1$. For example, the product $K_4 \times K_2$ is shown in Fig. 3.5. If $G = F \times H$, then the distance between two vertices (u_1, v_1) and (u_2, v_2) in G is the sum of the distances $d_F(u_1, u_2) + d_H(v_1, v_2)$.

The n-cube Q_n ($n \geq 1$) is defined using Cartesian products of graphs as follows. $Q_1 \cong K_2$; and for $k \geq 1$, $Q_{k+1} \cong Q_k \times K_2$, i.e., Q_{k+1} is the Cartesian product of Q_k and K_2. Mulder [56] showed that the n-cube possesses interesting structural properties that can be described using distance notions. To this end, let u, v be vertices in a graph G. Then the *geodesic interval* between u and v, denoted by $I_g[u, v]$, is the collection of all vertices that belong to some $u - v$-geodesic (shortest path). A graph G is a *median graph* if for any three vertices u, v, w of G, $|I_g[u, v] \cap I_g[u, w] \cap I_g[v, w]| = 1$. For example, the 3-cube, shown in Fig. 3.4, is a median graph.

The next result was established in [56].

Theorem 5. *A graph G is isomorphic to Q_n if and only if G is a median graph with maximum degree n such that G contains two vertices at distance $diam(G)$ apart at least one of which has degree n.*

Hamming graphs are Cartesian products of complete graphs. Suppose $\{a_1, a_2, \ldots, a_n\}$ are positive integers and that $A_i = \{0, 1, \ldots, a_i-1\}$. Then the Hamming graph $H_{a_1, \ldots, a_n}$ is the graph with vertex set the Cartesian product $A_1 \times A_2 \times \cdots \times A_n$ in which two vertices are joined by an edge if and only if the corresponding vectors differ in exactly one coordinate. So $H_{a_1, a_2 \ldots, a_n} = K_{a_1} \times K_{a_2} \times \cdots \times K_{a_n}$ and the n-cube is the Hamming graph with $a_1 = a_2 = \cdots = a_n = 2$. The Hamming graph $H_{4,2}$ is shown in Fig. 3.5. The distance between two vertices is equal to the *Hamming distance* between these vertices, namely, the number of coordinates in which the corresponding vectors differ. Hamming graphs are used in coding theory and have applications to other distance invariants as discussed in Sect. 3.8. Mulder [55] extended Theorem 5 to characterize Hamming graphs in terms of intervals and forbidden subgraphs.

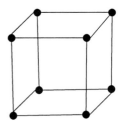

Fig. 3.4 The 3-cube: a median graph

3 Distance in Graphs

Fig. 3.5 The Hamming graph $H_{4,2}$

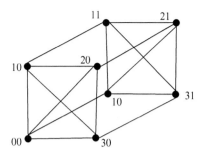

3.4.3 Distance Hereditary Graphs

Howorka [41] defined a graph G to be *distance hereditary* if for each connected induced subgraph H of G and every pair u, v of vertices in H, $d_H(u, v) = d_G(u, v)$. Howorka [41] provided several conditions that characterize distance hereditary graphs.

Theorem 6. *For a graph G the following are equivalent:*

1. *G is distance hereditary.*
2. *Every induced path of G is a geodesic.*
3. *Every subpath of a cycle C of more than half C's length is induced.*
4. *Every cycle C of length at least 5 has a pair of chords e_1, e_2 of C such that $C + \{e_1, e_2\}$ is homeomorphic with K_4.*

Distance hereditary graphs form a subclass of the perfect graphs (graphs where every induced subgraph has equal clique and chromatic numbers), and several NP-hard problems have efficient solutions for distance hereditary graphs. One such problem is discussed in Sect. 3.10. Another characterization of distance hereditary graphs that lends itself well to algorithmic applications was given independently by Bandelt and Mulder [2] and Hammer and Maffray [37]. A graph H is said to be obtained from a graph G by (1) *adding a leaf v* to some vertex v' of G, if v is added to G and joined to v' by an edge; and (2) by *adding a twin v* to some vertex v' of G, if v is added to G and if v is joined to the vertices in the open or closed neighbourhood of v' in G, i.e., v is joined to the vertices in $N_G(v')$ or $N_G(v') \cup \{v'\}$, respectively.

Theorem 7. *A graph G is distance hereditary if and only if there is a sequence of graphs $G_1, G_2, \ldots, G_{n-1}$ such that $G_1 \cong K_2$, $G \cong G_{n-1}$ and for $2 \leq i \leq n-1$, G_i is obtained from G_{i-1} by adding a vertex as a leaf or a twin to some vertex of G_{i-1}.*

Another polynomial recognition algorithm for distance hereditary graphs that is useful for solving the "Steiner problem" (see Sect. 3.10) for these graphs is

58   W. Goddard and O.R. Oellermann

described in [20]. Bandelt and Mulder [2] gave a characterization of these graphs in terms of forbidden subgraphs.

3.4.4 Random Graphs

The random graph G_p^n is obtained by starting with n vertices, and then for every pair of distinct vertices, making them adjacent with probability p (each decision independent).

Fact 8 *For any fixed p, G_p^n has diameter 2 with high probability (meaning the limit as $n \to \infty$ is 1).*

There is another version of random graphs, called power-law or scale-free graphs, which models the Web better (see, for example, Bonato [6]).

3.5 Average Distance

The *average distance* of a graph $G = (V, E)$ of order n, denoted by $\mu(G)$, is the expected distance between a randomly chosen pair of distinct vertices; that is,

$$\mu(G) = \frac{1}{\binom{n}{2}} \sum_{u,v \subset V} d(u, v).$$

The study of the average distance began with the chemist Wiener [72], who noticed that the melting point of certain hydrocarbons is proportional to the sum of all distances between unordered pairs of vertices of the corresponding graph. This sum is now called the *Wiener number* or *Wiener index* of the graph and is denoted by $\sigma(G)$. (Note that $\sigma(G) = \binom{n}{2}\mu(G)$.) The average distance of a graph has been used for comparing the compactness of architectural plans [52]. Doyle and Graver [22] were the first to define $\mu(G)$ as a graph parameter.

Here are the values for some simple graphs.

Fact 9 *1. Complete graphs: $\mu(K_n) = 1$ (for $n \geq 2$).*
2. Complete bipartite graph: $\mu(K_{a,b}) = (ab + 2a^2 + 2b^2)/((a + b)(a + b - 1))$.
3. Path on n vertices: $\mu(P_n) = (n + 1)/3$.
4. Cycle on n vertices: $\mu(C_n) = (n + 1)/4$ if n is odd, and $n^2/(4(n - 1))$ if n is even.

The result on the path is a discrete version of the fact that a pair of randomly chosen points on a line of length 1 have expected distance $1/3$.

3 Distance in Graphs   59

One might expect some relationship between the radius or diameter of the graph and its average distance. However, Plesník [62] showed that, apart from the trivial inequality $\mu(G) \leq \text{diam}(G)$, no such relationship exists.

3.5.1 Bounds on the Average Distance

The following upper bound was established independently in [22, 24, 51].

Theorem 8. *If G is a connected graph of order n, then*

$$1 \leq \mu(G) \leq \frac{n+1}{3}.$$

Equality holds if and only if G is a complete graph or a path.

Plesník [62] improved this bound for two-edge-connected graphs to approximately $n/4$. Mahéo and Thuillier (see [30]) showed that $\mu(n) \leq n/(2\kappa) + o(n)$ if G is κ-connected.

A straightforward lower bound in terms of the order n and number of edges m follows from the fact that there are exactly m pairs of vertices distance 1 apart and the remaining pairs are distance at least 2 apart; see [24].

Fact 10 *If G is a connected graph of n vertices and m edges, then*

$$\mu(G) \geq 2 - \frac{2m}{n(n-1)}.$$

Equality holds if and only if $\text{diam}(G) = 2$.

Finding the upper bound for $\mu(G)$ in terms of n and m is much more difficult. Šoltés [69] found a sharp upper bound:

Theorem 9. *The maximum average distance of a connected graph of order n and m edges, is achieved by the path-complete graph.*

There are a few other graphs achieving the maximum average distance; the extremal graphs were characterized in [35].

Plesník [62] found a sharp lower bound for the average distance of a graph in terms of the order and diameter. However, the problem of finding the exact maximum average distance among all graphs of a given order and diameter remains unsolved.

In Sect. 3.2.2 a bound on the diameter in terms of the distinct eigenvalues of the adjacency matrix and of the Laplacian was given. Rodriguez and Yebra [64] obtained a similar result for the average distance of G.

Theorem 10. *Let G be a connected graph of n vertices and m edges, and let b be the number of distinct eigenvalues of G. Then*

$$\mu(G) \leq b - \frac{2(b-1)m}{n(n-1)}.$$

The same result holds if b is the number of distinct Laplacian eigenvalues of G.

The spectrum (set of eigenvalues) does not necessarily determine a graph, not even if the graph is a tree. McKay (see [54]) and Merris [53] showed, however, that the average distance of a tree is determined by its spectrum.

Theorem 11. *Let T be a tree of order n and let $0 = \lambda_1 \leq \lambda_2 \leq \cdots \leq \lambda_n$ be the Laplacian eigenvalues of T. Then*

$$\mu(T) = \frac{2}{n-1} \sum_{i=2}^{n} \frac{1}{\lambda_i}.$$

Mohar [54] determined both upper and lower bounds on the average distance of a graph in terms of some of its Laplacian eigenvalues.

3.5.2 The Average Distance and Other Graph Invariants

Considerable interest in average distance was generated by conjectures made by the computer program GRAFFITI, developed by Fajtlowicz [27, 28]. One of these was that $\mu(G)$ is at most the independence number $\beta(G)$ of the graph. Chung [14] proved this conjecture.

Theorem 12. *For any connected graph G, $\mu(G) \leq \beta(G)$.*

This bound is attained if and only if G is a complete graph.

The computer program GRAFFITI also conjectured in [28] that every δ-regular connected graph of order n has average distance at most n/δ. GRAFFITI later restated this conjecture for graphs with minimum degree δ. It took 10 years before Kouider and Winkler [46] proved the following.

Theorem 13. *Let G be a connected graph of order n and minimum degree δ. Then*

$$\mu(G) \leq \frac{n}{\delta + 1} + 2.$$

While this result is stronger than the GRAFFITI conjecture when n is much larger than δ, it does not imply it. The conjecture was finally proven by Beezer, Riegsecker and Smith [3].

3.5.3 Edge Removal and the Average Distance

In this section we look at the effect of edge removal on the average distance of a graph. Since the removal of a bridge disconnects the graph, we consider only cyclic edges e. We can measure the effect of e's removal from G by considering either the difference $\mu(G - e) - \mu(G)$ or the ratio $\mu(G - e)/\mu(G)$. In both cases the *best* edge is one whose removal minimizes the quantity in question, and the *worst* edge is one that maximizes the quantity. These questions are of importance in network design: how badly would an edge failure affect the network, or how much would the network suffer if we omitted a particular link to save costs.

If an edge $e = ab$ is removed, the distance between a and b increases and no other distances decrease. Thus, the difference $\mu(G - e) - \mu(G)$ for the best edge is at least $1/\binom{n}{2}$. Finding attainable upper bounds is more difficult. Favaron et al. [30] found the maximum value for the difference.

Theorem 14. *Let G be a graph of order n and e a cyclic edge of G. Then*

$$\mu(G - e) - \mu(G) \leq \frac{1}{3}\left(\sqrt{2} - 1\right)n + O(1).$$

We now consider the ratio $\mu(G - e)/\mu(G)$. If G is a cycle, then the removal of an edge increases the average distance by a factor of about $4/3$. Winkler [73, 74] conjectured that for two-edge-connected graphs this is the maximum possible ratio for the best edge. This became known as the "four-thirds conjecture", and was eventually proven by Bienstock and Győri [5]. Soon thereafter, Győri [36] extended it and proved:

Theorem 15. *The four-thirds conjecture holds for all connected graphs that are not trees.*

The ratio $\mu(G - e)/\mu(G)$ can be arbitrarily large if a worst edge e is removed (see [17]). To see this, consider the graph G obtained from a cycle C_r : $v_0 v_1 v_2 \ldots v_{r-1} v_0$, where $r \geq 6$ is a fixed integer, by replacing vertices v_0 and v_3 by complete graphs of order $\lfloor (n-r)/2 \rfloor$ and $\lceil (n-r)/2 \rceil$ whose vertices are adjacent to v_{r-1}, v_1 and v_2, v_4, respectively. If n is large, then two randomly chosen vertices are almost certainly in the union of the two complete graphs and the probability that they are in the same (different) complete graph(s) and thus have distance 1 (3) tends to $\frac{1}{2}$. Hence $\mu(G) = \frac{1}{2}3 + \frac{1}{2}1 + o(1)$. By a similar argument $\mu(G-e) = (r-3)/2 + o(1)$, where e is the edge $v_1 v_2$. Choosing $r = \lfloor 2\sqrt{n} \rfloor$ gives $\mu(G - e)/\mu(G) = O(\sqrt{n})$. Favaron et al. [30] showed that this achieves the order of magnitude of the maximum possible value.

Theorem 16. *Let G be a connected graph of order n and e a cyclic edge of G. Then*

$$\frac{\mu(G - e)}{\mu(G)} \leq \frac{\sqrt{n}}{2\sqrt{3}} + O(1).$$

The *minimum average distance spanning tree* (or MAD tree) of a connected graph is a spanning tree having minimum average distance. Such a tree is also referred to as a minimum routing cost tree. It is surprising that the removal of a single best edge can increase the average distance by a factor of $4/3$, but the removal of $m - n + 1$ best edges (where m is the number of edges in the graph) increases the average distance by a factor less than 2. This fact was established by Entringer et al. in [25] (and a related result is discussed in [76]).

Theorem 17. *Let G be a connected graph of order n. Then there exists a vertex v and spanning tree T_v that is distance preserving from v, such that*

$$\mu(T_v) \le 2\frac{n-1}{n}\mu(G).$$

Johnson et al. [43] showed that the problem of finding a MAD tree in a graph is NP-hard.

3.5.4 Vertex Removal and Average Distance

We now turn our attention to the effect of vertex removal on the average distance. Unlike edge removal, vertex removal can both decrease or increase the average distance. For convenience, we express our results in terms of the Wiener index of the graph. Swart [70] showed that the maximum possible decrease occurs when an end-vertex is removed from a path.

Theorem 18. *Let G be a graph of order $n \ge 2$ and let v be a non-cut vertex of G. Then*

$$\sigma(G) - \sigma(G - v) \le \frac{n(n-1)}{2},$$

with equality if and only if G is a path and v is an end-vertex of G.

Šoltés [69] showed that the path-complete graphs are extremal for the ratio:

Theorem 19. *Let G be a graph of order n and $m \ge n - 1$ edges, and let v be a non-cut-vertex of G. Then*

$$\frac{\sigma(G - v)}{\sigma(G)} \ge \frac{\sigma(PK_{n-1,m-1})}{\sigma(PK_{n,m})}.$$

Šoltés [69] also gave sharp upper bounds for $\sigma(G - v) - \sigma(G)$ in terms of the order and number of edges of G.

In some instances the removal of any vertex increases the average distance. For example, the cycle C_n leaves P_{n-1} after the removal of any vertex. Thus the average distance increases by a factor of nearly $4/3$. Winkler [73, 74] conjectured that this is the worst increase. This vertex version of the "four-thirds conjecture" was proven asymptotically by Bienstock and Győri [5].

Theorem 20. *Every connected graph has a vertex whose removal increases the average distance by a factor of at most $\frac{4}{3} + O(n^{-5})$.*

Althöfer [1] proved the four-thirds conjecture for four-connected graphs, and improved on it for graphs of higher connectivity.

3.6 Directed Graphs

Directed graphs behave somewhat differently to undirected graphs. In general we assume that the digraph D is strongly connected; that is, there is a directed path from each vertex to each other vertex. Note, however, that the radius of D might well exist even though the digraph is not strongly connected and the diameter does not exist. Further, it is possible that $\mathrm{diam}(D) \gg 2\,\mathrm{rad}(D)$.

An *oriented* graph is one obtained by assigning directions to each edge of an undirected graph. That is, it is a digraph without two-cycles. Füredi et al. [33] provided a lower bound on the number of arcs in an oriented graph of diameter 2.

Theorem 21. *If an oriented graph has diameter 2, then $m \geq (1 - o(1))\,n\log n$, where n is the order and m the number of arcs.*

Chvátal and Thomassen [15] studied the problem of taking an undirected graph and finding an orientation of minimum diameter. Earlier, Robbins [63] had shown that an undirected graph has a strong orientation if and only if it is bridgeless. Chvátal and Thomassen showed that a bridgeless graph of diameter d has an orientation of diameter at most $2d^2 + d$. They also showed that determining whether a graph has an orientation of diameter 2 is NP-complete.

3.6.1 The Average Distance of Digraphs

The average distance of digraphs has not received as much attention as the average distance of graphs. Ng and Teh [57] gave a lower bound for the average distance of a strong digraph D in terms of its order n and number of arcs m similar to the one given for undirected graphs; they showed that $\mu(D) \geq 2 - m/(n(n-1))$ with equality if and only if $\mathrm{diam}(D) = 2$. For $n \geq 3$ this bound is sharp if $m \geq 2n - 2$, the smallest number of arcs for which there exists a digraph of order n and diameter 2.

Plesník [62] proved the following upper bound on the average distance of a strong digraph.

Theorem 22. *Let D be a strong digraph of order n. Then $\mu(D) \leq n/2$. Equality holds if and only if D is a directed cycle.*

It is natural to ask whether there are upper bounds on the average distance of a digraph in terms of the order and minimum degree that are analogous to the ones for graphs, but it turns out that in general these bounds do not carry over for general digraphs.

In [19] the problem of finding an orientation of a two-edge-connected graph that minimizes the average distance is studied.

3.6.2 Tournaments

A *tournament* is an oriented complete graph. Landau [47] showed that every tournament has radius at most 2. It is easy to construct a tournament on n vertices with diameter $n - 1$. For $n \neq 4$, it is also easy to construct a tournament with diameter 2.

Plesník [62] gave bounds for average distance in strongly connected tournaments.

Theorem 23. *Let T_n be a strongly connected tournament of order $n \geq 3$. Then*

$$\frac{3}{2} \leq \mu(T_n) \leq \frac{n+4}{6}.$$

Equality holds if and only if T_n has diameter 2 (this is possible only if $n \neq 4$) or if T_n is the unique tournament of diameter $n - 1$, respectively.

3.7 Convexity

Interval notions in graphs have led to the study of abstract convexity in graphs and structural characterizations of several interesting graph classes.

Suppose V is a collection of points and $\mathcal{M}$ a collection of subsets of V. Then $\mathcal{M}$ is a *convexity* if it contains both $\emptyset$ and V and it is closed under intersections. The elements of $\mathcal{M}$ are called *convex sets*. If $T \in \mathcal{M}$, then a point v of T is an *extreme point* of T if $T \setminus \{v\} \in \mathcal{M}$. If $S \subseteq V$, then the smallest convex set containing S is called the *convex hull* of S. A *convex geometry* is a convexity with the additional property that every convex set is the convex hull of its extreme points.

The most well-known graph convexity is defined in terms of geodesic intervals, which were introduced in Sect. 3.4.2. Suppose $G = (V, E)$ is a connected graph. Then a set $S \subseteq V$ is *g-convex* if $I_g[u, v] \subseteq S$ for all pairs $u, v \in S$. Let $\mathcal{M}_g(G)$ be the collection of all g-convex sets of G. Then $\mathcal{M}_g(G)$ is a convexity.

It is not difficult to see that a vertex v of a g-convex set S is an extreme vertex of S if and only if v is simplicial in $\langle S \rangle$, i.e., the neighbourhood of v in S induces a complete graph. Farber and Jamison [29] characterized the class of graph for which the g-convex sets form a convex geometry.

3 Distance in Graphs

Fig. 3.6 The 3-fan

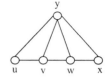

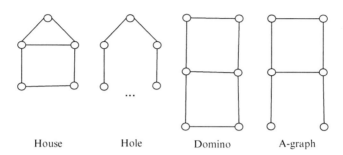

      House   Hole   Domino   A-graph

Fig. 3.7 Forbidden graphs

Theorem 24. *Let G be a connected graph. Then $\mathcal{M}_g(G)$ is a convex geometry if and only if G is chordal without an induced 3-fan (see Fig. 3.6).*

Every shortest path is necessarily induced, but not conversely. This leads to another type of graph interval. The *monophonic interval* between a pair u, v of vertices in a graph G, denoted by $I_m[u, v]$, is the collection of all vertices that lie on an induced $u - v$ path in G. A set S of vertices in a graph is m-convex if $I_m[u, v] \subseteq S$ for all pairs $u, v \in S$. It is not difficult to see that the collection $\mathcal{M}_m(G)$ of all m-convex sets is a convexity, and that the extreme points of an m-convex set S are precisely the simplicial vertices of $\langle S \rangle$. Farber and Jamison [29] characterized those graphs for which the m-convex sets form a convex geometry.

Theorem 25. *Let G be a connected graph. Then $\mathcal{M}_m(G)$ is a convex geometry if and only if G is chordal.*

Dragan et al. [23] defined another type of graph interval. If u, v is a pair of vertices in a connected graph G, then the m^3-interval, denoted by $I_{m^3}[u, v]$, between u and v is the collection of all vertices that lie on some induced $u - v$ path of length at least 3. A set S of vertices in G is m^3-convex if $I_{m^3}[u, v] \subseteq S$ for all pairs u, v of vertices in S. For a graph G, let $\mathcal{M}_{m^3}(G)$ be the collection of all m^3-convex sets. This collection of sets is certainly a convexity. Further, it can be shown that a vertex v is an extreme point of an m^3-convex set S if and only if v is *semisimplicial*, i.e., not the centre of an induced P_4. The class of graphs for which the m^3-convex sets form a convex geometry was characterized in [23].

Theorem 26. *Let G be a connected graph. Then $\mathcal{M}_{m^3}(G)$ is a convex geometry if and only if G is (house, hole, domino, A)-free (see Fig. 3.7).*

3.8 Metric Dimension

Distances in graphs have interesting applications. One such application is to uniquely locate the position of a vertex in a network using distances. A vertex v *resolves* a pair u, w of vertices in a connected graph G if $d(u, v) \neq d(w, v)$. A set of vertices S is a *resolving set* of G, if every pair of vertices in G is resolved by some vertex of S. A resolving set of minimum cardinality is called a *metric basis* of G and its cardinality the *metric dimension, $dim(G)$*, of G.

The metric dimension of a graph was introduced by Slater [66] and independently by Harary and Melter [39]. Slater referred to the metric dimension as the location number, and motivated its study by its application to the placement of a minimum number of sonar/loran detecting devices in a network so that the position of every vertex in the network could be uniquely described in terms of its distances to the detecting devices. A problem in pharmaceutical chemistry once again led to an independent discovery of the notion of a resolving set of a graph [12]. In [34] it was noted that the metric dimension of a graph is NP-hard.

The formula for the metric dimension of trees has been discovered independently by several authors (see [12, 39, 66]). The metric dimension of a nontrivial path is 1 since a leaf resolves the path. Suppose now that T is a tree that contains vertices of degree at least 3. A vertex v of degree at least 3 is an *exterior vertex* if there is some leaf u in T such that the $v - u$ path of T contains no vertices of degree exceeding 2 except for v. Let $ex(T)$ denote the number of exterior vertices of T and $\ell(T)$ the number of leaves of T. It turns out that a metric basis for a tree can be found by selecting, for each exterior vertex, all but one of its exterior leaves. That is, the metric dimension of T is given in the following.

Theorem 27. *For a tree T, $\dim(T) = \ell(T) - ex(T)$.*

Apart from trees, very few exact results for the metric dimension of graphs are known unless the graphs are highly structured (usually vertex transitive). It was claimed in [45] that the metric dimension of the Cartesian product of k paths is k; but indeed it was only verified that k is an upper bound in this case. In [65] a connection between the metric dimension of the n-cube and the solution to a coin weighing problem was noted. This observation and results by Lindström [50] and Erdös and Rényi [26] show that $\lim_{n \to \infty} \dim(Q_n) \log n / n = 2$, thereby disproving the claim about n-cubes made in [45].

Motivated by the connection between coin weighing problems/strategies for the Mastermind game and resolving sets in Cartesian products of certain classes of graphs, the metric dimension of Cartesian products of graphs was investigated in [10]. This paper introduces "doubly resolving sets" as a useful tool for obtaining upper bounds on the metric dimension of graphs, particularly in Cartesian products of graphs.

3 Distance in Graphs   67

3.9 Algorithms and Complexity

3.9.1 Shortest Paths

To compute the distance between two vertices in an unweighted graph, one can use
a breadth-first search. To compute the distance between two vertices in a weighted
graph, one can use Dijkstra's algorithm (which is in some sense a generalization of
breadth-first search). Note that the algorithm actually finds the distance from a given
start vertex to all other vertices.

```
ShortestPath (G:graph, a:vertex)
    for all vertices v do currDis(v) := infinity
    currDis(a) := 0
    remainder := [all vertices]
  while remainder nonempty do {
    let w be vertex in remainder with minimum value of currDis
    remainder -= [w]
    for all vertices v in remainder do
        currDis (v) := min (currDis(v), currDis(w)+length(w,v))
  }
```

The running time of the above implementation of Dijkstra's algorithm is $O(n^2)$.
By using suitable data structures this can be brought down for sparse graphs to
$O(m + n \log n)$, where m is the number of edges.

3.9.2 All Pairs Shortest Paths

Suppose we wanted instead to calculate the shortest path between every pair of
vertices, for example, in order to compute the average distance. One idea would be to
run Dijkstra with every vertex as a start vertex. This takes $O(n^3)$ time. There are two
dynamic programming algorithms with similar running times. One is due to Bellman
and Ford and the other to Floyd and Warshall. A variant of the former is used in
routing protocols in networks. We describe here the Floyd–Warshall algorithm [31].

Suppose the vertices are ordered 1 up to n. Then define

$d_m(u, v)$ as the length of the shortest path between u and v that uses only the vertices
numbered 1 up to m as intermediates.

The desired value is $d_n(u, v)$ for all u and v.

There is a formula for d_m in terms of d_{m-1}. For, the shortest u to v path that uses
only vertices labeled up to m, either uses vertex m or it doesn't. Thus:

$$d_m(u, v) = \min \begin{cases} d_{m-1}(u, v) \\ d_{m-1}(u, m) + d_{m-1}(m, v) \end{cases}$$

The resultant program iterates m from $m = 0$ to $m = n - 1$.

3.10 Steiner Distances

Up to this point we have considered distance invariants that hinge on shortest paths between pairs of vertices. In this section we give a brief overview of related invariants that arise by considering the "cheapest" subgraph that connects a given set of vertices.

3.10.1 Extending Distance Measures

Suppose G is a (weighted) graph and S a set of vertices in G. Then the *Steiner distance* for S, denoted by $d_G(S)$, is the smallest weight of a connected subgraph of G containing S. Such a subgraph is necessarily a tree, called a *Steiner tree* for S. The problem of finding a Steiner tree for a given set S of vertices is called the Steiner problem. In its two extremes, namely if $|S| = 2$ or $|S| = n$, the Steiner problem is solved efficiently by well-known algorithms, for example, Dijkstra's algorithm and Kruskal's minimum spanning tree algorithm, respectively. In general, however, this problem is NP-hard (see [34]), even for unweighted bipartite graphs. Winter's survey [75] provides a good overview of different heuristics that have been developed for the problem as well as exact solutions for various graph classes.

The radius, diameter and average distance have a natural extension. For a given vertex v in a connected (weighted) graph G and integer k $(2 \leq k \leq n)$, the k-eccentricity of v, denoted by $e_k(v)$, is the maximum Steiner distance among all k-sets of vertices in G that contain v. The k-radius, $rad_k(G)$, of G is the minimum k-eccentricity of the vertices of G, and the k-diameter, $diam_k(G)$, of G is the maximum k-eccentricity. The *average Steiner k-distance*, $\mu_k(G)$ of G, is the average Steiner distance among all k-sets of vertices of G. The k-distance of a vertex v, denoted by $e_k(v)$, is the sum of the Steiner distances of k-sets of vertices containing v. The subgraph induced by vertices of minimum k-eccentricity is called the k-centre of G and is denoted by $C_k(G)$; the subgraph induced by the vertices of minimum k-distance is called the k-median and is denoted by $M_k(G)$.

The k-diameter is clearly an upper bound for the k-radius. No upper bound for the k-diameter as a function of k and the k-radius of an (unweighted) graph is known. For trees, the following generalization of Fact 2 was established in [13].

Theorem 28. *For a tree T of order n and integer $k \leq n$, $diam_k(T) \leq \frac{n}{n-1} rad_k(T)$.*

It was shown in [60], that the k-centre of a tree T can be found by successively pruning leaves. If T has at most $k - 1$ leaves, then T is its own k-centre. If T has at least k leaves, then the k-centre is the ith derivative of T where i is the smallest integer such that $T^{(i)}$ has at most $k - 1$ leaves. This also shows that the k-centre of a tree is contained in the $(k + 1)$-centre of a tree. (This containment does not hold in general graphs). Moreover, it follows that k-centres of trees are connected.

The k-median of trees was shown in [4] to be connected. In the same paper it was shown that a tree H of order p is the k-median of a tree if and only if $p = 1, 2$

or k or if H has at most $k - p + 1$ leaves. An algorithm for finding the k-median of a tree T was also described. If a tree has order at least $2k - 1$, then its k-median consists of a single vertex or a pair of adjacent vertices. The k-centre and k-median of a tree may be arbitrarily far apart (see [59]). Centrality structures that connect the k-centre and k-median of a tree are introduced and studied in [59].

The average Steiner k-distance of a graph was first defined in [18]. In the same paper it was shown that $\mu_k(G) \leq \mu_l(G) + \mu_{k+1-l}(G)$ for $2 \leq l \leq k - 1$, and that the range of average Steiner k-distance of a graph is given by:

Theorem 29. *If G is a connected graph of order n and $2 \leq k \leq n$, then*

$$k - 1 \leq \mu_k(G) \leq \frac{k-1}{k+1}(n+1),$$

with equality on the left if and only if G is $(n + 1 - k)$-connected or $k = n$, and equality on the right if and only if G is a path or $n = k$.

An efficient procedure that finds the average Steiner k-distance of a tree is described in [18]. This algorithm counts the number of k-sets such that a given edge belongs to a Steiner tree for the k-set. Moreover, it is shown that for a tree T, $\mu_k(T) \leq k\mu_l(T)/l$ for $2 \leq l \leq k - 1$ with equality if and only if T is a star, and the lower bound given in Theorem 29 is improved to $k(1 - 1/n)$ for trees.

3.10.2 Steiner Intervals and Graph Convexity

The *Steiner interval* of a set X of vertices in a connected graph G, denoted by $I(X)$, is the collection of all vertices that belong to some Steiner tree for X. A set S of vertices is *k-Steiner convex*, denoted by g_k-convex, if $I(X) \subseteq S$ for all subsets X of S with $|X| = k$. Thus a g_2-convex set is a g-convex set. We call the extreme vertices of a g_k-convex set a *k-Steiner simplicial* vertex and abbreviate this by kSS. The $3SS$ vertices are characterized in [11] as the vertices that are **not** the centre of an induced claw, paw or P_4; see Fig. 3.8.

Thus the $3SS$ vertices are semisimplicial. Several graph convexities related to the g_3-convexity are introduced in [58] and those graphs for which these graph

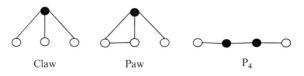

Claw   Paw P_4

● Indicates a centre vertex

Fig. 3.8 Characterizing $3SS$ vertices

Fig. 3.9 The replicated twin C_4's

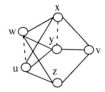

convexities form convex geometries are characterized in the same paper. We state here a characterization of those graphs for which the g_3-convex sets form a convex geometry. A replicated-twin C_4 is any one of the four graphs shown in Fig. 3.9 where any subset of the dotted edges belongs to the graph. The collection of replicated-twin C_4's is denoted by $\mathcal{R}_{C_4}$.

Theorem 30. *Let G be a connected graph and $\mathcal{M}_{g_3}(G)$ the collection of all g_3-convex sets of G. Then $\mathcal{M}_{g_3}(G)$ is a convex geometry if and only if $diam(G) \leq 2$ and if G is (house, hole, 3-fan, $\mathcal{R}_{C_4}$)-free.*

Acknowledgments We would like to thank Peter Dankelmann for sharing his thoughts on average distance with us.

References

1. Althöfer I (1990) Average distances in undirected graphs and the removal of vertices. J Combin Theory B 48(1):140–142
2. Bandelt H-J, Mulder HM (1986) Distance–hereditary graphs. J Combin Theory B 41:182–208
3. Beezer RA, Riegsecker JE, Smith BA (2001) Using minimum degree to bound average distance. Discrete Math 226:365–371
4. Beineke LW, Oellermann OR, Pippert RE (1996) On the Steiner median of a tree. Discrete Appl Math 68:249–258
5. Bienstock D, Györi E (1988) Average distance in graphs with removed elements. J Graph Theory 12(3):375–390
6. Bonato A (2005) A survey of models of the web graph. In: Combinatorial and algorithmic aspects of networking, vol 3405, Lecture notes in computer science. Springer, Berlin, pp 159–172
7. Buckley F, Harary F (1990) Distance in graphs. Addison-Wesley Publishing Company Advanced Book Program, Redwood City, CA
8. Buckley F, Lewinter M (1988) A note on graphs with diameter-preserving spanning trees. J Graph Theory 12(4):525–528
9. Buckley F, Miller Z, Slater PJ (1981) On graphs containing a given graph as center. J Graph Theory 5:427–432
10. Cáceres J, Hernando C, Mora M, Pelayo IM, Puertas ML, Seara C, Wood DR (2007) On the metric dimension of the cartesian product of graphs. SIAM J Discrete Math 21(2):423–441
11. Cáceres J, Oellermann OR (2009) On 3-Steiner simplicial orderings. Discrete Math 309: 5828–5833
12. Chartrand G, Eroh L, Johnson MA, Oellermann OR (2000) Resolvability in graphs and the metric dimension of a graph. Discrete Appl Math 105:99–113
13. Chartrand G, Oellermann OR, Tian S, Zou HB (1989) Steiner distance in graphs. Časopis Pro Pěstování Matematiky 114(4):399–410

3 Distance in Graphs

14. Chung FRK (1988) The average distance and the independence number of a graph. J Graph Theory 12:229–235
15. Chvátal V, Thomassen C (1978) Distances in orientations of graphs. J Combin Theory B 24:61–75
16. Cockayne EJ, Hedetniemi SM, Hedetniemi ST (1981) Linear algorithms for finding the Jordan center and path center of a tree. Transport Sci 15:98–114
17. Dankelmann P Average distance in graphs – a survey, (submitted)
18. Dankelmann P, Oellermann OR, Swart HC (1996) The average Steiner distance of a graph. J Graph Theory 22(1):15–22
19. Dankelmann P, Oellermann OR, Wu J-L (2004) Minimum average distance of strong orientations of graphs. Discrete Appl Math 143:204–212
20. D'Atri A, Moscarini M (1988) Steiner trees and connected domination. SIAM J Discrete Math 17:521–538
21. Dirac GA (1961) On rigid circuit graphs. Abh Math Sem Univ Hamburg 25:71–76
22. Doyle JK, Graver JE (1977) Mean distance in a graph. Discrete Math 7(2):147–154
23. Dragan FF, Nicolai F, Brandstädt A (1999) Convexity and HHD-free graphs. SIAM J Discrete Math 12:119–135
24. Entringer RC, Jackson DE, Snyder DA (1976) Distance in graphs. Czech Math J 26:283–296
25. Entringer RC, Kleitman DJ, Székely LA (1996) A note on spanning trees with minimum average distance. Bull Inst Combinator Appl 17:71–78
26. Erdös P, Rényi A (1963) On two problems of information theory. Magyar Tud Akad Mat Kutató Int Közl 8:229–243
27. Fajtlowicz S (1986) On two conjectures of GRAFFITI. Congr Numer 55:51–56
28. Fajtlowicz S (1987) On conjectures of GRAFFITI II. Congr Numer 60:187–197
29. Farber M, Jamison RE (1986) Convexity in graphs and hypergraphs. SIAM J Algebra Discrete Methods 7:433–444
30. Favaron O, Kouider M, Mahéo M (1989) Edge-vulnerability and mean distance. Networks 19(5):493–504
31. Floyd RW (1962) Algorithm 97: shortest path. Commun ACM 5:345
32. Freeman LC (1978/1979) Centrality in social networks: conceptual clarification. Soc Networks 1:215–239
33. Füredi Z, Horak P, Pareek CM, Zhu X (1998) Minimal oriented graphs of diameter 2. Graph Combinator 14:345–350
34. Garey MR, Johnson DS (1979) Computers and intractibility: a guide to the theory of NP-completeness. W.H. Freeman and Company, New York
35. Goddard W, Swart CS, Swart HC (2005) On the graphs with maximum distance or k-diameter. Math Slovaca 55(2):131–139
36. Györi E (1988) On Winkler's four thirds conjecture on mean distance in graphs. Congr Numer 61:259–262
37. Hammer PL, Maffray F (1990) Completely separable graphs. J Discrete Appl Math 27:85–99
38. Harary F (1962) The maximum connectivity of a graph. Proc Natl Acad Sci USA 48:1142–1146
39. Harary F, Melter RA (1976) On the metric dimension of a graph. Ars Combinatoria 2:191–195
40. Harary F, Norman RZ (1953) The dissimilarity characteristic of Husimi trees. Ann Math (2) 58:134–141
41. Howorka E (1977) A characterization of distance-hereditary graphs. Quart J Math Oxford 28:417–420
42. Jeong H, Mason SP, Barabási A-L, Oltvai ZN (2001) Lethality and centrality in protein networks. Nature 411:41–42
43. Johnson DS, Lenstra JK, Rinnooy-Kan AHG (1978) The complexity of the network design problem. Networks 8:279–285
44. Jordan C (1869) Sur les assembalges des lignes. J Reine Angew Math 70:185–190
45. Khuller S, Raghavachari B, Rosenfeld A (1996) Landmarks in graphs. Discrete Appl Math 70(3):217–229

46. Kouider M, Winkler P (1997) Mean distance and minimum degree. J Graph Theory 25:95–99
47. Landau HG (1953) On dominance relations and the structure of animal societies. III. The condition for a score structure. Bull Math Biophys 15:143–148
48. Laskar R, Shier D (1983) On powers and centers of chordal graphs. Discrete Appl Math 6(2):139–147
49. Lekkerkerker CG, Boland J (1962) Representation of a finite graph by a set of intervals on the real line. Fund Math 51:45–64
50. Lindström B (1964) On a combinatory detection problem. I. Magyar Tud Akad Mat Kutató Int Közl 9:195–207
51. Lovász L (1979) Combinatorial problems and exercises. Akadémiai Kiadó, Budapset
52. March L, Steadman P (1974) The geometry of environment: an introduction to spatial organization in design. MIT Press, Cambridge, MA
53. Merris R (1989) An edge version of the matrix-tree theorem and the Wiener index. Linear Multilinear Algebra 25(4):291–296
54. Mohar B (1991) Eigenvalues, diameter, and mean distance in graphs. Graph Combinator 7(1):53–64
55. Mulder HM (1980) The interval function of a graph. Mathematical Centre Tracts, Amsterdam
56. Mulder HM (1980) n-cubes and median graphs. J Graph Theory 4:107–110
57. Ng CP, Teh HH (1966/1967) On finite graphs of diameter 2. Nanta Math 1:72–75
58. Nielsen M, Oellermann OR (2009) Steiner trees and convex geometries. SIAM J Discrete Math 23:680–693
59. Oellermann OR (1995) From Steiner centers to Steiner medians. J Graph Theory 20(2):113–122
60. Oellermann OR, Tian S (1990) Steiner centers in graphs. J Graph Theory 14(5):585–597
61. Plesník J (1975) Note on diametrically critical graphs. In: Recent advances in graph theory (Proceedings of the 2nd Czechoslovak symposium, Prague, 1974). Academia, Prague, pp 455–465
62. Plesník J (1984) On the sum of all distance in a graph or digraph. J Graph Theory 8:1–24
63. Robbins HE (1939) Questions, discussions, and notes: a theorem on graphs, with an application to a problem of traffic control. Am Math Mon 46(5):281–283
64. Rodriguez JA, Yebra JLA (1999) Bounding the diameter and the mean distance of a graph from its eigenvalues: Laplacian versus adjacency matrix methods. Discrete Math 196(1–3):267–275
65. Sebö A, Tannier E (2004) On metric generators of graphs. Math Oper Res 29(2):383–393
66. Slater PJ (1975) Leaves of trees. Congr Numer 14:549–559
67. Slater PJ (1978) Centers to centroids in graphs. J Graph Theory 2:209–222
68. Slater PJ (1981) Centrality of paths and vertices in a graph: cores and pits. In: The theory and applications of graphs. Wiley, New York, pp 529–542
69. Šoltés L (1991) Transmission in graphs: a bound and vertex removing. Math Slovaca 41(1):11–16
70. Swart CS (1996) Distance measures in graphs and subgraphs. Master's thesis, University of Natal, Durban
71. Vizing VG (1967) On the number of edges in a graph with a given radius. Dokl Akad Nauk SSSR 173:1245–1246
72. Wiener H (1947) Structural determination of paraffin boiling points. J Am Chem Soc 69(1):17–20
73. Winkler P (1986) Mean distance and the four thirds conjecture. Congr Numer 54:53–61
74. Winkler P (1989) Graph theory in memory of G.A. Dirac. In: Proceedings of meeting in Sandbjerg, Denmark, 1985, Ann Discrete Math, North-Holland, Amsterdam
75. Winter P (1987) Steiner problem in networks: a survey. Networks 17:129–167
76. Wong R (1980) Worst-case analysis of network design problem heuristics. SIAM J Algebra Discrete Methods 1:51–63
77. Zelinka B (1968) Medians and peripherians of trees. Arch Math (Brno) 4:87–95

Chapter 4
Domination in Graphs

Nawarat Ananchuen, Watcharaphong Ananchuen, and Michael D. Plummer

Abstract A set of vertices S in a graph G dominates G if every vertex in G is either in S or adjacent to a vertex in S. The size of any smallest dominating set is called the domination number of G. Two variants on this concept that have attracted recent interest are total domination and connected domination. A set of vertices S is a total dominating set if every vertex in the graph is adjacent to a vertex of S and S is a connected dominating set if it is dominating and, in addition, induces a connected subgraph. The size of any smallest total dominating set in G is called the total domination number of G and the size of a smallest connected dominating set is the connected domination number of G. These simple, yet wide-ranging, graph-theoretic concepts have a multitude of real-world applications. There are already in print several surveys of results on domination; therefore, in this chapter we adopt a slightly different approach. We begin by surveying results on bounding the three domination numbers. We then focus on criticality concepts for domination. The two types of criticality most widely studied to date are graphs for which the domination number decreases upon the addition of any missing edge and the graphs for which the domination number decreases upon the deletion of any vertex. Recently, there has been increased activity in the study of these critical concepts and we survey these new results, focusing especially upon matching in critical graphs.

Keywords Domination · Total domination · Connected domination · Edge critical · Vertex critical · Matching

MSC2000: Primary 05C69; Secondary 05C70, 05C35

N. Ananchuen (✉)
Department of Mathematics, Faculty of Science, Silpakorn University,
Nakorn Pathom 73000, Thailand
e-mail: nawarat@su.ac.th

M. Dehmer (ed.), *Structural Analysis of Complex Networks*,
DOI 10.1007/978-0-8176-4789-6_4, © Springer Science+Business Media, LLC 2011

4.1 Introduction

The study of domination and related topics is one of the fast-developing areas in graph theory. To date, at least 2000 published research papers on domination have appeared in various journals. Excellent comprehensive collections of results together with open problems in this area were published in 1998 in two books [71,72].

The aim of this chapter is to provide a fundamental understanding of three types of domination and to present some recent results, most of which concern graphs edge- (or vertex-) critical with respect to each. We cannot, however, include all such results due to space limitations, but instead provide a guide to the relevant references.

For the most part, our terminology follows that of Bondy and Murty [22]. Thus G is a graph with the vertex set $V(G)$, edge set $E(G)$, $v(G)$ vertices, $\varepsilon(G)$ edges, (vertex) connectivity $\kappa(G)$, and independence number $\alpha(G)$. We denote the *complement* of a graph G by $\overline{G}$. A graph G is said to be even (odd) if $v(G)$ is even (odd). For $S \subseteq V(G), G[S]$ denotes the subgraph induced by S. For $v \in V(G), N_G(v)$ denotes the set of all vertices adjacent to vertex v and is called the *neighborhood* of v. The *closed neighborhood* of v, $N_G[v]$ is defined by $N_G[v] = N_G(v) \cup \{v\}$. If $S \subseteq V(G)$, then $N_G(S)$ denotes $\bigcup_{x \in S} N_G(x)$ and $N_G[S]$ denotes $\bigcup_{x \in S} N_G[x]$. The degree of $v \in V(G)$ denoted by $d_G(v)$ is defined to be $|N_G(v)|$. Let $\delta(G)$ and $\Delta(G)$ denote the minimum degree and maximum degree of G, respectively. As usual, K_n, C_n, and P_n denote a complete graph, a cycle, and a path of order n, respectively, and $K_{m,n}$ denotes a complete bipartite graph with bipartite sets of size m and n. We denote the number of components (odd components) of a graph G by $\omega(G)$ ($\omega_o(G)$).

A vertex $v \in V(G)$ is called an *endvertex* (or *leaf* or *pendant vertex*) if $d_G(v) = 1$. If $\{u\} = N_G(v)$, then u is a *support vertex* of v and the edge uv is called a *pendant edge*. For vertices u and v of $G, d(u, v)$ denotes the distance from u to v. The *eccentricity* of a vertex v denoted by $e(v)$ is $\max\{d(u, v)|u \in V(G)\}$. The *radius* of G denoted by $\mathrm{rad}(G)$ and the *diameter* of G denoted by $\mathrm{diam}(G)$ are $\min\{e(v)|v \in V(G)\}$ and $\max\{e(v)|v \in V(G)\}$, respectively.

If S and T are subsets of $V(G)$, we say that S *dominates* T if $T \subseteq N_G[S]$. In particular, if S dominates $V(G)$ then S is said to be a *dominating set* for G; that is, $S \subseteq V(G)$ is a dominating set for G if every vertex of G not in S is adjacent to one in S. The cardinality of any smallest dominating set in G is denoted by $\gamma(G)$. We call a dominating set with cardinality $\gamma(G)$, a $\gamma(G)$-set. It is easy to see that for $n \geq 1, \gamma(K_n) = \gamma(K_{1,n}) = 1$ and if $m, n \geq 2$, then $\gamma(K_{m,n}) = 2$. Further, for $n \geq 3, \gamma(P_n) = \gamma(C_n) = \left\lceil \frac{n}{3} \right\rceil$.

The concept of domination, in both the theoretical and applied sense, has received the attention of many researchers. It has been used to study the optimal location of facilities such as radar stations (see [17]), hardware or software resources (see [61,62]), and communication networks (see [97]). Typical problems are concerned with the minimum cardinality of a subset of a set of facilities, such as radar stations and communication centers, which collectively monitor all the facilities or perhaps can transmit messages to every facility in the network. In graph-theoretic

4 Domination in Graphs

terms, the minimum cardinality of such a set of monitoring facilities is called the domination number. In the real world, it might be additionally required that such a set of monitors be able to communicate among themselves in case of a security breach or perhaps each of these monitors must be reached by at least one other monitor. In these two special cases, the minimum cardinality of a monitor set is called the connected domination number and the total domination number, respectively.

This chapter is divided into six sections. Section 4.2 contains some fundamental results concerning upper bounds for the classical domination number. Two of the more intensely studied variants of classical domination – connected domination and total domination – are the focus of Sect. 4.3.

The concept of criticality has proved very useful in the past when applied to such important graph parameters as connectivity, chromatic number, and independence number. Only very recently has this idea been applied to domination. For the first time, a number of results concerning criticality of various types of domination are assembled in Sects. 4.4 and 4.5. Finally, we conclude the chapter with Sect. 4.6 in which we suggest some directions for future study.

4.2 Bounds on the Domination Number

Clearly, $V(G)$ is a dominating set for G. Thus $\gamma(G) \leq \nu(G)$. It is easy to see that $\gamma(G) = \nu(G)$ if $\overline{G}$ is complete. If the graph under consideration has no isolated vertices, then a bound on its domination number can be improved considerably. Many researchers have been concerned with establishing upper bounds on the domination number since determining the domination number of a graph is an NP-complete problem (see [31, 65] or [71], page 34). Various bounds on the domination number are given in terms of other graph parameters, for example, in terms of order and minimum degree (see [16, 29, 30, 40, 78, 104, 111, 117]), diameter (see [48, 67, 71, 103]), size and the minimum degree (see [79, 119]), and others (see [90, 102, 116]). The reader is directed to [42, 44, 71, 77, 96] for more on these bounds.

We now state some fundamental results on bounding the domination number. The first result was given by Ore in 1962.

Theorem 1 ([111]). *Let G be a graph with $\delta(G) \geq 1$. Then $\gamma(G) \leq \frac{1}{2}\nu(G)$.*

Payan and Xuong [114] characterized the graphs that achieve the bound in Theorem 1, as we show in Theorem 2. Before stating this theorem we need a new definition. The *corona* of a graph H denoted by $H \circ K_1$ is the graph obtained from H by adding a pendant edge to each vertex of H.

Theorem 2 ([114]). *Let G be a graph with $\delta(G) \geq 1$. Then $\gamma(G) = \frac{1}{2}\nu(G)$ if and only if each component of G is isomorphic to C_4 or to $H \circ K_1$ for some connected graph H.*

The bound in Theorem 1 may be improved if the minimum degree of a graph is increased as shown in the following theorems.

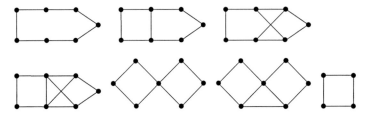

Fig. 4.1 Graphs G with $\delta(G) \geq 2$ and $\gamma(G) > \frac{2}{5}\nu(G)$

Theorem 3 ([104]). *Suppose G is a connected graph with $\delta(G) \geq 2$. Then G is either one of seven graphs in Fig. 4.1 or $\gamma(G) \leq \frac{2}{5}\nu(G)$.*

When certain cycle lengths are forbidden, the bound in Theorem 3 was improved by Frendrup et al. [60] and Harant and Rautenbach [70].

Theorem 4 ([70]). *If G is a graph with $\delta(G) \geq 2$ and G does not contain cycles of length $4, 5, 7, 10$, or 13, then $\gamma(G) \leq \frac{3}{8}\nu(G)$.*

As a corollary of Theorem 4, an upper bound on the domination number for bipartite graphs follows.

Corollary 1 ([70]). *If G is a bipartite graph with $\delta(G) \geq 2$ and G does not contain cycles of length 4 or 10, then $\gamma(G) \leq \frac{3}{8}\nu(G)$.*

Theorem 5 ([117]). *If G is a graph with $\delta(G) \geq 3$, then $\gamma(G) \leq \frac{3}{8}\nu(G)$.*

Beside establishing Theorem 5, Reed [117] conjectured that if G is 3-regular, $\gamma(G) \leq \left\lceil \frac{\nu(G)}{3} \right\rceil$. This conjecture is true for Hamiltonian 3-regular graphs (see [46]), but it was recently shown to be false in general by Kostochka and Stodolsky [92]. They constructed a connected 3-regular graph G of order 60 having $\gamma(G) = 21$. Further, they found a sequence of connected 3-regular graphs $\{G_k\}_{k=1}^{\infty}$ with $\nu(G_k) = 46k$ and $\gamma(G_k) \geq 16k$ such that $\lim_{k \to \infty} \frac{\gamma(G_k)}{\nu(G_k)} \geq \frac{8}{23} = \frac{1}{3} + \frac{1}{69}$. However, as they pointed out, their examples contain cutedges. But even more recently, the two authors, and, independently, Kelmans (unpublished private communication), have obtained 2-connected examples. However, it is unknown at this point if Reed's conjecture holds for 3-connected 3-regular graphs. In a related work, Kawarabayashi et al. [90] proved that a 2-edge-connected 3-regular graph G of girth at least $3k$ satisfies $\gamma(G) \leq \frac{3k+2}{9k+3}\nu(G)$.

Kostochka and Stodolsky [93] have also proved the following result.

Theorem 6 ([93]). *If G is a connected 3-regular graph with $\nu(G) > 8$, then $\gamma(G) \leq \frac{4}{11}\nu(G)$.*

The next two results provide bounds on the domination number of graphs with minimum degree 4 and 5.

Theorem 7 ([98]). *If G is 4-regular, then $\gamma(G) \leq \frac{4}{11}\nu(G)$.*

4 Domination in Graphs

Theorem 8 ([140]). *If G is a graph with $\delta(G) \geq 5$, then $\gamma(G) \leq \frac{5}{14}\nu(G)$.*

In 1998, Clark et al. [40] established an upper bound on the domination number based on the degree sequence of a graph. Their result stated in Theorem 9 has proved to be a useful tool in establishing Theorem 10 by Hellwig and Volkmann [78].

Theorem 9 ([40]). *Suppose G is a graph with $\delta(G) \geq 1$. Then*

$$\gamma(G) \leq \left[1 - \prod_{k=1}^{\delta(G)+1} \frac{k}{k + \frac{1}{\delta(G)}}\right]\nu(G).$$

Theorem 10 ([78]). *Suppose G is a connected graph:*

(i) *If $\delta(G) = 6$, then $\gamma(G) \leq \frac{6}{17}\nu(G)$.*
(ii) *If $\delta(G) \geq 7$, then $\gamma(G) \leq \lfloor \frac{1}{3}\nu(G) \rfloor$.*

The upper bound on the domination number can be very small if the graphs under consideration are planar or are of small diameter.

Theorem 11 ([71], page 55). *Let G be a graph with $diam(G) = 2$. Then $\gamma(G) \leq \delta(G)$.*

The bound in Theorem 11 follows immediately from the fact that, in a graph G of diameter 2, a neighborhood of a vertex $v \in V(G)$ dominates G for any v. This bound was improved by Hellwig and Volkmann [78] as follows.

Theorem 12 ([78]). *Let G be a graph with $diam(G) = 2$. Then*

(i) *$\gamma(G) \leq \kappa(G)$.*
(ii) *If G contains an induced complete graph $K_{\delta(G)}$ with all its vertices of degree $\delta(G)$, then $\gamma(G) \leq 3$.*

Hellwig and Volkmann [78] also established an upper bound on the domination number of graphs having diameter 2 in terms of their order. Further, they presented several sufficient conditions for $\gamma(G) \leq \delta(G) - 1$ to hold in graphs G having diameter 2. They also posed the problem of characterizing graphs G having diameter 2 with $\gamma(G) = \delta(G)$.

MacGillivray and Seyffarth [103] established an upper bound for planar graphs having small diameter.

Theorem 13 ([103]). *Suppose G is a planar graph. Then*

(i) *If $diam(G) = 2, \gamma(G) \leq 3$.*
(ii) *If $diam(G) = 3, \gamma(G) \leq 10$.*

MacGillivray and Seyffarth [103] also showed that the bound in Theorem 13(i) is best possible since the planar graph G in Fig. 4.2 is of diameter 2 and $\gamma(G) = 3$. They pointed out that the bound in Theorem 13(ii) might not be best possible. However, they have no example of a planar graph having diameter 3 and domination number greater than 6. They also gave examples of planar graphs of diameter 4 with arbitrarily large domination numbers.

In 2002, Goddard and Henning [67] improved Theorem 13 as follows.

Fig. 4.2 The planar graph G with diam$(G) = 2$ and $\gamma(G) = 3$

G :

Theorem 14 ([67]). *Suppose G is a planar graph:*

(i) *If $diam(G) = 2$, then either $\gamma(G) \leq 2$ or G is isomorphic to the graph in Fig. 4.2.*
(ii) *If $diam(G) = 3$ and $rad(G) = 2$, then $\gamma(G) \leq 6$.*
(iii) *If $diam(G) = 3$ and G is sufficiently large, then $\gamma(G) \leq 7$.*

The bounds in Theorem 14(ii) and (iii) were slightly improved by Dorfling et al. [48].

Theorem 15 ([48]). *Suppose G is a planar graph with $diam(G) = 3$. Then*

(i) $\gamma(G) \leq 9$.
(ii) *If $rad(G) = 2$, then $\gamma(G) \leq 5$.*
(iii) *If G is sufficiently large, then $\gamma(G) \leq 6$.*

4.3 Two Other Types of Domination

The concept of domination can be extended to yield new concepts by combining it with other graph parameters. For example, suppose that the dominating set S has an additional property such as being connected (it is then called a connected dominating set), being independent (it is then called an independent dominating set), having no isolated vertex (it is then called a total dominating set), or being a clique (it is then called a dominating clique), etc. Hence, there are lots of variations on domination. In this section we concentrate on two of the more widely studied variants of domination: total domination and connected domination. The reader is directed to [71, 72] and also [42, 77, 96] which deal with other types of domination.

A dominating set S of a graph G is said to be a *total dominating set* if $G[S]$ has no isolated vertex; that is, a subset S of $V(G)$ is a total dominating set for G if every vertex in $V(G)$ is adjacent to a vertex of S. The *total domination number* of a graph G, denoted by $\gamma_t(G)$, is the minimum size of any total dominating set of G. We call a total dominating set with cardinality $\gamma_t(G)$ a $\gamma_t(G)$-set. Note that since only graphs with no isolated vertex can contain a total dominating set, henceforth when referring to total domination, we assume all graphs under consideration have no isolates. Clearly, a graph with no isolated vertex contains at least one total dominating set, namely the set of all its vertices. It follows by definition that for any graph G with no isolated vertex, $\gamma(G) \leq \gamma_t(G)$ and $\gamma_t(G) \geq 2$. It is easy to see that for $n \geq 2, \gamma_t(K_n) = 2$ and for $m, n \geq 1, \gamma_t(K_{m,n}) = 2$. Zwierzchowski [144] computed the total domination numbers of a path P_n and a cycle C_n.

4 Domination in Graphs

Lemma 1 ([144]). *For integer $n \geq 3$,*

$$\gamma_t(P_n) = \gamma_t(C_n) = \begin{cases} \frac{n}{2}, & for\ n \equiv 0(mod4), \\ \frac{n+1}{2}, & for\ n \equiv 1(mod4), \\ \frac{n+2}{2}, & for\ n \equiv 2(mod4), \\ \frac{n+1}{2}, & for\ n \equiv 3(mod4). \end{cases}$$

Total domination was first studied by Cockayne et al. [43] in 1980. Since then this topic has received much attention. An excellent survey of this topic may be found in the two books by Haynes et al. [71,72]. A more recent survey is due to Henning [82]. As is the case with domination number, determining the total domination number of a graph is also NP-complete (see [82]), so bounds on total domination number have been a natural subject for investigation. We begin with a property of minimum dominating sets in graphs established by Bollobás and Cockayne [20] which leads to a relationship between domination and total domination numbers in a graph with no isolated vertex.

Theorem 16 ([20] or see [82]). *Every graph G with no isolated vertex has a minimum dominating set D in which each vertex $v \in D$ has the property that there exists a vertex $v' \in V(G) - D$ that is adjacent to v, but to no other vertex of D.*

Theorem 17 ([82]). *For every graph G with no isolated vertex, $\gamma(G) \leq \gamma_t(G) \leq 2\gamma(G)$.*

The next result concerns an upper bound on the total domination number in terms of the order of the graph.

Theorem 18 ([43]). *If G is a connected graph with $v(G) \geq 3$, then $\gamma_t(G) \leq \frac{2}{3}v(G)$.*

The *2-corona* of a graph H, denoted by $H \circ P_2$, is the graph obtained from H by adding a path of length 2 to each vertex of H. Brigham et al. [23] characterized graphs achieving the bound in Theorem 18 as follows.

Theorem 19 ([23]). *Let G be a connected graph with $v(G) \geq 3$. Then $\gamma_t(G) = \frac{2}{3}v(G)$ if and only if G is C_3, C_6 or $H \circ P_2$ for some connected graph H.*

Henning [82] established an upper bound on the total domination number in terms of minimum degree and order.

Theorem 20 ([82]). *If G is a graph with $\delta(G) \geq 1$, then*

$$\gamma_t(G) \leq \left(\frac{1 + ln(\delta(G))}{\delta(G)} \right) v(G).$$

When the minimum degree is increased, the bound on the total domination number in Theorems 18 and 20 can be improved.

Theorem 21 ([80]). *If G is a connected graph with $\delta(G) \geq 2$ and $G \notin \{C_3, C_5, C_6, C_{10}\}$, then $\gamma_t(G) \leq \frac{4}{7}v(G)$.*

The bound in Theorem 21 was improved by Lam and Wei [94] by imposing an extra condition on the graph.

Theorem 22 ([94]). *Suppose G is a graph with $\delta(G) \geq 2$. Then $\gamma_t(G) \leq \frac{1}{2}\nu(G)$ if either $d_G(u) + d_G(v) \geq 5$ for every two adjacent vertices u and v of G or every component of the subgraph of G induced by its set of vertices having degree 2 has size at most one.*

Recently, Henning and Yeo [87] improved the bound in Theorem 21 for a connected graph G with $\delta(G) \geq 2$ and $\Delta(G) \geq 3$ to $\frac{1}{2}(\nu(G) + p)$ where p is defined in terms of the sum of the number of paths having certain specified properties. For more details, the reader is directed to [87]. The next two results give upper bounds on the total domination number when the minimum degree is at least 3 or 4, respectively. Note that Theorem 23 can also be regarded as a corollary of Theorem 22.

Theorem 23 ([15,94,127]). *If G is a graph with $\delta(G) \geq 3$, then $\gamma_t(G) \leq \frac{1}{2}\nu(G)$.*

Theorem 24 ([127]). *If G is a graph with $\delta(G) \geq 4$, then $\gamma_t(G) \leq \frac{3}{7}\nu(G)$.*

If the graph under consideration is planar, the next result is an immediate consequence of Theorem 14(i).

Theorem 25 ([67]). *If G is a planar graph with $diam(G) = 2$, then $\gamma_t(G) \leq 3$.*

Theorem 26 ([48]). *Suppose G is a planar graph with $diam(G) = 3$. Then*

(i) $\gamma_t(G) \leq 10$.
(ii) If $rad(G) = 2$, then $\gamma_t(G) \leq 5$.
(iii) If G is sufficiently large, then $\gamma_t(G) \leq 7$.

Upper bounds on the total domination number have also been established in terms of the size of the graph ([81,121]), the girth ([47,74,86]), and the size of a maximum matching ([83,85]). Bounds in the presence of forbidden subgraphs such as $K_{1,3}$ ([53–55,85]) have been studied as well.

We now turn our attention to connected domination. A dominating set S for G is a *connected dominating set* if $G[S]$ is connected. The minimum cardinality of a connected dominating set is called the *connected domination number* of G and is denoted by $\gamma_c(G)$. We call a connected dominating set with cardinality $\gamma_c(G)$, a $\gamma_c(G)$-set. Note that since a graph must be connected to have a connected dominating set, henceforth when referring to connected domination, we assume all graphs under consideration are connected. It follows directly from the definitions that for a connected graph G, $\gamma(G) \leq \gamma_c(G)$ and if $\gamma(G) = 1$, then $\gamma_c(G) = \gamma(G)$. Further, a connected dominating set of size at least 2 is a total dominating set for G. Hence, $\gamma(G) \leq \gamma_t(G) \leq \gamma_c(G)$ for a connected graph G with $\Delta(G) < \nu(G) - 1$. It then follows that for a connected graph G and for $k \in \{2, 3\}$, $\gamma_t(G) = k$ if and only if $\gamma_c(G) = k$.

Connected domination seems to have been studied first by Sampathkumar and Walikar [118] who attribute the terminology to Hedetniemi (see [71, 72, 77]). The

4 Domination in Graphs

algorithmic aspects of connected domination were first discussed by Garey and Johnson in [65] where they claimed that determining the connected domination number of a graph is NP-complete, even when the graph is planar and regular of degree 4. For a summary of the algorithmic status of connected domination for various special graph classes, see [45, 50, 108, 138]. For relations between connected domination and other graph parameters (see [19, 56, 76, 100, 110, 120]).

We begin with the connected domination number for some special graph classes.

Theorem 27 ([118]).

(i) $\gamma_c(K_n) = 1$.

(ii) $\gamma_c(K_{m,n}) = \begin{cases} 1, & \text{if either } m = 1 \text{ or } n = 1, \\ 2, & \text{if } m, n \geq 2. \end{cases}$

(iii) $\gamma_c(P_n) = \gamma_c(C_n) = n - 2$ for $n \geq 3$.

(iv) For any tree T with $v(T) \geq 3$, $\gamma_c(T) = v(T) - p(T)$, where $p(T)$ denotes the number of endvertices in T.

Sampathkumar and Walikar [118] pointed out that the connected domination number of a graph is bounded above by the connected domination number of any spanning subgraph. They then used this fact together with Theorem 27(iv) to establish an upper bound on the connected domination number of graphs.

Theorem 28 ([118]). For a connected graph G with $v(G) \geq 3$, $\gamma_c(G) \leq v(G) - 2$.

The same two authors [118] also provided a bound for the connected domination number in terms of the maximum degree, the order, and the size of the graph.

Theorem 29 ([118]). For a connected graph G, $\left\lfloor \frac{v(G)}{\Delta(G)+1} \right\rfloor \leq \gamma_c(G) \leq 2\varepsilon(G) - v(G)$.

Note that a graph G containing a vertex of degree $v(G) - 1$ satisfies the lower bound and a path P_n of order n satisfies the upper bound. Hence, both these bounds are sharp.

In 1984, Hedetniemi and Laskar [76] observed a relationship between the connected domination number and the number of endvertices in a spanning tree. They then gave an upper bound on the connected domination number in terms of the maximum degree of the graph.

Theorem 30 ([76]). Let G be a connected graph. Then

(i) $\gamma_c(G) + p(T) = v(G)$ where $p(T)$ is the maximum number of endvertices in any spanning tree of G.

(ii) $\gamma_c(G) \leq v(G) - \Delta(G)$.

Kleitman and West [91] provided a result concerning the minimum number of leaves that a spanning tree in a connected graph with specified minimum degree can have. Combining their result together with Theorem 30(i), an upper bound on the connected domination number of graphs can be obtained.

Theorem 31 ([91]). *Let G be a connected graph with $\delta(G) \geq k$. Then G contains a spanning tree with at least $v(G) - 3 \left\lfloor \frac{v(G)}{k+1} \right\rfloor + 2$ leaves.*

Corollary 2 ([71], page 163). *If G is a connected graph with $\delta(G) \geq k$, then $\gamma_c(G) \leq 3 \left\lfloor \frac{v(G)}{k+1} \right\rfloor - 2$.*

Bounds on the connected domination number have been obtained by Duchet and Meyniel [49] (in terms of the domination number) and by Favaron and Kratsch [56] (in terms of the total domination number).

Theorem 32 ([49]). *For a connected graph G, $\gamma_c(G) \leq 3\gamma(G) - 2$.*

Theorem 33 ([56]). *For a connected graph G, $\gamma_c(G) \leq 2\gamma_t(G) - 2$.*

Before closing this section, we must mention the famous conjecture on domination due to Vizing [135] which remains unsettled.

Conjecture 1. For any graphs G and H, if $G \square H$ denotes the Cartesian product of G and H, then $\gamma(G)\gamma(H) \leq \gamma(G \square H)$.

The best bound known in this sense is due to Clark and Suen [41].

Theorem 34 ([41]). *For any graphs G and H, $\gamma(G)\gamma(H) \leq 2\gamma(G \square H)$.*

Vizing's conjecture may also be made for total and for connected domination. For total domination, Ho [88] has established the same bound as that of Clark and Suen.

Theorem 35 ([88]). *For any graphs G and H, $\gamma_t(G)\gamma_t(H) \leq 2\gamma_t(G \square H)$.*

As far as the authors know, no analogous bound has been established for the connected domination version of Vizing's conjecture.

4.4 Results on Criticality

In studying a graph parameter P, it is often useful to study a more restricted class of graphs, the so-called *critical graphs*. That is, a graph with parameter P, but in which an alteration of a certain type (such as edge addition, edge deletion, or vertex deletion) results in a graph no longer having property P.

In this section, we study graphs edge-critical and vertex-critical with respect to graph parameter P where $P \in \{\gamma, \gamma_t, \gamma_c\}$. Edge-critical graphs are graphs in which P decreases upon the addition of any missing edge while vertex-critical graphs are graphs in which P decreases when any vertex is removed. More precisely, we say that a graph G is k-P-*edge-critical* if $P(G) = k$ and $P(G + e) < k$ for every edge $e \in E(\overline{G})$ and G is k-P-*vertex-critical* if $P(G) = k$ but for each vertex v of G, $P(G - v) < k$. If we do not specify a value k, we simply say that a graph is P-edge-critical or P-vertex-critical. Of the two concepts of edge-criticality and

4 Domination in Graphs

vertex-criticality, edge-criticality has received more attention as we show in the coming subsections. Although the class of k-P-edge-critical graphs and the class of k-P-vertex-critical graphs are not the same, note that every k-P-vertex-critical graph can be extended to one that is both vertex- and edge-critical by successively adding edges that do not decrease P (see [24]). For each $P \in \{\gamma, \gamma_c\}$, it is well known that 1-P-edge-critical graphs are precisely the complete graphs K_n for all positive integers n, whereas the only 1-P-vertex-critical graph is K_1.

Remark 1. Suppose G is a k-P-edge-critical graph where $P \in \{\gamma, \gamma_c, \gamma_t\}$ and u and v are nonadjacent vertices of G. Then at least one of u and v is in a $P(G)$-set for $G + uv$. Further, if $P = \gamma$, then exactly one of u and v is in a $\gamma(G)$-set for $G + uv$.

4.4.1 k-γ-Edge-Critical Graphs

k-γ-edge-critical graphs were first studied by Sumner and Blitch [126] in 1983. Most of their results were concerned with the cases $2 \le k \le 3$. They gave a characterization of 2-γ-edge-critical graphs and 3-γ-edge-critical disconnected graphs as we show in Theorem 36. Since then the concept of connected k-γ-edge-critical graphs for $k \ge 3$ has received considerable attention. Most of the known results are confined to the case $k = 3$. These graphs have been studied with respect to graph parameters such as toughness and matching, [2, 3, 5, 6, 8, 10, 36] Hamiltonicity [32, 34, 37, 38, 51, 58, 107, 109, 128, 139, 141–143], and more [35, 57, 59, 73, 89, 112, 113, 122, 125, 132, 143]. We do not intend to list all of these results, but give a brief survey on connected 3-γ-edge-critical graphs. For $k \ge 4$, connected k-γ-edge-critical graphs are far from completely understood. In fact, even for $k = 3$, no characterization of such graphs is known.

The following lemma follows immediately from the definition of k-γ-edge-criticality.

Lemma 2. *Suppose G is k-γ-edge-critical and u and v are nonadjacent vertices of G. Then*

(i) $\gamma(G + uv) = k - 1$.
(ii) There is a subset S of $V(G) - \{u, v\}$ of size $k - 2$ such that either $S \cup \{u\}$ dominates $G - v$ or $S \cup \{v\}$ dominates $G - u$. Further, if $S \cup \{v\}$ dominates $G - u$, then $S \cap N_G(u) = \emptyset$.

The following characterizations of 2-γ-edge-critical graphs and disconnected 3-γ-edge-critical graphs were established by Sumner and Blitch [126].

Theorem 36 ([125, 126]).

(i) A graph G is 2-γ-edge-critical if and only if $\overline{G} = \bigcup_{i=1}^{n} K_{1,r_i}$ where r_i and n are positive integers.

(ii) A graph G is a 3-γ-edge-critical disconnected graph if and only if G is a disjoint union of a 2-γ-edge-critical graph and a complete graph K_1 or a disjoint union of a complete graph with a perfect matching removed and a complete graph K_n where $n \geq 1$.

Before stating the next result, we need some new notation. If u, v, and w are vertices of G and $\{u, v\}$ dominates $G - w$, then we write $[u, v] \to w$. This so-called *arrow notation* made its first appearance in [126]. Note that if G is 3-γ-edge-critical and u and v are nonadjacent vertices of G, then $\gamma(G + uv) = 2$ and so there is a vertex $z \in V(G) - \{u, v\}$ such that $[u, z] \to v$ or $[v, z] \to u$.

Sumner and Blitch [126] proved the next theorem for the case $n \geq 4$. The cases $n = 2$ and 3 were added by Flandrin et al. [59]. This result is the most useful tool in studying 3-γ-edge-critical graphs yet available.

Theorem 37 ([59, 126]). *Let G be a connected 3-γ-edge-critical graph and let S be an independent set of $n \geq 2$ vertices in $V(G)$:*

(i) Then the vertices of S can be ordered as $a_1, a_2, \ldots, a_n$ in such a way that there exists a sequence of distinct vertices $x_1, x_2, \ldots, x_{n-1}$ so that $[a_i, x_i] \to a_{i+1}$ for $i = 1, 2, \ldots, n - 1$.
(ii) If, in addition, $n \geq 4$, then the x_i's can be chosen so that $x_1, x_2, \ldots, x_{n-1}$ span a path and $S \cap \{x_1, x_2, \ldots, x_{n-1}\} = \emptyset$.

The next result concerns the diameter and number of components of 3-γ-edge-critical graphs.

Theorem 38 ([126]). *Suppose G is a connected 3-γ-edge-critical graph. Then*

(i) $2 \leq diam(G) \leq 3$.
(ii) If S is a vertex cutset in G, then $\omega(G - S) \leq |S| + 1$.

Ananchuen and Plummer [2, 6, 8] sharpened Theorem 38(ii) as follows.

Theorem 39 ([2,6,8]). *Let G be a connected 3-γ-edge-critical graph and let S be a vertex cutset in G. Then*

(i) If $|S| \geq 6, \omega(G - S) \leq |S| - 2$.
(ii) If $|S| \geq 4, \omega(G - S) \leq |S| - 1$.
(iii) If $4 \leq |S| \leq 5$ and each component of $G - S$ has at least three vertices, then $\omega(G - S) \leq |S| - 2$.
(iv) If $|S| = 3$, then $\omega(G - S) \leq 3$ and if $G - S$ has exactly three components, then each component is complete and at least one of them is a singleton.
(v) If $|S| = 2$, then $\omega(G - S) \leq 3$ and if $G - S$ has exactly three components, then G must be the graph shown in Fig. 4.3 with $n \geq 2$.
(vi) If $|S| = 1$, then $\omega(G - S) = 2$ and exactly one of the components of $G - S$ is a singleton. Furthermore, G has at most three cutvertices. If it has two, then G is a graph of the type shown in Fig. 4.3 with $n \geq 2$, and if it has three, it is the graph shown in Fig. 4.3 with $n = 1$.

Fig. 4.3 A 3-γ-edge-critical graph with a cutvertex

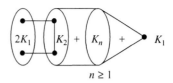

Theorem 38(ii) together with Theorem 39 might be considered a result on toughness. Both are useful for establishing results on matchings in Sect. 4.5. The *toughness* of a graph G denoted by $\tau(G)$ is min $\{|S|/\omega(G - S) : S$ is a vertex cutset of $G\}$. A graph G is said to be *t-tough* if $\tau(G) \geq t$.

Remark 2. By Theorem 38(ii), if G is a connected 3-γ-edge-critical graph, then G is 1/2-tough. Further, if $\delta(G) \geq 2$, then G is 2-connected, by Theorem 39, and thus $\tau(G) \geq 1$ (see also [36]). In fact, Blitch [18] proved that if G is a connected 3-γ-edge-critical graph containing v as a cutvertex, then v is adjacent to an endvertex of G. Hence, if $\delta(G) = 1$, then $\tau(G) < 1$. Chen et al. [36] gave a characterization of connected 3-γ-edge-critical graphs with toughness 1.

We now turn our attention to Hamiltonicity. In 1990, Wojcicka [139] proved that every connected 3-γ-edge-critical graph of order at least 7 has a Hamiltonian path. (A shorter proof was given in 2002 by Zhang and Tian [143].) Wojcicka then posed the following conjecture.

Conjecture 2. Every connected 3-γ-edge-critical graph having minimum degree at least 2 has a Hamiltonian cycle.

Moreover, she conjectured that, in general, for $k \geq 3$, $(k - 1)$-connected k-γ-edge-critical graphs are Hamiltonian. Yuansheng et al. [142] showed that this conjecture is not true for $k = 4$ by constructing a class of 3-connected 4-γ-edge-critical non-Hamiltonian graphs. Conjecture 2 was settled via the following two results and is now referred to as "Wojcicka's theorem" by Mynhardt [109].

Theorem 40 ([58]). *If G is a connected 3-γ-edge-critical graph with $\delta(G) \geq 2$, then $\alpha(G) \leq \delta(G) + 2$.*

Theorem 41 ([58, 128]). *Suppose G is a connected 3-γ-edge-critical graph with $\delta(G) \geq 2$:*

(i) If $\alpha(G) \leq \delta(G) + 1$, then G is Hamiltonian.
(ii) If $\alpha(G) = \delta(G) + 2$, then G is Hamiltonian.

Part (i) of the above theorem was proved by Favaron et al. [58] and Tian et al. [128] established part (ii).

The next corollary is a consequence of Wojcicka's theorem and Remark 2, together with the observation that if a graph G is Hamiltonian, then $\tau(G) \geq 1$.

Corollary 3. *Suppose G is a connected 3-γ-edge-critical graph. Then G is Hamiltonian if and only if $\tau(G) \geq 1$.*

Recently, Chen and Tian [34] obtained an even simpler proof of Wojcicka's theorem by using Hanson's [68] and Bondy–Chvátal's [21] closure operations.

A result closely related to Wojcicka's theorem was proved by Xue and Chen [141]. They established that if G is a connected 3-γ-edge-critical with $\delta(G) = 1$, then $G - A$ is Hamiltonian where $A = \{v \in V(G)|d_G(v) = 1\}$.

A graph G is Hamiltonian-connected if every two distinct vertices are joined by a Hamiltonian path. Chen et al. [37] posed a conjecture similar to that of Wojcicka.

Conjecture 3. A connected 3-γ-edge-critical graph is Hamiltonian-connected if and only if $\tau(G) > 1$.

This conjecture was proved in a series of four papers [32, 37, 38, 51] by Chen, Cheng, Ng, Tian, Wei, and Zhang.

4.4.2 k-γ_t-Edge-Critical Graphs

The concept of a k-γ_t-edge-critical graph was first introduced by van der Merwe et al. [130] in 1998. Since $\gamma_t(G) \geq 2$ for any graph G with no isolates, the first nontrivial case for k-γ_t-edge-critical graphs is $k = 2$. But it is easy to show that the only 2-γ_t-edge-critical graphs are the complete graphs K_n with $n \geq 2$. Most of the nontrivial results on k-γ_t-edge-critical graphs are concerned with the case $k = 3$ and were established by van der Merwe et al. [130, 131, 133, 134], Hanson and Wang [69], and more recently by Simmons [123].

In [130] van der Merwe et al. showed that the addition of an edge to a graph may decrease the total domination number of the resulting graph by at most two.

Theorem 42 ([130]). *For any edge* $e \in E(\overline{G})$, $\gamma_t(G) - 2 \leq \gamma_t(G + e) \leq \gamma_t(G)$.

It then follows that if G is a k-γ_t-edge-critical graph, then for an edge $e \in E(\overline{G}), \gamma_t(G + e) = k - 1$ or $k - 2$.

Observe that if G is k-γ_t-edge-critical, then for any pair of nonadjacent vertices u and v of G, at least one of u and v must be in a minimum total dominating set for $G + uv$. Further, if $\gamma_t(G + uv) = k - 2$, then u and v share a minimum total dominating set for $G + uv$.

A k-γ_t-edge-critical graph G with $\gamma_t(G + e) = k - 2$ for every edge $e \in E(\overline{G})$ is called *supercritical* (cf. [130]). These graphs were characterized by van der Merwe et al. [131].

Theorem 43 ([131]). *A graph G is supercritical if and only if G is the union of two or more nontrivial complete graphs.*

We now turn our attention to 3-γ_t-edge-critical graphs.

Theorem 44 ([130]). *Suppose G is a 3-γ_t-edge-critical graph. Then*

(i) $2 \leq diam(G) \leq 3$.
(ii) Every vertex of G is adjacent to at most one endvertex.
(iii) G has at most one cutvertex.

4 Domination in Graphs   87

Note that Theorem 44(ii) also holds for k-γ_t-edge-critical graphs for $k \geq 3$. Further, it follows by Theorem 44(iii) that no tree is 3-γ_t-edge-critical.

The next two results provide information on 3-γ_t-edge-critical graphs containing a cutvertex.

Theorem 45 ([130]). *Suppose G is a 3-γ_t-edge-critical containing x as a cutvertex. Then*

(i) $x \in \gamma_t(G)$-set *for every* $\gamma_t(G)$-set *of* G.
(ii) $diam(G) = 3$.
(iii) $G - x$ *has exactly two components.*
(iv) x *is adjacent to an endvertex.*

Theorem 46 ([130]). *Let G be a graph with a pendent edge uv (where v is a cutvertex and u an endvertex) and let $A = N(v) - \{u\}$ and $B = V(G) - N[v]$. Then G is 3-γ_t-edge-critical if and only if:*

(i) $G[A]$ *is complete and* $|A| \geq 2$.
(ii) $G[B]$ *is complete and* $|B| \geq 2$.
(iii) *Every vertex in A is adjacent to $|B| - 1$ vertices in B and every vertex in B is adjacent to at least one vertex in A.*

The next corollary follows immediately.

Corollary 4 ([130]). *If G is a 3-γ_t-edge-critical with an endvertex u, then*

(i) $\gamma(G) = 2$.
(ii) $G - u$ *is Hamiltonian.*

van der Merwe et al. [130] proved that for any graph G, there is a 3-γ_t-edge-critical graph H such that G is an induced subgraph of H. Consequently, it is not possible to characterize 3-γ_t-edge-critical graphs in terms of forbidden subgraphs.

Simmons [123] defined the closure of a 3-γ_t-edge-critical graph G, denoted by $D^t(G)$, to be the graph obtained from G by adding the edge uv to G for each pair of nonadjacent vertices u and v if $\{u, v\}$ dominates G. It was then shown that:

Theorem 47 ([123]). *For any 2-connected 3-γ_t-edge-critical graph G, G is Hamiltonian if and only if $D^t(G)$ is Hamiltonian.*

Simmons [123] also established several results for $D^t(G)$ and showed that these results also hold for G when G is 3-γ_t-edge-critical.

Theorem 48 ([123]). *Let G be a 3-γ_t-edge-critical graph and I an independent set in G with $|I| = m \geq 3$. Then the vertices in I can be ordered as $a_1, a_2, \ldots, a_m$ in such a way that there exists a path $x_1, x_2, \ldots, x_{m-1}$ in $G - I$ where $\{x_i, a_i\}$ is a minimum total dominating set for $G - a_{i+1}$ for $i = 1, 2, \ldots, m - 1$.*

Theorem 49 ([123]). *Let G be a 2-connected 3-γ_t-edge-critical graph. If the independence number $\alpha(G) \geq 3$, then $\alpha(G) \leq \delta(G) + 2$. Moreover, if $\alpha(G) = \delta(G) + 2$, then every maximum independent set contains all the vertices of degree $\delta(G)$.*

Theorem 50 ([123]). *Let G be a connected 3-γ_t-edge-critical graph. If S is a vertex cutset of G, then $G - S$ has at most $|S| + 1$ components.*

Note that Theorems 48 and 50 are analogous to Theorems 37 and 38(ii), respectively.

4.4.3 k-γ_c-Edge-Critical Graphs

k-γ_c-edge-critical graphs were first introduced by Chen et al. [33] in 2004. They gave a characterization of 2-γ_c-edge-critical graphs and provided conditions for some particular classes of graphs to be critical. They also established some results concerning the case $k = 3$, most of which have previous analogues for ordinary domination critical graphs. Ananchuen [1] studied k-γ_c-edge-critical graphs with cutvertices and gave a characterization for such graphs when $k = 3$. Ananchuen et al. [11] studied matching properties in 3-γ_c-edge-critical graphs as we show in Sect. 4.5.

The first result in this section shows that the addition of an edge to a graph may decrease the connected domination number of the resulting graph by at most two. Thus this result is analogous to Theorem 42 for total domination number.

Theorem 51 ([33]). *Let G be a connected graph. For any edge $e \in E(\overline{G})$, $\gamma_c(G) - 2 \leq \gamma_c(G + e) \leq \gamma_c(G)$.*

It then follows that if G is k-γ_c-edge-critical, then for each $e \in E(\overline{G})$, $\gamma_c(G + e) = k - 1$ or $k - 2$.

The next result provides a characterization of 2-γ_c-edge-critical graphs.

Theorem 52 ([33]). *A connected graph G is 2-γ_c-edge-critical if and only if $\overline{G} = \bigcup_{i=1}^{n} K_{1,r_i}$ for $r_i \geq 1$ and $n \geq 2$.*

By Theorem 36(i), it is easy to see that G is a connected 2-γ-edge-critical graph if and only if G is a 2-γ_c-edge-critical graph.

Chen et al. [33] proved that for an integer $n \geq 3$, no path or tree of order n is γ_c-edge-critical. Further, neither $K_{1,k}$, for a positive integer $k \geq 2$, nor $K_{r,s}$, for $\max\{r, s\} \geq 3$, where r and s are positive integers, is γ_c-edge-critical. However, C_n, a cycle of order n, is γ_c-edge-critical for every integer $n \geq 4$.

For $k = 3$, Chen et al. [33] established a bound on the diameter and the number of components of k-γ_c-edge-critical graphs analogous to Theorem 38. (Just replace γ with γ_c.)

Ananchuen [1] studied k-γ_c-edge-critical graphs with cutvertices and established the following.

Theorem 53 ([1]). *For $k \geq 3$, let G be a k-γ_c-edge-critical graph with a cutvertex x. Then*

(i) $x \in S$ for every $\gamma_c(G)$-set S of G.
(ii) $G - x$ contains exactly two components.

4 Domination in Graphs

(iii) If C is a component of $G - x$, then $\gamma_c(C) \le k - 1$ and $G[N_C(x)]$ is complete. Further, if $\gamma_c(C) = k - 1$, then C is $(k - 1)$-γ_c-edge-critical.

Theorem 54 ([1]). *For $k \ge 3$, let G be a k-γ_c-edge-critical graph with a cutvertex x. Suppose C_1 and C_2 are the components of $G - x$. Let $A = G[V(C_1) \cup \{x\}]$ and $B = G[V(C_2) \cup \{x\}]$. Then*

(i) $k - 1 \le \gamma_c(A) + \gamma_c(B) \le k$.
(ii) $\gamma_c(A) + \gamma_c(B) = k$ if and only if exactly one of C_1 and C_2 is a singleton.

For 3-γ_c-edge-critical graphs, Ananchuen [1] improved the result of Chen et al. [33] on a number of components.

Theorem 55 ([1]). *Let G be a 3-γ_c-edge-critical graph and S a vertex cutset of G with $|S| \ge 2$. Then $\omega(G - S) \le |S|$.*

One might expect a result analogous to Theorem 39 for 3-γ_c-edge-critical graphs. But this is not the case. In [1] it was shown that the bound in Theorem 55 is best possible.

4.4.4 k-γ-Vertex-Critical Graphs

Recall that a graph G is k-γ-vertex-critical if $\gamma(G) = k$ and $\gamma(G - v) < k$ for each vertex $v \in V(G)$. It is not difficult to see, for example, that C_{3k+1} is $(k + 1)$-γ-vertex-critical. Clearly, a disconnected graph G is γ-vertex-critical if and only if each component of G is γ-vertex-critical. So in the rest of this section, we concern ourselves with connected graphs only.

The concept of k-γ-vertex-criticality was first introduced by Brigham et al. [24,25]. Clearly, the only 1-γ-vertex-critical graph is K_1. Brigham et al. [25] pointed out that the 2-γ-vertex-critical graphs are precisely those obtained from the complete graphs K_{2n} by deleting a perfect matching. For $k > 2$, an understanding of the structure of k-γ-vertex-critical graphs is far from complete. However, some structural properties have been established in terms of an upper bound on the order [24, 25] and the diameter [63, 64].

In what follows, D_v denotes any minimum dominating set for $G - v$. The first result provides basic properties that have proved very useful in studying k-γ-vertex-critical graphs.

Lemma 3 ([24, 25]). *Let G be a k-γ-vertex-critical graph. Then*

(i) For each $v \in V(G), |D_v| = k - 1$ and $N_G[v] \cap D_v = \emptyset$.
(ii) For every pair of distinct vertices u and v, $D_u \ne D_v$.
(iii) If $d_G(v) \ge 1$, $N_G[v]$ is not complete.
(iv) There exists no pair of vertices u and v of G such that $N_G[u] \subseteq N_G[v]$.

The next result establishes an upper bound on the order of a k-γ-vertex-critical graph.

Theorem 56 ([24, 25]). *Let G be a k-γ-vertex-critical graph. Then*

(i) $v(G) \leq (\Delta(G) + 1)(k - 1) + 1$.
(ii) $v(G) \leq \frac{1}{3}(2\varepsilon(G) + 3k - \Delta(G))$.

Brigham et al. [24] conjectured that if G is a k-γ-vertex-critical graph with $v(G) = (\Delta(G) + 1)(k - 1) + 1$, then G is regular. This conjecture was settled in the affirmative by Fulman et al. [64]. In the same paper [24], Brigham et al. also posed a conjecture on an upper bound of the diameter of k-γ-vertex-critical connected graphs which they proved for $k \leq 5$ and which Fulman et al. [64] later proved for $k \geq 2$.

Theorem 57 ([64]). *The diameter of a k-γ-vertex-critical connected graph is at most $2(k - 1)$.*

Fulman et al. [64] showed that the bound in Theorem 57 is best possible. They also characterized 3-γ-vertex-critical graphs with diameter 4 and 4-γ-vertex-critical graphs with diameter 6 described in Theorem 58. Recall that a block of a graph is a maximal connected subgraph of G with no cutvertices of itself. An end block of G is a block containing exactly one cutvertex of G.

Theorem 58 ([64]).

(i) A graph G with diameter 4 is 3-γ-vertex-critical if and only if G has two blocks, each of which is 2-γ-vertex-critical.
(ii) A graph G with diameter 6 is 4-γ-vertex-critical if and only if it has three blocks, two of which are end blocks and all of which are 2-γ-vertex-critical.

Before proceeding, we need a new definition. Let G and H be disjoint graphs. $H \cdot G$, the *coalescence* of H and G, is the graph obtained by identifying one vertex of H with one vertex of G. (Note that the identified vertices become a cutvertex of $H \cdot G$.) Brigham et al. [24, 25] showed that:

Theorem 59 ([24, 25]). *Let H and G be graphs $\neq K_1$. Then*

(i) $\gamma(H) + \gamma(G) - 1 \leq \gamma(H \cdot G) \leq \gamma(H) + \gamma(G)$.
(ii) If both H and G are γ-vertex-critical or if $H \cdot G$ is γ-vertex-critical, then $\gamma(H \cdot G) = \gamma(H) + \gamma(G) - 1$.
(iii) $H \cdot G$ is γ-vertex-critical if and only if both H and G are γ-vertex-critical.

They then used the above result and induction to establish the following.

Theorem 60 ([24,25]). *A graph G is γ-vertex-critical if and only if each block of G is γ-vertex-critical. Further, if G is γ-vertex-critical and contains $G_1, G_2, \ldots, G_n$ as its blocks, then $\gamma(G) = \sum_{i=1}^{n} \gamma(G_i) - (n - 1)$.*

4 Domination in Graphs 91

Brigham et al. [24, 25] also used the idea of coalescence to show that any graph G with $\gamma(G) = k \geq 3$ can be embedded in a k-γ-vertex-critical graph. They first showed that for a given graph G with $\gamma(G) \geq 3$, there exists a 3-γ-vertex-critical graph H containing G as an induced subgraph. The graph H can be constructed as follows. Let $V(G) = \{v_1, v_2, \ldots, v_p\}$. Note that $p \geq 3$ since $\gamma(G) \geq 3$. Now let $V(H) = \{v_1, v_2, \ldots, v_p, w_1, w_2, \ldots, w_p, x_1, x_2, \ldots, x_p\}$ and $E(H) = E(G) \cup \{w_i x_j, w_i v_j, x_i v_j | 1 \leq i, j \leq p$ and $j \neq i\}$. It is not difficult to show that H is 3-γ-vertex-critical. Clearly, if $\gamma(G) = k = 3$, then H is the required graph in which G can be embedded. If $\gamma(G) = k \geq 3$, then they showed that $H \cdot C_{3(k-3)+1}$ is the desired graph in which G can be embedded. Note that $\gamma(H \cdot C_{3(k-3)+1}) = \gamma(H) + \gamma(C_{3(k-3)+1}) - 1 = 3 + (k - 2) - 1 = k$.

Hence we have the following theorem.

Theorem 61 ([**24, 25**]). *Let G be any graph with $\gamma(G) = k \geq 3$. Then*

(i) *There exists a 3-γ-vertex-critical graph H containing G as an induced subgraph.*
(ii) *G can be embedded in a k-γ-vertex-critical graph.*

Note that for any graph G, $\gamma(G \cup 2K_1) \geq 3$, thus Theorem 61(i) also holds for any graph G with $\gamma(G) < 3$.

It follows immediately by Theorem 61 that it is not possible to characterize k-γ-vertex-critical graphs in terms of forbidden subgraphs.

A graph is *claw-free* if it does not contain the graph $K_{1,3}$ as an induced subgraph. Ananchuen and Plummer [7] studied the connectivity of claw-free 3-γ-vertex-critical graphs in terms of minimum degree. They showed that:

Theorem 62 ([**7**]). *Let G be a connected claw-free 3-γ-vertex-critical graph. Then*

(i) *G is 2-connected.*
(ii) *If G is of even order or if $\delta(G) \geq 3$, then G is 3-connected.*
(iii) *If $\delta(G) \geq 5$, then G is 4-connected.*

We conclude this section by reminding the reader that the concept of vertex criticality may also be applied to other variations of domination such as distance domination (see [72] Chap. 12, and [84, 129]), as well as independent domination [14]. (See also [26–28, 39, 75, 115].)

4.4.5 k-P-Vertex-Critical Graphs Where $P \in \{\gamma_t, \gamma_c\}$

Observe that a graph containing an isolated vertex cannot be totally dominated with any vertex set. Similarly, a disconnected graph cannot be dominated by any connected dominating set. Hence, it makes sense to add an extra assumption for a vertex to be deleted when we want to study vertex deletion with respect to total domination and connected domination.

A graph G with no isolated vertex is k-γ_t-*vertex-critical* if $\gamma_t(G) = k$ and for each $v \in V(G)$ that is not adjacent to any vertex of degree one, $\gamma_t(G - v) < k$. A graph G with $\kappa(G) \geq 2$ is k-γ_c-*vertex-critical* if $\gamma_c(G) = k$ but $\gamma_c(G - v) < k$ for every vertex $v \in V(G)$. Note that if we assume that the connectivity of G is at least 2, then G is 3-γ_c-vertex-critical if and only if G is 3-γ_t-vertex-critical. Hence, for graphs with connectivity at least 2, results dealing with 3-γ_c-vertex-critical graphs may be interpreted as results pertaining to 3-γ_t-vertex-critical graphs and vice versa since for any graph G, $\gamma_c(G) = 3$ if and only if $\gamma_t(G) = 3$.

k-γ_t-vertex-critical graphs were first studied by Goddard et al. [66] in 2004 and later on by Mojdeh and Rad [105, 106]. It is easy to see that G is γ_t-vertex-critical if and only if each component of G is γ_t-vertex-critical. So we restrict our attention to connected graphs only. The following basic result is similar to Lemma 3.

Lemma 4 ([66]). *Suppose G is a k-γ_t-vertex-critical graph. Let $S(G) = \{u \in V(G) | u$ is a support vertex in $G\}$. Then*

(i) *For each $v \in V(G) - S(G)$, if D_v is a minimum total dominating set for $G - v$, then $|D_v| = k - 1$ and $D_v \cap N_G[v] = \emptyset$.*
(ii) *For each $v \in V(G) - S(G)$, there is no vertex u of G with $uv \notin E(G)$ and $N_G(u) \subseteq N_G(v)$.*

Theorem 63 ([66]). *Let G be a connected graph of order at least 3 with at least one endvertex. Then, G is k-γ_t-vertex-critical if and only if $G = H \circ K_1$ for some connected graph H of order k with $\delta(H) \geq 2$.*

It follows immediately by Theorem 63 that no tree is γ_t-vertex-critical.

A graph H is *vertex diameter k-critical* if diam$(H)=k$ and diam$(H-v) > k$. Hanson and Wang [69] observed that for a graph G, $\gamma_t(G) = 2$ if and only if diam$(\overline{G}) > 2$. By combining this result together with a concept of vertex diameter k-criticality, Goddard et al. [66] obtained the following characterization.

Theorem 64 ([66]). *A connected graph G is 3-γ_t-vertex-critical if and only if $\overline{G}$ is vertex diameter 2-critical or G is $K_3 \circ K_1$.*

Goddard et al. [66] also gave an upper bound for the diameter of k-γ_t-vertex-critical graphs which is sharp for small k.

Theorem 65 ([66]). *Let G be a k-γ_t-vertex-critical graph. Then*

(i) *diam$(G) \leq 2k - 3$.*
(ii) *For $k \leq 8$, the diameter of G is at most the value given by Table 4.1.*

Table 4.1 Upper bounds on the diameter of k-γ_t-vertex-critical graphs with $3 \leq k \leq 8$

k	3	4	5	6	7	8
$diam(G)$	3	4	6	7	9	11

4 Domination in Graphs   93

Finally, they posed several open problems, for example, the characterization of a k-γ_t-vertex-critical graph G of order $\Delta(G)(k-1)+1$ and $k+\Delta(G)$, respectively. These two problems were addressed by Mojdeh and Rad [105, 106] and more recently by Wang et al. [136].

We now turn our attention to k-γ_c-vertex-critical graphs. It is easy to see that the only 1-γ_c-vertex-critical graph is K_1 and the 2-γ_c-vertex-critical graphs are obtained from the even complete graphs K_{2n}, with $n \geq 2$, by deleting a perfect matching. For $k \geq 3$, the structure of k-γ_c-vertex-critical graphs is much more complicated. k-γ_c-vertex-critical graphs were studied in [12, 13], where most of the results are concerned with 3-γ_c-vertex-critical graphs. Some basic properties of 3-γ_c-vertex-critical graphs especially with respect to connectivity are established which were used to study matching properties in 3-γ_c-vertex-critical graphs as we show in the next section. The following result is analogous to Lemma 3(i, ii, iv).

Lemma 5 ([12]). *Suppose G is k-γ_c-vertex-critical. Let D_v be a minimum connected dominating set for $G - v$. Then*

 (i) *If $v \in V(G)$, $D_v \cap N_G[v] = \emptyset$.*
 (ii) *If $u, v \in V(G)$ and $u \neq v$, then $D_u \neq D_v$.*
 (iii) *If $v \in V(G)$, then $|D_v| = k - 1$.*
 (iv) *If $v \in V(G)$, then there is no $u \neq v$ such that $N_G[v] \subseteq N_G[u]$.*

Theorem 66 ([12]). *If G is a 3-γ_c-vertex-critical graph, then either $G = C_5$ or G is 3-connected.*

In the case when G is exactly 3-connected, we can say more about its structure.

Theorem 67 ([12]). *Suppose G is a 3-γ_c-vertex-critical graph and S is a vertex cutset in G with $|S| = 3$. Then*

 (i) *$|E(G[S])| \leq 1$.*
 (ii) *$G - S$ consists of precisely two components.*

Theorem 67 can then be used to establish the following results.

Theorem 68 ([12]). *Suppose G is a 3-γ_c-vertex-critical graph and S is a minimum vertex cutset in G with $|S| = 3$. Let C_1 and C_2 be the two components of $G - S$ where $|V(C_1)| \leq |V(C_2)|$. Then*

 (i) *Either $|V(C_1)| \leq 2$ or $|V(C_2)| \leq 2$.*
 (ii) *If $|V(C_1)| = 2$, then $|V(C_2)| = 2$, and G is isomorphic to the graph G_0 in Fig. 4.4.*
 (iii) *If $|V(C_1)| = 1$ and S is independent, then G is isomorphic to one of G_0 or G_0', both of which are shown in Fig. 4.4.*
 (iv) *If $|V(C_1)| = 1$ and S contains an edge, then G has G_1 or G_2 as a spanning subgraph, where G_1 and G_2 are shown in Fig. 4.5.*

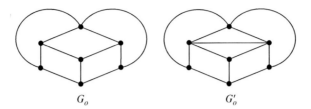

Fig. 4.4 The 3-γ_c-vertex-critical graphs G_0 and G'_0

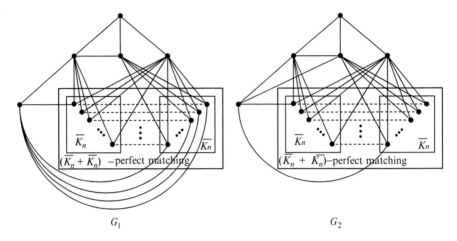

Fig. 4.5 The graphs G_1 and G_2

4.5 Matching Properties

Recall that a *perfect matching* in a graph G is a matching that covers all of the vertices of G while a *near-perfect matching* is a matching that covers all but one of the vertices of G.

A graph G is *k-factor-critical* if and only if for every set $S \subseteq V(G)$ with $|S| = k$, the graph $G - S$ contains a perfect matching. Note that if G is k-factor-critical then $v(G) \equiv k \pmod{2}$. k-factor-critical graphs are called *factor-critical* if $k = 1$ and *bicritical* if $k = 2$. Factor critical and bicritical graphs play an important role in a canonical decomposition theory for arbitrary graphs in term of their matchings. The interested reader is referred to [101] for much more on this subject.

In this section, we are concerned with the existence of a perfect matching, a near-perfect matching, and being k-factor-critical for the classes of 3-P-edge-critical graphs and 3-P-vertex-critical graphs where $P \in \{\gamma, \gamma_c\}$.

4.5.1 3-P-Edge-Critical Graphs and Matching Properties

We begin by recalling Tutte's famous theorem on perfect matchings.

4 Domination in Graphs

Theorem 69 (Tutte's theorem (see [22] page 76)). *A graph G has a perfect matching if and only if $\omega_o(G - S) \leq |S|$, for all $S \subset V(G)$.*

By combining Theorem 38(ii) and Tutte's theorem, the first part of Theorem 70 below follows immediately for $P = \gamma$. This was established by Sumner and Blitch [126]. For $P = \gamma_c$, Chen et al. [33] also established a result similar to Theorem 38(ii). So the first part of Theorem 70 also holds for $P = \gamma_c$, (see [33]). The second part of Theorem 70 was established in [3] for $P = \gamma$ and in [11] for $P = \gamma_c$.

Theorem 70 ([3, 11, 33, 126]). *For $P \in \{\gamma, \gamma_c\}$, let G be a connected 3-P-edge-critical graph. Then*

(i) *If $v(G)$ is even, G contains a perfect matching.*
(ii) *If $v(G)$ is odd, G contains a near-perfect matching.*

Sufficient conditions for 3-γ-edge-critical graphs to be k-factor-critical for $k = 1, 2, 3$ are presented next.

Theorem 71 ([3, 10]). *Suppose G is a 3-γ-edge-critical graph.*

(i) *If G is 2-connected of odd order, then G is factor-critical.*
(ii) *Suppose further that G is 3-connected of even order and if G is either planar or having minimum degree at least 4, then G is bicritical.*
(iii) *If G is 4-connected of odd order and $\delta(G) \geq 5$, then G is 3-factor-critical.*

The next theorem shows that the hypotheses on both the connectivity and minimum degree in Theorem 71 can be relaxed if the graph under consideration is claw-free.

Theorem 72 ([3, 10]). *Suppose G is a 3-γ-edge-critical claw-free graph. Then, for $k \in \{2, 3\}$, if G is k-connected with $\delta(G) \geq k + 1$ and $v(G) \equiv k(mod 2)$, then G is k-factor-critical.*

In [10] it was conjectured that Theorem 72 holds for $k \geq 2$. The following conjecture was also posed.

Conjecture 4. Suppose G is a graph with $k \geq 2$ and suppose $k - 1$ and $v(G)$ have the same parity. Then if G is k-connected and 3-γ-edge-critical with $\delta(G) \geq k + 1$, G is $(k - 1)$-factor-critical.

Now we turn our attention to sufficient conditions for 3-γ_c-edge-critical graphs to be k-factor-critical graphs for $k = 1, 2, 3$.

Theorem 73 ([11]). *Suppose G is a 3-γ_c-edge-critical graph.*

(i) *If $v(G) = 2n + 1 \geq 5$ and $\delta(G) \geq 2$, then G is factor-critical.*
(ii) *If G is 3-connected with $v(G) = 2n \geq 8$ and $\delta(G) \geq n-1$, then G is bicritical.*
(iii) *If G is 4-connected of odd order and $K_{1,4}$-free, then G is 3-factor-critical.*

Note that the above minimum degree requirement of 3-γ_c-edge-critical graphs to be bicritical is much stronger than that for 3-γ-edge-critical graphs. However, this bound is sharp and it cannot be lowered even if the connectivity is increased.

4.5.2 3-P-Vertex-Critical Graphs and Matching Properties

From the point of view of matchings, the properties of 3-P-vertex-critical graphs, for $P \in \{\gamma, \gamma_c\}$, differ quite dramatically from those of the 3-P-edge-critical graphs. For example, by Theorem 70, every connected even 3-P-edge-critical graph must have a perfect matching and every connected odd 3-P-edge-critical graph must have a near-perfect matching. These conclusions do not hold for general 3-P-vertex-critical graphs as shown below.

We first present an infinite family $H_{k,\binom{k}{2}-k}$ of 3-γ-vertex-critical graphs (see [4]) and an infinite family J_k of 3-γ_c-vertex-critical graphs (see [12]).

Let k be any positive integer with $k \geq 5$. We proceed to construct the graph which we call $H_{k,\binom{k}{2}-k}$. The vertex set consists of two disjoint subsets of vertices called *central* and *peripheral*, respectively. Let $\{v_1, v_2, \ldots, v_k\}$ denote the set of central vertices. The subgraph induced by these central vertices will be the complete graph K_k with the Hamiltonian cycle $v_1 v_2 \cdots v_k v_1$ deleted. The peripheral vertices will be $\binom{k}{2} - k$ in number and will be denoted by the symbol $\sim \{i, j\}$ where the (unordered) pair $\{i, j\}$ ($i \neq j$) ranges over all the $\binom{k}{2} - k$ subsets of size 2 of the set $1, \ldots, k$, *except* those having $j = i + 2$ where $i + 2$ is read modulo k. The neighbor set of peripheral vertex $\sim \{i, j\}$ will be precisely the set of all central vertices, except i and j. There are no edges joining pairs of peripheral vertices.

Let $k \geq 6$ be a positive integer. We construct the graph J_k as follows. Let the integers $1, 2, \ldots, k$ span a complete k-graph with the Hamiltonian cycle $12 \cdots k1$ removed. These k vertices are called *central* vertices. Consider a second set of $k(k - 3)/2 - k$ *peripheral* vertices labeled with unordered pairs of distinct integers $\sim \{i, j\}$, $1 \leq i < j \leq k$, except for exactly those pairs of the form $\{i, i + 1\}$ and $\{i, i + 2\}$ modulo k. Now join the pair labeled $\sim \{i, j\}$ to all central vertices except i and j. Figure 4.6 shows graphs $H_{6,9}$ and J_6, respectively.

It is not difficult to see that the graphs $H_{k,\binom{k}{2}-k}$ for $k \geq 6$ and J_k for $k \geq 8$ do not contain a perfect (or near-perfect) matching. Hence, 3-P-vertex-critical graphs, for $P \in \{\gamma, \gamma_c\}$, need not have a perfect (or near-perfect) matching. So what might be some reasonable conditions one might place on a 3-P-vertex-critical graph sufficient to guarantee the existence of a perfect (or near-perfect) matching? One of the classical theorems about matching is the following, due independently to Sumner [124] and Las Vergnas [95].

Theorem 74 ([95, 124]). *Every connected claw-free graph of even order has a perfect matching.*

The next result can be considered as a variation on this theme.

Theorem 75 ([4, 9]). *Suppose G is a $K_{1,5}$-free 3-γ-vertex-critical graph. Then*

(i) If $\nu(G)$ is even, then G has a perfect matching.
(ii) If $\nu(G) \geq 11$ is odd with $\delta(G) \geq 1$, then G has a near-perfect matching.

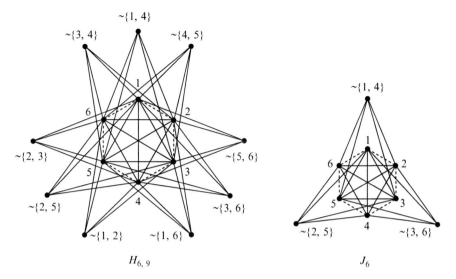

Fig. 4.6 The 3-γ-vertex-critical graph $H_{6,9}$ and the 3-γ_c-vertex-critical graph J_6

The lower bound on $\nu(G)$ in part (ii) is best possible.

The following conjecture is found in [4, 9].

Conjecture 5. Suppose G is a 3-γ-vertex-critical graph. Then

(i) If $\nu(G)$ is even $K_{1,7}$-free, then G contains a perfect matching.
(ii) If $\nu(G)$ is odd $K_{1,5}$-free 2-connected with $\delta(G) \geq 3$, then G is factor-critical.

Wang and Yu [137] showed that Conjecture 5(ii) is true with two small exceptions.

If a 3-γ-vertex critical graph is also claw-free, more can be said. The following theorem due independently to Favaron et al. [52] and to Liu and Yu [99] is useful here.

Theorem 76 ([52,99]). *If G is a $(k+1)$-connected claw-free graph of order n and if $n - k$ is even, then G is k-factor-critical.*

The next theorem follows immediately by Theorems 62 and 76.

Theorem 77 ([7]). *Let G be a connected claw-free 3-γ-vertex-critical graph. Then*

(i) *If G has odd order, then G is factor-critical.*
(ii) *If G has even order, then G is bicritical.*
(iii) *If G has odd order and $\delta(G) \geq 5$, then G is 3-factor-critical.*

We now turn our attention to 3-γ_c-vertex-critical graphs. We saw above that 3-γ_c-vertex-critical graphs need not contain a perfect (or near-perfect) matching. Sufficient conditions for 3-γ_c-vertex-critical graphs to contain a perfect (or near-perfect) matching are given in the following theorem. This theorem can be considered as another variation of Theorem 74.

Theorem 78 ([13]). *Suppose G is a $K_{1,7}$-free 3-γ_c-vertex-critical graph. Then*

(i) If $v(G)$ is even, then G contains a perfect matching.
(ii) If $v(G)$ is odd, then G contains a near-perfect matching.

For k-factor-critical results, we have:

Theorem 79 ([13]). *Suppose G is a 3-γ_c-vertex-critical graph of even order. If G is either (i) $K_{1,4}$-free, or (ii) $K_{1,5}$-free and 5-connected, then G is bicritical.*

Theorem 80 ([13]). *Suppose G is a 3-γ_c-vertex-critical graph of odd order. Then*

(i) If G contains a vertex cutset S with $|S| = 3$, then G is factor-critical.
(ii) If G is $K_{1,6}$-free, then G is factor-critical.
(iii) If G is $K_{1,3}$-free and $\delta(G) \geq 4$, then G is 3-factor-critical.

In closing, we would like to remind the reader that if the connectivity of a graph is at least 2, then G is 3-γ_c-vertex-critical if and only if G is 3-γ_t-vertex-critical. Hence, the results dealing with 3-γ_c-vertex-critical graphs presented in this section may be interpreted as results pertaining to 3-γ_t-vertex-critical graphs as well.

4.6 Summary and Conclusions

So what are the possible future directions in the study of domination? Domination and its many variations form a vast subject. Already two entire books have been written on the subject [71,72]. And since their publication in 1998, nearly 1000 new research papers on domination have appeared in scientific journals worldwide! It is therefore beyond the scope of this chapter to speculate individually on the future of all these variants. Certainly, with much of the world preoccupied with security issues these days, domination in all its forms will attract ever more attention.

It seems clear that on the applied side, researchers will continue to classify domination problems confined to special classes of graphs as to whether they are polynomially solvable or NP-complete. In Chap. 8 of [72], chapter author Kratsch provides a nice table of algorithmic complexity results for five of the more widely studied variants of domination as applied to 15 different special classes of graphs. His table will no doubt continue to be enlarged.

On the theoretical side, many unsolved problems will no doubt continue to occupy the efforts of researchers. In closing, we mention just two of our favorites. The first is the study of well-dominated graphs; i.e., those graphs in which every minimal dominating set is minimum. Graphs in this class have a trivially polynomial procedure to determine their domination number, namely the greedy algorithm. Clearly, one can decide in polynomial time if a graph is *not* well-dominated; just exhibit

4 Domination in Graphs

two minimal dominating sets of differing cardinalities. Hence the well-dominated problem is in the class co-NP. But can a well-dominated graph be so certified in polynomial time? That is, is the well-dominated problem in NP? This is unknown.

Finally, in line with our emphasis on criticality issues in this chapter, we mention the following problem. Is there a polynomial algorithm to decide if a graph is edge-critical or vertex-critical with respect to domination (connected domination, total domination, etc.)? Although most domination problems for general graphs have been shown to be NP-complete, that does not automatically make the answer to this question "no." As far as we know, it has not been proven that to recognize a domination critical graph one must know its domination number.

References

1. Ananchuen N (2007) On domination critical graphs with cutvertices having connected domination number 3. Int Math Forum 2:3041–3052
2. Ananchuen N, Plummer MD (2003) Some results related to the toughness of 3-domination critical graphs. Discrete Math 272:5–15
3. Ananchuen N, Plummer MD (2004) Matching properties in domination critical graphs. Discrete Math 277:1–13
4. Ananchuen N, Plummer MD (2005) Matching in 3-vertex-critical graphs: the even case. Networks 45:210–213
5. Ananchuen N, Plummer MD (2006) Erratum to: Matching properties in domination critical graphs. Discrete Math 306:291–291
6. Ananchuen N, Plummer MD (2006) Erratum to: Some results related to the toughness of 3-domination critical graphs. Discrete Math 306:292
7. Ananchuen N, Plummer MD (2006) On the connectivity and matchings in 3-vertex-critical claw-free graphs. Utilitas Math 69:85–96
8. Ananchuen N, Plummer MD (2006) Some results related to the toughness of 3-domination critical graphs II. Utilitas Math 70:11–32
9. Ananchuen N, Plummer MD (2007) Matchings in 3-vertex-critical graphs: the odd case. Discrete Math 307:1651–1658
10. Ananchuen N, Plummer MD (2007) 3-Factor-criticality in domination critical graphs. Discrete Math 307:3006–3015
11. Ananchuen N, Ananchuen W, Plummer MD (2008a) Matching properties in connected domination critical graphs. Discrete Math 308:1260–1267
12. Ananchuen W, Ananchuen N, Plummer MD (2008b) Vertex criticality for connected domination. Utilitas Math (to appear)
13. Ananchuen W, Ananchuen N, Plummer MD (2008c) Connected domination: vertex criticality and matchings, (submitted) (to appear)
14. Ao S (1994) Independent domination critical graphs. M.Sc. thesis, Department of Mathematics and Statistics, University of Victoria
15. Archdeacon D et al. (2004) Some remarks on domination. J Graph Theory 46:207–210
16. Baogen X, Cockayne EJ, Haynes TW, Hedetniemi ST, Shangchao Z (2000) Extremal graphs for inequalities involving domination parameters. Discrete Math 216:1–10
17. Berge C (1973) Graphs and hypergraphs. North-Holland, Amsterdam
18. Blitch PM (1983) Domination in graphs. Ph.D. thesis, Department of Mathematics and Statistics, University of South Carolina
19. Bo C, Liu B (1996) Some inequalities about connected domination number. Discrete Math 159:241–245

20. Bollobás B, Cockayne EJ (1979) Graph theoretic parameters concerning domination, independence and irredundance. J Graph Theory 3:241–250
21. Bondy JA, Chvátal V (1976) A method in graph theory. Discrete Math 15:111–135
22. Bondy JA, Murty USR (1976) Graph theory with applications. Macmillan, London
23. Brigham RC, Carrington JR, Vitray RP (2000) Connected graphs with maximum total domination number. J Combin Math Combin Comput 34:81–96
24. Brigham RC, Chinn PZ, Dutton RD (1984) A study of vertex domination critical graphs. Tech. Report M-2, Department of Mathematics, University of Central Florida
25. Brigham RC, Chinn PZ, Dutton RD (1988) Vertex domination-critical graphs. Networks 18:173–179
26. Brigham RC, Haynes TW, Henning MA, Rall DF (2005) Bicritical domination. Discrete Math 305:18–32
27. Burton T, Sumner DP (2006) Domination dot-critical graphs. Discrete Math 306:11–18
28. Burton T, Sumner DP (2007) γ-Excellent, critically dominated, end-dominated, and dot-critical trees are equivalent. Discrete Math 307:683–693
29. Caro Y, Roditty Y (1985) On the vertex-independence number and star decomposition of graphs. Ars Combinatoria 20:167–180
30. Caro Y, Roditty Y (1990) A note on the k-domination number of a graph. Int J Math Sci 13:205–206
31. Chang GJ (1998) Algorithmic aspects of domination in graphs. In: Du D-Z, Pardalos PM (eds) Handbook of combinatorial optimization, vol 3. Kluwer, Boston, pp 339–405
32. Chen Y, Edwin Cheng TC, Ng CT (2008) Hamilton-connectivity of 3-domination critical graphs with $\alpha = \delta + 1 \geq 5$. Discrete Math 308:1296–1307
33. Chen XG, Sun L, Ma D-X (2004) Connected domination critical graphs. Appl Math Lett 17:503–507
34. Chen Y, Tian F (2003) A new proof of Wojcicka's conjecture. Discrete Appl Math 127:545–554
35. Chen Y, Tian F, Wei B (2002) Codiameters of 3-connected 3-domination critical graphs. J Graph Theory 39:76–85
36. Chen Y, Tian F, Wei B (2002) The 3-domination-critical graphs with toughness one. Utilitas Math 61:239–253
37. Chen Y, Tian F, Wei B (2003) Hamilton-connectivity of 3-domination-critical graphs with $\alpha \leq \delta$. Discrete Math 271:1–12
38. Chen Y, Tian F, Zhang Y (2002) Hamilton-connectivity of 3-domination critical graphs with $\alpha = \delta + 2$. Eur J Combinator 23:777–784
39. Chengye Z, Yuansheng Y, Linlin S (2008) Domination dot-critical graphs with no critical vertices. Discrete Math 308:3241–3248
40. Clark WE, Shekhtman B, Suen S (1998) Upper bounds for the domination number of a graph. Congr Numer 132:99–123
41. Clark WE, Suen S (2000) An inequality related to Vizing's conjecture. Electron J Combinator 7(4):3
42. Cockayne EJ (1978) Domination of undirected graphs – a survey. In: Alavi, Y., Lick, D.R. (eds) Theory and applications of graphs in America's bicentennial year. Springer, Berlin, pp 141–147
43. Cockayne EJ, Dawes RM, Hedetniemi ST (1980) Total domination in graphs. Networks 10:211–219
44. Cockayne EJ, Hedetniemi ST (1977) Towards a theory of domination in graphs. Networks 7:247–261
45. Colbourn CJ, Stewart LK (1990) Permutation graphs: connected domination and Steiner trees. Discrete Math 86:179–189
46. Cropper M, Greenwell D, Hilton AJW, Kostochka AV (2005) The domination number of cubic Hamiltonian graphs. AKCE Int J Graph Combinator 2:137–144
47. DeLaViña E, Liu Q, Pepper R, Waller B, West DB (2007) Some conjectures of Graffiti.pc on total domination. Congr Numer 185:81–95

4 Domination in Graphs

101

48. Dorfling M, Goddard W, Henning MA (2006) Domination in planar graphs with small diameter II. Ars Combinatoria 78:237–255
49. Duchet P, Meyniel H (1982) On Hadwiger's number and the stability number. In: Bollobás B (ed) Graph theory. Annals of Discrete Mathematics, vol 13. North-Holland, Amsterdam, pp 71–74
50. Duckworth W, Mans B (2002) On the connected domination number of random regular graphs. In: Ibarra OH, Zhang L (eds) Computing and combinatorics. Lecture notes in computer science, vol 2387. Springer, Berlin, pp 210–219
51. Edwin Cheng TC, Chen Y, Ng CT (2009) Codiameters of 3-domination critical graphs with toughness more than one. Discrete Math 309:1067–1078
52. Favaron O, Flandrin E, Ryjáček Z (1997) Factor-criticality and matching extension in DCT-graphs. Discuss Math Graph Theory 17:271–278
53. Favaron O, Henning MA (2004) Paired-domination in claw-free cubic graphs. Graph Combinator 20:447–456
54. Favaron O, Henning MA (2008) Total domination in claw-free graphs with minimum degree 2. Discrete Math 308:3213–3219
55. Favaron O, Henning MA (2008) Bounds on total domination in claw-free cubic graphs. Discrete Math 308:3491–3507
56. Favaron O, Kratsch D (1991) Ratios of domination parameters. In: Kulli VR (ed) Advances in graph theory. Vishwa, Gulbarga, pp 173–182
57. Favaron O, Sumner DP, Wojcicka E (1994) The diameter of domination k-critical graphs. J Graph Theory 18:723–734
58. Favaron O, Tian F, Zhang L (1997) Independence and hamiltonicity in 3-domination-critical graphs. J Graph Theory 25:173–184
59. Flandrin E, Tian F, Wei B, Zhang L (1999) Some properties of 3-domination-critical graphs. Discrete Math 205:65–76
60. Frendrup A, Henning MA, Randerath B, Vestergaard PD (2009) An upper bound on the domination number of a graph with minimum degree 2. Discrete Math 309:639–646
61. Fujita S, Kameda T, Yamashita M (1995) A resource assignment problem on graphs. In: Staples J, Eades P, Katoh N, Moffat A (eds) Algorithms and computation. Lecture notes in computer science, vol 1004. Springer, London, pp 418–427
62. Fujita S, Kameda T, Yamashita M (2000) A study on r-configurations-a resource arrangement problem. SIAM J Discrete Math 13:227–254
63. Fulman J (1994) Domination in vertex and edge critical graphs. Manuscript, Department of Mathematics, Harvard University
64. Fulman J, Hanson D, MacGillivray G (1995) Vertex domination-critical graphs. Networks 25:41–43
65. Garey MR, Johnson DS (1979) Computers and intractability: a guide to the theory of NP-completeness. W.H. Freeman, San Francisco, CA
66. Goddard W, Haynes TW, Henning MA, van der Merwe LC (2004) The diameter of total domination vertex critical graphs. Discrete Math 286:255–261
67. Goddard W, Henning MA (2002) Domination in planar graphs with small diameter. J Graph Theory 40:1–25
68. Hanson D (1993) Hamilton closures in domination critical graphs. J Combin Math Combin Comput 13:121–128
69. Hanson D, Wang P (2003) A note on extremal total domination edge critical graphs. Utilitas Math 63:89–96
70. Harant J, Rautenbach D (2009) Domination in bipartite graphs. Discrete Math 309:113–122
71. Haynes TW, Hedetniemi ST, Slater PJ (1998) Fundamentals of domination in graphs. Marcel Dekker, New York
72. Haynes TW, Hedetniemi ST, Slater PJ (eds) (1998) Domination in graphs – advanced topics. Marcel Dekker, New York
73. Haynes TW, Henning MA (1998) Domination critical graphs with respect to relative complements. Australas J Combinator 18:115–126

74. Haynes TW, Henning MA (2009) Upper bounds on the total domination number. Ars Combinatoria 91:243–256
75. Haynes TW, Phillips JB, Slater PJ (1999) Realizability of (j, t)-critical graphs for sets of values. Congr Numer 137:65–75
76. Hedetniemi ST, Laskar R (1984) Connected domination in graphs. In: Bollobás B (ed) Graph theory and combinatorics. Academic, London, pp 209–217
77. Hedetniemi ST, Laskar RC (1990) Bibliography on domination in graphs and some basic definitions of domination parameters. Discrete Math 86:257–277
78. Hellwig A, Volkmann L (2006) Some upper bounds for the domination number. J Combin Math Combin Comput 57:187–209
79. Henning MA (1999) A characterisation of graphs with minimum degree 2 and domination number exceeding a third their size. J Combin Math Combin Comput 31:45–64
80. Henning MA (2000) Graphs with large total domination number. J Graph Theory 35:21–45
81. Henning MA (2005) A linear Vizing-like relation relating the size and total domination number of a graph. J Graph Theory 49:285–290
82. Henning MA (2009) A survey of selected recent results on total domination in graphs. Discrete Math 309:32–63
83. Henning MA, Kang L, Shan E, Yeo A (2008) On matching and total domination in graphs. Discrete Math 308:2313–2318
84. Henning MA, Oellermann OR, Swart HC (2003) Distance domination critical graphs. J Combin Math Combin Comput 44:33–45
85. Henning MA, Yeo A (2006) Total domination and matching numbers in claw-free graphs. Electron J Combinator 13, # R 59, 28
86. Henning MA, Yeo A (2008) Total domination in graphs with given girth. Graph Combinator 24:333–348
87. Henning MA, Yeo A (2007) A new upper bound on the total domination number of a graph. Electron J Combinator 14, # R 65, 10
88. Ho PT (2008) A note on the total domination number. Utilitas Math 77:97–100
89. Jinquan D (1995) Some results on 3-domination critical graphs. Acta Sci Nat Univ NeiMonggol 26:39–42
90. Kawarabayashi K, Plummer MD, Saito A (2006) Domination in a graph with a 2-factor. J Graph Theory 52:1–6
91. Kleitman DJ, West DB (1991) Spanning trees with many leaves. SIAM J Discrete Math 4: 99–106
92. Kostochka AV, Stodolsky BY (2005) On domination in connected cubic graphs. Discrete Math 304:45–50
93. Kostochka AV, Stodolsky BY (2009) An upper bound on the domination number of n-vertex connected cubic graphs. Discrete Math 309:1142–1162
94. Lam PCB, Wei B (2007) On the total domination number of graphs. Utilitas Math 72:223–240
95. Las Vergnas M (1975) A note on matchings in graphs, Colloque sur la Théorie des Graphes (Paris, 1974) Cahiers Centre Études Rech Oṕer 17:257–260
96. Laskar R, Walikar HB (1981) On domination related concepts in graph theory. In: Rao SB (ed) Combinatorics and graph theory. Lecture notes in mathematics, vol 885. Springer, Berlin, pp 308–320
97. Liu CL (1968) Introduction to combinatorial mathematics. McGraw-Hill, New York
98. Liu H, Sun L (2004) On domination number of 4-regular graphs. Czechoslovak Math J 54:889–898
99. Liu G, Yu Q (1998) On n-edge-deletable and n-critical graphs. Bull Inst Combinator Appl 24:65–72
100. Liying K, Erfang S (1995) On connected domination number of a graph. In: Ku T-H (ed) Combinatorics and graph theory'95, vol 1. World Science, River Edge, NJ, pp 199–204
101. Lovász L, Plummer MD (1986) Matching theory. Annals of Discrete Mathematics, vol 29. North-Holland, Amsterdam
102. Löwenstein C, Rautenbach D (2008) Domination in graphs of minimum degree at least two and large girth. Graph Combinator 24:37–46

4 Domination in Graphs

103. MacGillivray G, Seyffarth K (1996) Domination numbers of planar graphs. J Graph Theory 22:213–229
104. McCuaig W, Shepherd B (1989) Domination in graphs with minimum degree two. J Graph Theory 13:749–762
105. Mojdeh DA, Rad NJ (2006) On the total domination critical graphs. Electron Notes Discrete Math 24:89–92
106. Mojdeh DA, Rad NJ (2007) On an open problem concerning total domination critical graphs. Expositiones Math 25:175–179
107. Moodley L (2000) Wojcicka's theorem: complete, consolidated proof. J Combin Math Combin Comput 33:129–179
108. Moscarini M (1993) Doubly chordal graphs, Steiner trees, and connected domination. Networks 23:59–69
109. Mynhardt CM (1998) On two conjectures concerning 3-domination-critical graphs. Congr Numer 135:119–138
110. Newman-Wolfe RE, Dutton RD, Brigham RC (1988) Connecting sets in graphs – a domination related concept. Congr Numer 67:67–76
111. Ore O (1962) Theory of graphs. Am Math Soc Colloq Publ 38
112. Paris M (1994) Note: the diameter of edge domination critical graphs. Networks 24:261–262
113. Paris M, Sumner DP, Wojcicka E (1999) Edge-domination-critical graphs with cut-vertices. Congr Numer 141:111–117
114. Payan C, Xuong NH (1982) Domination-balanced graphs. J Graph Theory 6:23–32
115. Phillips JB, Haynes TW (2000) A generalization of domination critical graphs. Utilitas Math 58:129–144
116. Rautenbach D (2008) A note on domination, girth and minimum degree. Discrete Math 308:2325–2329
117. Reed B (1996) Paths, stars and the number three. Comb Probab Comput 5:277–295
118. Sampathkumar E, Walikar HB (1979) The connected domination number of a graph. J Math Phys Sci 13:607–613
119. Sanchis LA (1997) Bounds related to domination in graphs with minimum degree two. J Graph Theory 25:139–152
120. Sanchis LA (2000) On the number of edges in graphs with a given connected domination number. Discrete Math 214:193–210
121. Shan E, Kang L, Henning MA (2007) Erratum to: A linear Vizing-like relation relating the size and total domination number of a graph. J Graph Theory 54:350–353
122. Shiu WC, Zhang L-Z (2008) Pancyclism of 3-domination-critical graphs with small minimum degree. Utilitas Math 75:175–192
123. Simmons J (2005) Closure operations and hamiltonian properties of independent and total domination critical graphs. Ph.D. thesis, Department of Mathematics and Statistics, University of Victoria
124. Sumner DP (1976) 1-Factors and anti-factor sets. J Lond Math Soc 13:351–359
125. Sumner DP (1990) Critical concepts in domination. Discrete Math 86:33–46
126. Sumner DP, Blitch P (1983) Domination critical graphs. J Combin Theory B 34:65–76
127. Thomassé S, Yeo A (2007) Total domination of graphs and small transversals of hypergraphs. Combinatorica 27:473–487
128. Tian F, Wei B, Zhang L (1999) Hamiltonicity in 3-domination-critical graphs with $\alpha = \delta + 2$. Discrete Appl Math 92:57–70
129. Tian F, Xu J-M (2008) Distance domination-critical graphs. Appl Math Lett 21:416–420
130. van der Merwe LC, Mynhardt CM, Haynes TW (1998) Total domination edge critical graphs. Utilitas Math 54:229–240
131. van der Merwe LC, Mynhardt CM, Haynes TW (1998) Criticality index of total domination. Congr Numer 131:67–73
132. van der Merwe LC, Mynhardt CM, Haynes TW (1999) 3-domination critical graphs with arbitrary independent domination numbers. Bull ICA 27:85–88
133. van der Merwe LC, Mynhardt CM, Haynes TW (2001) Total domination edge critical graphs with maximum diameter. Discuss Math Graph Theory 21:187–205

134. van der Merwe LC, Mynhardt CM, Haynes TW (2003) Total domination edge critical graphs with minimum diameter. Ars Combinatoria 66:79–96
135. Vizing VG (1968) Some unsolved problems in graph theory. Uspekhi Mat Nauk 23:117–134
136. Wang C, Hu Z, Li X (2009) A constructive characterization of total domination vertex critical graphs. Discrete Math 309:991–996
137. Wang T, Yu Q (2009) Factor-critical property in 3-dominating-critical graphs. Discrete Math 309:1079–1083
138. White K, Farber M, Pulleyblank W (1985) Steiner trees, connected domination and strongly chordal graphs. Networks 15:109–124
139. Wojcicka E (1990) Hamiltonian properties of domination-critical graphs. J Graph Theory 14:205–215
140. Xing H-M, Sun L, Chen X-G (2006) Domination in graphs of minimum degree five. Graph Combinator 22:127–143
141. Xue Y, Chen Z (1991) Hamiltonian cycles in domination-critical graphs. J Nanjing Univ 27:58–62
142. Yuansheng Y, Chengye Z, Xiaohui L, Yongsong J, Xin H (2005) Some 3-connected 4-edge-critical non-Hamiltonian graphs. J Graph Theory 50:316–320
143. Zhang L-Z, Tian F (2002) Independence and connectivity in 3-domination-critical graphs. Discrete Math 259:227–236
144. Zwierzchowski M (2007) Total domination number of the conjunction of graphs. Discrete Math 307:1016–1020

Chapter 5
Spectrum and Entropy for Infinite Directed Graphs

Jun Ichi Fujii

Abstract From the viewpoint of operator theory, we discuss spectral properties for infinite directed graphs that have bounded valences. Graphs may have selfloops, but they are assumed not to have multiedges. Note that we use the transpose adjacency operator throughout this chapter by reason of this viewpoint. As a subsidiary effect, one may read this as a visual introduction to operator theory.

Keywords Infinite directed graph · Spectrum · Entropy · Numerical range · Fractal · Coding theory

MSC2000: Primary 05C50; Secondary 94A17, 94A24, 05C20, 05C38, 05C90, 47A10, 47A12

In this chapter, we observe spectral properties for infinite directed graphs and discuss entropies for them from the viewpoint of operator theory. First we give an exact definition of the adjacency operator for an infinite directed graph and then its representation by the Shatten operator also called dyad. In Sect. 5.2, we give invariants for graphs related to their spectra, e.g., the spectral radii, numerical ones, and norms. These are monotone increasing quantities as graphs are growing. Also, coloring problems for directed graphs are closely related to locations for spectra in the complex plane. In Sect. 5.3, we introduce various products for two graphs and discuss these spectral invariants. In Sect. 5.4, we consider (infinite) directed trees, which are often generated by (finite) graphs. Infinite (generating) directed trees are interesting objects related to other mathematical ones, e.g., Fibonacci sequences, self-similar fractals or entropies. Based on this consideration, we introduce two types of entropies for graphs. In infinite graphs, the topological entropy differs from the spectral entropy although these coincide for finite graphs. In an extreme

---

J.I. Fujii (✉)
Department of Arts and Sciences (Information Science), Osaka Kyoiku University, Asahigaoka, Kashiwara, Osaka 582-8582, Japan
e-mail: fujii@cc.osaka-kyoiku.ac.jp

M. Dehmer (ed.), *Structural Analysis of Complex Networks*,
DOI 10.1007/978-0-8176-4789-6_5, © Springer Science+Business Media, LLC 2011

case, the former is twice as large as the latter. As an application for graphs in the last section, we use Ziv's entropy and show a source coding theorem which is a generalization of Ziv's one.

5.1 Adjacency Operator and Dyadic Representation

For a directed graph G with the *vertices* $V(G)$ of G, an *arrow* (or *arc*) from u to v is denoted by (u, v) or $u \to v$ and the set of all arrows by $E = E(G)$. We exclude multigraphs here and $V(G)$ is countable. The set of vertices $V = V(G)$ corresponds to an orthonormal basis $\{e_v\}$ of a (complex) Hilbert space

$$\mathcal{H} = \ell^2(V(G)) = \ell^2(G) = \left\{ x = \sum_{v \in V} x_v e_v \,\middle|\, \sum_{v \in V} |x_v|^2 < \infty \right\}$$

with the inner product $\langle x, y \rangle = \sum_{v \in V} x_v \overline{y_v}$ and the norm $\|x\| = \sqrt{\sum_{v \in V} |x_v|^2}$.

Following Mohar's adjacency operator for undirected graphs in [20], Fujii et al. [12] defined the *adjacency operator* $A = A(G)$ for a directed graph G as the closed operator with the domain

$$\text{Dom } A = \left\{ x = \sum_{v \in V} x_v e_v \in \mathcal{H} \,\middle|\, \sum_{u \in V} \left| \sum_{v \to u} x_v \right|^2 < \infty \right\}$$

and the value for the above x

$$Ax = \sum_{u \in V} \left(\sum_{v \to u} x_v \right) e_u. \tag{5.1}$$

Here we remark that this definition is not a usual adjacency operator but its transpose. If $A(G)$ is selfadjoint (or transpose invariant), then G is considered as an undirected graph. Define the *outdegree* $d^+(v)$ (resp. *indegree* $d^-(v)$) for a vertex v as the cardinal number of the arrows from (resp. to) v. It is easy to see that

$$d^+(v) = \|Ae_v\| = \sum_{u \leftarrow v} 1 \quad \text{and} \quad d^-(v) = \|A^* e_v\| = \sum_{u \to v} 1.$$

For example, $\begin{pmatrix} 1 & 1 \\ 0 & 0 \end{pmatrix}$ is the adjacency operator for the graph $\overset{1}{\bigcirc}\hspace{-0.3em}\circ\!\!\leftarrow\!\!\overset{2}{\circ}$ and

$$d^+(1) = d^+(2) = 1, \ d^-(1) = 2 \quad \text{and} \quad d^-(2) = 0.$$

5 Spectrum and Entropy for Infinite Directed Graphs

A graph G is said to have a *bounded valency* k if $d^{\pm}(v) \leq k$ for all $v \in V(G)$ (One might think $A(G)$ is an infinite matrix with each row-sum and each column-sum are not greater than k). Then we have (see [12, Theorem 2]):

Proposition 1. *An infinite directed graph G has a bounded valency if and only if $A(G)$ is a bounded operator on $\mathcal{H}$. In this case, the valency k implies*

$$\|A(G)\| \equiv \inf \left\{ t \geq 0 \,\middle|\, \|A(G)x\| \leq t\|x\| \ (\forall x \in \mathcal{H}) \right\} \leq k.$$

Proof. Suppose G has the valency k. Then the property

$$n(|a_1|^2 + \cdots + |a_n|^2) - |a_1 + \cdots + a_n|^2 = \sum_{1 \leq i < j \leq n} |a_i - a_j|^2 \geq 0$$

shows

$$\begin{aligned}
\|A(G)x\|^2 = \sum_{u \in V} \Big| \sum_{v \to u} x_v \Big|^2 &\leq \sum_{u \in V} d^-(u) \sum_{v \to u} |x_v|^2 \\
&\leq k \sum_{u \in V} \sum_{v \to u} |x_v|^2 = k \sum_{v \in V} d^+(v)|x_v|^2 \\
&\leq k^2 \sum_{v \in V} |x_v|^2 = k^2 \|x\|^2,
\end{aligned}$$

so that $\|A(G)\| \leq k$. Conversely suppose G does not have any finite valency. We may assume that for any N there exists e_{v_N} with $d^+(e_{v_N}) > N$. Then

$$\|A(G)e_{v_N}\| = d^+(e_{v_N}) > N = N\|e_{v_N}\|$$

implies that $A(G)$ cannot be bounded. $\qquad\square$

Throughout this chapter, we assume G has a bounded valency; that is, $A(G)$ is a bounded operator. So we define the *norm* of the graph G by $\|G\| = \|A(G)\|$. Since $A(G)^* = {}^t A(G)$ is the adjacency operator for the *converse directed graph* for G, the graph G^* denotes it.

Finally in this section, we give a representation of $A(G)$. Each arrow (u, v) is expressed as a *dyad* $e_v \otimes \overline{e_u}$ for $u, v \in V(G)$, where $(e_v \otimes \overline{e_u})x = \langle x, e_u \rangle e_v$ for $x \in \mathcal{H}$. Then the adjacency operator $A(G)$ of G has a dyadic representation

$$A(G) \equiv \sum_{u \in V} \sum_{v \to u} e_u \otimes \overline{e_v} \quad \left(= \sum_{v \to u} e_u \otimes \overline{e_v} \quad \text{in short} \right),$$

where the sum is in the *strong operator topology*; that is,

$$A_n \to A \iff \|(A_n - A)x\| \to 0 \quad \text{for all vectors } x \in \mathcal{H}.$$

Then we obtain (5.1) again: For $x = \sum_{w \in V} x_w e_v$,

$$A(G)x = \sum_{u,w \in V} \sum_{v \to u} \langle e_w, e_v \rangle x_w e_u = \sum_{u,w \in V} \sum_{w \to u} x_w e_u = \sum_{u \in V} \left(\sum_{w \to u} x_w \right) e_u.$$

5.2 Spectral Invariants

Since $A(G)$ represents G itself, we adopt definitions for notions of G by those of $A(G)$ in the operator theory (as a beginner's book, see [17]):

$$\begin{aligned}
\textbf{spectrum} \quad & \sigma(G) = \sigma(A(G)) = \{\lambda \mid A - \lambda I \text{ is not invertible}\}, \\
\textbf{numerical range} \ & W(G) = W(A(G)) = \{\langle A(G)x, x \rangle \mid \|x\| = 1\}, \\
\textbf{spectral radius} \ & r(G) = r(A(G)) = \sup\{\lambda \mid \lambda \in \sigma(G)\}, \\
\textbf{numerical radius} \ & w(G) = w(A(G)) = \sup\{\lambda \mid \lambda \in W(G)\}.
\end{aligned}$$

The spectrum is a nonempty closed set and included by $\overline{W(G)}$, the closure of $W(G)$, which is convex (the Toeplitz–Hausdorff theorem). If the closed convex hull of $\sigma(A)$ coincides with $\overline{W(A)}$, then A is called *convexoid*. Recall G is *normally symmetric* if G satisfies

$$\{u \in V(G) \mid v \to u, w \to u \in E(G)\} = \{u \in V(G) \mid v \leftarrow u, w \leftarrow u \in E(G)\}.$$

Then we can show as in [12] that G is normally symmetric if and only if $A(G)$ is normal; i.e., $A(G)^* A(G) = A(G)A(G)^*$. Thus normally symmetric (and hence undirected) graphs are convexoid. An example for convexoid graphs which is not normally symmetric is the backward shift U^*, which is discussed in the preceding paragraph of Theorem 3. Moreover, the following relations are well known in operator theory:

$$r(G) = \lim_{n \to \infty} \|A(G)^n\|^{1/n} \leq w(G) \leq \|G\| \leq 2w(G), \tag{5.2}$$

where $r(G) = w(G) = \|G\|$ holds if G is normally symmetric. In operator theory, each equality in the inequalities (5.2) determines a class of operators, respectively (cf. [17]). If G is *purely directed*, i.e., $A(G) + A(G)^*$ defines the undirected graph $\overline{G}$ which is called the *symmetric cover* of G, then $w(\overline{G}) = 2w(G)$; see [24].

Since the triangle inequality

$$w(A + B) \leq w(A) + w(B)$$

holds for all operators, the numerical radius determines an equivalent norm to the usual one.

5 Spectrum and Entropy for Infinite Directed Graphs

Example 1. Take a *nilpotent* graph $N_2 : \overset{v}{\circ} \to \overset{w}{\circ}$ with the adjacency matrix $A(N_2) = e_w \otimes \overline{e_v}$. The term "nilpotent" is derived from the property $A(N)^2 = O$ (We discuss a general case later around Theorem 9). Then

$$r(N_2) = 0, \quad w(N_2) = \frac{1}{2} \quad \text{and} \quad \|N_2\| = \|e_v\|\|e_w\| = 1.$$

Another typical example that shows the difference between finite and infinite graphs is given by Mohar and Woess [21]: A graph G is called k-regular if $d^{\pm}(v) = k$ for all $v \in V(G)$. If G is finite, then $r(G) = k$ [3]. But it does not hold for the following tree, which motivated us to study entropies for graphs.

Example 2. Let T be the 3-regular homogeneous undirected tree (Fig. 5.1); then $r(T) = 2\sqrt{2}$ is shown in [21, (7.6)] in spite of $k = 3$.

First we observe properties for the spectrum and the numerical range in the complex plane [13, 14].

Theorem 1. *The spectrum $\sigma(G)$ and the numerical range $W(G)$ of a graph G are symmetric with respect to the real axis.*

Proof. For a unit vector x, we put its conjugate $\overline{x} = \sum_v \overline{x_v} e_v$. Then, for $\lambda = \langle A(G)x, x \rangle \in W(G)$, via dyadic representation, we have

$$\lambda = \sum_{v,w \in V} \sum_{v \to u} x_v \overline{x_w} \langle e_u, e_w \rangle = \sum_{v \in V} \sum_{v \to u} x_v \overline{x_u},$$

and hence

$$\overline{\lambda} = \sum_{v \in V} \sum_{u \leftarrow v} \overline{x_v} x_u = \langle A(G)\overline{x}, \overline{x} \rangle \in W(G).$$

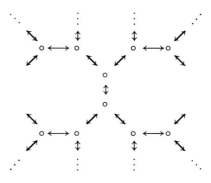

Fig. 5.1 Regular homogeneous undirected tree

Consider the *approximate point spectrum* $\pi(G)$;

$$\pi(G) \equiv \pi(A(G)) = \{\lambda \in \mathbb{C} \,|\, \exists \|x_n\| = 1; \ \|(A(G) - \lambda)x_n\| \to 0\}.$$

Similarly we can show $\pi(G)$ is symmetric. Since $\sigma(G) = \pi(G) \cup \pi(G^*)$, we have $\sigma(G)$ is also symmetric. $\quad\square$

Recall that G is called a *bipartite graph* if $V(G)$ can be divided into two disjoint sets V_0 and V_1 such that there is no arrow between two vertices in the same set. More generally, if $V(G)$ can be divided into m disjoint sets V_k ($k = 0, \ldots, m - 1$) such that all the arrows from V_k strike in $V_{(k+1) \bmod m}$, then G is called *cyclically m-partite*.

Example 3. The following graph is cyclically 3-partite:

$$
\begin{array}{l}
\cdots \circledcirc \leftarrow \odot \rightarrow \circledcirc \leftarrow \odot \rightarrow \circledcirc \cdots \\
\quad \downarrow \nearrow \downarrow \searrow \downarrow \nearrow \downarrow \searrow \downarrow \\
\cdots \ominus \leftarrow \circledcirc \rightarrow \ominus \leftarrow \circledcirc \rightarrow \ominus \cdots
\end{array}
\qquad
\left\{
\begin{array}{l}
\odot \in V_0 \\
\circledcirc \in V_1 \\
\ominus \in V_2
\end{array}
\right.
$$

Theorem 2. *The spectrum $\sigma(G)$ and the numerical range $W(G)$ of a cyclically m-partite graph G are invariant under the $\frac{2\pi}{m}$-rotation in the complex plane.*

Proof. Rearranging the vertices according to the direct sum decomposition

$$\mathcal{H} \equiv \ell^2(V) = \ell^2(V_0) \oplus \cdots \oplus \ell^2(V_{m-1}),$$

we may assume that the adjacency operator is of the form for adjacency suboperators A_j on $\ell^2(V_j)$:

$$
A(G) = \begin{pmatrix}
O & \cdots & & A_{m-1} \\
A_0 & O & \cdots & O \\
& \ddots & \ddots & \vdots \\
O & & A_{m-2} & O
\end{pmatrix}.
$$

Then, for a unit vector $x = (x_k) \in \ell^2(V_0) \oplus \cdots \oplus \ell^2(V_{m-1})$, we have

$$\lambda \equiv \langle A(G)x, x \rangle = \langle A_{m-1}x_{m-1}, x_0 \rangle + \langle A_0 x_0, x_1 \rangle + \cdots + \langle A_{m-2}x_{m-2}, x_{m-1} \rangle.$$

For $\mu = \exp(2\pi i/m)$, take

$$y = {}^t(x_0, \mu^{m-1}x_1, \mu^{m-2}x_2, \ldots, \mu x_{m-1}).$$

Note that $\mu^{m+k} = \mu^k$ and

$$\overline{y} = {}^t(\overline{x_0}, \mu\,\overline{x_1}, \mu^2\overline{x_2}, \ldots, \mu^{m-1}\overline{x_{m-1}}).$$

Thereby, $\|y\| = 1$ and $\langle A(G)y, y \rangle = \mu \langle A(G)x, x \rangle = \mu\lambda$. Thus $\mu\lambda \in W(G)$, that is, $W(G)$ is invariant for the rotation. Similarly we have $\sigma(G)$ is invariant. $\quad\square$

5 Spectrum and Entropy for Infinite Directed Graphs

Corollary 1. *The spectrum and the numerical range of a bipartite graph are symmetric at the origin.*

Next we see the radii of graphs. It is known that the spectrum is closed but the numerical range is not: For example, let U^* be the backward shift graph

$$\circ \leftarrow \circ \leftarrow \circ \leftarrow \circ \leftarrow \cdots .$$

Then $W(U^*)$ is open unit disc [18, Sol. 168(2)] and $\sigma(U^*)$ is the closed one [18, Sol. 67], and therefore $w(U^*) = 1 \notin W(U^*)$. In this case, $r(U^*) = 1 \in \sigma(U^*)$, which implies $\sigma(U^*) \not\subset W(U^*)$ and $\sigma(A) \subset \overline{W(A)}$ holds for all operators A. In general, the statements $r(A) \in \sigma(A)$ and $w(A) \in \overline{W(A)}$ are false. But $r(G) \in \sigma(G)$ and $w(G) \in \overline{W(G)}$ hold for any graphs G as we later. We show in [14] the former, which is a natural infinite version of the Perron–Frobenius theorem. To see the latter, for a unit vector $x = \sum_v x_v e_v$, define $|x| = \sum_v |x_v| e_v$. Then

$$|\langle A(G)x, x \rangle| \leq \langle A(G)|x|, |x| \rangle \quad \text{and} \quad \||x|\| = 1.$$

Thus $w(G) \in \overline{W(G)}$ holds [9].

Theorem 3. *For any graph G, the relations $r(G) \in \sigma(G)$ and $w(G) \in \overline{W(G)}$ hold. Moreover, there exist positive unit vectors $x_n \in \ell^2(G)$ with $\|(r(G) - A(G))x_n\| \to 0$ and consequently $r(G) \in \pi(G)$.*

To show the above results on the spectral radius, we need a notation and a lemma as in Bonsall's method in [23]. For $\mathcal{H} = \ell^2(G)$, define an equivalence relation $\sim$ on the direct sum $\mathcal{H} \oplus \mathcal{H}$ by $i(x, y) \sim (-y, x)$ and the equivalence class of (x, y) is denoted by $[x, y]$ as we identify $\mathbb{C}$ with $\mathbb{R} \oplus \mathbb{R}$. Then the quotient space $\mathcal{B} = \mathcal{H} \oplus \mathcal{H} / \sim$ is a Banach space with a norm

$$\|[x, y]\| = \sup_t \|(\cos t)x + (\sin t)y\|,$$

where $\|x\| + \|y\| \geq \|[x, y]\| \geq \max\{\|x\|, \|y\|\}$ hold. For $A \in B(\mathcal{H})$, define $\tilde{A}$ on $\mathcal{B}$ by $\tilde{A}[x, y] = [Ax, Ay]$. Then the map $A \mapsto \tilde{A}$ is an isometrical isomorphism and hence $r(A) = r(\tilde{A})$. Then we have

Lemma 1. *For a complex number λ with $|\lambda| > r(G)$ for some graph G,*

$$\|(\lambda - A(\tilde{G}))^{-1}[x, y]\| \leq \|(|\lambda| - A(G))^{-1}(x + y)\|$$

for all nonnegative vectors x (i.e., $x_v \geq 0$ for all $v \in V$) and y in $\mathcal{H}$.

Proof. Put $A = A(G)$. For $\lambda = re^{it}$, we have $r = |\lambda| > r(A) = r(\tilde{A})$ and

$$(\lambda - \tilde{A})^{-1}[x, y] = \sum_{n=1}^{\infty} \lambda^{-n} \tilde{A}^{n-1}[x, y] = \sum_{n=1}^{\infty} r^{-n} e^{-int} \tilde{A}^{n-1}[x, y]$$

$$= \sum_{n=1}^{\infty} r^{-n} \tilde{A}^{n-1}[(\cos nt)x + (\sin nt)y, (\cos nt)y - (\sin nt)x].$$

For a vector function

$$P(s) = \sum_n r^{-n} A^{n-1} \big((\cos(nt + s))x + (\sin(nt + s))y \big),$$

the nonnegativity for $r^{-n} A^{n-1}$, x, and y shows

$$\|(\lambda - \tilde{A})^{-1}[x, y]\| = \sup_s \|P(s)\| \leq \|(r - A)^{-1}(x + y)\|. \qquad \square$$

Proof of Theorem 3. We have only to show the results on the spectral radius. Suppose $r = r(G) \notin \sigma(G)$. Put $A = A(G)$. For $re^{it} \in \sigma(A) = \sigma(\tilde{A})$, take positive numbers $\{r_n\} \downarrow r$ and a vector $z \in \mathcal{B}$ with

$$\lim_{n \to \infty} \|(r_n e^{it} - \tilde{A})^{-1} z\| = \infty$$

via the uniform boundedness theorem. Here we can take positive vectors v, w, x, y with $z = [v - w, x - y]$. In fact, take positive vectors x_k and y_k with

$$x = x_1 - x_2 + i(x_3 - x_4) \quad \text{and} \quad y = y_1 - y_2 + i(y_3 - y_4).$$

Then we can rearrange

$$\begin{aligned}
z = [x, y] &= [x_1 - x_2, y_1 - y_2] + [i(x_3 - x_4), i(y_3 - y_4)] \\
&= [x_1 - x_2, y_1 - y_2] + [-(y_3 - y_4), x_3 - x_4] \\
&= [(x_1 + y_4) - (x_2 + y_3), (y_1 + x_3) - (y_2 + x_4)].
\end{aligned}$$

It follows from Lemma 1 that

$$\begin{aligned}
\|(r_n e^{it} - \tilde{A})^{-1} z\| &\leq \|(r_n e^{it} - \tilde{A})^{-1}[v, x]\| + \|(r_n e^{it} - \tilde{A})^{-1}[w, y]\| \\
&\leq \|(r_n - A)^{-1}(v + x)\| + \|(r_n - A)^{-1}(w + y)\| \\
&\leq \|(r_n - A)^{-1}\| (\|(v + x)\| + \|(w + y)\|),
\end{aligned}$$

which contradicts the choice of z. Thus $r(G) \in \sigma(G)$.

Next we see $r(G) \in \pi(G)$. Then we can take positive numbers $t_n \downarrow r(G)$ and a positive vector $z \in \mathcal{H}$ with

$$\lim_{n \to \infty} \|(t_n - A)^{-1} z\| = \infty.$$

Moreover, putting a strictly positive unit vector y by

$$y = \frac{z + w}{\|z + w\|} \quad \text{where} \quad w = \sum_k 2^{-k} e_{v(k)},$$

we also have

$$\lim_{n \to \infty} \|(t_n - A)^{-1} y\| = \infty.$$

5   Spectrum and Entropy for Infinite Directed Graphs   113

Take unit vectors $x_n = (t_n - A)^{-1} y / \|(t_n - A)^{-1} y\|$. It follows that

$$\|(r(G) - A)x_n\| \leq \|(r(G) - t_n)x_n\| + \|(t_n - A)x_n\|$$

$$= |r(G) - t_n| \, \|x_n\| + \frac{\|y\|}{\|(t_n - A)^{-1} y\|}$$

$$= t_n - r(G) + \frac{1}{\|(t_n - A)^{-1} y\|} \longrightarrow 0,$$

which shows $r(G) \in \pi(G)$ and gives the required vectors x_n.   □

Recall the *spanning subgraph* G' of G is a graph such that $V(G) = V(G')$ and $E(G) \supset E(G')$. Then $\|G\|, r(G)$ and $w(G)$ are monotone increasing:

Theorem 4. *If G' is a spanning subgraph of G, then*

$$\|G'\| \leq \|G\|, \quad r(G') \leq r(G) \quad and \quad w(G') \leq w(G).$$

Proof. For nonnegative unit vector x, we have $0 \leq \langle A(G')x, x \rangle \leq \langle A(G)x, x \rangle$ and

$$\|A(G')x\|^2 = \langle A(G')^* A(G')x, x \rangle \leq \langle A(G)^* A(G)x, x \rangle = \|A(G)x\|^2.$$

Thus the numerical radius and the norm are monotone increasing. So is the spectral radius by

$$r(G') = \lim_{n \to \infty} \|A(G')^n\|^{1/n} \leq \lim_{n \to \infty} \|A(G)^n\|^{1/n} = r(G).$$ □

Although one might conjecture that $r(G)$ (or else) is approximated by finite spanning subgraphs G_n, it is false: Consider the unilateral shift graph:

$$U : \circ \to \circ \to \circ \to \cdots .$$

Then all the finite subgraphs of U are nilpotent, and $r(U) = 1$. But Y. Seo gave the following interesting example in [15] which has the approximating finite subgraphs slightly related to Chebyshev polynomials T_n with $T_n(\cos x) = \cos nx$. (We give another proof here based on Seo's idea.)

Example 4. Let an infinite graph G be

$$
\begin{array}{cccccccc}
1 & & 3 & & 5 & \cdots & 2n-1 & \cdots \\
\circ & \leftarrow & \circ & \leftarrow & \circ & \leftarrow \cdots & \circ & \leftarrow \cdots \\
G : \updownarrow & & \updownarrow & & \updownarrow & & \updownarrow & \\
\circ & \to & \circ & \to & \circ & \to \cdots & \circ & \to \cdots \\
2 & & 4 & & 6 & \cdots & 2n & \cdots .
\end{array}
$$

Then we have

$$A(G) = \begin{pmatrix} 0\,1\,1\,0 & \cdots \\ 1\,0\,0\,0\,0\cdots \\ 0\,0\,0\,1\,1 \\ 0\,1\,1\,0\,0 \\ 0\,0\,0\,0\,0 \\ 0\,0\,0\,1\,1\, \ddots \\ \vdots\;\vdots\;\vdots\,0\,0\,\ddots \end{pmatrix} \quad \text{and} \quad A(G)^*A(G) = \begin{pmatrix} 1\,0\,0\,0 & \cdots \\ 0\,2\,2\,0 \\ 0\,2\,2\,0\,0 \\ 0\,0\,0\,2\,2\,0 \\ 0\,0\,0\,2\,2\,0 \\ 0\,0\,0\,0\,0\,\ddots \\ \vdots\;\vdots\;\vdots\,0\,0\,\ddots \end{pmatrix},$$

which shows $r(G) \leq \|G\| = \sqrt{\|A(G)^*A(G)\|} = 2$. Take corresponding finite graphs for each natural numbers n:

$$G_n : \quad \begin{array}{ccccccc} 1 & 3 & 5 & \cdots\, 2n-1 \\ \circ \leftarrow \circ \leftarrow \circ \leftarrow \cdots & \circ \\ \updownarrow & \updownarrow & \updownarrow & & \updownarrow \\ \circ \rightarrow \circ \rightarrow \circ \rightarrow \cdots & \circ \\ 2 & 4 & 6 & \cdots & 2n \end{array}$$

Considering the characteristic polynomial (of order $2n$) $S_n(x)$ for G_n,

$$S_n(x) \equiv \det(A(G_n) - xI) = \det \begin{pmatrix} -x & 1 & 1 & 0 & & & & \\ 1 & -x & 0 & 0 & & & & \\ 0 & 0 & -x & 1 & & & & \\ 0 & 1 & 1 & -x & & & & \\ & & & & \ddots & & & \\ & & & & & -x & 1 & 1 & 0 \\ & & & & & 1 & -x & 0 & 0 \\ & & & & & 0 & 0 & -x & 1 \\ & & & & & 0 & 1 & 1 & -x \end{pmatrix}.$$

Putting the determinant (polynomial of order $2n - 1$) for the following $(2n - 1) \times (2n - 1)$ matrix (and expanding it at the 1st row to get S_{n-1}):

$$R_n(x) = \det \begin{pmatrix} 1 & 1 & 0 & & & & & \\ 0 & -x & 1 & & & & & \\ 1 & 1 & -x & & & & & \\ & & & \ddots & & & & \\ & & & & -x & 1 & 1 & 0 \\ & & & & 1 & -x & 0 & 0 \\ & & & & 0 & 0 & -x & 1 \\ & & & & 0 & 1 & 1 & -x \end{pmatrix} = S_{n-1}(x) + (-1)^2 R_{n-1}(x),$$

5 Spectrum and Entropy for Infinite Directed Graphs 115

we can expand S_n:

$$S_n(x) = x^2 S_{n-1}(x) - R_n(x).$$

It follows that

$$S_n(x) - S_{n-1}(x) = x^2(S_{n-1} - S_{n-2}) - (R_n(x) - R_{n-1}(x))$$
$$= x^2(S_{n-1} - S_{n-2}) - S_{n-1}(x),$$

and consequently

$$S_n(x) - x^2 S_{n-1}(x) + x^2 S_{n-2} = 0.$$

Putting

$$\lambda = \frac{x^2 + i\sqrt{4x^2 - x^4}}{2},$$

which is a solution of $t^2 - x^2 t + x^2 = 0$, we have

$$S_n(x) = \frac{\lambda^n - \bar{\lambda}^n}{\lambda - \bar{\lambda}} S_1(x) = \frac{(\lambda^n - \bar{\lambda}^n)(x^2 - 1)}{i\sqrt{4x^2 - x^4}}.$$

Let $x = 2\cos y$. Then

$$S_n(2\cos y) = \frac{\left(\frac{4\cos^2 y + 4i\sin y\cos y}{2}\right)^n - \left(\frac{4\cos^2 y - 4i\sin y\cos y}{2}\right)^n}{4i\sin y\cos y}(4\cos^2 y - 1)$$

$$= 2^n \cos^n y \left((\cos y + i\sin y)^n - (\cos y - i\sin y)^n\right)\frac{4\cos^2 y - 1}{2i\sin 2y}$$

$$= -2^{n+1}(\cos^n y)(\sin ny)\frac{4\cos^2 y - 1}{2i\sin 2y}.$$

For $y = \pi/n$, we have

$$S_n\left(2\cos\frac{\pi}{n}\right) = -2^{n+1}\left(\cos^n\frac{\pi}{n}\right)(\sin\pi)\frac{4\cos^2\frac{\pi}{n} - 1}{2i\sin\frac{2\pi}{n}} = 0.$$

Thus $2\cos\frac{\pi}{n} \in \sigma(G_n)$, so that

$$2\cos\frac{\pi}{n} \leqq r(G_n) \leqq r(G_{n+1}) \leqq \cdots \leqq r(G) \leqq 2 = \lim_{n\to\infty} 2\cos\frac{\pi}{n},$$

which implies $r(G_n) \uparrow r(G)$ as $n \to \infty$.

5.3 Products for Graphs

If $V(G) = V(G')$, then the product $A(G)A(G')$ (in particular, $A(G)^n$) makes sense, but it does not always represent the adjacency operator of some graph in discourse. This usual product is discussed in the next section. Here we investigate some "products" for graphs.

First we introduce the tensor product for graphs [10]. The tensor product $\ell^2(V) \otimes \ell^2(W)$ for Hilbert spaces $\ell^2(V)$ and $\ell^2(W)$ can be defined as $\ell^2(V \times W)$ where $e_v \otimes e_w$ denotes (e_v, e_w) which forms the basis of $\ell^2(V) \otimes \ell^2(W)$. The inner product of it means

$$\langle x_1 \otimes y_1, x_2 \otimes y_2 \rangle = \langle x_1, x_2 \rangle \langle y_1, y_2 \rangle.$$

The *tensor product* $A \otimes B$ for operators A and B is defined by

$$(A \otimes B)(x \otimes y) = Ax \otimes By,$$

which is an extension of the Kronecker product for matrices. It follows from the definition

$$(A \otimes B)(C \otimes D) = AB \otimes CD.$$

It is known that

$$\|A \otimes B\| = \|A\|\|B\| \quad \text{and} \quad \sigma(A \otimes B) = \sigma(A)\sigma(B).$$

Note that $W(A \otimes B) \supset W(A)W(B)$ and the equality does not always hold. In fact, let

$$A = \begin{pmatrix} 0 & 1 \\ 0 & 0 \end{pmatrix}, \quad B = {}^tA = \begin{pmatrix} 0 & 0 \\ 1 & 0 \end{pmatrix}.$$

Then $W(A) = W(B) = W(A \otimes B) = \{\lambda \mid |\lambda| \leqq 1/2\}$. In particular,

$$w(A)w(B) = \frac{1}{4} < \frac{1}{2} = w(A \otimes B),$$

and $r(A \otimes B) = r(A)r(B)$. Indeed,

$$r(A \otimes B) = \lim_{n \to \infty} \|(A \otimes B)^n\|^{1/n} = \lim_{n \to \infty} \|A^n \otimes B^n\|^{1/n}$$
$$= \lim_{n \to \infty} (\|A^n\|\|B^n\|)^{1/n} = r(A)r(B).$$

Now, we define the *tensor product* $G \otimes H$ for graphs G and H by

$$A(G \otimes H) = A(G) \otimes A(H) = \sum_{(v \to u) \in V(G)} \sum_{(w \to x) \in V(H)} (e_u \otimes \overline{e_v}) \otimes (e_x \otimes \overline{e_w})$$
$$= \sum_{(v \to u) \in V(G)} \sum_{(w \to x) \in V(H)} (e_u \otimes e_x) \otimes \overline{(e_v \otimes e_w)}.$$

5 Spectrum and Entropy for Infinite Directed Graphs

So, summing up, we have:

Theorem 5. *For graphs G and H,*

$$\|G \otimes H\| = \|G\|\|H\|, \ \sigma(G \otimes H) = \sigma(G)\sigma(H), \ r(G \otimes H) = r(G)r(H),$$
$$W(G \otimes H) \supset W(G)W(H) \quad and \quad w(G \otimes H) \geq w(G)w(H).$$

Example 5. Let $U = \sum_{k=1}^{\infty} e_{k+1} \otimes \overline{e_k}$ be the (unilateral) *shift*:

$$U: \quad \overset{1}{\circ} \longrightarrow \overset{2}{\circ} \longrightarrow \overset{3}{\circ} \longrightarrow \cdots \qquad \left(r(U) = 1\right)$$

and $SC(2)$ the *supercomplete* graph:

$$SC(2): \quad \overset{v}{\circlearrowleft \circ} \longleftrightarrow \overset{w}{\circ \circlearrowright} \qquad \left(r(SC(2)) = 2\right).$$

Since

$$A(SC(2)) = e_v \otimes \overline{e_v} + e_v \otimes \overline{e_w} + e_w \otimes \overline{e_v} + e_w \otimes \overline{e_w},$$

we have

$$A(SC(2) \otimes U) = \sum_{k=1}^{\infty} e_{v(k+1)} \otimes \overline{e_{vk}} + e_{v(k+1)} \otimes \overline{e_{wk}} + e_{w(k+1)} \otimes \overline{e_{vk}} + e_{w(k+1)} \otimes \overline{e_{wk}}$$

where the new vertices vk are determined by $e_{vk} = e_v \otimes e_k$. Thus the tensor product $SC(2) \otimes U$ is

$$SC(2) \otimes U: \quad \begin{matrix} \overset{v1}{\circ} \to \overset{v2}{\circ} \to \overset{v3}{\circ} \to \overset{v4}{\circ} \\ \times \quad \times \quad \times \quad \cdots \\ \underset{w1}{\circ} \to \underset{w2}{\circ} \to \underset{w3}{\circ} \to \underset{w4}{\circ} \end{matrix} \qquad \left(r(SC(2) \otimes U) = 2\right).$$

In general, we have $r\left(SC(n) \otimes U\right) = n$.

Next we mention the Hadamard (Schur) product for graphs. Let G and H be infinite directed graphs with $V(G) = V(H)$. Then the *Hadamard product $G * H$* is defined by $A(G) * A(H)$;

$$\langle A(G) * A(H) e_v, e_w \rangle = \langle A(G) e_v, e_w \rangle \langle A(H) e_v, e_w \rangle.$$

Take isometric operator U (i.e., $U^*U = I$) from $\ell^2(V)$ to $\ell^2(V) \otimes \ell^2(V)$ determined by $Ue_v = e_v \otimes e_v$ for all $v \in V$, namely, as a dyadic representation,

$$U = \sum_{v \in V} (e_v \otimes e_v) \otimes \overline{e_v}.$$

Then we have the Toyama–Marcus–Khan theorem [6, 10, 22]:

$$A(G) * A(H) = U^*(A(G) \otimes A(H))U,$$

which is a useful tool to investigate properties of the Hadamard product.

Since we have $E(G * H) = E(G) \cap E(H)$, the Hadamard product $G * H$ is a spanning subgraph of G and H. Thus Theorem 5 implies

$$\|G * H\| \leq \|G\|, \|H\|, \quad r(G * H) \leq r(G), r(H) \text{ and } w(G * H) \leq w(G), w(H).$$

Noting that $A \otimes B = (A \otimes I)(I \otimes B) = (I \otimes B)(A \otimes I)$, we have

$$\sigma(A \otimes I + I \otimes B) = \sigma(A) + \sigma(B)$$

for all bounded operators. But, even for graphs G and H, the operator $A(G) \otimes I + I \otimes A(H)$ is not always the adjacency one of some graph. Here we adopt Mohar's term [21]: A graph G is called *simple* if there is no selfloop.

Then, for simple graphs G and H, the graph $G \oplus H$ defined by

$$A(G \oplus H) = A(G) \otimes I_{\ell^2(V(H))} + I_{\ell^2(V(G))} \otimes A(H) \tag{5.3}$$

is called the *Cartesian product*. Although the Cartesian product can be defined for all simple graphs, we also define it for graphs if (5.3) determines some graph. In the case of Example 5, we have:

Example 6. Although the shift U in Example 5 is simple, a supercomplete graph $SC(n)$ is not. But the Cartesian product $SC(2) \oplus U$ can be defined as

$$SC(2) \oplus U: \qquad \qquad \left(r\big(SC(2) \oplus U\big) = 3 \right)$$

The definition (5.3) implies $r\big(SC(n) \oplus U\big) = n + 1$ in general.

In general, spectral relations for the Cartesian products for simple graphs are:

Theorem 6. *For simple graphs G and H, the following hold:*

$$\sigma(G \oplus H) = \sigma(G) + \sigma(H), \quad and \quad W(G \oplus H) \supset W(G) + W(H)$$

and

$$\|G \oplus H\| \leq \|G\| + \|H\|, \quad r(G \oplus H) = r(G) + r(H), \quad w(G \oplus H) = w(G) + w(H).$$

Proof. For unit vectors x and y

$$\langle A(G)x, x \rangle + \langle A(H)y, y \rangle = \langle A(G)x, x \rangle \langle y, y \rangle + \langle x, x \rangle \langle A(H)y, y \rangle$$
$$= \langle A(G) \otimes I + I \otimes B)x \otimes y, x \otimes y \rangle \in W(G \oplus H),$$

5 Spectrum and Entropy for Infinite Directed Graphs

which implies the inclusion for numerical ranges. Moreover it shows

$$w(G \oplus H) \geqq w(G) + w(H)$$

and the reversed inequality is the triangle one, so that we have the equality. Other relations are clear. □

For simple undirected graphs, Mohar and Woess [21] introduced the NEPS (i.e., the noncomplete extended p-sum) which is a general concept including tensor and Cartesian products. We extend it to the directed case in [10]: Let $\mathcal{C}$ be a nonempty subset of the two-dimensional binary space $\{0, 1\}^2$. For simple graphs G and H, the NEPS $G \underset{\mathcal{C}}{\times} H$ is defined as the graph with vertices $V = V(G) \times V(H)$ satisfying

$$A(G \underset{\mathcal{C}}{\times} H) = \sum_{(c_1,c_2)\in\mathcal{C}} A(G)^{c_1} \otimes A(H)^{c_2} \tag{5.4}$$

with convention $X^0 = I$. If the right-hand side of (5.4) determines a graph, it can be defined even for the nonsimple case.

If $\mathcal{C} = \{(1, 1)\}$, then NEPS $\underset{\mathcal{C}}{\times}$ means tensor product and $\mathcal{C} = \{(1,0), (0, 1)\}$ gives the Cartesian one. Now we have general spectral results by (5.4).

Theorem 7. *For simple graphs G and H,*

$$\sigma(G \underset{\mathcal{C}}{\times} H) = \left\{ \sum_{(c_1,c_2)\in\mathcal{C}} \lambda^{c_1}\mu^{c_2} \;\middle|\; \lambda \in \sigma(G), \mu \in \sigma(H) \right\}$$

and

$$r(G \underset{\mathcal{C}}{\times} H) = \sum_{(c_1,c_2)\in\mathcal{C}} r(G)^{c_1} r(H)^{c_2}.$$

Remark 1. Recently, from the viewpoint of quantum information theory, Accardi et al. discussed other products of *rooted* undirected simple graphs G (i.e., there is a unique vertex e_G called the root. Rooted graphs are extended easily to the directed case; in particular, we will discuss the "rooted directed tree" to consider entropy for all graphs). The *orthogonal product* for G and H is the graph obtained by attaching a copy of H by its root e_H to each vertex of G except its root e_G. If a copy of H is attached also to e_G in addition to the orthogonal product, then it is called the *comb product*. The *star product* is obtained by attaching a copy of H by e_H only to e_G. Their main interest is in the *free product* (V, E) of a finite set of rooted graphs G_k: Define the set of vertices as the words by

$$V = \{e\} \cup \left\{ v_1 v_2 \cdots v_m \;\middle|\; v_k \in V(G_{j_k}) \setminus \{e_{G_{j_k}}\} \; j_k \neq j_{k+1} \right\},$$

where e is newly added as the *empty word*. The empty word means $ev = ve = e$ for all vertices v and the original roots e_k are identified with e. Define the edge by

$$E = \left\{ \{vu, v'u\} \mid \{v, v'\} \in \cup_j E(G_j), \, u, vu, v'u \in V \right\}.$$

If the length ℓ of each word in E is restricted to $\ell \leq m$ for some m, then the free product is called an *m-free* one. (for details, see [1]).

All these notions can be extended to the directed case; for example, the above E is redefined by

$$E = \{\{vu, v'u\} \mid v \to v' \in \cup_j E(G_j), u, vu, v'u \in V\},$$

so that we have the definition of the free product for directed rooted graphs. Here we give each example for directed graphs. Let

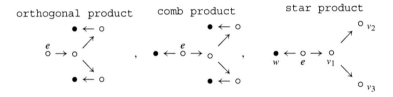

then we have

orthogonal product   comb product   star product

and

2-free product

5.4 Directed Path and Tree

To see the spectral structure of graphs, we consider directed trees and paths. Here a *k-path* from v to w in G means a set of connected k arrows from v to w in G whose existence corresponds to the (v, w)-element of $A(G)^k$. If $A(G)^k$ determines some graph, then G is called *k-powerable* (for $k = 2$, *squarable*). If $A(G)$ is *k-idempotent* (i.e., $A(G)^k = A(G)$) or *k-nilpotent* (i.e., $A(G)^k = O$), then G is *k*-powerable. The permutation graphs and shift ones are *k*-powerable for all $k \in \mathbb{R}$.

Algebraic operators, such as idempotent and nilpotents, have a specific spectral property: Equation $p(A) = O$ implies $p(\lambda) = 0$ for $\lambda \in \sigma(A)$ by the spectral mapping theorem. Thus, for *k*-idempotent (resp. nilpotent) A, we have

5 Spectrum and Entropy for Infinite Directed Graphs

$$\sigma(A) \subset \left\{e^{\frac{2\pi i j}{k-1}} \,\middle|\, j = 1,\ldots,k-1\right\} \cup \{0\} \quad (\text{resp. } \sigma(A) = \{0\}),$$

and hence $r(A) = 1$ (resp. 0). Moreover it has a standard block matrix form

$$A = \begin{pmatrix} I & T \\ O & O \end{pmatrix} \quad \left(\text{resp. } \begin{pmatrix} O & T \\ O & O \end{pmatrix}\right),$$

so that we have

$$W(A) = \{1 + \langle Ty, x \rangle \mid \|x\|^2 + \|y\|^2 = 1\} \ (\text{resp. } \{\langle Ty, x \rangle \mid \|x\|^2 + \|y\|^2 = 1\})$$

$$w(A) = 1 + \frac{\|T\|}{2} \quad \left(\text{resp. } \frac{\|T\|}{2}\right) \quad \text{and} \quad \|A\| = \sqrt{1 + \|T\|^2} \ (\text{resp. } \|T\|).$$

Example 7. A typical connected k-idempotent finite graph I_k is defined by the $k \times k$ adjacency matrix

$$A(I_k) = \begin{pmatrix} 0 & \cdots & \cdots & 0 & 1 & 1 \\ 1 & 0 & \cdots & \cdots & 0 & 1 \\ 0 & 1 & \ddots & \vdots & \vdots & \vdots \\ \vdots & \ddots & \ddots & 0 & 0 & 1 \\ 0 & \cdots & \cdots & 1 & 0 & 1 \\ 0 & \cdots & \cdots & 0 & 0 & 0 \end{pmatrix} = \begin{pmatrix} R_{k-1} & 1_{k-1} \\ 0_{k-1} & 0 \end{pmatrix},$$

where R_{k-1} is the $k-1$-dimensional cyclic permutation and 1_{k-1} (resp. 0_{k-1}) is the $k-1$-dimensional column (resp. row) vector with entries 1 (resp. 0). Then $r(I_k) = 1$, $w(I_k) = \frac{1+\sqrt{k-1}}{2}$ and $\|I_k\| = \sqrt{k}$. These graphs are:

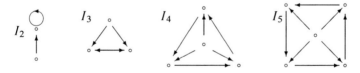

Clearly, the above notions can be rephrased in terms of k-paths.

Theorem 8. *The following equivalences hold.*

(1) *A graph G is k-powerable if and only if G has at most one k-path between each pair of vertices.*
(2) *A graph G is k-idempotent if and only if the condition that G has a unique k-path from v to w is equivalent to $v \to w \in E(G)$.*
(3) *A graph G is k-nilpotent if and only if G has no k-path.*

Similarly to (1) in the above theorem, we give a condition that the usual product makes sense.

Corollary 2. *For graphs* G *and* H *with* $V = V(G) = V(H)$, *the product* $A(G)A(H)$ *determines some graph* GH *if and only if*

$$\#\left(\{u \in V \mid v \to u \in E(H), u \to w \in E(G)\}\right) \leqq 1$$

for all pairs (v, w) *of vertices in* V.

A quasinilpotent operator is the one with $\sigma(N) = \{0\}$, or equivalently $r(N) = 0$. For example, the backward weighted shift

$$U^*_{\{2^{-k}\}} = \begin{pmatrix} 0 & 2^{-1} & 0 & \cdots & \\ 0 & 0 & 2^{-2} & 0 & \cdots \\ 0 & 0 & 0 & 2^{-3} & \cdots \\ \vdots & & \ddots & \ddots & \ddots \end{pmatrix}$$

satisfies $U_{\{2^{-k}\}}U^*_{\{2^{-k}\}} = \mathrm{diag}(0, 2^{-2}, 2^{-4}, 2^{-6}, \dots)$ and thereby

$$\|U^*_{\{2^{-k}\}}\| = 2^{-1} \quad \text{and} \quad \|(U^*_{\{2^{-k}\}})^n\| = 2^{-\frac{n(n-1)}{2}}.$$

It follows that

$$r(U^*_{\{2^{-k}\}}) = \lim_{n \to \infty} \|(U^*_{\{2^{-k}\}})^n\|^{1/n} = \lim_{n \to \infty} 2^{-\frac{n-1}{2}} = 0.$$

Thus the quasinilpotent operator is not always nilpotent. Nevertheless all quasinilpotent graphs are nilpotent and $r(G) \notin (0, 1)$ for all graphs G.

Theorem 9. *The following statements are equivalent:* (i) G *is nilpotent,* (ii) G *is quasinilpotent, and* (iii) $r(G) < 1$.

Proof. It suffices to show (iii) implies (i). In fact, suppose that $r(G) < 1$. Then there exists n with $\|A(G)^n\| < 1$. Since all the entries of $A(G)^n$ are nonnegative integers, its norm is not less than 1 unless it is a zero operator. Thus $A(G)^n$ must be zero; that is, G is nilpotent. $\square$

Next we consider trees generated by a graph. Here a (*mono*) *rooted tree* means a nontrivial directed tree graph G with a unique origin; that is, a graph G with the valency k has no selfloops and there is a vertex v_0 called the *origin* such that

$$d^+(v) \leqq k \quad \text{for all} \quad v \in V, \qquad d^+(v_0) \neq 0$$
$$d^-(v_0) = 0 \quad \text{and} \quad d^-(v) = 1 \quad \text{for} \quad v \neq v_0.$$

Thus, the minimum graph in the rooted trees is $\circ \longrightarrow \circ$. Note also that a rooted tree T is not normally symmetric but k-powerable for all k. Note that we can take T' naturally with $A(T') = A(T)^2$ and $V(T') = V(T)$, which is denoted by T^2. Then we have that T^2 is a rooted tree again and so is T^n for each n.

5 Spectrum and Entropy for Infinite Directed Graphs 123

For a rooted tree T, let $T[n; v]$ be the number of n-paths starting from the vertex v. In particular, $T[n]$ denotes $T[n; v_0]$. Thus, $T[n; v] = d_{(n)}^+(v)$ where $d_{(n)}^+$ is the degree function in T^n. So we can obtain the norm, the spectral radius, and afterwards the (prototype of) entropy of a rooted tree.

Theorem 10. *If T is a rooted tree, then*

$$r(T) = \lim_{n \to \infty} \left(\max_{v \in V} T[n; v]^{1/(2n)} \right) \quad and \quad \|T\| = \max_{v \in V} \sqrt{d^+(v)}.$$

Proof. Since $d^-(v) \equiv 1$ except v_0, then $A(T)^* A(T)$ is the diagonal matrix $\mathrm{diag}(d^+(v_1), d^+(v_2), \cdots)$, so that we obtain the formula for norm $\|T\|$. Thereby

$$r(T) = \lim_{n \to \infty} \|T^n\|^{\frac{1}{n}} = \lim_{n \to \infty} \left(\max_{v \in V} \sqrt{T[n; v]} \right)^{\frac{1}{n}} = \lim_{n \to \infty} \left(\max_{v \in V} T[n; v]^{\frac{1}{2n}} \right). \quad \square$$

Example 8. Let T_k be the (rooted) *k-regular (directed) tree*, that is, a rooted tree with $d^+(v) \equiv k$. The 1-regular tree T_1 is also called the *unilateral shift* and denoted also by U. Then we have $r(T_k) = \sqrt{k}$.

Example 9. For a monotone-increasing sequence $c = (c(n))$ in $\mathbb{N}$, let $\mathcal{M}_c$ be the class of rooted trees T with $T[c(n)] = 1$ for all n. Define a *locally k-regular tree* $F_c \in \mathcal{M}_c$ as a rooted tree with valency $k \geq 2$ such that $F_c[m] = k F_c[m-1]$ for all $m \in \mathbb{N} \setminus \{c(n) | n \in \mathbb{N}\}$. So we have

$$\liminf_{n \to \infty} \frac{\log F_c[n]}{n} = \liminf_{n \to \infty} \frac{\log F_c[c(n)]}{c(n)} = 0.$$

In this tree F_c, $d^+(v) = k$ except the vertices that are the terminals of paths from the origin whose lengths are $c(n) - 1$. For j with $0 < j \leq c(n) - c(n-1)$, we have

$$\frac{\log F_c[c(n) - j]}{c(n) - j} = \frac{(c(n) - c(n-1) - j) \log k}{c(n) - j} = \left(1 - \frac{c(n-1)}{c(n) - j}\right) \log k$$
$$\leq \left(1 - \frac{c(n-1)}{c(n) - 1}\right) \log k = \frac{\log F_c[c(n) - 1]}{c(n) - 1}.$$

It follows that

$$\limsup_{n \to \infty} \frac{\log F_c[n]}{n} = \limsup_{n \to \infty} \frac{\log F_c[c(n) - 1]}{c(n) - 1} = \left(1 - \limsup_{n \to \infty} \frac{c(n-1)}{c(n) - 1}\right) \log k.$$

Thus, $\limsup(\log F_c[n])/n = 0$ if $c(n)$ is a polynomial of n. However, it is equal to $(\log k)(M-1)/M$ for $c(n) = M^n$ and $\log k$ for $c(n) = M^{N^n}$.

Considering this example, we define $H_0(T)$ as

$$H_0(T) = \limsup_{n \to \infty} \frac{1}{n} \log T[n],$$

which is a prototype of entropy. Then, in Example 8, we have

$$H_0(T_k) = \log k = 2 \log r(T_k).$$

For a rooted tree T with the valency k, it holds for any n that $T[n] \leq T_k[n] = k^n$. So we have an upper bound of $H_0(T)$ by Example 9 and a lower bound.

Lemma 2. *Let T be a rooted tree with the valency k. If T is infinite, then $0 \leq H_0(T) \leq \log k$, and otherwise $H_0(T) = -\infty$.*

For a nontrivial directed graph G with the valency k and a fixed vertex w in G, define the *free-generating tree* $T(G; w)$ as the rooted tree satisfying (T_0)–(T_2).

- (T_0) There is a surjection Φ from $V(T(G; w))$ onto $V(G)$,
- (T_1) $d^+(v) = d^+(\Phi(v))$ for all $v \in V(T(G; w))$,
- (T_2) $\Phi(v_0) = w$ for the origin v_0 of $T(G; w)$.

For the sake of convenience, we identify v_0 with w. All the directed paths in $T(G; w)$ correspond one-to-one to those in G via Φ. Thus, the number of the n-paths from w in G is exactly $T(G; w)[n]$. For example:

Example 10. For all the following graphs G_k, the free-generating rooted tree $T(G_k; v)$ for any vertex v is a unilateral shift U:

$$G_1 : \circ\!\circlearrowright \qquad G_2 : \circlearrowright\!\circ \quad \circ\!\circlearrowright \qquad G_3 : \circ \longleftrightarrow \circ \qquad G_4 : \circ \to \circ\!\circlearrowright$$

$$G_5 : \overset{\circ}{\underset{\circ \longrightarrow \circ}{\swarrow \quad \nwarrow}} \qquad G_6 : \circ \to \circ \to \circ\!\circlearrowright \qquad G_7 = U : \circ \to \circ \to \circ \to \cdots .$$

Example 11. For the *homogeneous k-regular undirected tree* U_k (i.e., $d^\pm(v) \equiv k$), the free-generating tree $T(U_k; v)$ is the k-regular tree T_k for all v.

The following lemma is a characterization of nilpotent graphs in terms of the free-generating trees. Recall that the length of a finite tree T is the maximum of lengths of the paths in T.

Lemma 3. *A graph G is nilpotent if and only if all free-generating trees of G are finite and their lengths are bounded.*

It is easy to see that $T(T; v_0) = T$ for a rooted tree T with the origin v_0. For a k-regular tree T_k, we have $T(T_k; v) = T_k$ for all v in T_k. In the following simple example, we see how the entropy of the free-generating tree is changing.

5 Spectrum and Entropy for Infinite Directed Graphs

Example 12. Consider a rooted tree $G = T_{2,4}$ defined as the star product graph for $\underset{e_1}{\circ} \to T_2$ and $\underset{e_2}{\circ} \to T_4$ attaching to the root to each tree T_2 and T_4:

$$T_{2,4}: \qquad \overset{v_0}{\underset{T_2 \qquad T_4}{\circ}}$$

Then we have $H_0(T(G;v_1)) = \log 2$ and $H_0(T(G;v_2)) = \log 4$. In order to obtain $H_0(T(G;v_0))$, let us see the number of n-paths $T(G;v_0)[n]$ from the origin v_0. Since

$$T(G;v_0)[n] = 2^{n-1} + 4^{n-1},$$

we have

$$H_0(T(G;v_0)) = \limsup_{n \to \infty} \frac{\log(2^{n-1} + 4^{n-1})}{n} = \lim_{n \to \infty} \frac{\log(2^{n-1} + 4^{n-1})}{n} = \log 4.$$

Finally in this section, we give an interesting example for free-generating trees, which shows clearly the property of the original graph.

Example 13. It is known that the following undirected graph G

$$\circ \longleftrightarrow \circ \circlearrowright,$$

which is frequently used as an elementary automaton, is closely related to the Fibonacci sequence

$$(a_0, a_1, a_2, \ldots) = (1, 1, 2, 3, 5, 8, 13, 21, \ldots) \tag{5.5}$$

and the golden ratio $\frac{1+\sqrt{5}}{2}$. In fact, its adjacency matrix is

$$A = A(G) = \begin{pmatrix} 0 & 1 \\ 1 & 1 \end{pmatrix}$$

and its spectrum (in this case, eigenvalues) is

$$\frac{1 \pm \sqrt{5}}{2} \in \sigma(A),$$

which shows $r(G)$ is the golden ratio. Also the transition

$$\begin{pmatrix} a_n \\ a_{n+1} \end{pmatrix} = A \begin{pmatrix} a_{n-1} \\ a_n \end{pmatrix} = \begin{pmatrix} 0 & 1 \\ 1 & 1 \end{pmatrix} \begin{pmatrix} a_{n-1} \\ a_n \end{pmatrix}$$

gives the Fibonacci relation, cf. [19]:

$$a_{n+1} = a_n + a_{n-1}.$$

Then, the initial vector ${}^t(a_0, a_1) = {}^t(1,1)$ yields the above sequences (5.5). Moreover, by the Jordan decomposition, we have

$$A^n = \frac{1}{4\sqrt{5}} \begin{pmatrix} 2 & 2 \\ 1-\sqrt{5} & 1+\sqrt{5} \end{pmatrix} \begin{pmatrix} \left(\frac{1-\sqrt{5}}{2}\right)^n & 0 \\ 0 & \left(\frac{1+\sqrt{5}}{2}\right)^n \end{pmatrix} \begin{pmatrix} 1+\sqrt{5} & -2 \\ \sqrt{5}-1 & 2 \end{pmatrix}$$

which shows

$$a_n = \frac{1}{\sqrt{5}} \left(\left(\frac{1+\sqrt{5}}{2}\right)^{n+1} - \left(\frac{1-\sqrt{5}}{2}\right)^{n+1} \right).$$

But it is not easy to see that the above graph G gives (5.5). So we consider the free-generating tree of $T = T(G; v_0)$. (To make it easy, we label and color vertices.)

$$G: \quad \underset{v_0}{\circ} \longleftrightarrow \underset{v_1}{\bullet} \circlearrowright.$$

Then (5.5) appears as each number a_n of the nth node vertices in T and the relation between G and the Fibonacci sequence is visually clear.

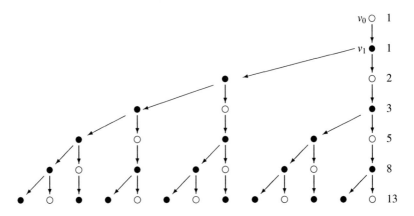

Thereby, we have

$$H_0(T(G; v_0)) = \limsup_{n \to \infty} \frac{1}{n} \log a_n$$

$$= \limsup_{n \to \infty} \log \left(\frac{1}{\sqrt{5}}\right)^{1/n} \left(\left(\frac{1+\sqrt{5}}{2}\right)^{n+1} - \left(\frac{1-\sqrt{5}}{2}\right)^{n+1} \right)^{1/n}$$

$$= \lim_{n \to \infty} \log \left(\frac{1+\sqrt{5}}{2}\right)^{(n+1)/n} = \log \frac{1+\sqrt{5}}{2}.$$

5 Spectrum and Entropy for Infinite Directed Graphs

Thus we have the logarithm of the golden ratio. This gives us much information to consider entropies for infinite directed graphs.

5.5 Two Entropies for Graphs

Based on the preceding section, we define two entropies of graphs. First, according to the case of finite graphs, we defined an entropy $h(G)$ for a graph G.

Definition 1. For a nonnilpotent graph G, the *spectral entropy* $h(G)$ is defined by $h(G) = \log r(G)$.

This entropy for finite graphs was observed again as the complexity in the sense of Kolmogorov by Fujii, Nakamura, Seo, and Watatani [16]. (This is extended to the complexity for positive operators in [8].)

$$h(G) = \lim_{n \to \infty} \frac{\log \langle A(G)^n u, u \rangle}{n}$$

where $u = \begin{pmatrix} 1 \\ \vdots \\ 1 \end{pmatrix}$. Here, for each natural number n, the number

$$a_n(G) = \langle A(G)^n u, u \rangle$$

is called the *Schwarz constant* for G and is equal to the number of n-paths in G. To consider the distribution of spectral entropies for graphs, we give the following example.

Example 14. For $k \geq 2$, $n \geq 1$, there exist finite graphs $\{G_n(k)\}$ with $r(G_n(k)) = k^{1/n}$. For example, for the *cyclic permutation* $U(n)$ with n vertices, i.e.,

$$A(U(n)) = R_n = \sum_{k=0}^{n-1} e_{(k+1) \bmod n} \otimes \overline{e_k} = \begin{pmatrix} 0 & 0 & \cdots & 0 & 1 \\ 1 & 0 & & & 0 \\ 0 & 1 & \ddots & & \vdots \\ \vdots & & \ddots & \ddots & 0 \\ 0 & \cdots & 0 & 1 & 0 \end{pmatrix},$$

let $G_n(k)$ be a graph including exactly k $U(n)$'s as subgraphs with only one common vertex called the *center,* as in the following examples, see Fig. 5.2.

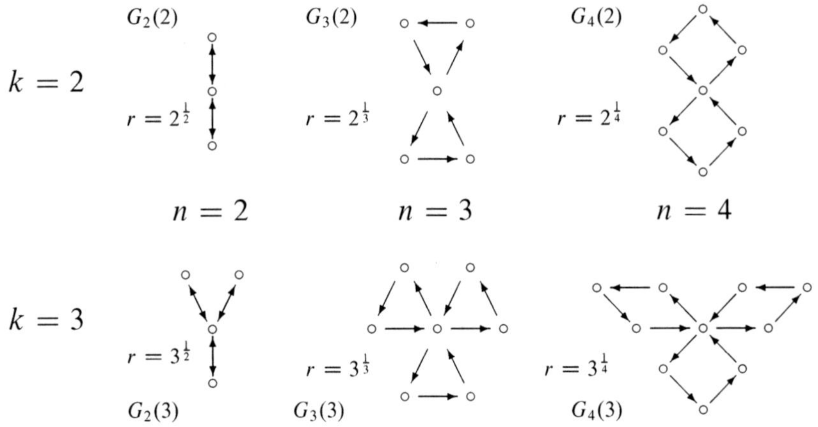

Fig. 5.2 $G_n(k)$

Then the number of m-paths in $G_n(k)$ is

$$a_m(G_n(k)) = k\left(\sum_{j=0}^{n-1} k^{\lfloor \frac{m+j}{n} \rfloor}\right) = \left(\sum_{j=0}^{n-1} k^{\lfloor \frac{m+j}{n} \rfloor + 1}\right),$$

where $\lfloor\ \rfloor$ means Gauss' symbol. Then complexity formula shows

$$r(G_n(k)) = \lim_{m\to\infty}\left(\sum_{j=0}^{n-1} k^{\lfloor \frac{m+j}{n} \rfloor + 1}\right)^{1/m} = \lim_{m\to\infty}\left(k^{\lfloor \frac{m+n-1}{n} \rfloor + 1}\right)^{1/m}$$

$$= \lim_{m\to\infty} k^{\frac{m+1}{mn-n+1}} = k^{1/n}$$

by replacing m with $mn - n + 1$.

Considering the tensor products for the above examples, we have the distribution for the spectral radii or entropies for graphs by Theorem 5.

Theorem 11. *The entropies $h(G)$ for graphs G are densely distributed in the half-line $[0, \infty)$.*

Examples 12–14 suggest another entropy for graphs as in [5]. Let $a_n(v)$ be the number of n-paths in G started from the vertex v:

$$a_n(v) = \sum_{u\in V(G)} \langle A(G)^n e_v, e_u \rangle.$$

5 Spectrum and Entropy for Infinite Directed Graphs

Definition 2. For a nonnilpotent graph G, the *topological (or combinatorial) entropy* $H(G)$ of a graph G is defined by

$$H(G) = \sup_{v \in V(G)} \lim_{n \to \infty} \frac{\log a_n(v)}{n}.$$

Then we showed in [5] that

Theorem 12. *For all nonnilpotent graphs G, the topological entropy is estimated by the spectral entropy $h(G)$:*

$$h(G) \leq H(G) \leq 2h(G).$$

Moreover if G is a finite graph, then $h(G) = H(G)$.

Proof. By Lemma 3, $T(G; v)[n]$ may be assumed to be nonzero. Let $p(v; n) \subset V$ be the vertices that are the terminals of n-paths from v. For the adjacency operator A of G, we have

$$T(G; v)[n]^2 = \left| \sum_w \langle A^n e_v, e_w \rangle \right|^2 \leq \sum_{w \in p(v;n)} |\langle A^n e_v, e_w \rangle|^2$$

$$\leq \|A^n e_v\|^2 \sum_{w \in p(v;n)} \|e_w\|^2 \leq \|A^n\|^2 T(G; v)[n].$$

Thus, $T(G; v)[n]^{1/n} \leq \|A^n\|^{2/n} \to r(G)^2$ for all v, hence $H(G) \leq 2h(G)$.

To show $h(G) = \log r(G) \leq H(G)$, put $A = A(G)$, $B = A^*$, and $r = r(G) = r(A) = r(B)$. Theorem 3 shows there exist unit positive approximate proper vectors $\{x_n\}$ of r for B such that $\|z_n\| \leq 1/(nk^{n-1})$ where $z_n = (B - r)x_n$. Here we may assume that there is a vertex v with $\delta \equiv \inf_n \langle e_v, x_n \rangle > 0$. Putting $\alpha_n = ((A^{n-1} + rA^{n-2} + \cdots + r^{n-1})e_v, z_n)$, we have

$$T(G : v)[n] = \sum_{w \in V} \langle A^n e_v, e_w \rangle = \sum_{w \in V} \langle e_v, B^n e_w \rangle$$

$$\geq \langle e_v, B^n x_n \rangle = \langle e_v, r^n x_n \rangle + \langle e_v, (B^{n-1} + rB^{n-2} + \cdots + r^{n-1})z_n \rangle$$

$$= r^n \langle e_v, x_n \rangle + \alpha_n \geq \delta r^n + \alpha_n.$$

Note that α_n is bounded by the following inequalities:

$$0 \leq \alpha_n \leq (\|A^{n-1}e_v\| + r\|A^{n-2}e_v\| + \cdots + r^{n-1}\|e_v\|)\|z_n\|$$

$$\leq (k^{n-1} + rk^{n-2} + \cdots + r^{n-1})\|z_n\| \leq nk^{n-1}\|z_n\| \leq 1$$

since $r \leq k$. Thereby we have

$$H(G) \geq \limsup_{n \to \infty} \frac{1}{n} \log T(G; v)[n] \geq \limsup_{n \to \infty} \frac{1}{n} \log(\delta r^n + \alpha_n) = \log r = h(G).$$

Next we show the equality for the finite case. The above equation holds when $r(G) < 1$. So suppose G is a finite nonnilpotent graph. Then G has an infinite free-generating tree and consequently G has a cycle. Considering (strongly) connected components, we may assume that the adjacency matrix $A(G)$ is irreducible. Then the Perron–Frobenius theorem shows that there exists the unit positive eigenvector p for an eigenvalue $r(G)$. Let e be a vector whose entries are all 1. Then

$$T(G;v)[n] = \langle G^n e_v, e \rangle$$

where e_v is the vector corresponding to the vertex v. It follows from the positivity of p that there exists a positive number α such that $\alpha p - e$ is a positive vector. Then

$$r(G)^n \langle p, e \rangle = \langle G^n p, e \rangle \leqq \langle G^n e, e \rangle \leqq \alpha \langle G^n p, e \rangle = \alpha \, r(G)^n \langle p, e \rangle$$

for all n. By $\langle e, e \rangle = \dim \ell^2(V) = \sharp V$, we have

$$H(G) = \sup_v \limsup_{n \to \infty} \frac{1}{n} \log T(G;v)[n]$$

$$\geqq \limsup_{n \to \infty} \frac{1}{n} \log \frac{\sum_v T(G;v)[n]}{\langle e, e \rangle} = \limsup_{n \to \infty} \frac{1}{n} \log \frac{\langle G^n e, e \rangle}{\langle e, e \rangle}$$

$$\geqq \lim_{n \to \infty} \frac{1}{n} \log \frac{r(G)^n \langle p, e \rangle}{\langle e, e \rangle} = \log r(G) = h(G).$$

Conversely, we have

$$H(G) \leqq \limsup_{n \to \infty} \frac{1}{n} \log \langle G^n e, e \rangle$$

$$\leqq \limsup_{n \to \infty} \frac{1}{n} \log(\alpha \, r(G)^n \langle p, e \rangle) = \log r(G) = h(G)$$

so that $h(G) = H(G)$ holds. $\qquad\square$

The above theorems show the distribution for $H(G)$.

Corollary 3. *The entropies $H(G)$ for nonnilpotent graphs G are densely distributed in the half-line $[0, \infty)$.*

We have that the estimation in Theorem 12 is best possible by the following example.

Example 15. As in Fig. 5.3, let $T_{k,m}$ be the directed tree such that in the nth nodes, $\frac{(k-m)w_n}{k}$ vertices have k arrows and others have none, where w_n is the number of

5 Spectrum and Entropy for Infinite Directed Graphs

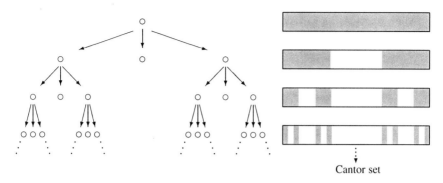

Fig. 5.3 The tree $T_{3,1}$ and the Cantor set

the *n*th node vertices and the *root* (i.e., 0th node) vertex has k arrows, or $w_1 = k$. (In particular, $T_k = T_{k,0}$ is the k-ary directed tree.) For a tree $T_{k,m}$, we have

$$w_n = k(k-m)^{n-1}$$

and $T_{k,m}$ is *squarable* in the sense that $A(T_{k,m})^2$ is also an adjacency operator for some graph, say $T_{k,m}^2$. In fact,

$$T_{k,m}^2 = T_{k(k-m),m(k-m)} \text{ and in general } T_{k,m}^n = T_{k(k-m)^{n-1},m(k-m)^{n-1}}.$$

So the topological entropy is

$$H(T_{k,m}) = \lim_{n\to\infty} \frac{\log k(k-m)^{n-1}}{n}$$
$$= \lim_{n\to\infty} \frac{\log k + (n-1)\log(k-m)}{n} = \log(k-m).$$

On the other hand, by $A(T_{k,m})^*A(T_{k,m}) = kP$ for some projection P, we have the norm

$$\|T_{k,m}\| = \sqrt{k}$$

and hence the spectral entropy is

$$h(T_{k,m}) = \log r(T_{k,m}) = \lim_{n\to\infty} \frac{\log \|T_{k,m}^n\|}{n}$$
$$= \lim_{n\to\infty} \frac{\log \sqrt{(k-m)^n}}{n} = \frac{\log(k-m)}{2}.$$

The above examples $T_{k,m}$ are attractive ones which represent fractals called self-similar sets (see, e.g., [4]). For example, $T_{3,1}$ represents the Cantor set as in Fig. 5.3.

Other examples for self-similar sets are in the following table.

Self-similar set	k	m	J	Scale	Similarity dimension
Cantor dust	3	1	1	1/3	$\log 2/\log 3$
Sierpiński gasket	4	1	2	1/2	$\log 3/\log 2$
Menger sponge	27	7	3	1/3	$\log 20/\log 3$

Here J is the dimension of the space on which the figure in discourse lies. In these fractals, the ratio is $k^{1/J}$ and the number of self-similar maps is $k - m$. Thus the similarity dimension is $d = J \log(k - m)/\log k$ and the relation to the entropies is

$$d = \frac{J \log(k - m)}{\log k} = \frac{J h(T_{k,m})}{\log \|T_{k,m}\|} = \frac{J H(T_{k,m})}{2 \log \|T_{k,m}\|}.$$

5.6 Ziv's Entropy and Coding

Finally, we apply our discussion to coding theory. If we put a letter on each arrow in G, we may consider an infinite path in G as a *message* on G. Hereafter suppose G is a finite directed graph, and consequently $h(G) = H(G)$.

For a message $x = (v_0, v_1, v_2, \ldots)$ on a graph G, let $x[n]$ be the *n-path* $(v_0, v_1, \ldots, v_n)$ in x, say *the first n-words in x*. For a shift S on the message $S(\ldots, v_i, \ldots) = (\ldots, v_{i+1}, \ldots)$, put

$$X_n = \sharp \left\{ S^j(x)[n] \mid j = 1, 2, \ldots \right\},$$

the number of *n-words* in x. Then a kind of complexity of x is defined; see [25].

Definition 3. For a message x on G, *Ziv's entropy* $h(x)$ is defined by

$$h(x) = \lim_{n \to \infty} \frac{\log X_n}{n}.$$

By the definition, we have

Theorem 13. *For all messages x on a finite directed graph G, $h(x) \leq h(G)$. If G is irreducible, then there exists a message x such that $h(x) = h(G)$.*

Proof. Since X_n is not greater than the Schwarz constant $a_n = a_n(G)$, we have $h(x) \leq h(G)$.

Next suppose G is irreducible. For any integer $m \geq 1$, let $x_1^m, x_2^m, \ldots, x_{a_m}^m$ be all the distinct m-words on G. For two words $x_i^m = v_1 v_2 \cdots v_m$ and $x_{i+1}^n = u_1 u_2 \cdots u_n$, there exists a path $(v_m, w_1, \ldots, w_k, u_1)$ from v_m to u_1 in G since G is irreducible. Let $w_{i,j}^{m,n}$ be $w_1 w_2 \cdots w_k$ if $x_i^m x_{i+1}^n$ is not the word on G, and otherwise the empty word. Then $x_i^m w_{i,j}^{m,n} x_{i+1}^m$ is the word on G. Thus we put a message x by

5 Spectrum and Entropy for Infinite Directed Graphs

$$x = x_1^1 w_{1,2}^{1,1} x_2^1 \cdots x_{a_1}^1 w_{a_1,1}^{1,2} x_1^2 w_{1,2}^{2,2} x_2^2 \cdots ,$$

and hence we have $X_m = a_m(G)$ for all m. Therefore $h(x) = h(G)$. $\qquad\square$

Remark 2. The above argument for entropy or complexity is in the nonstochastic situation. Here we mention the relation between the entropy of graphs and that of a probability matrix. By setting probabilities $\mathcal{P}$ on all arrows in G, we obtain a transition matrix $G(\mathcal{P})$ of a Markov chain. Then it is known that

$$h(G)(= H(G)) = \max_{\mathcal{P}} H(G(\mathcal{P})),$$

where $H(G(\mathcal{P}))$ is the entropy as a Markov source. In fact, for $A(G) = (a_{ij})$ and its Frobenius vector $f = (r_i)$, the above maximum is attained at $B = (b_{ij})$ with

$$b_{ij} = \frac{r_i a_{ij}}{r_j r(G)}.$$

Here we deal with block source encoding and decoding without noise, which is used for data compression. The (finite state) *block encoder* (resp. *decoder*) with length N means the map $\phi : x \mapsto y$ (resp. $\psi : y \mapsto \hat{x}$) for a message x, the encoded message y, and the *replica* (i.e., decoded message) $\hat{x}$ such that the ith block $S^{iN}(y)[N]$ (resp. $S^{iN}(\hat{x})[N]$) depends only on $S^{iN}(x)[N]$ (resp. $S^{iN}(y)[N]$) for all i. Then, Ziv showed the following lemma in [25].

Lemma 4. (Ziv's Data processing lemma) $h(\hat{x}) \leqq h(y) \leqq h(x)$ *for every encoder-decoder pair.*

Here we assume that the messages x and $\hat{x}$ are written by an alphabet of M letters and y by that of M' letters with $M \geqq M'$. Considering the case that the encoded message is on a finite directed graph, we can show the estimation by the entropy of the graph, which is better than by $\log M'$, and considered as a generalization of Ziv's result [25].

Theorem 14. (Bound for invertibility) *Let F be a finite directed graph and suppose that a message x is encoded to the message on F. If $h(x) > h(F)$, then $x \neq \hat{x}$.*

Proof. To show the contraposition, suppose that there exists a block encoder φ and a block decoder ψ such that $x = \hat{x}$. Then Theorem 13 implies $h(y) \leqq h(F)$. It follows from Ziv's data processing lemma that $h(\hat{x}) \leqq h(y) \leqq h(F)$. By $x = \hat{x}$, we have $h(x) = h(\hat{x})$, so that $h(x) \leqq h(F)$. $\qquad\square$

Finally in this section, we try to generalize Ziv's block source coding theorem in [25], which, however, included a somewhat ambiguous part. In [2], S. Arimotro reformulated it to a mathematically accurate statement. Now we generalize Ziv–Arimoto's coding theorem by a finite graph. Recall that an irreducible finite graph G is *aperiodic* if there is a positive integer m such that there exists an m-path from v to u for all vertices v and u.

Theorem 15. (Coding theorem of graph version) *For an aperiodic finite directed graph F, let x be the message on the alphabet of M letters and encoded messages on F. If $h(x) < h(F)$ for all x, then there exists a pair of N-block encoder ϕ and decoder ψ with $x = \hat{x} \equiv \psi(\phi(x))$.*

Proof. To estimate length for codewords, suppose that we can take N and its largest divisor ℓ with $\ell^2 M^\ell \leqq N$. For all the distinct ℓ-words $x_1^\ell, x_2^\ell, \cdots, x_s^\ell$ in the ith block $S^{iN}(x)[N]$, we have $s = X_\ell \leqq M^\ell$. Since F is aperiodic, there exists a set of m-paths from u and v for all vertices u and v of F. Taking an integer k with $b_{k-m} \geqq M^\ell \geqq b_{k-m-1}$ where b_n is the Schwarz constant of F, we can translate the words x_i^ℓ over A into the codewords y_i^{k-m} on F. As in the proof of Theorem 13, there exist m-words z_i^m with $y_1^{k-m} z_1^m y_2^{k-m} z_2^m \cdots y_M^{k-m} z_s^m$ being also the words on F, which is called the *list of the casts* and defines the first L letters in the codeword $S^{iN}(y)[N]$. Then we have

$$L = (k - m + m)s = kX_\ell \leqq X_\ell \left(m + 1 + \frac{k - m - 1}{\log b_{k-m-1}} \log M^\ell \right)$$

$$\leqq N \left(\frac{m + 1}{\ell^2} + \frac{k - m - 1}{\log b_{k-m-1}} \frac{\log M}{\ell} \right).$$

Now parse $S^{iN}(x)[N]$ as

$$S^{iN}(x)[N] = S^{iN}(x)[\ell] S^{iN+\ell}(x)[\ell] \cdots S^{i(N+1)-\ell}(x)[\ell].$$

There are at most $\frac{N}{\ell}$ vectors in the parsed $S^{iN}(x)[N]$. Then the remainder of the ith block codeword $S^{iN}(y)[N]$ is taken to be the *list of the addresses*: Each word $S^{iN+j\ell}(x)[\ell]$ is encoded into a word on F pointing to the place of the corresponding ℓ-word in the list part. Thus the length $Q = q - m$ of this codeword is large enough if

$$b_{Q-1} < X_\ell \leqq b_Q,$$

or equivalently

$$q < m + 1 + \frac{q - m - 1}{\log b_{q-m-1}} \log X_\ell.$$

For the addresses $w_1^{q-m}, \ldots, w_P^{q-m}$ for

$$S^{iN}(x)[\ell], \ S^{iN+\ell}(x)[\ell], \ \ldots, \ S^{i(N+1)-\ell}(x)[\ell],$$

respectively, we can take m-words z_i' such that $w_1^{q-m} z_1' w_2^{q-m} z_2' \cdots w_P^{q-m} z_P'$ is also a word on F of length $(\frac{N}{\ell})q$. If $K = L + (\frac{N}{\ell})q$ turns out to be less than N, prolong it by adding $(N - K)$-words whose letters are all equal to the first letter in F (say 0). For $K \geqq N$, summing up the list of the casts and the addresses, we have

5 Spectrum and Entropy for Infinite Directed Graphs

$$K = L + \left(\frac{N}{\ell}\right) q$$

$$\leq N \left(\frac{m+1}{\ell^2} + \frac{k-m-1}{\log b_{k-m-1}} \frac{\log M}{\ell} + \frac{m+1}{\ell} + \frac{q-m-1}{\log b_{q-m-1}} \frac{\log X_\ell}{\ell}\right).$$

Now we show that we can take N. Since

$$\lim_{\ell \to \infty} \frac{1}{\ell} \log X_\ell = h(x) < \log r(F) = \lim_{q \to \infty} \frac{1}{q-m-1} \log b_{q-m-1}$$

and $\lim_{\ell \to \infty} \dfrac{N}{\ell} = 0$, we can take N large enough to make $K \leq N$. $\qquad \square$

Remark 3. If $F = SC(M')$ is the supercomplete graph with vertices M', then we can obtain the coding theorem by Ziv and Arimoto from Theorem 13.

Corollary 4. *Let G be a finite directed graph with M arrows and F an aperiodic finite directed graph with M' arrows. If $h(G) < h(F)$ and $M \geq M'$, then there exists a pair of a block encoder ϕ and a decoder ψ with length N such that $x = \widehat{x} \equiv \psi(\phi(x))$ for all messages x on G.*

Proof. It follows from Theorem 13 that $h(x) \leq h(G) < h(F)$ for all messages x on G. Thereby Theorem 15 implies the required result. $\qquad \square$

5.7 Notes

The adjacency operator for graphs was introduced by Mohar [20] for an undirected case and then by Fujii et al. [12] for a general case. The latter is the starting point of our study of infinite directed graphs from the viewpoint of operator theory [5, 7, 9–11, 13–16, 19, 24]. In particular, Seo and the author investigated spectral properties and various operations for graphs. Based on these studies, the author introduced two entropies of infinite graphs and discussed the difference between them [5]. On the other hand, considering S. Arimoto's idea for Ziv's coding theorem in a Japanese text [2], Y. Seo and the author obtained a generalization of his theorem in [11].

References

1. Accardi L, Staszewski R, Salacity R (2007) Decompositions of the free product of graphs. Infix Dampens Anal Quantum Probed Relax Top 10:303–334. http://arxiv.org/pdf/math/0609329.pdf
2. Arimoto S (1980) Probability, information and entropy (in Japanese). Morikita, Tokyo
3. Cvetković DM, Doob M, Sachs H (1980) Spectra of graphs. Academic, New York

4. Edgar GA (1990) Measure, topology and fractal geometry. Springer, New York
5. Fujii JI (1993) Entropy of graphs. Math Japon 38:39–46
6. Fujii JI (1995) The Marcus-Khan theorem for Hilbert space operators. Math Japon 41:531–535
7. Fujii JI, Fujii M (1993) Theorems of Williams and Pasadena. Math Japon 38:35–37
8. Fujii JI, Fujii M (2002) Kolmogorov's complexity for positive definite matrices. Lin Algebra Appl 341:171–180
9. Fujii JI, Seo Y (1993) Graphs and numerical center of mass. Math Japon 38:351–359
10. Fujii JI, Seo Y (1995) Graphs and tensor products of operators. Math Japon 41:245–252
11. Fujii JI, Seo Y (1997) Entropy and coding for graphs. Int J Math Stat Sci 6(1):63–77
12. Fujii M, Sasaoka H, Watatani Y (1989) Adjacency operators of infinite directed graphs. Math Japon 34:727–735
13. Fujii M, Sasaoka H, Watatani Y (1990) The numerical range of an infinite directed graph. Math Japon 35:577–582
14. Fujii JI, Sasaoka H, Watatani Y (1991) The spectrum of an infinite directed graph. Math Japon 36:607–625
15. Fujii M, Nakamoto M, Seo Y (1996) Graphs and Gersgorin's theorem. Math Japon 44:517–523
16. Fujii M, Nakamoto M, Seo Y, Watatani Y (1996) Graphs and Kolmogorov's complexity. Math Japon 44:113–117
17. Furuta T (2001) Invitation to linear operators. Taylor & Francis, New York
18. Halmos PR (1982) A Hilbert space problem book, 2nd edn. Springer, Berlin, New York
19. Matsumoto A, Seo Y (1996) Graphs and Fibonacci numbers. Math Japon 44:317–322
20. Mohar B (1982) The spectrum of an infinite graph. Lin Algebra Appl 48:245–256
21. Mohar B, Woess W (1989) A survey on spectra of infinite graphs. Bull Lond Math Soc 21:209–234
22. Paulsen I (1986) Completely bounded maps and dilations (Pitman Res Notes Math 146). Longman Scientific and Technical, Essex, and John Wiley and Sons, New York
23. Schaefer HH (1980) Topological vector spaces. Springer, Berlin, New York
24. Seo Y (1993) Graphs and nilpotent operators. Math Japon 38:1089–1093
25. Ziv J (1978) Coding theorems for individual sequences. IEEE Trans Inf Theory IT-24:405–412

Chapter 6
Application of Infinite Labeled Graphs to Symbolic Dynamical Systems

Kengo Matsumoto

Abstract We apply a theory of infinite labeled graphs to studying presentations and classifications of symbolic dynamical systems, by introducing a class of infinite labeled graphs, called λ-graph systems. Its matrix presentation is called a symbolic matrix system. The notions of a λ-graph system and symbolic matrix system are generalized notions of a finite labeled graph and symbolic matrix for sofic subshifts to general subshifts. Strong shift equivalence and shift equivalence between symbolic matrix systems are formulated so that two subshifts are topologically conjugate if and only if the associated canonical symbolic matrix systems are strong shift equivalent. We construct several kinds of shift equivalence invariants for symbolic matrix systems. They are the dimension groups, the K-groups, and the Bowen–Franks groups that are generalizations of the corresponding notions for nonnegative matrices. They yield topological conjugacy invariants of subshifts. The entropic quantities called λ-entropy and volume entropy for λ-graph systems are also studied related to the topological entropy of symbolic dynamics.

Keywords Subshifts · Symbolic dynamics · λ-Graph systems · Strong shift equivalence · Bowen–Franks group · K-theory · Topological entropy

MSC2000: Primary 37B10; Secondary 28D20, 46L80

6.1 Introduction

Graph theory has intersections with other branches of mathematics. The theory of symbolic dynamics is one of them. It has significant uses for coding theory and formal language theory in information sciences. The class of symbolic dynamical systems is a basic part of topological dynamical systems. Graph-theoretical

---

K. Matsumoto (✉)
Department of Mathematics, Joetsu University of Education, Joetsu 943-8512, Japan
e-mail: kengo@juen.ac.jp

M. Dehmer (ed.), *Structural Analysis of Complex Networks*,
DOI 10.1007/978-0-8176-4789-6_6, © Springer Science+Business Media, LLC 2011

techniques and linear algebraic techniques are very useful to study symbolic dynamical systems. Let Σ be a finite set, called an alphabet. Let $\Sigma^{\mathbb{Z}}$ be the infinite product spaces $\prod_{i=-\infty}^{\infty} \Sigma_i$ where $\Sigma_i = \Sigma$, endowed with the product topology. The transformation σ on $\Sigma^{\mathbb{Z}}$ given by $\sigma((x_i)_{i\in\mathbb{Z}}) = (x_{i+1})_{i\in\mathbb{Z}}$ is called the (full) shift. Let Λ be a shift-invariant closed subset of $\Sigma^{\mathbb{Z}}$ i.e. $\sigma(\Lambda) = \Lambda$. The topological dynamical system $(\Lambda, \sigma|_\Lambda)$ is called a subshift. We denote $\sigma|_\Lambda$ by σ and write the subshift as Λ for short. We denote by $X_\Lambda (\subset \prod_{i=1}^{\infty} \Sigma_i)$ the set of all right-infinite sequences that appear in Λ.

Symbolic dynamical systems have several subclasses by viewing the underlying graphs. The class of Markov shifts, which are often called SFT (shifts of finite type) comes from finite directed graphs, and the class of sofic shifts, which are a generalization of Markov shifts, comes from finite directed labeled graphs. In the study of Markov shifts and sofic shifts, graph-theoretical techniques are very useful. More generally we apply a theory of infinite labeled graphs to studying general symbolic dynamical systems, by introducing a class of infinite labeled graphs, called λ-graph systems. Its matrix presentation is called a symbolic matrix system. The notions of the λ-graph system and symbolic matrix system are generalized notions of a finite labeled graph and symbolic matrix for sofic shifts to general subshifts. Strong shift equivalence and shift equivalence between symbolic matrix systems are formulated such that two subshifts are topologically conjugate if and only if the canonical symbolic matrix systems are strong shift equivalent. We construct several kinds of shift equivalence invariants for symbolic matrix systems. They are the dimension groups, the Bowen–Franks groups, and the nonzero spectra that are generalizations of the corresponding notions for nonnegative matrices. The K-groups for symbolic matrix systems are introduced. They are also shift equivalence invariants and stronger than the Bowen–Franks groups but weaker than the dimension triples. These kinds of shift equivalence invariants naturally induce topological conjugacy invariants for subshifts. In particular the K-groups and the Bowen–Franks groups induce not only topological conjugacy invariants for subshifts, but also flow equivalence invariants. As entropic quantities, the volume entropy and the λ-entropy for λ-graph systems are introduced. They are shift equivalence invariants and hence induce topological conjugacy invariants of subshifts. The λ-entropy measures a distance from sofic shifts. The volume entropy does not necessarily coincide with classical topological entropy unless the subshift is sofic.

A finite sequence $\mu = (\mu_1, ..., \mu_k)$ of elements $\mu_j \in \Sigma$ is called a word. We denote by $|\mu|$ the length k of μ. A word $\mu = (\mu_1, ..., \mu_k)$ is said to appear in $x = (x_i)_{i\in\mathbb{Z}} \in \Sigma^{\mathbb{Z}}$ if $x_m = \mu_1, ..., x_{m+k-1} = \mu_k$ for some $m \in \mathbb{Z}$. For a subshift Λ, we denote by $B_k(\Lambda)$ the set of all words of length k appearing in some $x \in \Lambda$, where $B_0(\Lambda)$ denotes the empty word. We set $B_*(\Lambda) = \cup_{k=0}^{\infty} B_k(\Lambda)$. A matrix with entries in nonnegative integers is called a nonnegative matrix. Throughout the chapter, $\mathbb{Z}_+$ and $\mathbb{N}$ denote the set of all nonnegative integers, and the set of all positive integers, respectively. For an introduction to the theory of symbolic dynamical systems, see [21, 33] (cf. [11, 27]).

The chapter is organized as follows. In Sects. 6.2–6.4, basic classes of subshifts called topological Markov shifts that are defined by finite directed graphs, sofic shifts that are defined by finite labeled graphs, and coded shifts that are defined

6 Application of Infinite Labeled Graphs to Symbolic Dynamical Systems

by codes, respectively, are briefly explained. In Sect. 6.5, λ-graph systems and symbolic matrix systems are introduced as a presentation of subshifts. In Sect. 6.6, strong shift equivalences and shift equivalences of symbolic matrix systems are defined and studied. In Sect. 6.7, nonnegative matrix systems are introduced to define K-theoretic invariants of symbolic matrix systems. In Sect. 6.8, K-theoretic invariants for symbolic matrix systems called dimension groups, K-groups, and Bowen–Franks groups are introduced. They yield topological conjugacy invariants of subshifts. In Sect. 6.9, spectrum, λ-entropy, and volume entropy for λ-graph systems and symbolic matrix systems are introduced. They are closely related to the topological entropy of subshifts. In Sect. 6.10, the K-groups and the Bowen–Franks groups for a class of nonsofic shifts is presented. In Sect. 6.11, a relation of the K-theoretic invariants for λ-graph systems and symbolic matrix systems to K-theory for C^*-algebras is remarked. Finally, in Sect. 6.12, conclusions and further works are described.

6.2 Markov Shifts and Finite Directed Graphs

The classification of symbolic dynamical systems has been very important and one of the central problems in the theory of topological dynamical systems. The classification problem was first examined for a class of symbolic dynamical systems called topological Markov shifts. Each dynamical system of the class is determined by a single square nonnegative matrix. Hence the behavior of such a dynamical system depends on the underlying matrix. Let A be an $n \times n$ nonnegative matrix. Put $V_A = \{1, \ldots, n\}$: the vertex set. Write $A(i, j)$ edges from $i \in V_A$ to $j \in V_A$. We have a directed graph denoted by G_A. Let Σ be the set E_A of all edges of G_A. Let s_A and r_A be the maps from E_A to V_A that assign the source vertex and the range vertex of an edge respectively. Let Λ_A be the set of all bi-infinite sequences $(e_i)_{i \in \mathbb{Z}}$ of $e_i \in E_A$ with $r_A(e_i) = s_A(e_{i+1}), i \in \mathbb{Z}$. Then Λ_A becomes a subshift, called the topological Markov shift defined by nonnegative matrix A.

In [52], R. F. Williams introduced the notions of strong shift equivalence and shift equivalence between nonnegative matrices and showed that two topological Markov shifts are topologically conjugate if and only if the underlying matrices are strong shift equivalent. Strong shift equivalence implies shift equivalence. Although the converse implication had been a long-standing open problem, Kim and Roush [20] have solved negatively for even irreducible matrices.

Two nonnegative matrices M and N are said to be *strong shift equivalent in 1-step*, written as $M \underset{1}{\approx} N$, if there exist rectangular nonnegative matrices R and S such that

$$ M = RS, \qquad N = SR. $$

Definition 6.2.1 ([52]). Two matrices A and B are said to be *strong shift equivalent in l-step*, written as $A \underset{l}{\approx} B$, if there exist nonnegative square matrices $A_1, A_2, \ldots, A_{l-1}$ such that

$$A \underset{1}{\approx} A_1 \underset{1}{\approx} A_2 \underset{1}{\approx} \cdots \underset{1}{\approx} A_{l-1} \underset{1}{\approx} B.$$

We say A and B are *strong shift equivalent* if $A \underset{l}{\approx} B$ for some l, and write it as $A \approx B$.

Williams has proved the following fundamental theorem of Markov shifts.

Theorem 6.2.2 ([52]). *Topological Markov shifts Λ_A and Λ_B are topologically conjugate if and only if A and B are strong shift equivalent.*

Definition 6.2.3 ([52]). Two matrices A and B are said to be *shift equivalent of lag l*, written as $A \underset{l}{\sim} B$, if there exist nonnegative rectangular matrices R and S such that

$$A^l = RS, \qquad B^l = SR, \qquad SA = BS, \qquad RB = AR.$$

We say A and B are *shift equivalent* if $A \underset{l}{\sim} B$ for some l, and write it as $A \sim B$. Both strong shift equivalence and shift equivalence are equivalence relations, and strong shift equivalence implies shift equivalence.

6.3 Sofic Shifts and Finite Labeled Graphs

We fix a finite set Σ called an alphabet. We write the empty symbol $\emptyset$ in Σ as 0. We denote by $\mathfrak{S}_\Sigma$ the set of all finite formal sums of elements of Σ. A square matrix with entries in $\mathfrak{S}_\Sigma$ is called a symbolic matrix over Σ. It is an equivalent object to a labeled graph called a λ-graph [47]. There is a class of subshifts called sofic shifts that are a generalized class of Markov shifts determined by symbolic matrices (see [14, 25, 26, 51], etc.).

Let $\mathcal{A}(= [\mathcal{A}(i, j)]_{i,j=1,\dots,n})$ be an $n \times n$ symbolic matrix over Σ. We have a directed graph $G_\mathcal{A}$ from the matrix $\mathcal{A}$ with labeled edges by the symbols in Σ. We denote by $\lambda(e) = \alpha \in \Sigma$ the label α of edge e. Let $\Lambda_\mathcal{A}$ be the set of all bi-infinite sequences $(\lambda(e_i))_{i \in \mathbb{Z}}$ of labels of edges $e_i \in E_\mathcal{A}$ with $r_\mathcal{A}(e_i) = s_\mathcal{A}(e_{i+1}), i \in \mathbb{Z}$, where $r_\mathcal{A}(e_i)$ and $s_\mathcal{A}(e_{i+1})$ denote the range vertex of e_i and the source vertex of e_{i+1}, respectively. Then $\Lambda_\mathcal{A}$ becomes a subshift, called the *sofic shift* defined by symbolic matrix $\mathcal{A}$.

M. Nasu in [47,48] generalized the notion of strong shift equivalence to symbolic matrices. He showed that two sofic shifts are topologically conjugate if and only if their canonical symbolic matrices are strong shift equivalent ([47, 48]; see also [16]). M. Boyle and W. Krieger in [4] introduced the notion of shift equivalence for symbolic matrices and studied topological conjugacy for sofic shifts. For two symbolic matrices $\mathcal{A}$ over alphabet Σ and $\mathcal{A}'$ over alphabet Σ' and a bijection ϕ from a subset of Σ onto a subset of Σ', $\mathcal{A}$ and $\mathcal{A}'$ are said to be specified equivalent

6 Application of Infinite Labeled Graphs to Symbolic Dynamical Systems 141

under specification ϕ if $\mathcal{A}'$ can be obtained from $\mathcal{A}$ by replacing every symbol a appearing in $\mathcal{A}$ by $\phi(a)$ [47,48]. We write it as $\mathcal{A} \overset{\phi}{\simeq} \mathcal{A}'$, or simply $\mathcal{A} \simeq \mathcal{A}'$. We call ϕ a specification from Σ to Σ'. Two symbolic matrices $\mathcal{M}$ and $\mathcal{M}$ are said are *strong shift equivalent in 1-step*, written as $\mathcal{M} \underset{1}{\approx} \mathcal{M}$, if there exist rectangular symbolic matrices $\mathcal{R}$ and $\mathcal{S}$ such that

$$\mathcal{M} \simeq \mathcal{R}\mathcal{S}, \qquad \mathcal{M} \simeq \mathcal{S}\mathcal{R}.$$

Definition 6.3.1 ([47]). Two symbolic matrices $\mathcal{A}$ and $\mathcal{B}$ are said to be *strong shift equivalent in l-step*, written as $\mathcal{A} \underset{l}{\approx} \mathcal{B}$, if there exist symbolic matrices $\mathcal{A}_1, \mathcal{A}_2, \ldots, \mathcal{A}_{l-1}$ such that

$$\mathcal{A} \underset{1}{\approx} \mathcal{A}_1 \underset{1}{\approx} \mathcal{A}_2 \underset{1}{\approx} \cdots \underset{1}{\approx} \mathcal{A}_{l-1} \underset{1}{\approx} \mathcal{B}.$$

We say $\mathcal{A}$ and $\mathcal{B}$ are *strong shift equivalent* if $\mathcal{A} \underset{l}{\approx} \mathcal{B}$ for some l, and write it as $\mathcal{A} \approx \mathcal{B}$.

For a sofic shift Λ and $x = (x_n)_{n \in \mathbb{N}} \in X_\Lambda$, we put

$$P(x) = \{(\mu_1, \ldots, \mu_k) \in B_*(\Lambda) \mid (\mu_1, \ldots, \mu_k, x_1, x_2, \ldots) \in X_\Lambda\}.$$

Two $x = (x_n)_{n \in \mathbb{N}}, y = (y_n)_{n \in \mathbb{N}} \in X_\Lambda$ are said to be past equivalent if $P(x) = P(y)$. Let V_Λ be the equivalence classes of X_Λ under the past equivalence. As Λ is sofic, the set V_Λ is a finite set. We write an edge labeled α from $[x]$ to $[y]$ if $[y] = [\alpha x]$, where $\alpha x = (\alpha, x_1, x_2, \ldots)$. We have a finite labeled graph G_Λ with vertex set V_Λ. The labeled graph is called the *left Krieger cover graph* for the sofic shift Λ [25,26]. Let us denote by $\mathcal{A}^\Lambda$ its symbolic matrix. We call the matrix $\mathcal{A}^\Lambda$ the canonical symbolic matrix for the sofic shift Λ.

Nasu has proved the following fundamental theorem of sofic shifts.

Theorem 6.3.2 ([47]). *Sofic shifts Λ and Λ' are topologically conjugate if and only if their canonical symbolic matrices $\mathcal{A}^\Lambda$ and $\mathcal{A}^{\Lambda'}$ are strong shift equivalent.*

Definition 6.3.3 ([4]). Two symbolic matrices $\mathcal{A}$ and $\mathcal{B}$ are said to be *shift equivalent of lag l*, written as $\mathcal{A} \underset{l}{\sim} \mathcal{B}$, if there exist symbolic rectangular matrices $\mathcal{R}$ and $\mathcal{S}$ such that

$$\mathcal{A}^l \simeq \mathcal{R}\mathcal{S}, \qquad \mathcal{B}^l \simeq \mathcal{S}\mathcal{R}, \qquad \mathcal{S}\mathcal{A} \simeq \mathcal{B}\mathcal{S} \qquad \mathcal{R}\mathcal{B} \simeq \mathcal{A}\mathcal{R}.$$

We say $\mathcal{A}$ and $\mathcal{B}$ are *shift equivalent* if $\mathcal{A} \underset{l}{\sim} \mathcal{B}$ for some l, and write it as $\mathcal{A} \sim \mathcal{B}$. Both strong shift equivalence and shift equivalence are equivalence relations, and strong shift equivalence implies shift equivalence.

6.4 Coded Shifts

Denote by Σ^* the set of all words of an alphabet Σ. A (finite or infinite) collection $\mathcal{C}(\subset \Sigma^*)$ of words over Σ is said to be uniquely decipherable if whenever $\alpha_1 \alpha_2 \cdots \alpha_n = \gamma_1 \gamma_2 \cdots \gamma_m$ with $\alpha_i, \gamma_j \in \mathcal{C}$, then $n = m$ and $\alpha_i = \gamma_i$ for $i = 1, \ldots, n$. A uniquely decipherable set $\mathcal{C}$ is called a code. Blanchard and Hansel [2] have introduced the notion of coded shift. A subshift Λ is called a *coded shift* if Λ is the closure of the set of bi-infinite sequences obtained by freely concatenating the words in a code $\mathcal{C}$. It is denoted by $\Lambda_{\mathcal{C}}$.

Proposition 6.4.1 ([2]). *For a subshift Λ over Σ, the following four conditions are equivalent:*

1. *Λ is a coded shift.*
2. *Λ is presented by an irreducible countable labeled graph.*
3. *Λ is the closure of the set of sequences obtained by freely concatenating the words in a subset of Σ^*.*
4. *Λ contains an increasing sequence of irreducible Markov shifts whose unions are dense in Λ.*

Irreducible sofic shifts are coded shifts, and any coded shift is irreducible.

The following are examples of coded shifts.

1. Synchronizing counter shift (cf. [9, 17, 28]). Let $\Sigma = \{a, b, c\}$. Put $\mathcal{C} = \{a \overbrace{b \cdots b}^{n} \overbrace{c \cdots c}^{n} \mid n = 0, 1, \ldots\}$. It is easy to see that $\mathcal{C}$ is a code, and the coded shift $\Lambda_{\mathcal{C}}$ is not sofic. Its language is accepted by a synchronizing counter automaton. It is named the context-free shift in [33, Example 1.2.9].

2. Dyck shifts ([22]). For $2 \leq N \in \mathbb{N}$, let $\Sigma = \{(_1, \ldots, (_N,)_N, \ldots,)_1\}$ be the N-kinds of brackets. Define a semi-group structure in Σ^* by setting

$$(_i \cdot)_j = \begin{cases} 1 & \text{if } i = j, \\ 0 & \text{otherwise.} \end{cases}$$

Let $\mathcal{C}_N$ be the set of all words in Σ whose products are reduced to **1**. It is easy to see that $\mathcal{C}_N$ is a code, and the resulting coded shift $\Lambda_{\mathcal{C}_N}$ is not sofic. The coded shift is called the Dyck shift of order N and written as D_N. Its language is accepted by a push-down automaton and forms a context-free language (cf. [9, 17]).

6.5 λ-Graph Systems and Symbolic Matrix Systems

A symbolic matrix corresponds to a labeled graph, called a λ-graph, that is a presentation of a sofic shift. We consider a generalization of λ-graphs to present any subshift. Details of this section are seen in [36].

6 Application of Infinite Labeled Graphs to Symbolic Dynamical Systems 143

We first explain the notion of the Bratteli diagram that appears in the theory of operator algebras (see [5, 12]). A Bratteli diagram consists of a vertex set $V = V_1 \cup V_2 \cup \cdots$ and an edge set $E = E_{1,2} \cup E_{2,3} \cup \cdots$ where each V_l and $E_{l,l+1}$ are finite and nonempty. We have maps $s, r : E \rightarrow V$ such that $s(E_{l,l+1}) = V_l$, $r(E_{l,l+1}) = V_{l+1}$. They are called a source map and a range map, respectively. For $u \in V_l, v \in V_{l+1}$, put

$$E_{l,l+1}(u, v) = \{e \in E_{l,l+1} | s(e) = u, r(e) = v\}.$$

A labeled Bratteli diagram (V, E, λ) over alphabet Σ consists of a Bratteli diagram (V, E) with a map λ from E to Σ.

Definition 6.5.1. A λ-*graph system* (V, E, λ, ι) over alphabet Σ consists of a labeled Bratteli diagram (V, E, λ) over Σ and a surjective map ι from $V \setminus V_1$ to V satisfying the following two conditions:

1. $\iota(V_{l+1}) = V_l$ for $l \in \mathbb{N}$.
2. For $u \in V_l, w \in V_{l+2}$, there exists a bijective correspondence between

$$E_{l,l+1}(u, \iota(w)) \quad \text{and} \quad \bigcup_{v \in V_{l+1}, \iota(v) = u} E_{l+1,l+2}(v, w)$$

that is compatible with the labeling λ.

By the above condition 1, the cardinality of the set V_{l+1} is greater than or equal to that of the set V_l. The above condition 2 is called the local property of the λ-graph system. It yields the following two conditions:

1. For $e \in E_{l+1,l+2}$, there exists $e' \in E_{l,l+1}$ such that

$$\iota(s(e)) = s(e'), \quad \iota(r(e)) = r(e') \quad \text{and} \quad \lambda(e) = \lambda(e').$$

2. For $f \in E_{l,l+1}, v \in V_{l+2}$ with $\iota(v) = r(f)$, there exists $e \in E_{l+1,l+2}$ such that

$$\iota(s(e)) = s(f), \quad r(e) = v \quad \text{and} \quad \lambda(e) = \lambda(f).$$

Two λ-graph systems (V, E, λ, ι) over Σ and $(V', E', \lambda', \iota')$ over Σ' are said to be isomorphic if there exist bijections $\Phi_V : V \rightarrow V'$, $\Phi_E : E \rightarrow E'$, and a specification $\phi : \Sigma \rightarrow \Sigma'$ such that

1. $\Phi_V(V_l) = V_l'$ and $\Phi_E(E_{l,l+1}) = E_{l,l+1}'$ for $l \in \mathbb{N}$,
2. $\Phi_V(s(e)) = s(\Phi_E(e))$ and $\Phi_V(r(e)) = r(\Phi_E(e))$ for $e \in E$,
3. $\iota'(\Phi_V(v)) = \Phi_V(\iota(v))$ for $v \in V$,
4. $\lambda'(\Phi_E(e)) = \phi(\lambda(e))$ for $e \in E$.

We construct subshifts from λ-graph systems. Let $\mathfrak{L} = (V, E, \lambda, \iota)$ be a λ-graph system over Σ. For $k < l$, set

$$P_{k,l} = \{(e_k, e_{k+1}, \ldots, e_{l-1}) | e_i \in E_{i,i+1}, r(e_i) = s(e_{i+1}) \text{ for } i = k, k+1, \ldots, l-2\}$$

the set of all paths from V_k to V_l, and

$$L_{k,l} = \{\lambda(e_k)\lambda(e_{k+1}) \cdots \lambda(e_{l-1}) \in \overbrace{\Sigma \times \cdots \times \Sigma}^{l-k \text{ times}} \mid (e_k, e_{k+1}, \ldots, e_{l-1}) \in P_{k,l}\}$$

the set of all labeled paths from V_k to V_l. Put $L_l = L_{1,l+1}$ and endow it with discrete topology. We set

$$X_{\mathfrak{L}} = \{(\lambda(e_1), \lambda(e_2), \ldots) \in \Sigma^{\mathbb{N}} | e_i \in E_{i,i+1}, r(e_i) = s(e_{i+1}) \text{ for } i \in \mathbb{N}\}$$

the set of all right infinite sequences consisting of labels along infinite paths. The topology on $X_{\mathfrak{L}}$ is defined from open sets of the form

$$U_{(\mu_1, \ldots, \mu_k)} = \{(\alpha_1, \alpha_2, \ldots) \in X_{\mathfrak{L}} | \alpha_i = \mu_i \text{ for } i = 1, \ldots, k\}$$

for $(\mu_1, \ldots, \mu_k) \in L_k$. By the local property of the λ-graph system, one sees:

1. if $(\alpha_1, \alpha_2, \ldots) \in X_{\mathfrak{L}}$, we have $(\alpha_2, \alpha_3, \ldots) \in X_{\mathfrak{L}}$,
2. if $(\alpha_k, \ldots, \alpha_{l-1}) \in L_{k,l}$ for $l > k$, we have $(\alpha_k, \ldots, \alpha_{l-1}) \in L_{k+1,l+1}$.

As in [33, Definition 1.3.1], a set $\mathcal{L}$ of words of Σ is called a language if it satisfies the following conditions:

(a) Every subword of a word w in $\mathcal{L}$ belongs to $\mathcal{L}$.
(b) For a word w in $\mathcal{L}$, there are nonempty words u, v in $\mathcal{L}$ such that uwv belongs to $\mathcal{L}$.

Let $\mathcal{L}(\mathfrak{L})$ be the set of all words appearing in $X_{\mathfrak{L}}$. That is, $\mathcal{L}(\mathfrak{L}) = \cup_{k \leq l} L_{k,l}$. Then

Proposition 6.5.2. $\mathcal{L}(\mathfrak{L})$ *is a language.*

By [33, Proposition 1.3.4], we see

Theorem 6.5.3. *There exists a subshift Λ over alphabet Σ whose language is $\mathcal{L}(\mathfrak{L})$. Namely the set $B_*(\Lambda)$ of all admissible words of the subshift Λ is $\mathcal{L}(\mathfrak{L})$.*

The subshift is realized as

$$\Lambda = \{(\ldots, \alpha_{-2}, \alpha_{-1}, \alpha_0, \alpha_1, \alpha_2, \ldots) | (\alpha_n, \alpha_{n+1}, \ldots) \in X_{\mathfrak{L}} \text{ for all } n \in \mathbb{Z}\}.$$

We denote by $\Lambda_{\mathfrak{L}}$ the subshift Λ in the above theorem and call it the subshift presented by the λ-graph system $\mathfrak{L}$.

We next construct λ-graph systems from subshifts. For a subshift Λ over Σ, denote by X_{Λ} its right one-sided subshift. Set

$$\Lambda^l(x) = \{\mu \in B_l(\Lambda) | \mu x \in X_{\Lambda}\} \qquad \text{for} \quad x \in X_{\Lambda}, \quad l \in \mathbb{N}.$$

6 Application of Infinite Labeled Graphs to Symbolic Dynamical Systems 145

We define a nested sequence of equivalence relations in the space X_Λ. Two points $x, y \in X_\Lambda$ are said to be l-*past equivalent*, written as $x \sim_l y$, if $\Lambda^l(x) = \Lambda^l(y)$. Denote by $\Omega_l = X_\Lambda / \sim_l$ the quotient space. For $x, y \in X_\Lambda$ and $\mu \in B_k(\Lambda)$, one sees that

1. if $x \sim_l y$, we have $x \sim_m y$ for $m < l$,
2. if $x \sim_l y$ and $\mu x \in X_\Lambda$, we have $\mu y \in X_\Lambda$ and $\mu x \sim_{l-k} \mu y$ for $l > k$.

We have the following sequence of surjections in a natural way:

$$\Omega_1 \leftarrow \Omega_2 \leftarrow \cdots \leftarrow \Omega_l \leftarrow \Omega_{l+1} \leftarrow \cdots .$$

The subshift Λ is a sofic shift if and only if $\Omega_l = \Omega_{l+1}$ for some $l \in \mathbb{N}$. For a fixed $l \in \mathbb{N}$, let $F_i^l, i = 1, 2, \ldots, m(l)$ be the set of all l-past equivalence classes of X_Λ. Hence X_Λ is a disjoint union of $F_i^l, i = 1, 2, \ldots, m(l)$. The vertex set V_l at level l consist of the sets $F_i^l, i = 1, \ldots, m(l)$. We write an edge with label a from the vertex $F_i^l \in V_l$ to $F_j^{l+1} \in V_{l+1}$ if $ax \in F_i^l$ for some $x \in F_j^{l+1}$. We denote by $E_{l,l+1}$ the set of all edges from V_l to V_{l+1}. There exists a natural map ι_l^Λ from V_{l+1} to V_l by mapping F_j^{l+1} to F_i^l when F_i^l contains F_j^{l+1}. Set $V^\Lambda = \cup_{l=1}^\infty V_l$ and $E^\Lambda = \cup_{l=1}^\infty E_{l,l+1}$. The labeling of edges is denoted by $\lambda^\Lambda : E \to \Sigma$. We then see

Theorem 6.5.4. *For a subshift* Λ, $(V^\Lambda, E^\Lambda, \lambda^\Lambda, \iota^\Lambda)$ *becomes a λ-graph system which we denote by* $\mathfrak{L}^\Lambda$. *Moreover the subshift* $\Lambda_{\mathfrak{L}^\Lambda}$ *presented by* $\mathfrak{L}^\Lambda$ *coincides with the original subshift* Λ.

We call $\mathfrak{L}^\Lambda$ the *canonical λ-graph system* for Λ. For a λ-graph system $\mathfrak{L}$, let $\Lambda_{\mathfrak{L}}$ be the presented subshift by $\mathfrak{L}$. Then its canonical λ-graph system $\mathfrak{L}^{\Lambda_{\mathfrak{L}}}$ does not necessarily coincide with the original λ-graph system $\mathfrak{L}$. We indeed see that the canonical λ-graph system $\mathfrak{L}^\Lambda$ for a subshift Λ is left-resolving; i.e., the incoming edges to each vertex carry different labels, and is predecessor-separated; i.e., distinct vertices at each level have distinct predecessor sets of labels. If in particular, Λ is a sofic shift, its canonical λ-graph system is eventually realized as the left Krieger cover graph for Λ.

Let us consider matrix presentation of λ-graph systems.

Definition 6.5.5. Let $(\mathcal{M}_{l,l+1}, I_{l,l+1}), l \in \mathbb{N}$ be a pair of sequences of rectangular matrices such that the following conditions for each $l \in \mathbb{N}$ are satisfied:

1. $\mathcal{M}_{l,l+1}$ is an $m(l) \times m(l+1)$ rectangular matrix with entries in $\mathfrak{S}_\Sigma$.
2. $I_{l,l+1}$ is an $m(l) \times m(l+1)$ rectangular matrix with entries in $\{0, 1\}$ satisfying the relation:

$$I_{l,l+1} \mathcal{M}_{l+1,l+2} = \mathcal{M}_{l,l+1} I_{l+1,l+2}, \qquad l \in \mathbb{N}. \tag{6.1}$$

We further assume that both the matrices $\mathcal{M}_{l,l+1}$ and $I_{l,l+1}$ have no zero columns and no zero rows. For j, there uniquely exists i such that $I_{l,l+1}(i, j) \neq 0$. By the above conditions one sees $m(l) \leq m(l+1)$. The pair $(\mathcal{M}, I)$ is called a *symbolic matrix system* over Σ.

Two symbolic matrix systems $(\mathcal{M}, I)$ over Σ and $(\mathcal{M}', I')$ over Σ' are said to be isomorphic if $m(l) = m'(l)$ for $l \in \mathbb{N}$ and there exist a specification ϕ from Σ to Σ' and an $m(l) \times m(l)$ permutation matrix P_l for each $l \in \mathbb{N}$ such that

$$P_l \mathcal{M}_{l,l+1} \overset{\phi}{\simeq} \mathcal{M}'_{l,l+1} P_{l+1}, \qquad P_l I_{l,l+1} = I'_{l,l+1} P_{l+1} \qquad \text{for} \quad l \in \mathbb{N}.$$

The notion of a symbolic matrix system is a generalized notion of a symbolic matrix. For an $n \times n$ symbolic matrix $\mathcal{A}$, we set

$$\mathcal{M}_{l,l+1} = \mathcal{A}, \qquad I_{l,l+1} = E_n \qquad \text{for} \quad l \in \mathbb{N}$$

where E_n denotes the identity matrix of size n. Then $(\mathcal{M}_{l,l+1}, I_{l,l+1}), l \in \mathbb{N}$ is a symbolic matrix system. The following proposition shows that symbolic matrix systems are nothing but the matrix presentation of λ-graph systems.

Proposition 6.5.6. *There exists a bijective correspondence between the set of all isomorphism classes of symbolic matrix systems and the set of all isomorphism classes of λ-graph systems.*

We say a symbolic matrix system is canonical for a subshift Λ if its corresponding λ-graph system is canonical. It is denoted by $(\mathcal{M}^\Lambda, I^\Lambda)$.

For example, set for each $l \in \mathbb{N}$, $\mathcal{M}_{l,l+1} = \begin{bmatrix} a & b \\ b & 0 \end{bmatrix}$ and $I_{l,l+1} = \begin{bmatrix} 1 & 0 \\ 0 & 1 \end{bmatrix}$. The λ-graph system for the symbolic matrix system gives rise to the even shift, denoted by Y. Its canonical symbolic matrix system is given by the following matrices:

$$\mathcal{M}^Y_{1,2} = \begin{bmatrix} a & a+b & b \\ b & 0 & 0 \end{bmatrix}, \qquad I^Y_{1,2} = \begin{bmatrix} 1 & 1 & 0 \\ 0 & 0 & 1 \end{bmatrix}$$

and

$$\mathcal{M}^Y_{l,l+1} = \begin{bmatrix} a & a & b \\ 0 & b & 0 \\ b & 0 & 0 \end{bmatrix}, \qquad I^Y_{l,l+1} = \begin{bmatrix} 1 & 0 & 0 \\ 0 & 1 & 0 \\ 0 & 0 & 1 \end{bmatrix} \qquad \text{for} \quad l \geq 2.$$

6.6 Strong Shift Equivalences and Shift Equivalences

In this section, we define two kinds of strong shift equivalences between symbolic matrix systems. One is called the properly strong shift equivalence that exactly reflects a bipartite decomposition of the associated λ-graph systems. The other one is called the strong shift equivalence that is weaker than the former strong shift equivalence. They coincide at least among symbolic matrix systems whose λ-graph systems are predecessor-separated and left-resolving, and hence between canonical symbolic matrix systems for subshifts. The latter is more easily defined and treated

6 Application of Infinite Labeled Graphs to Symbolic Dynamical Systems 147

than the former. The main purpose in this section is to see that topological conjugacy between two subshifts is completely characterized by strong shift equivalence between their canonical symbolic matrix systems. We first define properly strong shift equivalence between two symbolic matrix systems as a generalization of strong shift equivalence between two nonnegative matrices defined by Williams in [52] and between two symbolic matrices defined by Nasu in [47]. For the details of this section see [36].

For alphabets C, D, put $C \cdot D = \{cd \mid c \in C, d \in D\}$. For $x = \sum_j c_j \in \mathfrak{S}_C$ and $y = \sum_k d_k \in \mathfrak{S}_D$, define $xy = \sum_{j,k} c_j d_k \in \mathfrak{S}_{C \cdot D}$.

Let $(\mathcal{M}, I)$ and $(\mathcal{M}', I')$ be symbolic matrix systems over Σ and Σ', respectively, where $\mathcal{M}_{l,l+1}, I_{l,l+1}$ are $m(l) \times m(l+1)$ matrices and $\mathcal{M}'_{l,l+1}, I'_{l,l+1}$ are $m'(l) \times m'(l+1)$ matrices.

Definition 6.6.1. Two symbolic matrix systems $(\mathcal{M}, I)$ and $(\mathcal{M}', I')$ are said to be *properly strong shift equivalent in 1-step*, written as $(\mathcal{M}, I) \underset{1-pr}{\approx} (\mathcal{M}', I')$, if there exist alphabets C, D and specifications

$$\varphi : \Sigma \to C \cdot D, \qquad \phi : \Sigma' \to D \cdot C$$

and increasing sequences $n(l), n'(l)$ on $l \in \mathbb{N}$ such that for each $l \in \mathbb{N}$, there exist an $n(l) \times n'(l+1)$ matrix $\mathcal{P}_l$ over C, an $n'(l) \times n(l+1)$ matrix $\mathcal{Q}_l$ over D, an $n(l) \times n(l+1)$ matrix X_l over $\{0, 1\}$, and an $n'(l) \times n'(l+1)$ matrix X'_l over $\{0, 1\}$ satisfying the following equations:

$$\mathcal{M}_{l,l+1} \overset{\varphi}{\simeq} \mathcal{P}_{2l} \mathcal{Q}_{2l+1}, \qquad \mathcal{M}'_{l,l+1} \overset{\phi}{\simeq} \mathcal{Q}_{2l} \mathcal{P}_{2l+1}, \tag{6.2}$$

$$I_{l,l+1} = X_{2l} X_{2l+1}, \qquad I'_{l,l+1} = X'_{2l} X'_{2l+1} \tag{6.3}$$

and

$$X_l \mathcal{P}_{l+1} = \mathcal{P}_l X'_{l+1}, \qquad X'_l \mathcal{Q}_{l+1} = \mathcal{Q}_l X_{l+1}. \tag{6.4}$$

It follows by (6.1) that $n(2l) = m(l)$ and $n'(2l) = m'(l)$ for $l \in \mathbb{N}$.

Two symbolic matrix systems $(\mathcal{M}, I)$ and $(\mathcal{M}', I')$ are said to be *properly strong shift equivalent in N-step*, written as $(\mathcal{M}, I) \underset{N-pr}{\approx} (\mathcal{M}', I')$, if there exists a sequence of symbolic matrix systems $(\mathcal{M}^{(i)}, I^{(i)}), i = 1, 2, \ldots, N-1$ such that

$$(\mathcal{M}, I) \underset{1-pr}{\approx} (\mathcal{M}^{(1)}, I^{(1)}) \underset{1-pr}{\approx} \cdots \underset{1-pr}{\approx} (\mathcal{M}^{(N-1)}, I^{(N-1)}) \underset{1-pr}{\approx} (\mathcal{M}', I').$$

We simply call it a *properly strong shift equivalence*.

Lemma 6.6.2. *Properly strong shift equivalence is an equivalence relation on symbolic matrix systems.*

148 K. Matsumoto

Proof. It suffices to show that $(\mathcal{M}, I) \underset{1-pr}{\approx} (\mathcal{M}, I)$. Put $C = \Sigma, D = \{0, 1\}$. Define $\varphi : a \in \Sigma \to a \cdot 1 \in C \cdot D$ and $\phi : a \in \Sigma \to 1 \cdot a \in D \cdot C$. Let E_k be the $k \times k$ identity matrix. Set

$$\mathcal{P}_{2l} = \mathcal{P}_{2l+1} = \mathcal{M}_{l,l+1}, \quad \mathcal{Q}_{2l} = E_{m(l)}, \quad \mathcal{Q}_{2l+1} = E_{m(l+1)},$$
$$X_{2l} = E_{m(l)}, \quad X_{2l+1} = I_{l,l+1}, \quad X'_{2l} = I_{l,l+1}, \quad X'_{2l+1} = E_{m(l+1)}.$$

They give a properly strong shift equivalence in 1-step between $(\mathcal{M}, I)$ and $(\mathcal{M}, I)$. $\square$

We now introduce the notion of a bipartite symbolic matrix system and a bipartite λ-graph system.

Definition 6.6.3. A symbolic matrix system $(\mathcal{M}, I)$ over Σ is said to be *bipartite* if there exist disjoint subsets $C, D \subset \Sigma$ and increasing sequences $n(l), n'(l)$ on $l \in \mathbb{N}$ with $m(l) = n(l) + n'(l)$ such that for each $l \in \mathbb{N}$, there exist an $n(l) \times n'(l + 1)$ matrix $\mathcal{P}_{l,l+1}$ over C, an $n'(l) \times n(l + 1)$ matrix $\mathcal{Q}_{l,l+1}$ over D, an $n(l) \times n(l + 1)$ matrix $X_{l,l+1}$ over $\{0, 1\}$ and an $n'(l) \times n'(l+1)$ matrix $X'_{l,l+1}$ over $\{0, 1\}$ satisfying the equations

$$\mathcal{M}_{l,l+1} = \begin{bmatrix} 0 & \mathcal{P}_{l,l+1} \\ \mathcal{Q}_{l,l+1} & 0 \end{bmatrix}, \qquad I_{l,l+1} = \begin{bmatrix} X_{l,l+1} & 0 \\ 0 & X'_{l,l+1} \end{bmatrix}.$$

Since the relations $I_{l,l+1} \mathcal{M}_{l+1,l+2} = \mathcal{M}_{l,l+1} I_{l+1,l+2}$ and hence

$$X_{l-1,l} \mathcal{P}_{l,l+1} = \mathcal{P}_{l-1,l} X'_{l,l+1}, \qquad X'_{l-1,l} \mathcal{Q}_{l,l+1} = \mathcal{Q}_{l-1,l} X_{l,l+1}.$$

hold, the following lemma is straightforward.

Lemma 6.6.4. *For a bipartite symbolic matrix system $(\mathcal{M}, I)$ as above, set*

$$\mathcal{P}_l = \mathcal{P}_{l,l+1}, \quad \mathcal{Q}_l = \mathcal{Q}_{l,l+1}, \quad X_l = X_{l,l+1}, \quad X'_l = X'_{l,l+1}$$

and

$$\mathcal{M}^{CD}_{l,l+1} = \mathcal{P}_{2l} \mathcal{Q}_{2l+1}, \qquad \mathcal{M}^{DC}_{l,l+1} = \mathcal{Q}_{2l} \mathcal{P}_{2l+1},$$
$$I^{CD}_{l,l+1} = X_{2l} X_{2l+1}, \qquad I^{DC}_{l,l+1} = X'_{2l} X'_{2l+1}.$$

Then both pairs $(\mathcal{M}^{CD}, I^{CD})$ and $(\mathcal{M}^{DC}, I^{DC})$ are symbolic matrix systems over $C \cdot D$ and $D \cdot C$, respectively, and they are properly strong shift equivalent in 1-step.

Definition 6.6.5. A λ-graph system (V, E, λ, ι) over Σ is said to be *bipartite* if there exist disjoint subsets $C, D \subset \Sigma$ such that $\Sigma = C \cup D$ and disjoint subsets $V_l^C, V_l^D \subset V_l$ for each $l \in \mathbb{N}$ such that $V_l^C \cup V_l^D = V_l$ and

6 Application of Infinite Labeled Graphs to Symbolic Dynamical Systems 149

1. for each $e \in E_{l,l+1}$

$$s(e) \in V_l^D, \quad r(e) \in V_{l+1}^C \quad \text{if and only if} \quad \lambda(e) \in C,$$
$$s(e) \in V_l^C, \quad r(e) \in V_{l+1}^D \quad \text{if and only if} \quad \lambda(e) \in D,$$

2. $\iota(V_{l+1}^D) = V_l^D, \quad \iota(V_{l+1}^C) = V_l^C.$

Lemma 6.6.6. *A symbolic matrix system is bipartite if and only if its corresponding λ-graph system is bipartite.*

Nasu introduced the notion of bipartite subshift in [47, 48]. A subshift Λ over alphabet Σ is said to be bipartite if there exist disjoint subsets $C, D \subset \Sigma$ such that any $(x_i)_{i \in \mathbb{Z}} \in \Lambda$ is either

$$x_i \in C \text{ and } x_{i+1} \in D \text{ for all } i \in \mathbb{Z} \quad \text{or} \quad x_i \in D \text{ and } x_{i+1} \in C \text{ for all } i \in \mathbb{Z}.$$

Let $\Lambda^{(2)}$ be the 2-higher power shift for Λ. Put

$$\Lambda_{CD} = \{(c_i d_i)_{i \in \mathbb{Z}} \in \Lambda^{(2)} | c_i \in C, d_i \in D\},$$
$$\Lambda_{DC} = \{(d_i c_i)_{i \in \mathbb{Z}} \in \Lambda^{(2)} | c_i \in C, d_i \in D\}.$$

They are subshifts over alphabets $C \cdot D$ and $D \cdot C$, respectively. Hence $\Lambda^{(2)}$ is partitioned into the two subshifts Λ_{CD} and Λ_{DC}.

Lemma 6.6.7. *A subshift Λ is bipartite if and only if its canonical symbolic matrix system $(\mathcal{M}^\Lambda, I^\Lambda)$ is bipartite.*

Proof. It is clear that a bipartite canonical symbolic matrix system gives rise to a bipartite subshift from the preceding lemma. Suppose that Λ is bipartite with respect to alphabets C, D. Denote by $(V^\Lambda, E^\Lambda, \lambda^\Lambda, \iota^\Lambda)$ the canonical λ-graph system $\mathfrak{L}^\Lambda$ for Λ. As in the construction of the canonical λ-graph system, the vertex set V_l^Λ is the set of all l-past equivalence classes $\{F_i^l\}_{i=1,\ldots,m(l)}$. Put

$$V_l^C = \{F_i^l | x_1 \in D \text{ for all } (x_1, x_2, \ldots) \in F_i^l\},$$
$$V_l^D = \{F_i^l | x_1 \in C \text{ for all } (x_1, x_2, \ldots) \in F_i^l\}$$

so that we have a disjoint union $V_l^C \cup V_l^D = V_l^\Lambda$ to yield a bipartite decomposition of $\mathfrak{L}^\Lambda$. $\qquad\square$

Let Λ be a bipartite subshift over Σ with respect to alphabets C, D. We have two symbolic matrix systems $(\mathcal{M}^{CD}, I^{CD})$ and $(\mathcal{M}^{DC}, I^{DC})$ over $C \cdot D$ and $D \cdot C$ from the bipartite canonical symbolic matrix system $(\mathcal{M}^\Lambda, I^\Lambda)$ for Λ, respectively. They are naturally identified with the canonical symbolic matrix systems for the subshifts Λ_{CD} and Λ_{DC}, respectively.

Corollary 6.6.8. *For a bipartite subshift Λ with respect to alphabets C, D, we have*

$$(\mathcal{M}^{CD}, I^{CD}) \underset{1-pr}{\approx} (\mathcal{M}^{DC}, I^{DC})$$

a properly strong shift equivalence in 1-step.

The following notion of bipartite conjugacy has been introduced by Nasu in [47, 48]. The conjugacy from Λ_{CD} onto Λ_{DC} that maps $(c_i d_i)_{i \in \mathbb{Z}}$ to $(d_i c_{i+1})_{i \in \mathbb{Z}}$ is called the forward bipartite conjugacy. The conjugacy from Λ_{CD} onto Λ_{DC} that maps $(c_i d_i)_{i \in \mathbb{Z}}$ to $(d_{i-1} c_i)_{i \in \mathbb{Z}}$ is called the backward bipartite conjugacy. A topological conjugacy between subshifts is called a symbolic conjugacy if it is a 1-block map given by a bijection between the underlying alphabets of the subshifts. Nasu proved the following factorization theorem.

Lemma 6.6.9 ([47]). *Any topological conjugacy ψ between subshifts is factorized into a composition of the form*

$$\psi = \kappa_n \zeta_n \kappa_{n-1} \zeta_{n-1} \cdots \kappa_1 \zeta_1 \kappa_0$$

where $\kappa_0, \ldots, \kappa_n$ are symbolic conjugacies and $\zeta_1, \ldots, \zeta_n$ are either forward or backward bipartite conjugacies.

Thanks to the above Nasu's result, we reach the following theorem.

Theorem 6.6.10. *For two subshifts Λ and Λ', let $(\mathcal{M}, I)$, and $(\mathcal{M}', I')$ be their canonical symbolic matrix systems for Λ and Λ', respectively. If Λ and Λ' are topologically conjugate, the symbolic matrix systems $(\mathcal{M}, I)$ and $(\mathcal{M}', I')$ are properly strong shift equivalent.*

Conversely, assume that two symbolic matrix systems $(\mathcal{M}, I)$ over Σ and $(\mathcal{M}', I')$ over Σ' are properly strong shift equivalent in 1-step. Let $\mathfrak{L}$ and $\mathfrak{L}'$ be their associated λ-graph systems for $(\mathcal{M}, I)$ and $(\mathcal{M}', I')$, respectively.

Lemma 6.6.11. *Let $\varphi : \Sigma \to C \cdot D$ and $\phi : \Sigma' \to D \cdot C$ be specifications that give a properly strong shift equivalence in 1-step between $(\mathcal{M}, I)$ and $(\mathcal{M}', I')$. For a word $x_1 x_2 \in B_2(\Lambda_\mathfrak{L})$ of the presented subshift $\Lambda_\mathfrak{L}$, put $\varphi(x_i) = c_i d_i, i = 1, 2$ where $c_i \in C, d_i \in D$. Then there uniquely exists $y_0 \in \Sigma'$ such that $\phi(y_0) = d_1 c_2$.*

Theorem 6.6.12. *If two symbolic matrix systems are properly strong shift equivalent, their presented subshifts are topologically conjugate.*

Proof. Suppose $(\mathcal{M}, I) \underset{1-pr}{\approx} (\mathcal{M}', I')$. We use the same notation as in the definition of properly strong shift equivalence. Set $\Lambda = \Lambda_\mathfrak{L}$ and $\Lambda' = \Lambda_{\mathfrak{L}'}$. By the preceding lemma, we have a 2-block map Φ from $B_2(\Lambda)$ to Σ' defined by $\Phi(x_1 x_2) = y_0$ where $\phi(y_0) = d_1 c_2$ and $\varphi(x_i) = c_i d_i, i = 1, 2$. Let Φ_∞ be the sliding block code induced by Φ so that Φ_∞ is a map from Λ to $\Sigma'^{\mathbb{Z}}$. We also write as Φ the map from Λ^* to the set of all words of Σ' defined by

6 Application of Infinite Labeled Graphs to Symbolic Dynamical Systems 151

$$\Phi(x_1 x_2 \cdots x_n) = \Phi(x_1 x_2)\Phi(x_2 x_3)\cdots \Phi(x_{n-1} x_n).$$

To prove $\Phi_\infty(\Lambda) \subset \Lambda'$, it suffices to show that for any word w in Λ, $\Phi(w)$ is an admissible word in Λ'. For $w = w_1 w_2 \cdots w_n \in B_n(\Lambda)$ and any fixed $l \geq n + 1$, we find $j = 1, 2, \ldots, m(l + n)$ and $k = 1, 2, \ldots, m(l)$ such that w appears in $\mathcal{M}_{l,l+n}(k, j)$, where $\mathcal{M}_{l,l+n} = \mathcal{M}_{l,l+1} \cdots \mathcal{M}_{l+n-1,l+n}$. Take $i = 1, 2, \ldots, m(l - n)$ with $I_{l-n,l}(i, k) = 1$, where $I_{l-n,l} = I_{l-n,l-n+1} \cdots I_{l-1,l}$. Hence w appears in $I_{l-n,l}\mathcal{M}_{l,l+n}(i, j)$. Put $\varphi(w_i) = c_i d_i, i = 1, 2, \ldots, n$. By the equality

$$I_{l-1,l}\mathcal{M}_{l,l+n} \overset{\varphi}{\simeq} X_{2l-2}\mathcal{P}_{2l-1}\mathcal{Q}_{2l}\mathcal{P}_{2l+1}\mathcal{Q}_{2l+2}\cdots \mathcal{P}_{2l+2n-3}\mathcal{Q}_{2l+2n-2}X_{2l+2n-1},$$

the word $d_1 c_2 d_2 c_3 \cdots d_{n-1} c_n$ appears in a component of $\mathcal{Q}_{2l}\mathcal{P}_{2l+1}\mathcal{Q}_{2l+2}\cdots$ $\mathcal{P}_{2l+2n-3}$. Hence the word $\phi^{-1}(d_1 c_2)\phi^{-1}(d_2 c_3)\cdots \phi^{-1}(d_{n-1}c_n)$ appears in a component of $\mathcal{M}'_{l,l+1} \cdot \mathcal{M}'_{l+1,l+2} \cdots \mathcal{M}'_{l+n-2,l+n-1}$. Thus $\Phi(w)$ is an admissible word in Λ' so that the sliding block code Φ_∞ maps Λ to Λ'. Similarly, we can construct a sliding block code Ψ_∞ from Λ' to Λ that is an inverse of Φ_∞. $\qquad\square$

Properly strong shift equivalence exactly corresponds to a finite sequence of bipartite decompositions of symbolic matrix systems and λ-graph systems. The definition of properly strong shift equivalence for symbolic matrix systems, however, needs rather more complicated formulations than that of strong shift equivalence for nonnegative matrices. We next introduce the notion of strong shift equivalence between two symbolic matrix systems that is a simpler and weaker condition than properly strong shift equivalence. Let $(\mathcal{M}, I)$ and $(\mathcal{M}', I')$ be two symbolic matrix systems over alphabet Σ and Σ', respectively. Let $m(l)$ and $m'(l)$ be the sequences for which $\mathcal{M}_{l,l+1}, I_{l,l+1}$ are $m(l) \times m(l + 1)$ matrices and $\mathcal{M}'_{l,l+1}, I'_{l,l+1}$ are $m'(l) \times m'(l + 1)$ matrices, respectively.

Definition 6.6.13. Two symbolic matrix systems $(\mathcal{M}, I)$ and $(\mathcal{M}', I')$ are said to be *strong shift equivalent in 1-step*, written as $(\mathcal{M}, I) \underset{1-st}{\approx} (\mathcal{M}', I')$, if there exist alphabets C, D and specifications

$$\varphi : \Sigma \to C \cdot D, \qquad \phi : \Sigma' \to D \cdot C$$

such that for each $1 < l \in \mathbb{N}$, there exist an $m(l - 1) \times m'(l)$ matrix $\mathcal{H}_l$ over C and an $m'(l - 1) \times m(l)$ matrix $\mathcal{K}_l$ over D satisfying the following equations:

$$I_{l-1,l}\mathcal{M}_{l,l+1} \overset{\varphi}{\simeq} \mathcal{H}_l \mathcal{K}_{l+1}, \qquad I'_{l-1,l}\mathcal{M}'_{l,l+1} \overset{\phi}{\simeq} \mathcal{K}_l \mathcal{H}_{l+1}$$

and

$$\mathcal{H}_l I'_{l,l+1} = I_{l-1,l}\mathcal{H}_{l+1}, \qquad \mathcal{K}_l I_{l,l+1} = I'_{l-1,l}\mathcal{K}_{l+1}.$$

Two symbolic matrix systems $(\mathcal{M}, I)$ and $(\mathcal{M}', I')$ are said to be *strong shift equivalent in N-step*, written as $(\mathcal{M}, I) \underset{N-st}{\approx} (\mathcal{M}', I')$, if there exist symbolic matrix systems $(\mathcal{M}^{(i)}, I^{(i)}), i = 1, 2, \ldots, N - 1$ such that

$$(\mathcal{M}, I) \underset{1-st}{\approx} (\mathcal{M}^{(1)}, I^{(1)}) \underset{1-st}{\approx} \cdots \underset{1-st}{\approx} (\mathcal{M}^{(N-1)}, I^{(N-1)}) \underset{1-st}{\approx} (\mathcal{M}', I').$$

We simply call it a *strong shift equivalence*. Similarly to the case of properly strong shift equivalence, we see that strong shift equivalence on symbolic matrix systems is an equivalence relation.

Proposition 6.6.14. *Properly strong shift equivalence in 1-step implies strong shift equivalence in 1-step.*

Proof. Let $\mathcal{P}_l, \mathcal{Q}_l, X_l$, and X'_l be the matrices in the definition of properly strong shift equivalence in 1-step between $(\mathcal{M}, I)$ and $(\mathcal{M}', I')$. We set

$$\mathcal{H}_l = X_{2l-1}\mathcal{P}_{2l-1}, \qquad \mathcal{K}_l = X'_{2l-1}\mathcal{Q}_{2l-1}.$$

They give rise to a strong shift equivalence in 1-step between $(\mathcal{M}, I)$ and $(\mathcal{M}', I')$. $\square$

Conversely we have

Proposition 6.6.15 ([41]). *Let $(\mathcal{M}, I)$ and $(\mathcal{M}', I')$ be the symbolic matrix systems for λ-graph systems $\mathfrak{L}$ and $\mathfrak{L}'$, respectively. Suppose that both $\mathfrak{L}$ and $\mathfrak{L}'$ are left-resolving and predecessor-separated. The following conditions are equivalent:*

1. $(\mathcal{M}, I)$ *and* $(\mathcal{M}', I')$ *are strong shift equivalent in 1-step.*
2. $(\mathcal{M}, I)$ *and* $(\mathcal{M}', I')$ *are properly strong shift equivalent in 1-step.*

Hence the two notions, strong shift equivalence and properly strong shift equivalence, coincide with each other in the canonical symbolic matrix systems.

By a similar argument to the proof of Theorem 6.6.12, we obtain

Theorem 6.6.16. *If two symbolic matrix systems (not necessarily canonical) are strong shift equivalent, their presented subshifts are topologically conjugate.*

We next introduce the notion of shift equivalence between two symbolic matrix systems as a generalization of Williams's notion for nonnegative matrices [52] and Boyle–Krieger's notion for symbolic matrices [4]. Let $(\mathcal{M}, I)$ and $(\mathcal{M}', I')$ be two symbolic matrix systems over alphabets Σ and Σ', respectively. For $N \in \mathbb{N}$, we put $(\Sigma)^N = \Sigma \cdots \Sigma$ and $(\Sigma')^N = \Sigma' \cdots \Sigma'$: the N-times products.

Definition 6.6.17. For $N \in \mathbb{N}$, two symbolic matrix systems $(\mathcal{M}, I)$ and $(\mathcal{M}', I')$ are said to be *shift equivalent of lag N* if there exist alphabets C_N, D_N and specifications

$$\varphi_1 : \Sigma \cdot C_N \to C_N \cdot \Sigma', \qquad \varphi_2 : \Sigma' \cdot D_N \to D_N \cdot \Sigma$$

6 Application of Infinite Labeled Graphs to Symbolic Dynamical Systems 153

and

$$\psi_1 : (\Sigma)^N \to C_N \cdot D_N, \qquad \psi_2 : (\Sigma')^N \to D_N \cdot C_N$$

such that for each $l \in \mathbb{N}$, there exist an $m(l) \times m'(l + N)$ matrix $\mathcal{H}_l$ over C_N and an $m'(l) \times m(l + N)$ matrix $\mathcal{K}_l$ over D_N satisfying the following equations:

$$\mathcal{M}_{l,l+1}\mathcal{H}_{l+1} \overset{\varphi_1}{\simeq} \mathcal{H}_l \mathcal{M}'_{l+N,l+N+1}, \qquad \mathcal{M}'_{l,l+1}\mathcal{K}_{l+1} \overset{\varphi_2}{\simeq} \mathcal{K}_l \mathcal{M}_{l+N,l+N+1},$$

$$I_{l,l+N}\mathcal{M}_{l+N,l+2N} \overset{\psi_1}{\simeq} \mathcal{H}_l \mathcal{K}_{l+N}, \qquad I'_{l,l+N}\mathcal{M}'_{l+N,l+2N} \overset{\psi_2}{\simeq} \mathcal{K}_l \mathcal{H}_{l+N}.$$

and

$$I_{l,l+1}\mathcal{H}_{l+1} = \mathcal{H}_l I'_{l+N,l+N+1}, \qquad I'_{l,l+1}\mathcal{K}_{l+1} = \mathcal{K}_l I_{l+N,l+N+1}.$$

We denote this situation by

$$(\mathcal{M}, I) \underset{lag N}{\sim} (\mathcal{M}', I') \quad \text{or} \quad (\mathcal{H}, \mathcal{K}) : (\mathcal{M}, I) \underset{lag N}{\sim} (\mathcal{M}', I')$$

and simply call it a shift equivalence.

Similarly to the case of nonnegative matrices and symbolic matrices, we see that

1. $(\mathcal{M}, I) \underset{lag N}{\sim} (\mathcal{M}', I')$ implies $(\mathcal{M}, I) \underset{lag L}{\sim} (\mathcal{M}', I')$ for all $L \geq N$.
2. $(\mathcal{M}, I) \underset{lag N}{\sim} (\mathcal{M}', I')$ and $(\mathcal{M}', I') \underset{lag N'}{\sim} (\mathcal{M}'', I'')$ imply $(\mathcal{M}, I) \underset{lag N+N'}{\sim}$ $(\mathcal{M}'', I'')$.

Hence shift equivalence is an equivalence relation on symbolic matrix systems.

Proposition 6.6.18. *Strong shift equivalence in N-step implies shift equivalence of lag N in symbolic matrix systems.*

For a subshift (Λ, σ) over Σ, its n-higher power shift $(\Lambda^{(n)}, \sigma)$ is defined to be the subshift (Λ, σ^n) over $(\Sigma)^n$ (cf. [33]). Two subshifts are called eventually conjugate if their n-higher power shifts are topologically conjugate for all large enough n [19, 52]. Williams and Kim and Roush showed that two square nonnegative matrices are shift equivalent if and only if the associated topological Markov shifts are eventually conjugate. Boyle and Krieger generalized their result to symbolic matrices and sofic subshifts [4]. For a symbolic matrix system $(\mathcal{M}, I)$, let Λ be the presented subshift. We set for $l \in \mathbb{N}$,

$$I^n_{l,l+1} = I_{nl,nl+1} \cdots I_{nl+n-1,nl+n}, \qquad \mathcal{M}^n_{l,l+1} = \mathcal{M}_{nl,nl+1} \cdots \mathcal{M}_{nl+n-1,nl+n}.$$

Then $(\mathcal{M}^n, I^n)$ becomes a symbolic matrix system whose presented subshift is the n-higher power shift $\Lambda^{(n)}$ of Λ.

Proposition 6.6.19. *If symbolic matrix systems* $(\mathcal{M}, I)$ *and* $(\mathcal{M}', I')$ *are shift equivalent, their presented subshifts* Λ *and* Λ' *are eventually conjugate.*

Proof. Assume $(\mathcal{H}, \mathcal{K}) : (\mathcal{M}, I) \underset{lagN}{\sim} (\mathcal{M}', I')$. For a number $K \in \mathbb{N}$, put $n = K + N$. Let C_N, D_N be the alphabets as in the definition of shift equivalence. Set $C = C_N, D = D_N \cdot (\Sigma)^K$. There are natural specifications $(\Sigma)^n \to C \cdot D$ and $(\Sigma')^n \to D \cdot C$ by using the specifications in the shift equivalence between $(\mathcal{M}, I)$ and $(\mathcal{M}', I')$. Put the matrices

$$\mathcal{P}_l = \mathcal{H}_{nl-n} I'_{nl-K, nl-K+1} I'_{nl-K+1, nl-K+2} \cdots I'_{nl-1, nl},$$
$$\mathcal{Q}_l = \mathcal{K}_{nl-n} \mathcal{M}_{nl-K, nl-K+1} \mathcal{M}_{nl-K+1, nl-K+2} \cdots \mathcal{M}_{nl-1, nl}.$$

They are an $m(nl-n) \times m'(nl)$ matrix over C and an $m'(nl-n) \times m(nl)$ matrix over D, respectively. They yield a strong shift equivalence in 1-step between $(\mathcal{M}^n, I^n)$ and $(\mathcal{M}'^n, I'^n)$ so that their presented subshifts are topologically conjugate. $\square$

Properly shift equivalences $(\mathcal{M}, I) \underset{N-pr}{\sim} (\mathcal{M}', I')$ are defined in [36, 41]. They are slightly stronger than shift equivalence and weaker than properly strong shift equivalence.

6.7 Nonnegative Matrix Systems

In this section, we introduce the notion of a nonnegative matrix system that is also a generalization of nonnegative matrices. We then generalize strong shift equivalent and shift equivalence between nonnegative matrices to between nonnegative matrix systems. Let $(A_{l,l+1}, I_{l,l+1}), l \in \mathbb{N}$ be a pair of sequences of rectangular matrices such that the following conditions for each $l \in \mathbb{N}$ are satisfied:

1. $A_{l,l+1}$ is an $m(l) \times m(l+1)$ rectangular matrix with entries in nonnegative integers.
2. $I_{l,l+1}$ is an $m(l) \times m(l+1)$ rectangular matrix with entries in $\{0, 1\}$ satisfying the relation:

$$I_{l,l+1} A_{l+1,l+2} = A_{l,l+1} I_{l+1,l+2}, \qquad l \in \mathbb{N}. \tag{6.5}$$

We further assume that both the matrices $A_{l,l+1}$ and $I_{l,l+1}$ have no zero columns and no zero rows. For j, there uniquely exists i such that $I_{l,l+1}(i, j) \neq 0$. The above conditions imply that $m(l) \leq m(l+1)$. The pair (A, I) is called a *nonnegative matrix system*. The following is basic.

Lemma 6.7.1. *For a symbolic matrix system* $(\mathcal{M}, I)$, *let* $M_{l,l+1}$ *be the* $m(l) \times m(l+1)$ *nonnegative matrix obtained from* $\mathcal{M}_{l,l+1}$ *by setting all the symbols equal to* 1. *Then the resulting pair* (M, I) *becomes a nonnegative matrix system.*

6 Application of Infinite Labeled Graphs to Symbolic Dynamical Systems   155

We write the matrices above as $\text{supp}(\mathcal{M}_{l,l+1}) = M_{l,l+1}$ and call $M_{l,l+1}$ the support of $\mathcal{M}_{l,l+1}$. The pair (M, I) is also called the support of $(\mathcal{M}, I)$ or the nonnegative matrix system associated with $(\mathcal{M}, I)$.

Definition 6.7.2. Two nonnegative matrix systems (A, I) and (A', I') are said to be *strong shift equivalent in 1-step*, written as $(A, I) \underset{1-st}{\approx} (A', I')$, if for each $1 < l \in \mathbb{N}$, there exist an $m(l - 1) \times m'(l)$ nonnegative matrix H_l and an $m'(l - 1) \times m(l)$ nonnegative matrix K_l satisfying the equations:

$$I_{l-1,l} A_{l,l+1} = H_l K_{l+1}, \qquad I'_{l-1,l} A'_{l,l+1} = K_l H_{l+1}$$

and

$$H_l I'_{l,l+1} = I_{l-1,l} H_{l+1}, \qquad K_l I_{l,l+1} = I'_{l-1,l} K_{l+1}.$$

Two nonnegative matrix systems (A, I) and (A', I') are said to be *strong shift equivalent in N-step*, written as $(A, I) \underset{N-st}{\approx} (A', I')$, if there exist nonnegative matrix systems $(A^{(i)}, I^{(i)}), i = 1, 2, \ldots, N - 1$ such that

$$(A, I) \underset{1-st}{\approx} (A^{(1)}, I^{(1)}) \underset{1-st}{\approx} \cdots \underset{1-st}{\approx} (A^{(N-1)}, I^{(N-1)}) \underset{1-st}{\approx} (A', I').$$

We simply call it a strong shift equivalence.

This formulation is also a generalization of Williams's strong shift equivalent between nonnegative matrices [52]. Similarly to symbolic matrix systems, strong shift equivalence is an equivalence relation on nonnegative matrix systems. We directly have

Proposition 6.7.3. *If two symbolic matrix systems are strong shift equivalent (in N-step), then the associated nonnegative matrix systems are strong shift equivalent (in N-step).*

For a nonnegative matrix system (A, I), the transpose $I^t_{l,l+1}$ of the matrix $I_{l,l+1}$ naturally induces an ordered homomorphism from $\mathbb{Z}^{m(l)}$ to $\mathbb{Z}^{m(l+1)}$, where the positive cone $\mathbb{Z}^{m(l)}_+$ of the group $\mathbb{Z}^{m(l)}$ is defined by

$$\mathbb{Z}^{m(l)}_+ = \{(n_1, n_2, \ldots, n_{m(l)}) \in \mathbb{Z}^{m(l)} | n_i \geq 0, i = 1, 2 \ldots m(l)\}.$$

We put the inductive limits:

$$\mathbb{Z}_{I^t} = \varinjlim\{I^t_{l,l+1} : \mathbb{Z}^{m(l)} \to \mathbb{Z}^{m(l+1)}\},$$
$$\mathbb{Z}^+_{I^t} = \varinjlim\{I^t_{l,l+1} : \mathbb{Z}^{m(l)}_+ \to \mathbb{Z}^{m(l+1)}_+\}.$$

For each $l \in \mathbb{N}$, the homomorphism $I^t_{l,l+1} : \mathbb{Z}^{m(l)} \to \mathbb{Z}^{m(l+1)}$ is injective. Hence the canonical homomorphism $\iota_l : \mathbb{Z}^{m(l)} \to \mathbb{Z}_{I^t}$ is injective. By the relation (6.5), the sequence of the transposed matrices $A^t_{l,l+1}, l \in \mathbb{N}$ of $A_{l,l+1}, l \in \mathbb{N}$ yields an endomorphism of the ordered group $\mathbb{Z}_{I^t}$. We write it as $\lambda_{(A,I)}$.

Definition 6.7.4. For nonnegative matrix systems (A, I) and (A', I') and $L \in \mathbb{N}$, a homomorphism ξ from the group $\mathbb{Z}_{I^t}$ to the group $\mathbb{Z}_{I'^t}$ is called *a finite homomorphism of lag L* if it satisfies the condition $\xi(\mathbb{Z}^{m(l)}) \subset \mathbb{Z}^{m'(l+L)}$ for all $l \in \mathbb{N}$, where $\mathbb{Z}^{m(l)}$ and $\mathbb{Z}^{m'(l)}$ are naturally embedded into $\mathbb{Z}_{I^t}$ and $\mathbb{Z}_{I'^t}$, respectively.

Proposition 6.7.5. *Two nonnegative matrix systems (A, I) and (A', I') are strong shift equivalent in 1-step if and only if there exist order-preserving finite homomorphisms of lag 1: $\xi : \mathbb{Z}_{I^t} \to \mathbb{Z}_{I'^t}$ and $\eta : \mathbb{Z}_{I'^t} \to \mathbb{Z}_{I^t}$ such that*

$$\eta \circ \xi = \lambda_{(A,I)}, \qquad \xi \circ \eta = \lambda_{(A',I')}.$$

For a nonnegative matrix system (A, I), we set the $m(l) \times m(l+k)$ matrices:

$$I_{l,l+k} = I_{l,l+1} \cdots I_{l+k-1,l+k}, \qquad A_{l,l+k} = A_{l,l+1} \cdots A_{l+k-1,l+k}$$

for each $l, k \in \mathbb{N}$.

Definition 6.7.6. Two nonnegative matrix systems (A, I) and (A', I') are said to be *shift equivalent of lag N* if for each $l \in \mathbb{N}$, there exist an $m(l) \times m'(l+N)$ nonnegative matrix H_l and an $m'(l) \times m(l+N)$ nonnegative matrix K_l satisfying the equations:

$$\begin{aligned} A_{l,l+1} H_{l+1} &= H_l A'_{l+N,l+N+1}, & A'_{l,l+1} K_{l+1} &= K_l A_{l+N,l+N+1}, \\ H_l K_{l+N} &= I_{l,l+N} A_{l+N,l+2N}, & K_l H_{l+N} &= I'_{l,l+N} A'_{l+N,l+2N} \end{aligned}$$

and

$$I_{l,l+1} H_{l+1} = H_l I'_{l+N,l+N+1}, \qquad I'_{l,l+1} K_{l+1} = K_l I_{l+N,l+N+1}.$$

We write this situation as

$$(A, I) \underset{lagN}{\sim} (A', I') \quad \text{or} \quad (H, K) : (A, I) \underset{lagN}{\sim} (A', I')$$

and simply call it a shift equivalence. This formulation is a generalization of Williams's shift equivalence between nonnegative matrices ([52], see also [4]). Shift equivalence is an equivalence relation on nonnegative matrix systems. Similarly to symbolic matrix systems, strong shift equivalence in N-step implies shift equivalence of lag N in nonnegative matrix systems. As in the case of strong shift equivalence, we may describe shift equivalence on nonnegative matrix systems in terms of single homomorphisms between inductive limits of Abelian groups.

6 Application of Infinite Labeled Graphs to Symbolic Dynamical Systems 157

Proposition 6.7.7. *Two nonnegative matrix systems (A, I) and $(A'I')$ are shift equivalent of lag N if and only if there exist order-preserving finite homomorphisms of lag N: $\xi : \mathbb{Z}_{I'} \to \mathbb{Z}_{I''}$ and $\eta : \mathbb{Z}_{I''} \to \mathbb{Z}_{I'}$ such that*

$$\lambda_{(A',I')} \circ \xi = \xi \circ \lambda_{(A,I)}, \qquad \lambda_{(A,I)} \circ \eta = \eta \circ \lambda_{(A',I')} \tag{6.6}$$

and

$$\eta \circ \xi = \lambda^N_{(A,I)}, \qquad \xi \circ \eta = \lambda^N_{(A',I')}. \tag{6.7}$$

Let $(\mathcal{M}, I), (\mathcal{M}', I')$ be symbolic matrix systems and $(M, I), (M', I')$ be their supports, respectively. The following proposition is direct.

Proposition 6.7.8. *1. $(\mathcal{M}, I) \underset{n-st}{\approx} (\mathcal{M}', I')$ implies $(M, I) \underset{n-st}{\approx} (M', I')$.*
2. $(\mathcal{M}, I) \underset{lagN}{\sim} (\mathcal{M}', I')$ implies $(M, I) \underset{lagN}{\sim} (M', I')$.

6.8 Dimension Groups, K-Groups, and Bowen–Franks Groups

In this section, we study several shift equivalence invariants of nonnegative matrix systems. They are the dimension group, K-groups, and Bowen–Franks groups. The dimension group for a nonnegative matrix system is a generalization of the dimension group for a nonnegative matrix defined by W. Krieger in [23, 24, 35]. Krieger's idea to define dimension groups for nonnegative matrices is based on the K-theory for C^*-algebras (cf. [12, 50]). Let (A, I) be a nonnegative matrix system. We set $\mathbb{Z}_{I'}(k) = \mathbb{Z}_{I'}$ and $\mathbb{Z}_{I'}^+(k) = \mathbb{Z}_{I'}^+$ for $k \in \mathbb{N}$. We define an Abelian group and its positive cone by the inductive limits:

$$\Delta_{(A,I)} = \varinjlim_k \{\lambda_{(A,I)} : \mathbb{Z}_{I'}(k) \to \mathbb{Z}_{I'}(k+1)\},$$

$$\Delta_{(A,I)}^+ = \varinjlim_k \{\lambda_{(A,I)} : \mathbb{Z}_{I'}^+(k) \to \mathbb{Z}_{I'}^+(k+1)\}.$$

We call the ordered group $(\Delta_{(A,I)}, \Delta_{(A,I)}^+)$ the *dimension group for (A, I)*. Since the map $\delta_{(A,I)} : \mathbb{Z}_{I'}(k) \to \mathbb{Z}_{I'}(k+1)$ defined by $\delta_{(A,I)}([X,k]) = ([X,k+1])$ for $X \in \mathbb{Z}_{I'}$ yields an automorphism on $\Delta_{(A,I)}$ that preserves the positive cone $\Delta_{(A,I)}^+$. We also denote it by $\delta_{(A,I)}$ and call it the *dimension automorphism*. The triple $(\Delta_{(A,I)}, \Delta_{(A,I)}^+, \delta_{(A,I)})$ is called the *dimension triple for (A, I)*.

Proposition 6.8.1. *If two nonnegative matrix systems are shift equivalent, their dimension triples are isomorphic.*

Proof. Suppose $(A, I) \underset{lag N}{\sim} (A', I')$. By Proposition 6.7.7, there exist order-preserving finite homomorphisms $\xi : \mathbb{Z}_{I^l} \to \mathbb{Z}_{I'^l}$ and $\eta : \mathbb{Z}_{I'^l} \to \mathbb{Z}_{I^l}$ of lag N satisfying (6.6) and (6.7). Define the maps $\Phi_\xi : \mathbb{Z}_{I^l}(k) \to \mathbb{Z}_{I'^l}(k)$ and $\Phi_\eta : \mathbb{Z}_{I'^l}(k) \to \mathbb{Z}_{I^l}(k)$ by $\Phi_\xi([X, k]) = ([\xi(X), k])$ and $\Phi_\eta([Y, k]) = ([\eta(Y), k])$ for $X \in \mathbb{Z}_{I^l}$, $Y \in \mathbb{Z}_{I'^l}$. They induce homomorphisms from $\Delta_{(A,I)}$ to $\Delta_{(A',I')}$ and $\Delta_{(A',I')}$ to $\Delta_{(A,I)}$, respectively. We still denote them by Φ_ξ and Φ_η, respectively. Since the homomorphisms ξ, η are order preserving, the maps Φ_ξ, Φ_η also preserve order structures of the dimension groups. It then follows that $\delta_{(A,I)} \circ \Phi_\eta = \Phi_\eta \circ \delta_{(A'I')}$, $\delta_{(A',I')} \circ \Phi_\xi = \Phi_\xi \circ \delta_{(A,I)}$, and $\Phi_\eta \circ \Phi_\xi = \delta^{-N}_{(A,I)}$, $\Phi_\xi \circ \Phi_\eta = \delta^{-N}_{(A',I')}$. Both the maps Φ_ξ and Φ_η are isomorphisms and the corresponding dimension triples are isomorphic. $\qquad \square$

We define the dimension triple $(\Delta_{(\mathcal{M},I)}, \Delta^+_{(\mathcal{M},I)}, \delta_{(\mathcal{M},I)})$ for a symbolic matrix system $(\mathcal{M}, I)$ as the dimension triple $(\Delta_{(M,I)}, \Delta^+_{(M,I)}, \delta_{(M,I)})$ for its support (M, I). We may also define the dimension triple $(\Delta_\Lambda, \Delta^+_\Lambda, \delta_\Lambda)$ for subshift Λ as the dimension triple for its canonical symbolic matrix system.

Proposition 6.8.2. *The dimension triples for subshifts are topological conjugacy invariants.*

For an $n \times n$ nonnegative matrix A, the Bowen–Franks group $BF(A)$ is defined by the group $\mathbb{Z}^n/(1 - A)\mathbb{Z}^n$ [3]. This group was discovered in a study of suspension flows of topological Markov shifts by Bowen and Franks ([3, 15], cf. [49]). They showed that the groups are not only an invariant under shift equivalence but also an almost complete invariant under flow equivalence of topological Markov shifts. We introduce and study the notion of Bowen–Franks groups for nonnegative matrix systems as a generalization of the original Bowen–Franks groups for nonnegative matrices. Our Bowen–Franks groups for a nonnegative matrix system consist of a pair of Abelian groups. One corresponds to a generalization of the original Bowen–Franks group, called the Bowen–Franks group of degree zero, and the other one corresponds to its suspension, called the Bowen–Franks group of degree one. For matrices, the latter group is the torsion-free part of the original Bowen–Franks group. For a nonnegative matrix system, the group of degree one is not necessarily the torsion-free part of the group of degree zero (see Sect. 6.10).

We introduce two Abelian groups for nonnegative matrix systems, called K-groups, that are invariant under shift equivalence. Let (A, I) be a nonnegative matrix system. For $l \in \mathbb{N}$, we set the Abelian groups

$$K^l_0(A, I) = \mathbb{Z}^{m(l+1)}/(I^t_{l,l+1} - A^t_{l,l+1})\mathbb{Z}^{m(l)},$$

$$K^l_1(A, I) = \mathrm{Ker}(I^t_{l,l+1} - A^t_{l,l+1}) \text{ in } \mathbb{Z}^{m(l)}.$$

By the relation (6.5), the map $I^t_{l,l+1} : \mathbb{Z}^{m(l)} \to \mathbb{Z}^{m(l+1)}$ naturally induces homomorphisms: $i^l_* : K^l_*(A, I) \to K^{l+1}_*(A, I)$ for $* = 0, 1$.

Definition 6.8.3. The *K-groups for* (A, I) are defined as the inductive limits of the Abelian groups:

6 Application of Infinite Labeled Graphs to Symbolic Dynamical Systems   159

$$K_0(A, I) = \varinjlim_l \{i_0^l : K_0^l(A, I) \to K_0^{l+1}(A, I)\},$$

$$K_1(A, I) = \varinjlim_l \{i_1^l : K_1^l(A, I) \to K_1^{l+1}(A, I)\}.$$

For a symbolic matrix system $(\mathcal{M}, I)$, its K-groups $K_0(\mathcal{M}, I), K_1(\mathcal{M}, I)$ are defined to be the K-groups for its support. The groups $K_*(A, I)$ are also represented in the following way.

Proposition 6.8.4.

1. $K_0(A, I) = \mathbb{Z}_{I'}/(id - \lambda_{(A,I)})\mathbb{Z}_{I'} = \Delta_{(A,I)}/(id - \delta_{(A,I)}) \Delta_{(A,I)}$,
2. $K_1(A, I) = \mathrm{Ker}(id - \lambda_{(A,I)})$ in $\mathbb{Z}_{I'} = \mathrm{Ker}(id - \delta_{(A,I)})$ in $\Delta_{(A,I)}$.

Since the dimension triple $(\Delta_{(A,I)}, \Delta_{(A,I)}^+, \delta_{(A,I)})$ is invariant under shift equivalence of nonnegative matrix systems, we thus have

Proposition 6.8.5. *The groups $K_i(A, I), i = 0, 1$ are invariant under shift equivalence of nonnegative matrix systems.*

Set the Abelian group

$$\mathbb{Z}_I = \varprojlim_l \{I_{l,l+1} : \mathbb{Z}^{m(l+1)} \to \mathbb{Z}^{m(l)}\},$$

the projective limit of the system: $I_{l,l+1} : \mathbb{Z}^{m(l+1)} \to \mathbb{Z}^{m(l)}, l \in \mathbb{N}$. The sequence $A_{l,l+1}, l \in \mathbb{N}$ naturally acts on $\mathbb{Z}_I$ as an endomorphism, denoted by A. The identity on $\mathbb{Z}_I$ is denoted by I.

Definition 6.8.6. For a nonnegative matrix system (A, I), set

$$BF^0(A, I) = \mathbb{Z}_I/(I - A)\mathbb{Z}_I, \qquad BF^1(A, I) = \mathrm{Ker}(I - A) \text{ in } \mathbb{Z}_I.$$

We call $BF^0(A, I)$ the Bowen–Franks group for (A, I) of degree zero and $BF^1(A, I)$ the Bowen–Franks group for (A, I) of degree one.

Theorem 6.8.7. *The Bowen–Franks groups $BF^i(A, I), i = 0, 1$ are invariant under shift equivalence of nonnegative matrix systems.*

Proof. Suppose $(H, K) : (A, I) \underset{lag N}{\sim} (A', I')$. Put $\Phi_K((x_i)_{i \in \mathbb{N}}) = (K_i (x_{N+i})_{i \in \mathbb{N}})$ for $(x_i)_{i \in \mathbb{N}} \in \mathbb{Z}_I$. The map Φ_K yields a homomorphism from $\mathbb{Z}_I$ to $\mathbb{Z}_{I'}$. By the equality: $K_i \circ (I_{N+i,N+i+1} - A_{N+i,N+i+1}) = (I'_{i,i+1} - A'_{i,i+1}) \circ K_{i+1}$, the homomorphism induces a homomorphism from $BF^0(A, I)$ to $BF^0(A', I')$. We denote it by $\bar{\Phi}_K$. We similarly have a homomorphism $\bar{\Phi}_H$ from $BF^0(A', I')$ to $BF^0(A, I)$. Since we have $\Phi_H \circ \Phi_K = A^N$ on $\mathbb{Z}_I$ and $\Phi_K \circ \Phi_H = A'^N$ on $\mathbb{Z}_{I'}$, the homomorphisms $\bar{\Phi}_H$ and $\bar{\Phi}_K$ are inverses of each other. The homomorphisms Φ_H and Φ_K also induce isomorphisms between $BF^1(A, I)$ and $BF^1(A', I')$.   □

We see the following universal coefficient theorem by using elementary homological algebra. It says that the Bowen–Franks groups are determined by the K-groups.

Theorem 6.8.8. *1. There exists a short exact sequence*

$$0 \longrightarrow \mathrm{Ext}^1_{\mathbb{Z}}(K_0(A, I), \mathbb{Z}) \xrightarrow{\delta} BF^0(A, I) \xrightarrow{\gamma} \mathrm{Hom}_{\mathbb{Z}}(K_1(A, I), \mathbb{Z}) \longrightarrow 0$$

that splits unnaturally.
2. $BF^1(A, I) \cong \mathrm{Hom}_{\mathbb{Z}}(K_0(A, I), \mathbb{Z})$.

In the above theorem, $\mathrm{Ext}^1_{\mathbb{Z}}$ is the derived functor of the Hom-functor in homological algebra. The formulations above come from the universal coefficient theorem for K-theory of the C^*-algebra $\mathcal{O}_{\mathfrak{L}}$ associated with the λ-graph system $\mathfrak{L}$ ([40], cf. [6]).

Remark 1. As $\mathrm{Ext}^1_{\mathbb{Z}}(K_1(A, I), \mathbb{Z}) = 0$, the following short exact sequence clearly holds by Theorem 6.8.8(2).

$$0 \longrightarrow \mathrm{Ext}^1_{\mathbb{Z}}(K_1(A, I), \mathbb{Z}) \xrightarrow{\delta} BF^1(A, I) \xrightarrow{\gamma} \mathrm{Hom}_{\mathbb{Z}}(K_0(A, I), \mathbb{Z}) \longrightarrow 0.$$

Example. Let M be an $n \times n$ nonnegative matrix. Put for each $l \in \mathbb{N}$

$$A_{l,l+1} = M, \qquad I_{l,l+1} = \text{ the } n \times n \text{ identity matrix.}$$

Then (A, I) is a nonnegative matrix system that satisfies

$$K_0(A, I) = \mathbb{Z}^n / (1 - M^t)\mathbb{Z}^n, \qquad K_1(A, I) = \mathrm{Ker}(1 - M^t) \text{ in } \mathbb{Z}^n,$$
$$BF^0(A, I) = \mathbb{Z}^n / (1 - M)\mathbb{Z}^n, \qquad BF^1(A, I) = \mathrm{Ker}(1 - M) \text{ in } \mathbb{Z}^n.$$

Hence we have

$$K_0(A, I) \cong BF^0(A, I) = BF(M) : \text{ the original Bowen–Franks group for } M,$$
$$K_1(A, I) \cong BF^1(A, I) = \text{ the torsion-free part of } BF(M).$$

Let (A^Λ, I^Λ) be the canonical nonnegative matrix system for a subshift Λ. Define

$$K_i(\Lambda) = K_i(A^\Lambda, I^\Lambda), i = 0, 1, \quad : \text{ the K-groups for } \Lambda,$$
$$BF^i(\Lambda) = BF^i(A^\Lambda, I^\Lambda), i = 0, 1, \quad : \text{ the Bowen–Franks groups for } \Lambda.$$

Theorem 6.8.9. *The K-groups $K_i(\Lambda)$ and the Bowen–Franks groups $BF^i(\Lambda)$ for subshift Λ are Abelian groups that are topological conjugacy invariants of subshifts.*

6 Application of Infinite Labeled Graphs to Symbolic Dynamical Systems 161

Suppose that Λ is a sofic shift. We denote by $m(\Lambda)$ the cardinality of the vertices of the left Krieger cover graph for Λ and A_Λ its adjacency matrix. Then

$$BF^0(\Lambda) = \mathbb{Z}^{m(\Lambda)}/(1 - A_\Lambda)\mathbb{Z}^{m(\Lambda)}, \qquad BF^1(\Lambda) = \text{Ker}(1 - A_\Lambda) \text{ in } \mathbb{Z}^{m(\Lambda)}.$$

The Bowen–Franks group for the nonnegative matrix was first invented for use as an invariant of flow equivalence of the associated topological Markov shift rather than topological conjugacy [3, 15, 49]. For a general subshift, we have

Theorem 6.8.10. *The K-groups $K_*(\Lambda)$ and hence the Bowen–Franks groups $BF^*(\Lambda)$ for subshifts are also invariant under flow equivalence of subshifts.*

The proof is seen in [38, 39] by using a result of Parry–Sullivan [49].

6.9 Spectrum, λ-Entropy, and Volume Entropy

It is well known that the set of all nonzero eigenvalues of a nonnegative matrix M is a shift equivalence invariant. The set of M is called the nonzero spectrum of M and plays an important rôle for studying dynamical properties of the associated topological Markov shift (cf. [21, 33]). In this section, we introduce the notion of spectrum of nonnegative matrix system (A, I). It is an eigenvalue of (A, I) in the sense stated below. We denote by $Sp(A, I)$ the set of all eigenvalues of (A, I). As the sequence of the sizes of matrices $A_{l,l+1}, I_{l,l+1}, l \in \mathbb{N}$ is increasing, it seems to be natural to deal with eigenvalues of (A, I) with a certain boundedness condition defined below on the corresponding eigenvectors. We fix a nonnegative matrix system (A, I) for a while. A sequence $\{v^l\}_{l \in \mathbb{N}}$ of vectors $v^l = (v_1^l, \ldots, v_{m(l)}^l) \in \mathbb{C}^{m(l)}, l \in \mathbb{N}$ is called an I-*compatible vector* if it satisfies the conditions:

$$v^l = I_{l,l+1}v^{l+1} \qquad \text{for all} \quad l \in \mathbb{N}. \tag{6.8}$$

An I-compatible vector $\{v^l\}_{l \in \mathbb{N}}$ is said to be nonzero if v^l is a nonzero vector for some l. If $v_i^l \geq 0$ for all $i = 1, \ldots, m(l)$ and $l \in \mathbb{N}$, $\{v^l\}_{l \in \mathbb{N}}$ is said to be nonnegative. If there exists a number M such that $\sum_{i=1}^{m(l)} |v_i^l| \leq M$ for all $l \in \mathbb{N}$, $\{v^l\}_{l \in \mathbb{N}}$ is said to be bounded. We remark that, for an I-compatible vector $\{v^l\}_{l \in \mathbb{N}}$, $v^N \neq 0$ for some N implies $v^l \neq 0$ for all $l \geq N$.

Definition 6.9.1. For a complex number β, a nonzero I-compatible vector $\{v^l\}_{l \in \mathbb{N}}$ is called an *eigenvector* of (A, I) for *eigenvalue* β if it satisfies the conditions:

$$A_{l,l+1}v^{l+1} = \beta v^l \qquad \text{for all} \quad l \in \mathbb{N}. \tag{6.9}$$

An eigenvalue β is said to be bounded if it is an eigenvalue for a bounded eigenvector. We denote by $Sp^\times(A, I)$ the set of all nonzero eigenvalues of (A, I) and by $Sp_b^\times(A, I)$ the set of all nonzero bounded eigenvalues of (A, I), respectively. We call them the nonzero spectrum of (A, I) and the nonzero bounded spectrum of (A, I), respectively.

Theorem 6.9.2. *If two nonnegative matrix systems are shift equivalent, their non-zero spectra coincide.*

Proof. Suppose $(H, K) : (A, I) \underset{lagN}{\sim} (A', I')$. We show $Sp^\times(A, I) \subset Sp^\times (A', I')$. For $\beta \in Sp^\times(A, I)$ with nonzero eigenvector v^l, we set $u^l = K_l v^{l+N}$ for $l \in \mathbb{N}$. It is direct to see that

$$u^l = I'_{l,l+1} u^{l+1}, \qquad A'_{l,l+1} u^{l+1} = \beta u^l.$$

Now if the vectors u^l are zero for all $l \geq l_0$ for some l_0, by the equality $H_l K_{l+N} v^{l+2N} = I_{l,l+N} A_{l+N,l+2N} v^{l+2N}$, we see

$$0 = A_{l,l+N} I_{l+N,l+2N} v^{l+2N} = A_{l,l+N} v^{l+N} = \beta v^l.$$

Thus $v^l = 0$ for all $l \geq l_0$ and hence for all $l \in \mathbb{N}$, a contradiction. Therefore β is a nonzero eigenvalue of (A', I'). $\square$

We next show that the nonzero bounded spectrum of (A, I) is also invariant under shift equivalence. Put

$$N_A = \max_j \sum_{i=1}^{m(l)} A_{l,l+1}(i, j)$$

for $l \in \mathbb{N}$. By the relation (6.5), the right-hand side $\max_j \sum_{i=1}^{m(l)} A_{l,l+1}(i, j)$ does not depend on the choice of $l \in \mathbb{N}$. For an I-compatible vector $\{v^l\}_{l \in \mathbb{N}}$, we put $\|v^l\| = \sum_{i=1}^{m(l)} |v_i^l|$. The sequence $\{\|v^l\|\}_{l \in \mathbb{N}}$ is increasing. If $\{v^l\}_{l \in \mathbb{N}}$ is nonnegative, $\{\|v^l\|\}_{l \in \mathbb{N}}$ is constant and hence $\{v^l\}_{l \in \mathbb{N}}$ is bounded. We set

$$\|v\|_1 = \limsup_{l \to \infty} \|v^l\|.$$

Proposition 6.9.3. $Sp_b^\times(A, I) \subset \{z \in \mathbb{C} | |z| \leq N_A\}$.

Proof. For $\beta \in Sp(A, I)$ with a bounded eigenvector $\{v^l\}_{l \in \mathbb{N}}$, we have

$$\beta \sum_{i=1}^{m(l)} |v_i^l| \leq \sum_{j=1}^{m(l+1)} \left(\max_j \sum_{i=1}^{m(l)} A_{l,l+1}(i, j) \right) |v_j^{l+1}|.$$

Hence the inequality $\beta \|v^l\| \leq N_A \|v^{l+1}\|$ holds. As $\{v^l\}_{l \in \mathbb{N}}$ is bounded, $\lim_{l \to \infty} \|v^l\| = \|v\|_1$ exists so that we have a desired assertion. $\square$

6 Application of Infinite Labeled Graphs to Symbolic Dynamical Systems

We denote by $\mathfrak{B}_I$ the set of all bounded I-compatible vectors. It is a complex Banach space with norm $\| \cdot \|_1$. A nonnegative I-compatible vector $v = \{v^l\}_{l \in \mathbb{N}}$ is called a *state* if $\|v\|_1 = 1$. Let $\mathfrak{S}_I$ be the set of all states. It is a convex subset of $\mathfrak{B}_I$. Any bounded I-compatible vector $v \in \mathfrak{B}_I$ can be expressed as a finite linear combinations of states. For $v \in \mathfrak{B}_I$, we put

$$(L_A v)_i^l = \sum_{j=1}^{m(l+1)} A_{l,l+1}(i, j) v_j^{l+1} \qquad \text{for} \quad i = 1, \ldots, m(l), \quad l \in \mathbb{N}.$$

Then L_A gives rise to a bounded linear operator on the Banach space $\mathfrak{B}_I$ that satisfies $\|L_A\| = N_A$, where the norm of L_A is given by $\|L_A\| = \sup_{v \neq 0} \frac{\|L_A v\|_1}{\|v\|_1}$. Therefore we have

Proposition 6.9.4. *For a complex number β, it belongs to $Sp_b(A, I)$ if and only if it satisfies $L_A v = \beta v$ for some nonzero $v \in \mathfrak{B}_I$. That is, the bounded spectra of (A, I) are nothing but the eigenvalues of the bounded linear operator L_A on the Banach space $\mathfrak{B}_I$.*

By a similar manner to the proof of Theorem 6.9.2, we have

Theorem 6.9.5. *If two nonnegative matrix systems are shift equivalent, their nonzero bounded spectra coincide.*

Let $\sigma(L_A)$ be the set of all spectra of L_A as a bounded linear operator on the Banach space $\mathfrak{B}_I$. The general theory of bounded linear operators tells us that the set $\sigma(L_A)$ is not empty. Let r_A be the spectral radius of the operator L_A on $\mathfrak{B}_I$; that is, $r_A = \sup\{|r| : r \in \sigma(L_A)\}$. By a proof similar to [46, Lemma 4.1], we have

Proposition 6.9.6. *There exists a state $v \in \mathfrak{S}_I$ such that $L_A v = r_A v$. Hence we have $r_A \in Sp_b^{\times}(A, I)$.*

It is well known that the topological entropy $h_{\text{top}}(\Lambda)$ for subshift Λ is given by

$$h_{\text{top}}(\Lambda) = \lim_{k \to \infty} \frac{1}{k} \log \sharp B_k(\Lambda)$$

where $\sharp B_k(\Lambda)$ denotes the cardinality of the set $B_k(\Lambda)$ of all admissible words of length k in the subshift Λ (cf. [21, 33]). We say a symbolic matrix system $(\mathcal{M}, I)$ is left-resolving if a symbol appearing in $\mathcal{M}(i, j)$ cannot appear in $\mathcal{M}(i', j)$ for other $i' \neq i$; equivalently, its λ-graph system is left-resolving. As in Sect. 6.5, the canonical symbolic matrix system is left-resolving.

Proposition 6.9.7. *Let $(\mathcal{M}, I)$ be a left-resolving symbolic matrix system and (M, I) its support. For any $\beta \in Sp_b(M, I)$, we have the inequalities:*

$$\log |\beta| \leq \log r_M \leq h_{top}(\Lambda)$$

where r_M is the spectral radius of the operator L_M on $\mathfrak{B}_I$ and Λ is the presented subshift by $(\mathcal{M}, I)$.

Proof. The inequality $\log|\beta| \le \log r_M$ is clear. Take $v \in \mathfrak{S}_I$ such that $L_M v = r_M v$. We have for $k \in \mathbb{N}$,

$$r_M^k v_i^1 = \sum_{j=1}^{m(k+1)} M_{1,k+1}(i,j) v_j^{k+1}.$$

As $\sum_{i=1}^{m(1)} v_i^1 = 1$, it follows that

$$r_M^k \le \left(\max_j \sum_{i=1}^{m(1)} M_{1,k+1}(i,j) \right) \sum_{j=1}^{m(k+1)} v_j^{k+1} = \|L_M^k\|.$$

We may find j_0 such that $\|L_M^k\| = \sum_{i=1}^{m(1)} M_{1,k+1}(i,j_0)$. Since $(\mathcal{M}, I)$ is left-resolving, the number $\sum_{i=1}^{m(1)} M_{1,k+1}(i,j_0)$ is majorized by $\sharp B_k(\Lambda)$. Thus we obtain the inequalities

$$r_M^k \le \|L_M^k\| \le \sharp B_k(\Lambda).$$

As $\|L_M^k\|^{\frac{1}{k}} \to r_M$ for $k \to \infty$, we have the desired inequalities. $\square$

For a subshift Λ, define the nonzero spectrum $Sp^\times(\Lambda)$ and the nonzero bounded spectrum $Sp_b^\times(\Lambda)$ of Λ by the nonzero spectrum and the nonzero bounded spectrum of the canonical nonnegative matrix systems (M, I) for Λ, respectively. We thus have

Theorem 6.9.8. *Both the sets $Sp^\times(\Lambda)$ and $Sp_b^\times(\Lambda)$ are not empty and topological conjugacy invariants of subshifts. In particular, $Sp_b^\times(\Lambda)$ is bounded by the topological entropy of the subshift Λ.*

Other entropic quantities for a λ-graph system $\mathfrak{L}$ have been introduced in [30,42]; the first one is called λ-entropy, written as $h_\lambda(\mathfrak{L})$, which measures a growth rate of the cardinalities of the vertex sets $\{V_l\}_{l \in \mathbb{N}}$. The second one is called the volume entropy, written as $h_{vol}(\mathfrak{L})$, which measures a growth rate of the cardinalities of the sets of labeled paths. Their definitions are as follows:

$$h_\lambda(\mathfrak{L}) = \limsup_{l \to \infty} \frac{1}{l} \log \sharp V_l.$$

We denote by $P_l(\mathfrak{L})$ the set of all labeled paths starting at a vertex in V_1 and terminating at a vertex in V_l, and by $^\sharp P_l(\mathfrak{L})$ its cardinality. The volume entropy is defined by the formula:

$$h_{vol}(\mathfrak{L}) = \limsup_{l \to \infty} \frac{1}{l} \log \sharp P_l(\mathfrak{L}).$$

Theorem 6.9.9. *Both $h_\lambda(\mathfrak{L})$ and $h_{vol}(\mathfrak{L})$ are invariant under shift equivalence of λ-graph systems.*

6 Application of Infinite Labeled Graphs to Symbolic Dynamical Systems 165

This implies that they yield topological conjugacy invariants of subshifts. We see that $h_{vol}(\mathcal{L})$ is given by the topological entropy of a topological dynamical system $(\sigma_{\mathcal{L}}, X_{\mathcal{L}})$ associated with the λ-graph system $\mathcal{L}$ [42].

6.10 Example

We give examples of the K-groups and the Bowen–Franks groups for a class of certain nonsofic subshifts that include the synchronizing counter shift Λ_C defined in Sect. 6.4. Let Σ_N be the set of symbols $\{a_1, \ldots, a_N, b, c\}$ for a fixed $N \in \mathbb{N}$. The nonsofic subshift Λ_{C_N} is the coded shift over Σ_N whose forbidden words are

$$\mathcal{F}_{C_N} = \{a_i b^m c^k a_j \mid i, j = 1, \ldots, N \text{ with } m \neq k\},$$

where the word $a_i b^m c^k a_j$ means $a_i \overbrace{b \cdots b}^{m} \overbrace{c \cdots c}^{k} a_j$. In [45], the C^*-algebra $\mathcal{O}_{\Lambda_{C_N}}$ associated with the subshift Λ_{C_N} has been studied so that its K-groups and Bowen–Franks groups that are those of the canonical symbolic matrix system for Λ_{C_N} have been calculated. We have

Proposition 6.10.1.

$$K_0(\Lambda_{C_N}) = \mathbb{Z}/N\mathbb{Z} \oplus \mathbb{Z}, \quad K_1(\Lambda_{C_N}) = 0 \quad and$$
$$BF^0(\Lambda_{C_N}) = \mathbb{Z}/N\mathbb{Z}, \quad BF^1(\Lambda_{C_N}) = \mathbb{Z}.$$

These types of Bowen–Franks groups cannot be realized in sofic shifts because $BF^1(\Lambda_{C_N})$ is not the torsion-free part of $BF^0(\Lambda_{C_N})$ (cf. [34]).

In [18, 29, 43], the K-groups and the Bowen–Franks groups for β-shifts and the Dyck shifts are calculated.

6.11 Remark: Relation to K-Theory for C^*-Algebras

The author has constructed a C^*-algebra $\mathcal{O}_{\mathcal{L}}$ from the λ-graph system $\mathcal{L}$ ([40], cf. [10, 37]). The C^*-algebra $\mathcal{O}_{\mathcal{L}}$ has a canonical action α of the one-dimensional torus group $\mathbb{T}$, called gauge action. The fixed point algebra $\mathcal{F}_{\mathcal{L}}$ of $\mathcal{O}_{\mathcal{L}}$ under α is an AF-algebra which is stably isomorphic to the cross product $\mathcal{O}_{\mathcal{L}} \times_\alpha \mathbb{T}$. Let (M, I) be the nonnegative matrix system for $\mathcal{L}$. The invariants studied in this chapter are described in terms of the K-theory for the C^*-algebra in the following way:

$$(\Delta_{(M,I)}, \Delta^+_{(M,I)}, \delta_{(M,I)}) = (K_0(\mathcal{F}_{\mathcal{L}}), K_0(\mathcal{F}_{\mathcal{L}})_+, \hat{\alpha}_*),$$
$$K_i(M, I) = K_i(\mathcal{O}_{\mathcal{L}}), \quad i = 0, 1,$$
$$BF^i(M, I) = \text{Ext}^{i+1}(\mathcal{O}_{\mathcal{L}}), \quad i = 0, 1$$

where $\hat{\alpha}$ denotes the dual action of α and $\text{Ext}^1(\mathcal{O}_{\mathfrak{L}}) = \text{Ext}(\mathcal{O}_{\mathfrak{L}}), \text{Ext}^0(\mathcal{O}_{\mathfrak{L}}) = \text{Ext}(\mathcal{O}_{\mathfrak{L}} \otimes C_0(\mathbb{R}))$. The normalized nonnegative eigenvectors of (M, I) exactly correspond to the KMS-states for α on the C^*-algebra $\mathcal{O}_{\mathfrak{L}}$. Hence the set of all bounded spectra with nonnegative eigenvectors is the set of all inverse temperatures for the admitted KMS states [46], cf. [13].

6.12 Conclusions and Further Work

As a consequence, we see that infinite labeled graphs called λ-graph systems yield presentations of symbolic dynamical systems so that several kinds of computable topological conjugacy invariants of symbolic dynamical systems are defined. The invariants are related to invariants of the associated C^*-algebras.

Computations of these invariants for other kinds of symbolic dynamical systems are seen, for example, in [7, 8, 31]. Related works to C^*-algebras are, for example, [1, 32, 44].

Acknowledgments The author would like to deeply thank Matthias Dehmer and Jun Ichi Fujii for their invitation to the author to write this chapter and for their helpful suggestions in the presentation of this paper.

References

1. Bates T, Pask D (2007) C^*-algebras of labelled graphs. J Oper Theory 57:207–226
2. Blanchard F, Hansel G (1986) Systems codés. Theor Comput Sci 44:17–49
3. Bowen R, Franks J (1977) Homology for zero-dimensional nonwandering sets. Ann Math 106:73–92
4. Boyle M, Krieger W (1988) Almost Markov and shift equivalent sofic systems. In: Proceedings of Maryland special year in dynamics 1986–1987. Lecture Notes in Mathematics, vol 1342. Springer, pp 33–93
5. Bratteli O (1972) Inductive limits of finite-dimensional C^*-algebras. Trans Am Math Soc 171:195–234
6. Brown LG (1983) The universal coefficient theorem for Ext and quasidiagonality. Operator Algebras and Group Representation, vol 17. Pitmann Press, Boston, pp 60–64
7. Carlsen TM, Eilers S (2004) Matsumoto K-groups associated to certain shift spaces. Doc Math 9:639–671
8. Carlsen TM, Eilers S (2006) K-groups associated to substitutional dynamics. J Funct Anal 238:99–117
9. Chomsky N, Schützenberger MP (1963) The algebraic theory of context-free languages. In: Braffort P, Hirschberg D (eds) Computer programing and formal systems. North-Holland, Amsterdam, pp 118–161
10. Cuntz J, Krieger W (1980) A class of C^*-algebras and topological Markov chains. Invent Math 56:251–268
11. Denker M, Grillenberger C, Sigmund K (1976) Ergodic theory on compact spaces. Springer, Berlin, Heidelberg and New York
12. Effros EG (1981) Dimensions and C^*-algebras. In: AMS-CBMS Reg Conf Ser Math, vol 46. American Mathematical Society, Providence, RI

6 Application of Infinite Labeled Graphs to Symbolic Dynamical Systems   167

13. Enomoto M, Fujii M, Watatani Y (1984) KMS states for gauge action on $\mathcal{O}_A$. Math Japon 29:607–619
14. Fischer R (1975) Sofic systems and graphs. Monats für Math 80:179–186
15. Franks J (1984) Flow equivalence of subshifts of finite type. Ergod Theory Dyn Syst 4:53–66
16. Hamachi T, Nasu M (1988) Topological conjugacy for 1-block factor maps of subshifts and sofic covers. In: Proceedings of Maryland special year in dynamics 1986–1987. Lecture Notes in Mathematics, vol 1342. Springer, pp 251–260
17. Hopcroft JE, Ullman JD (2001) Introduction to automata theory, languages, and computation. Addison-Wesley, Reading, MA
18. Katayama Y, Matsumoto K, Watatani Y (1998) Simple C^*-algebras arising from β-expansion of real numbers. Ergod Theory Dyn Syst 18:937–962
19. Kim KH, Roush FW (1979) Some results on decidability of shift equivalence. J Combin Inf Syst Sci 4:123–146
20. Kim KH, Roush FW (1999) Williams conjecture is false for irreducible subshifts. Ann Math 149:545–558
21. Kitchens BP (1998) Symbolic dynamics. Springer, Berlin
22. Krieger W (1974) On the uniqueness of the equilibrium state. Math Syst Theory 8:97–104
23. Krieger W (1980) On dimension for a class of homeomorphism groups. Math Ann 252:87–95
24. Krieger W (1980) On dimension functions and topological Markov chains. Invent Math 56:239–250
25. Krieger W (1984) On sofic systems I. Isr J Math 48:305–330
26. Krieger W (1987) On sofic systems II. Isr J Math 60:167–176
27. Krieger W (2000) On subshifts and topological Markov chains. Numbers, Information and Complexity (Bielefeld 1998). Kluwer, Boston, MA, pp 453–472
28. Krieger W, Matsumoto K (2002) Shannon graphs, subshifts and lambda-graph systems. J Math Soc Jpn 54:877–900
29. Krieger W, Matsumoto K (2003) A lambda-graph system for the Dyck shift and its K-groups. Doc Math 8:79–96
30. Krieger W, Matsumoto K (2004) A class of topological conjugacy of subshifts. Ergod Theory Dyn Syst 24:1155–1172
31. Krieger W, Matsumoto K (2010) Subshifts and C^*-algebras from one-counter codes. Contemporary Math 503(2009):93–120 AMS
32. Kumjian A, Pask D, Raeburn I, Renault J (1997) Graphs, groupoids and Cuntz–Krieger algebras. J Funct Anal 144:505–541
33. Lind D, Marcus B (1995) An introduction to symbolic dynamics and coding. Cambridge University Press, Cambridge
34. Matsumoto K (1999) A simple C^*-algebra arising from certain subshift. J Oper Theory 42:351–370
35. Matsumoto K (1999) Dimension groups for subshifts and simplicity of the associated C^*-algebras. J Math Soc Jpn 51:679–698
36. Matsumoto K (1999) Presentations of subshifts and their topological conjugacy invariants. Doc Math 4:285–340
37. Matsumoto K (2000) Stabilized C^*-algebras constructed from symbolic dynamical systems. Ergod Theor Dyn Syst 20:821–841
38. Matsumoto K (2001) Bowen–Franks groups for subshifts and Ext-groups for C^*-algebras. K Theor 23:67–104
39. Matsumoto K (2001) Bowen–Franks groups as an invariant for flow equivalence of subshifts. Ergod Theory Dyn Syst 21:1831–1842
40. Matsumoto K (2002) C^*-algebras associated with presentations of subshifts. Doc Math 7:1–30
41. Matsumoto K (2003) On strong shift equivalence of symbolic matrix systems. Ergod Theory Dyn Syst 23:1551–1574
42. Matsumoto K (2005) Topological entropy in C^*-algebras associated with λ-graph systems. Ergod Theor Dyn Syst 25:1935–1951
43. Matsumoto K (2005) K-theoretic invariants and conformal measures on the Dyck shifts. Int J Math 16:213–248

44. Matsumoto K (2007) Actions of symbolic dynamical systems on C^*-algebras. J Reine Angew Math 605:23–49
45. Matsumoto K A class of simple C^*-algebras arising from certain nonsofic subshifts, to appear in Ergod Theory Dyn Syst, arxiv:0805.2767
46. Matsumoto K, Watatani Y, Yoshida M (1998) KMS-states for gauge actions on C^*-algebras associated with subshifts. Math Z 228:489–509
47. Nasu M (1986) Topological conjugacy for sofic shifts. Ergod Theory Dyn Syst 6:265–280
48. Nasu M (1995) Textile systems for endomorphisms and automorphisms of the shift. Mem Am Math Soc 114:546
49. Parry W, Sullivan D (1975) A topological invariant for flows on one-dimensional spaces. Topology 14:297–299
50. Tuncel S (1983) A dimension, dimension modules, and Markov chains. Proc Lond Math Soc 46:100–116
51. Weiss B (1973) Subshifts of finite type and sofic systems. Monatsh Math 77:462–474
52. Williams RF (1973) Classification of subshifts of finite type. Ann Math 98:120–153. Erratum (1974) Ann Math 99:380–381

Chapter 7
Decompositions and Factorizations
of Complete Graphs

Petr Kovář

Abstract Graph decompositions into isomorphic copies of a given graph are a well-established topic studied in both graph theory and design theory. Although spanning tree factorizations may seem to be just a special case of this concept, not many general results are known. We investigate necessary and sufficient conditions for a graph factorization into isomorphic spanning trees to exist.

Keywords Graph decomposition · Graph factorization · Graph labeling

MSC2000: Primary 05C70; Secondary 05C78, 05C05

In this chapter we investigate factorization of complete graphs into isomorphic spanning trees. The main goal is to present a recursive method that generalizes several previously introduced concepts. We start with a detailed overview of known methods. A brief summary of known results on spanning tree factorization of complete graphs follows. In the last part we describe a construction that unifies most of the presented methods in one general labeling called recursive labeling.

7.1 About Graph Decompositions and Factorizations

There are several real-life problems that lead to graph decompositions or constructing block designs. One of them – the design of ad hoc wireless networks – is described in Sect. 7.1.1. We proceed by giving a detailed overview of methods used for decompositions and spanning tree factorizations of complete graphs and in later sections we try to generalize them. Thus, this chapter provides a survey of techniques as well as a new result in this area.

P. Kovář (✉)
Department of Applied Mathematics, Technical University Ostrava, 17. listopadu,
708 33 Ostrava–Poruba, Czech Republic
e-mail: petr.kovar@vsb.cz

M. Dehmer (ed.), *Structural Analysis of Complex Networks*,
DOI 10.1007/978-0-8176-4789-6_7, © Springer Science+Business Media, LLC 2011

7.1.1 Ad Hoc Wireless Network Design

Various difficulties have to be resolved while designing ad hoc networks. After the network structure is changed locally (due to some failure) it can take a long time until this change is absorbed by the network, in particular, by the routing tables. Resolving such a failure by reconnecting nodes within the network may lead to huge diameter or high load for certain nodes in the network structure. Let us concentrate on wireless networks of mobile units of a firefighter department. The firefighters in action communicate with each other and headquarters broadcasts orders to all units. Their wireless networks consist of several mobile units, usually about two dozen with rather low traffic (compared with an Internet network) and with a limited number of stationary units. Each unit can communicate with every other unit but some pairs may temporarily lose connection since the firefighters are moving in the terrain. It is crucial to restore connections quickly, since blackouts can jeopardize the operation and firefighters' lives. We want both to quickly restore connectivity and keep appropriate network parameters: a low number of hops (small diameter, usually 3–10) and balance (low highest degree, up to 20). Each unit has its unique ID and responds only to this particular ID. Every unit operates as receiver, transmitter, and router to receive messages from one unit and forward it to another unit. Routing is maintained by the units itself.

Reformulating this problem as a graph theory problem we require network designs that are connected and do not contain circuits. The set of all possible connections forms a complete graph. Each particular network forms a spanning tree in this complete graph.

We present a method of decomposing a complete graph into spanning trees that are all isomorphic to the same structure. We obtain a collection of isomorphic factors (network structures), forming a pool of networks, all with the same diameter and the same highest degree. Moreover, all pairs of networks are edge disjoint, thus if one particular connection fails, no other factor from the pool will be affected by this failure. Instead of repairing one broken connection we swap to an entirely alternate network.

The list of networks can be hard-coded with the software on the units. In case of lost connection a broadcast message from the two separated units can resolve the problem immediately by switching to one of the backup networks. Once removed, a network structure can be put back into the list. We expect the missing connection to restore in time.

7.1.2 Description of the Graph Theory Problem

We are given a graph representing a network or a set of relations on n elements. Performing an operation on this net, one may want to parallelize or simply split one complex task into several smaller tasks. This immediately translates into decomposing the given graph into subgraphs. Each edge of the original graph is to appear in exactly one subgraph.

7 Decompositions and Factorizations of Complete Graphs   171

There may be further requirements put on the decomposition. It is natural to assume that all smaller graphs are not "too big," that they have the same number of vertices and edges, or that the diameter is bounded, or that they are all isomorphic. Further, one may require that all smaller subgraphs are spanning; i.e., each of them is connected is contains all vertices of the original graph. The main scope of this chapter is decompositions of complete graphs into small acyclic isomorphic spanning graphs: isomorphic spanning trees.

Decompositions of complete graphs are the subject of studies in both design theory and graph theory. Since design theory methods are suitable mainly for problems of decomposition into highly symmetric graphs (e.g., vertex transitive graphs such as complete graph, cycles or multipartite graphs [1,6]) and we examine decomposition into trees (nonregular graphs) the methods discussed further in this chapter are based almost exclusively on graph theory.

We use standard terminology that can be found in any textbook (see [35] or [7]). All graphs are finite and simple (without loops or multiple edges). A graph G is a pair (V, E), where the set of vertices V (often denoted by $V(G)$) is nonempty and the set of edges E (or $E(G)$) is a subset of $P_2(V)$, the set of all two-element subsets of V.

Definition 1. Let H be a graph on n vertices. A *decomposition of the graph H* is a set of pairwise edge disjoint subgraphs $G_1, G_2, \dots, G_s$ of H such that every edge of H belongs to exactly one of the subgraphs G_r. If each subgraph G_r is isomorphic to a graph G we speak about a *G-decomposition of H*. If G is a connected factor of H, then we call the G-decomposition a *G-factorization*.

7.1.3 Necessary Conditions for Decompositions

There are some obvious and some less obvious necessary conditions for a G-decomposition of a given graph H.

It is easy to see that the number of edges $|E(G)|$ has to divide the number of edges $|E(H)|$ since each edge of the graph H belongs to one copy of G. Moreover there are restrictions on the degree sequence of G. If $k = |E(H)|/|E(G)|$ is the number of copies of G in H, then the multiset $\mathcal{D}$ in which the degree of every vertex of G appears exactly k times has to be decomposable into multisets D_i, $i = 1, 2, \dots, |V(G)|$, such that the sum of degrees in each D_i corresponds to the degree of a vertex $w_i \in V(H)$. If H is r-regular this condition is simpler, the sum of elements in each D_i has to be r. In particular, if H is a complete graph K_n, then the sum of elements in each D_i has to be $n - 1$.

Example 1. We show that the graph $H \simeq K_r$ can be decomposed into $G \simeq K_4$ only if $r \equiv 1 \pmod{12}$. The number of edges $r(r - 1)/2$ has to be divisible by 6, thus $r \equiv 0, 1 \pmod{12}$. Moreover the degree 3 of every vertex in K_4 has to divide $r - 1$, thus $r \equiv 1 \pmod{3}$. Combining these conditions, the claim follows.

Example 2. Let t be an integer. The complete graph K_{2t} cannot be decomposed into cycles nor any even regular graphs, since it is odd-regular.

On the other hand there are also structural limitations that apply to the existence of a G-decomposition of H. A *caterpillar* is a tree such that by removing all leaves we obtain a path. Although the graphs R_1 and R_2 in Fig. 7.1 have the same number of edges and the same degree sequence, the caterpillar R_1 does and the caterpillar R_2 does not decompose the complete graph K_6 (for a small graph a brute force search can be performed; a general nonexistence result addressing certain caterpillars with diameter four is presented in [14]). The structural restrictions depend strongly on the particular class of graphs and are difficult to explore in general. Therefore such structural necessary conditions are known almost exclusively for highly symmetric graphs, such as vertex transitive graphs, and only sparsely for trees. Some results are presented in Sect. 7.3.

Definition 2. We say a G-decomposition of a graph H on $2n + 1$ vertices into $G_0, G_1, \ldots, G_{2n}$ is *cyclic* if there exists an ordering $(x_0, x_1, \ldots, x_{2n})$ of vertices of H and isomorphisms $\phi_i : G_1 \to G_i, i = 0, 1, \ldots, 2n$ such that $\phi_i(x_j) = x_{i+j}$ for every $j = 0, 1, \ldots, 2n$, where the subscripts are taken modulo $2n + 1$.

Example 3. In Fig. 7.2 is shown the factor G_1 of a cyclic C_3-decomposition of K_7. The next six subgraphs are obtained by isomorphisms "rotating" the graph G_1 clockwise.

If we have a graph G with m edges and we are looking for a cyclic G-decomposition of some K_n there have to be n copies of G in K_n. The number of edges in nG (n copies of G) has to divide $|E(K_n)|$,

$$nm | n(n-1)/2 \quad \text{thus} \quad n = mk + 1$$

for some positive integer k.

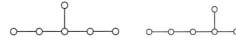

Fig. 7.1 Caterpillars R_1 and R_2

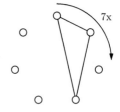

Fig. 7.2 Cyclic C_3-decomposition of K_7

7 Decompositions and Factorizations of Complete Graphs

7.1.4 Sufficient Conditions for Decompositions

There is no necessary and sufficient condition known for a graph G to decompose a graph H in general, not even in the case when H is a complete graph and G is arbitrary. Only for some classes of graph G was the sufficiency settled.

Now let us restrict only to the problem when $H \simeq K_n$ for some integer n. There are only a few sufficient conditions for a G-decomposition of a complete graph to exist. They include both design theory and graph theory techniques. Here we focus on the latter ones since only these also apply to the factorizations in Sect. 7.1.6. The problem of the existence of a G-decomposition of a complete graph is transformed into the existence of a certain labeling of a given graph G. To describe the methods of decompositions using a graph labeling we have to introduce some terms.

We denote the vertices of the complete graph K_n by $v_1, v_2, \ldots, v_n$. The *length* of an edge $v_i v_j$ is the smaller "distance" between v_i and v_j along the cycle $v_1, v_2, \ldots, v_n$. It can be counted by taking the smaller value of $|j - i|$ and $n - |j - i|$. This notion can also be extended to graphs that are labeled by the first n positive integers. In general, a *graph labeling* is a mapping that assigns integers (usually positive) to graph elements: vertices, edges, or both. Hence the following definition.

Definition 3. Let G be a graph with m edges and the vertex set $V(G)$ and let λ be an injection $\lambda : V(G) \to S$ where S is a subset of the set $\{0, 1, 2, \ldots, 2m\}$. The *length* of an edge xy is defined as $\ell(x, y) = \min\{|\lambda(x) - \lambda(y)|, 2m + 1 - |\lambda(x) - \lambda(y)|\}$. The set of vertex labels is denoted by $\lambda(V(G))$.

Additional requirements are put on the labelings to guarantee that in a G-decomposition of K_n each edge appears in precisely one copy of G. In 1967 Rosa (see [31]) defined the α-, β-, σ-, and ρ-valuations depending on which of the following conditions of a vertex labeling λ of a graph G with m edges are satisfied:

(a) the set of vertex labels $\lambda(V(G)) \subseteq \{0, 1, \ldots, m\}$
(b) the set of vertex labels $\lambda(V(G)) \subseteq \{0, 1, \ldots, 2m\}$
(c) the set $\{|\lambda(x) - \lambda(y)| : xy \in E(G)\} = \{1, 2, \ldots, m\}$
(d) the set of edge lengths $\{\ell(x, y) : xy \in E(G)\} = \{1, 2, \ldots, m\}$
(e) there exists $a \in \{0, 1, \ldots, m\}$, such that for all $xy \in E(G)$ either $\lambda(x) \leq a < \lambda(y)$ or $\lambda(x) > a \geq \lambda(y)$

If the labeling satisfies

- (a), (c), (e), then it is an α-valuation
- (a), (d), then it is a β-valuation
- (b), (c), then it is a σ-valuation
- (b), (d), then it is a ρ-valuation

The hierarchy of the valuations is α-, β-, σ-, and ρ-valuation which means that if G allows a certain valuation in the list, it also allows any latter one. Since 1967 hundreds of papers have been published extending the list of graphs known to allow some of these valuations. See Gallian [17] for an excellent survey updated every year. Through the years the names *α-labeling*, *ρ-labeling*, and *graceful labeling* for the *β*-valuation (see Golomb [18]) became more popular. The importance of the valuations for decompositions is apparent from the following paragraphs.

In the next paragraphs several other labelings are defined. There are plenty of definitions, some of them rather technical, that are used for spanning tree factorizations of complete graphs. Later, in Sect. 7.4 we try to find a general labeling that extends features of most of the labelings described here. Therefore the techniques are explained in detail.

α-Labeling

It is easy to observe that when a graph G allows an α-labeling then it has to be bipartite. In [31] Rosa proved the following sufficient condition for the existence of cyclic decompositions.

Theorem 1. *If a graph G with m edges has an α-labeling, then there exists a cyclic G-decomposition of the complete graph K_{2km+1}, where k is any natural number.*

We accept the following convention: if convenient the vertex name is unified with its label. For example, instead of saying "the vertex x labeled i" we say "the vertex i" and vice versa.

Example 4. In Fig. 7.3 is shown an example of an α-labeling. Notice that evaluating the edge lengths we get all consecutive integers $1, 2, \ldots, 5$. Moreover, taking $a = 3$ for every edge xy is either $x \leq 3$ and $y > 3$ or $x > 3$ and $y \leq 3$.

It should be pointed out that not all decompositions are necessarily cyclic as shown by the next example.

Example 5. K_9 can be decomposed into triangles C_3 (a Steiner triple system), but since there are 12 of them, this decomposition cannot be cyclic. Moreover, for K_{2t} no cyclic spanning tree decomposition into trees with t edges can exist, since there would have to be $2t - 1$ of the copies, not $2t$.

On the other hand, there are trees that have no α-labeling (see [2]), hence the property of being bipartite is not a sufficient condition for an α-labeling to exist.

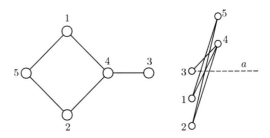

Fig. 7.3 α-Labeling of a graph G and a bipartite drawing of G

7 Decompositions and Factorizations of Complete Graphs

Fig. 7.4 p-Labeling of R_1 and its drawing as a subgraph in K_{11}

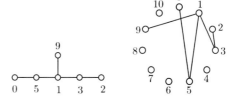

β-, σ-, and ρ-Labelings

Surprisingly most attention was devoted to graceful labelings (β-valuation) although it is the ρ-labeling that is the most general labeling sufficient for the existence of a G-decomposition of a complete graph. In [31] Rosa proved the following necessary and sufficient condition

Theorem 2. *Let G be a graph with m edges. A cyclic G-decomposition of the complete graph K_{2m+1} exists if and only if G has a ρ-labeling.*

The research in graceful labelings has been intensive during the past four decades; see [17].

Example 6. There is an example of a ρ-labeling of R_1 given in Fig. 7.4. Notice that evaluating the edge lengths we get all consecutive integers $1, 2, \ldots, 5$. This is easy to check in the same figure where R_1 is drawn as a subgraph G_1 in K_{11}.

7.1.5 Necessary Conditions for Factorizations

Recall that the main scope of this chapter is factorization of complete graphs. Since G-factorizations are a special case of G-decompositions, then besides the necessary conditions given in Sect. 7.1.3 more restrictions apply. For a G-factorization of K_n to exist, all the multisets D_i (defined on page 171) have to be of the same size, namely k where k is the number of factors isomorphic to G in K_n. Moreover, since the degree of every vertex x in each factor G_i has to be at least one, the largest degree $\Delta(G)$ has to be at most $n - k$. We call this the *degree condition*.

Let T be a spanning tree of K_n. Thus T is a tree on n vertices (and $m = n - 1$ edges) and since

$$|E(K_n)| = \binom{n}{2} = \frac{1}{2}n(n-1)$$

it follows that n has to be even, say $n = 2t$, and there have to be exactly $n/2 = t$ factors in a T-factorization of K_{2t} and the largest degree $\Delta(T) = n - k = 2t - t = t$.

Example 7. No star $K_{1,n-1}$ can factorize the complete graph K_n due to the degree condition. A double-star is a tree with only two vertices of degree t greater than 1

Fig. 7.5 A double-star factorizing K_6

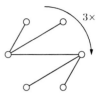

(called *nonleaves*). A double-star satisfies the degree condition for factorizing K_{2t} and it can easily be observed that it actually does decompose K_{2t} for every $t \geq 2$. See Fig. 7.5.

Let G be a graph with m edges and $m + 1$ vertices, with possibly some of them isolated. Since we are factorizing the complete graph (on $m + 1$ vertices) into isomorphic copies of G it follows that the order of the complete graph has to be even.

7.1.6 Sufficient Conditions for Factorizations

In comparison to decompositions, factorizations remained fairly unexplored. All methods known for spanning tree factorizations are based on graph labelings, save for the trivial Hamiltonian path or double-star factorizations. The methods include ρ-symmetric labeling, blended ρ-labeling, swapping labeling, $2n$-cyclic blended labeling, and fixing labeling. We describe these in the following paragraphs. To present clearly the idea of our main result in Sect. 7.4 we try to keep a general point of view by stressing the important facts and skipping some details. Then in Sect. 7.4 we combine these methods.

We start with a definition.

Definition 4. A G-decomposition of a graph H on $2n$ vertices into $G_1, G_2, \ldots, G_s$ is *bicyclic* if there exists an ordering $(x_0, x_1, \ldots, x_{n-1}, y_0, y_1, \ldots, y_{n-1})$ of vertices of H and isomorphisms $\phi_i : G_1 \to G_i, i = 1, 2, \ldots, s$ such that $\phi_i(x_j) = x_{j+i-1}$ and $\phi_i(y_j) = y_{j+i-1}$ for every $j = 0, 1, \ldots, n - 1$, where the subscripts are taken modulo n.

Factorizing Regular Complete Bipartite Graphs

For factorization of regular complete bipartite graphs $K_{n,n}$ Frončer introduced the *bipartite ρ-labeling*; see [12]. Although we are factorizing complete graphs, the notion of a, bipartite ρ-labeling is used in most of the subsequent labelings and plays an important role in Sect. 7.4. We give the definition here.

Definition 5. Let G be a bipartite graph with n edges and the vertex set $V(G) = V_0 \cup V_1$. Let λ be an injection $\lambda : V_i \to S_i$, where S_i is a subset of the set $\{0_i, 1_i, \ldots, (n - 1)_i\}, i = 0, 1$. The *length* of an edge (x_0, y_1) for $x_0 \in V_0$ and

$y_1 \in V_1$ with $\lambda(x_0) = a_0$ and $\lambda(y_1) = b_1$ is defined as $\ell_{01}(x_0, y_1) = b - a$ (mod n). If the set of all lengths of n edges is equal to $\{0, 1, 2, \ldots, n-1\}$, then λ is a *bipartite ρ-labeling*.

It was shown in [12] that the existence of a bipartite ρ-labeling of a graph G implies a bicyclic decomposition of the complete graph $K_{n,n}$.

Theorem 3. *Let a bipartite graph G with n edges have a bipartite ρ-labeling. Then there exists a bicyclic decomposition of $K_{n,n}$ into n copies of G.*

This idea plays an important role for all the blended, swapping, $2n$-cyclic blended, and fixing labelings.

ρ-Symmetric Labeling

A connected graph G with an edge (x, y) (called a *bridge*) is *symmetric* if there is an automorphism ψ of G such that $\psi(x) = y$ and $\psi(y) = x$. The isomorphic components of $G - (x, y)$ are called *banks* and denoted by H, H', respectively. See Fig. 7.6.

Definition 6. A labeling of a symmetric graph G with $2n - 1$ edges and banks H, H' is *ρ-symmetric graceful* if H has a ρ-labeling and $\psi(i) = i + n$ (mod $2n$) for each vertex i in H. A graph that admits a ρ-symmetric graceful labeling is called *ρ-symmetric graceful*.

This concept was introduced by Eldergill in [8]. The following theorem (proven in [8]) shows how this applies to spanning tree factorizations.

Theorem 4. *Let G be a symmetric graph with $2n - 1$ edges. Then there exists a cyclic G-decomposition of K_{2n} if and only if G is ρ-symmetric graceful.*

If G is an acyclic graph, it is a spanning tree of K_{2n}. Thus a symmetric tree on $2n$ vertices that allows a ρ-symmetric graceful labeling factorizes the complete graph K_{2n}. The requirement for a tree to be symmetric is too restrictive as we show in the next paragraph.

Example 8. In Fig. 7.6 is a symmetric graph with a ρ-symmetric graceful labeling drawn as a subgraph of K_{10}.

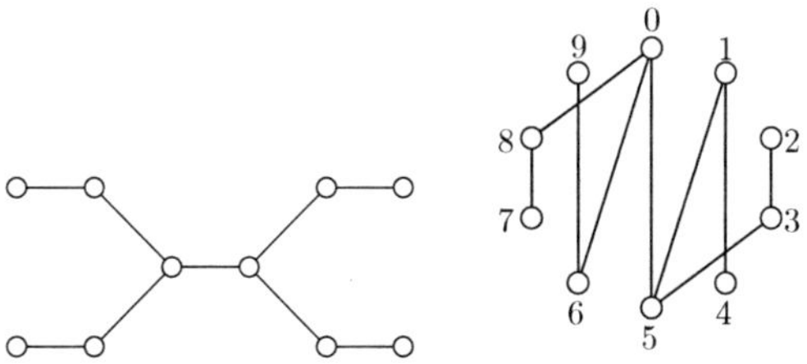

Fig. 7.6 Symmetric graph with a ρ-symmetric graceful labeling

178 P. Kovář

Blended ρ-Labeling

Blended ρ-labeling (or *blended labeling* for short) was introduced by Fronček [12] as a generalization of the ρ-symmetric graceful labeling.

Definition 7. Let G be a graph with $V(G) = V_0 \cup V_1$, $V_0 \cap V_1 = \emptyset$, and $|V_0| = |V_1| = r$. Let λ be an injection, $\lambda : V_i \longrightarrow \{0_i, 1_i, \ldots, (r-1)_i\}$, $i = 0, 1$.

The *pure length* of an edge (x_i, y_i) with $x_i, y_i \in V_i$, where $i \in \{0, 1\}$, for $\lambda(x_i) = p_i$ and $\lambda(y_i) = q_i$ is defined as

$$\ell_{ii}(x_i, y_i) = \min\{p - q \pmod{r}, \ q - p \pmod{r}\}.$$

The *mixed length* of an edge (x_0, y_1) with $x_0 \in V_0$, $y_1 \in V_1$, for $\lambda(x_0) = p_0$ and $\lambda(y_1) = q_1$, is defined as

$$\ell_{01}(x_0, y_1) = q - p \pmod{r},$$

where $p, q \in \{0, 1, \ldots, r-1\}$ are the vertex labels without subscripts. The edges (x_i, y_i) for $i = 0, 1$ with the pure length ℓ_{ii} are called *pure edges* and the edges (x_0, y_1) with the mixed length ℓ_{01} are called *mixed edges*.

This concept of lengths can be extended to any equal-sized partite set $V_0, V_1, \ldots, V_k$.

Definition 8. Let G be a graph with $4n + 1$ edges, $V(G) = V_0 \cup V_1$, $V_0 \cap V_1 = \emptyset$, and $|V_0| = |V_1| = 2n + 1$. Let λ be an injection, $\lambda : V_i \to \{0_i, 1_i, \ldots, (2n)_i\}$, $i = 0, 1$ and let the lengths be as defined in Definition 7.

We say G has a *blended ρ-labeling* if

(1) $\{\ell_{ii}(x_i, y_i) : (x_i, y_i) \in E(G)\} = \{1, 2, \ldots, n\}$ for $i = 0, 1$
(2) $\{\ell_{01}(x_0, y_1) : (x_0, y_1) \in E(G)\} = \{0, 1, \ldots, 2n\}$

Notice that in comparison to Definition 6 the requirement for T to be symmetric is lifted. However, there is a restriction on the order of G to be 2 modulo 4.

The vertex set of the graph G with a blended labeling is decomposed into two equal-sized partite sets V_0, V_1 and G can be split into three subgraphs as follows: two subgraphs H_0, H_1 induced on the vertices of V_0, V_1, respectively, and a bipartite graph H_{01} with partite sets V_0, V_1. The blended labeling restricted to the subgraphs H_0, H_1 can be viewed as a ρ-labeling (after having omitted the subscripts). Each of them guarantees a cyclic decomposition of a graph K_{2n+1} into n copies of H_0 or H_1, respectively. The labeling of the subgraph H_{01} is then a bipartite ρ-labeling that allows a bicyclic decomposition of $K_{2n+1,2n+1}$ into $2n + 1$ isomorphic copies of H_{01}. Hence the following theorem (for a proof see [12]).

Theorem 5. *Let G with $4n + 1$ edges have a blended ρ-labeling. Then there exists a bicyclic decomposition of K_{4n+2} into $2n + 1$ copies of G.*

Example 9. An example of a blended labeling of R_1 is shown in Fig. 7.7.

7 Decompositions and Factorizations of Complete Graphs

Fig. 7.7 Blended labeling of caterpillar R_1

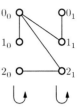

In a blended labeling all vertices in the "left" partite set V_0 are always $0_0, 1_0, \ldots, (2n-1)_0$ and in the "right" partite set V_1 always $0_1, 1_1, \ldots, (2n-1)_1$. Therefore we can omit the labels in the figures.

Swapping Labeling

Swapping labeling was introduced by Kovářová in her Ph.D. thesis [22]. Unlike blended labeling this labeling guarantees the factorization of complete graphs of order 0 modulo 4. For a proof see [22].

Definition 9. A graph G with $4n-1$ edges has a *swapping blended labeling* (briefly *swapping labeling*) if the following is satisfied. The vertex set $V(G) = V_0 \cup V_1$, $V_0 \cap V_1 = \emptyset$, and $|V_0| = |V_1| = 2n$. Let λ be an injection, $\lambda : V_i \to \{0_i, 1_i, \ldots, (2n-1)_i\}$ for $i = 0, 1$. Suppose the lengths are defined as in Definition 7, then

(1) $\{\ell_{ii}(x_i, y_i) : (x_i, y_i) \in E(G)\} = \{1, 2, \ldots, n\}$, for $i = 0, 1$,
(2) there exists an isomorphism φ such that G is isomorphic to G', where $V(G') = V(G)$ and $E(G') = E(G) \setminus \{(k_0, (k+n)_0), (l_1, (l+n)_1)\} \cup \{(k_0, (l+n)_1), ((k+n)_0, l_1)\}$,
(3) $\{\ell_{01}(x_0, y_1) : (x_0, y_1) \in E(G)\} = \{0, 1, \ldots, 2n-1\} \setminus \{\ell_{01}(k_0, (l+n)_1)\}$.

Again the vertex set of G with a swapping labeling can be split into three subgraphs: H_0 and H_1 induced on V_0 and V_1, respectively, and a bipartite subgraph H_{01} with the partite sets V_0 and V_1. The labelings of H_0 and H_1 induced by λ have all different pure lengths (Condition (1)). The labeling of H_{01} induced by λ gives edges of almost all different mixed lengths. Edges of mixed length $\ell_{01}(k_0, (l+n)_1) = l + n - k \pmod{2n}$ are missing (Condition (3)) among the first $n/2$ factors. In the next $n/2$ factors the longest pure edges are removed from the graphs H_0, H_1 and edges of mixed length $\ell_{01}(k_0, (l+n)_1)$ are added to the bipartite graph H_{01}.

Theorem 6. *Let G be a graph on $4n$ vertices with $4n-1$ edges that has a swapping blended labeling. Then there exists a G-decomposition of K_{4n} into $2n$ isomorphic copies of G.*

If G is a spanning tree in K_{4n} with a swapping labeling then by Theorem 6 it factorizes K_{4n}.

Fig. 7.8 Caterpillar R_3 and a swapping labeling of R_3

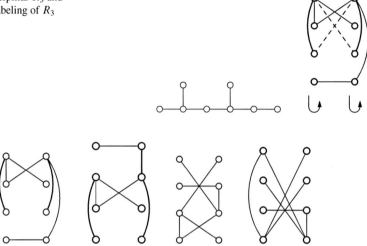

Fig. 7.9 R_3-factorization of K_8 based on the swapping labeling

Example 10. The caterpillar R_3 in Fig. 7.8 allows a swapping labeling. Figure 7.9 shows the four factors of K_8 given by this labeling. Again we omit the vertex labels, since they are implied by the drawing.

The *switching labeling* defined by Fronček and Kubesa in [16] also allows factorizations of K_{4n} into isomorphic spanning trees. In comparison to swapping labeling switching labeling is more restrictive since it requires certain strong automorphisms that exist only for very special classes of trees.

$2n$-Cyclic Labeling

So far the vertex set of G (as well as K_{2n}) was decomposed into two sets of size n. For ρ-symmetric labeling there is a strong automorphism required and for blended labeling n has to be odd, for the swapping labeling n has to be even. Kovářová in [22] approached the problem of factorizations of complete graphs into trees with a given diameter by decomposing the vertex set into $2n$ partite sets V_i of size k. This technique is designed for factorizations into "less dense" graphs with higher diameter. Hence the following definitions.

Definition 10. A tree T is called an *underlying tree of U* if the graph U arises from the tree T by blowing up two vertices r, s of T by K_k and all the remaining vertices by $\overline{K}_k = kK_1$. Every edge of the tree is replaced by the edges of $K_{k,k}$.

Definition 11. Let G be a graph with $2nk - 1$ edges, for k odd and $k, n > 1$, and the vertex set $V(G) = \bigcup_{i=0}^{2n-1} V_i$, where $|V_i| = k$ and $V_i \cap V_j = \emptyset$ for $i \neq j$. Let λ be an injection, $\lambda : V_i \to \{0_i, 1_i, 2_i, \ldots, (k-1)_i\}$, for $i = 0, 1, \ldots, 2n - 1$.

By H_{ij} we denote the bipartite subgraph of G induced on the vertices of the partite sets V_i and V_j with edges of mixed length ℓ_{ij}, and by H_i we denote the subgraph of G induced on the vertices of V_i with edges of pure length ℓ_{ii}.

We say that G has a *2n-cyclic blended labeling* (2n-cyclic labeling for short) if there exists an underlying tree T on $2n$ vertices with a ρ-symmetric graceful labeling such that the following hold:

(1) For some vertex $s \in T$ and its symmetric image $t = s + n \pmod{2n}$ we have

$$\{\ell_{ss}(x_s, y_s): (x_s, y_s) \in E(H_s)\} = \{1, 2, \ldots, (k-1)/2\}, \text{ and}$$
$$\{\ell_{tt}(x_t, y_t): (x_t, y_t) \in E(H_t)\} = \{1, 2, \ldots, (k-1)/2\},$$

(2) and for each edge $(i, j) \in E(T)$ we have

$$\{\ell_{ij}(x_i, y_j): (x_i, y_j) \in E(H_{ij})\} = \{0, 1, 2, \ldots, k-1\}.$$

Notice that after omitting subscripts the labelings induced on H_s and H_t are ρ-labelings and the labelings induced on all bipartite graphs H_{ij} are bipartite ρ-labelings. In [13] the following theorem is proven.

Theorem 7. *Let G with $2nk - 1$ edges be a graph that allows a 2n-cyclic blended labeling for k odd and $k, n > 1$. Then there exists a G-decomposition of K_{2nk} into nk copies of G.*

Fixing Labeling

By relaxing the condition of having a bipartite ρ-labeling in each H_{ij} (as introduced in the previous paragraph) and requiring that all vertices in one of the partite sets, say V_i, are always of degree one, we obtain *fixing labeling* as it was introduced by Kovářová in [21]; see Fig. 7.10. Therefore fixing labeling is more general than $2n$-cyclic labeling.

Definition 12. Let G be a graph with $2nk - 1$ edges, for k odd and $k, n > 1$, and vertex set $V(G) = \bigcup_{i=0}^{2n-1} V_i$, where $|V_i| = k$ and $V_i \cap V_j = \emptyset$ for $i \neq j$. Let λ be

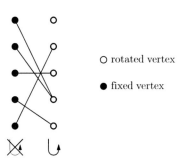

Fig. 7.10 Factorizing $K_{5,5}$ by fixing labeling

an injection, $\lambda : V_i \to \{0_i, 1_i, 2_i, \ldots, (k-1)_i\}$, for $i = 0, 1, \ldots, 2n - 1$. By H_{ij} we denote the bipartite subgraph of G induced on the vertices of the partite sets V_i and V_j with edges of the mixed length ℓ_{ij}, and by H_i we denote the subgraph of G induced on the vertices of V_i with edges of the pure length ℓ_{ii}. We say that G has a *fixing labeling* if there exists an underlying tree T on $2n$ vertices with a ρ-symmetric graceful labeling such that the following hold:

(1) For some vertex $s \in T$ and its symmetric image $t = s + n \pmod{2n}$ we have

$$\{\ell_{ss}(x_s, y_s) : (x_s, y_s) \in E(H_s)\} = \{1, 2, \ldots, (k-1)/2\}, \text{ and}$$
$$\{\ell_{tt}(x_t, y_t) : (x_t, y_t) \in E(H_t)\} = \{1, 2, \ldots, (k-1)/2\}.$$

(2) Let $F = \{i \in T : i \neq s, i \neq s + n ; \deg(x_i) = 1 \text{ for each } x_i \in H_{ij} \text{ and } j \in N(i)\}$, then F is the *set of fixable vertices* in T for given G, and each vertex i in F is called *fixable*. Let V_F be any independent set of fixable vertices in T called the *fixed set*. A vertex $i \in V_F$ is called a *fixed vertex*.

Then for every edge $(i, j) \in E(T)$ i or j is one of the fixed end-vertices or

$$\{\ell_{ij}(x_i, y_j) : (x_i, y_j) \in E(H_{ij})\} = \{0, 1, 2, \ldots, k-1\}.$$

Notice that if neither i nor j is a fixed vertex then the labeling of H_{ij} induced by λ is a bipartite ρ-labeling. Fixing labeling has been used in several papers (see Sect. 7.3) for spanning tree factorizations of complete graphs based on the next theorem proved in [21].

Theorem 8. *Let a graph G with $2nk - 1$ edges, for k odd and $k, n > 1$, have a fixing blended labeling. Then there exists a G-decomposition of K_{2nk} into nk copies of G.*

Example 11. In Fig. 7.11 is shown a symmetric tree T on 10 vertices with a symmetric ρ-labeling (T factorizes K_{10}). Fixable vertices and fixed vertices are highlighted in Fig. 7.12.

In Fig. 7.13 is a tree on 30 vertices that allows a fixing labeling. By adopting the convention, that the "level" i, $0 \le i \le 2$, stands for the vertex label and the partite set subscript stands for the label subscript, the labeling is implied by the drawing in Fig. 7.13.

Both $2n$-cyclic labeling and fixing labeling require the underlying tree to be symmetric and the number of vertices in each partite set V_i to be equal to the same odd number. The second restriction can be relaxed; see Sect. 7.2.

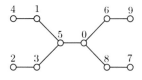

Fig. 7.11 Underlying tree T with a symmetric ρ-labeling

Fig. 7.12 Fixable and fixed vertices in T

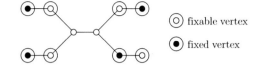

Fig. 7.13 Example of a fixing labeling of a tree on 30 vertices

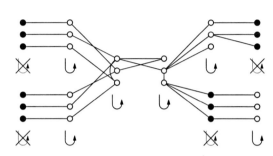

7.2 Advanced Methods

There is no necessary and sufficient condition for a T-factorization of a complete graph. Besides constructing additional classes of trees that factorize a complete graph it is natural to give a necessary condition for certain classes of trees. We say that a class of trees is classified if for each tree T_n in this class we know whether it factorizes the corresponding complete graph K_n (by labeling the tree) or it is shown that it cannot factorize K_n (such as paths or symmetric trees mentioned in Sect. 7.1.6). To achieve this classification more elaborate techniques are often needed. We present them here.

7.2.1 N-Z Construction

In [14] a recursive construction was used to obtain a classification of all caterpillars of diameter 5. This technique is based on factorizing K_{2n} into highly regular graphs and each of them further factorizing into a tree T. Here we give a short introduction to the method.

Definition 13. Let $G_1, G_2, \ldots, G_n$ be a G-factorization of K_{2n} induced by the isomorphisms $\phi_j, G_j = \phi_j(G)$ for $j = 1, 2, \ldots, n$. We call a vertex v *semisurjective* if there exists a partition $V_0, V_1, |V_0| = |V_1| = n$, of the vertex set $V(K_{2n})$ such that either $\{\phi_j(v): j = 1, 2, \ldots, n\} = V_0$ or V_1. We call the G-factorization of K_{2n} *semisurjective* if every vertex of G is semisurjective and the G-factorization is called *weakly semisurjective* if every nonleaf vertex of G is semisurjective.

Observe that if G has a blended labeling, then the G-factorization is semisurjective. This follows from the fact that the factorization is bicyclic. Fixing labeling of G gives in general a weakly semisurjective G-factorization. None of the above is true in general for swapping labeling, but for certain graphs it is possible to obtain swapping labelings that give weakly semisurjective factorizations.

The main idea is to *reduce* the graph G' (a caterpillar or a lobster; see [14,15]) on $2n + 2m$ vertices into two smaller graphs, one on $2n$ vertices and the second on $2m$ vertices, usually both with the same diameter. Let us focus on the first possibility reducing G' to G on $2n$ vertices; the second is done analogously. In the reduction we remove $2m$ leaves $v_1, v_2, \ldots, v_m$ and $u_1, u_2, \ldots, u_m$ to obtain a graph on $2n$ vertices. Suppose we have a semisurjective (or weakly semisurjective) G-factorization of K_{2n} with vertex partition V_0, V_1, both V_i of size n. Having all the vertices $v_1, v_2, \ldots, v_m$ in G' adjacent to vertices of V_0 and all the vertices $u_1, u_2, \ldots, u_m$ adjacent to vertices in V_1 we have reduced G' to G.

To achieve this reduction in general, we define the following.

Definition 14. Let N be the graph obtained from P_4 with vertices y_0, x_0, x_1, y_1 and edges $y_0 x_0, x_0 x_1, x_1 y_1$ by blowing up the inner vertices x_0, x_1 by K_n the end-vertices y_0, y_1 by $\overline{K_m} = m K_1$. The edge $x_0 x_1$ of P_4 is replaced by $K_{n,n}$ and the edges $y_0 x_0, x_1 y_1$ of P_4 are replaced by the edges of $K_{m,n}$ and $K_{n,m}$, respectively. The vertex set of N is then $V(N) = Y_0 \cup X_0 \cup X_1 \cup Y_1$, where $|X_0| = |X_1| = n$, $|Y_0| = |Y_1| = m$, and $|V(N)| = 2n + 2m$.

Similarly let Z be the graph obtained from P_4' with vertices x_0, y_1, y_0, x_1 and edges $x_0 y_1, y_1 y_0, y_0 x_1$ by blowing up the inner vertices y_0, y_1 of P_4' by K_m and the end-vertices x_0, x_1 by $\overline{K_n} = n K_1$. The edge $y_1 y_0$ of P_4' is replaced by $K_{m,m}$ and the edges $x_0 y_1, y_0 x_1$ of P_4' are replaced by the edges of $K_{n,m}$ and $K_{m,n}$, respectively.

The vertex set of Z is then the same as the vertex set of N, $V(Z) = V(N)$.

It is easy to observe that the graphs N and Z factorize the complete graph K_{2n+2m} into factors N and Z which may or may not be isomorphic (depending on whether $m = n$); see Fig. 7.14. We omit the proof.

Lemma 1. *If a graph G decomposes both graphs N and Z then G decomposes the complete graph K_{2n+2m}.*

Each of the graphs N and Z can be further factorized into isomorphic factors. The reduction is indeed the key step in a recursive method described in detail in [14]. The inductive step is based on m, therefore we try to keep m as small as possible. To prove the base for the induction a long list of graphs, called *starting cases*, of relatively small size up to a certain value had to be shown to factorize (weakly) semisurjectively the corresponding complete graph. This required a rather elaborate approach; see [14, 15].

7.2.2 Limitations

The use of the method for a G-factorization of complete graphs described in Sect. 7.2.1 is limited. The inductive step goes either by m or by n and naturally

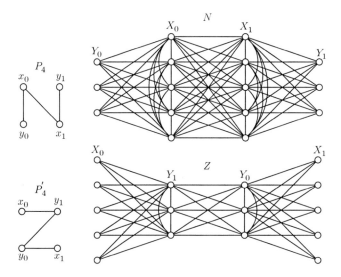

Fig. 7.14 Graphs N and Z for $n = 5$ and $m = 3$

we choose the smaller value, say m. Depending on the examined class of graphs the list of starting cases can already be, for $m = 3$ or $m = 5$, considerably long. For each of these graphs a labeling that guarantees factorization has to be found. Various labelings have to be used for various base graphs; the choice of labeling depends on the parity of n. In Sect. 7.4 we try to propose a unified and a more general approach.

Moreover, the recursion described in Sect. 7.2.1 strongly depends on the structure of the graphs N and Z. They are both of diameter 3; this may be a restriction on the structure of G.

7.3 Overview of Known Results

Among all graph factorizations we focus on spanning tree factorizations of complete graphs. For other classes of graphs we refer to [3, 6].

7.3.1 Trees

The decomposition of complete graphs has attracted attention for more than four decades. This research was definitely inspired by the famous conjecture by Ringel in 1963 saying that every tree with m edges decomposes the complete graph K_{2m+1}. Kotzig extended this conjecture saying that all trees are graceful; see, e.g., [35]. For an up-to-date survey of results on graceful and ρ-labelings see [17].

In this chapter we focus not on tree-decompositions in general, but on the special case of tree-factorizations. Decompositions into smaller trees or spanning trees of small diameters but not necessarily isomorphic were examined in [4,5,30].

In contrary to the search for graceful labelings of graphs the area of spanning tree factorization remained fairly unexplored, save the well-known Hamiltonian path factorization and the factorization into symmetric double-stars. In 1997, Eldergill [8] in his masters thesis, gave a necessary and sufficient condition for the existence of a cyclic factorization of K_{2n} into symmetric spanning trees (for a definition see page 177). He introduced ρ-symmetric graceful labelings and symmetric graceful labelings and classified all trees of order 10.

Another step was made by Fronček [10,12], who introduced blended ρ-labelings. By the method of blended ρ-labelings he found a wider class of trees on $4n + 2$ vertices that factorize the corresponding complete graph. In [16] the switching blended labeling was introduced for factorizations of complete graphs on $4n$ vertices by Fronček and Kubesa. Both switching blended labeling and ρ-symmetric graceful labeling require strong types of automorphisms of the factorizing graph. This narrows down the class of permissible trees.

More special classes of spanning trees were shown to factorize K_{4n+2} in [10,25] using the methods mentioned above. A spanning tree of any diameter that factorizes K_{4n+2} was found by Fronček in [12]. This result was completed in 2004 for K_{4n} by Kovářová in [20]. Among the most general results so far is the classification of caterpillars with diameter 4 (in a series of papers by Fronček [11], Kubesa [24,25]) that was completed later by Kovářová [21, 22]. The classification of caterpillars of diameter 5 was proved through the years in a series of papers by Kubesa [23, 24, 26–28] and finally completed in [14] by Fronček et al. The constructions of labelings of caterpillars with diameter 5 become technical; many subclasses need to be considered separately.

During the years 2002–2004 the methods (graph labelings) evolved. Fronček and Kovářová defined the $2n$-cyclic labeling (in [13]), then Kovářová defined fixing labeling (in [21]) and swapping labeling (in [22]) to cover the various cases.

The classification of all trees with at most four nonleaf vertices was completed in [15]. Vetrík in [34] found some caterpillars of diameter 6 that factorize the corresponding complete graph. The classification of caterpillars for diameters 3, 4, 5 and lobsters with four nonleaf vertices is complete since structural necessary conditions were proven by Fronček in [11] and by Fronček et al. in [15]. For example, for the trees with at most four nonleaf the following was shown in [15].

Theorem 9. *Let T be any tree with exactly four nonleaf vertices v_1, v_2, v_3 and v_4 such that* $\deg(v_1) \geq \deg(v_2) \geq \deg(v_3) \geq \deg(v_4) \geq 2$. *If T factorizes K_{2n} then*

(i) either $\deg(v_1) = n$ *and* $\deg(v_2) + \deg(v_3) + \deg(v_4) = n + 2$,
(ii) or $\deg(v_1) + \deg(v_4) = \deg(v_2) + \deg(v_3) = n + 1$.

The problem of spanning tree factorization of complete bipartite graphs $K_{m,n}$ was positively answered by Shibata and Seki in [32] and independently by El-Zanati and Vanden Eynden in [9]. For all positive integers m, n, where $m + n - 1$ divides mn, there exists a spanning tree factorization of $K_{m,n}$.

The general problem of finding a factorization for a given spanning tree remains far from being solved.

7.3.2 Disproof of Kubesa's Conjecture

Suppose there exists a T-factorization of K_{2n}. The observation that for symmetric graceful labeling, blended labeling, swapping labeling, and fixing labeling in general and for any known example it was always possible to split the vertex set of T into two equal-sized partite sets so that the sum of degrees was the same in each partite set led Kubesa to the conjecture that this is a necessary condition for a T-factorization. This conjecture was disproven by Meszka [29] who found an infinite class of trees factorizing a complete graph but not allowing such a partitioning of the vertex set. The smallest counterexample has 56 vertices.

On the other hand Tan [33] showed that the conjecture holds for trees such that: (1) the tree T has at most three different degrees or (2) the maximum degree $\Delta(T)$ is at most 4 or at least $n - 3$.

7.3.3 Packings, Coverings, and Orthogonal Double Covers

Both G-decompositions and G-factorizations require every edge of the decomposed complete graph to appear in precisely one copy of G. This restriction can be relaxed in two ways. Requiring every edge of K_{2n} to appear in at most one G results in a *packing* of G into K_{2n}. For a survey of results we refer to [36]. On the other hand if every edge of G has to appear in at least one copy of G we talk about *covering* K_{2n} by G. Moreover if every edge of K_n is covered by precisely two subgraphs and any two subgraphs share exactly one edge we talk about an *orthogonal double cover*; see [19]. Usually there are additional requirements on both the coverings or packings. There are a number of results; we refer the reader to [6].

7.4 Recursive Construction

In this section we describe a method that generalizes all of the methods for a G-factorization of K_{2n} presented in Sect. 7.1. We start by defining/redefining some terms. Later in this section the method of recursive labeling is presented and in the last part examples are given.

7.4.1 About the Method

First we introduce a few terms to simplify the description. A *permutation* π of a set A is a bijection $\pi : A \to A$. By ι we denote the identity permutation and by α_k the *cyclic permutation* of the set $V_i = \{0_i, 1_i, 2_i, \ldots, (k-1)_i\}$ given by $\alpha_k(a_i) = (a+1 \pmod{k})_i$ for any $a_i \in V_i$, where i is an integer and $0 \leq a \leq k-1$.

Definition 15. Given an (underlying) tree T on $2n$ vertices, two vertices r, s in T, and a positive integer k we can construct the *blown-up* graph $U[T; k; r, s]$. We replace the vertices r, s by K_k, and each vertex $i \in V(T)$, $i \neq r, s$ by $\overline{K_k}$. Such a blown-up vertex i we denote by H_i. Every edge $ij \in E(T)$ is replaced by a complete graph $K_{k,k}$ denoted by H_{ij}.

Example 12. In Fig. 7.15 is shown the caterpillar $T \simeq R_1$ with two vertices r, s chosen. Below is a blown-up graph $U[T; k; r, s]$ for $k = 3$.

The vertices r, s in the definition cannot be chosen arbitrarily. If T factorizes K_{2n} we require r, s to be mapped to each vertex of K_{2n} throughout the factors. In a bicyclic factorization such pairs of vertices always exist.

Definition 16. Let $G_1, G_2, \ldots, G_n$ be a G-factorization of K_{2n} induced by the isomorphisms ϕ_j, $G_j = \phi_j(G)$ for $j = 1, 2, \ldots, n$. A pair of vertices $x, y \in G$ is called a *semisurjective pair* if $\{\phi_j(x): j = 1, 2, \ldots, n\} = V_x$, $\{\phi_j(y): j = 1, 2, \ldots, n\} = V_y$, where $V_x \cap V_y = \emptyset$, and $V_x \cup V_y = V(K_{2n})$.

If λ is a vertex labeling of $U[T; k; r, s]$, it will be useful to work with a restriction of λ to the vertex set of H_i or H_{ij}, respectively.

Definition 17. Let λ be a vertex labeling of $U[T; k; r, s]$. We say the labeling $\lambda_i : H_i \to \{0_i, 1_i, \ldots, (k-1)_i\}$ is *induced* by λ if $\lambda_i(v) = \lambda(v)$ for all $v \in V(H_i)$. Similarly the labeling $\lambda_{ij} : H_{ij} \to \{0_i, 1_i, \ldots, (k-1)_i, 0_j, 1_j, \ldots, (k-1)_j\}$ is *induced* by λ if $\lambda_{ij}(v) = \lambda(v)$ for all $v \in V(H_{ij})$.

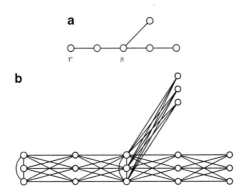

Fig. 7.15 (a) An underlying tree T; (b) a blown-up graph $U[T; k; r, s]$

7 Decompositions and Factorizations of Complete Graphs

The labeling introduced below extends features of the fixing labeling described in Sect. 7.1.6. We rewrite the definition of fixable and fixed vertices of T. By $N(v)$ we denote the set of vertices adjacent to v.

Definition 18. Let T be a tree on $2n$ vertices; let r, s be distinct vertices in T and k a positive integer. Let G be a subgraph in $U[T; k; r, s]$. If F is a set of vertices in T such that

$$F = \{i \in V(T): i \neq r, s \text{ where } \deg_G(x_i) = 1 \, \forall x_i \in V(H_{ij}) \text{ and } \forall j \in N(i)\},$$

then F is called a set of *fixable vertices in T* for a given G. All vertices in F are called *fixable*. Let $V_F \subseteq F$ be any set of vertices that are independent in T. We call V_F a *fixed set* and all vertices $i \in V_F$ we call *fixed vertices*.

We point out that in comparison to the previous Definition 12, in Definition 18, the requirement of T being symmetric has been lifted. For an example see Fig. 7.16.

Now we can define a new labeling that we will use to factorize complete graphs into spanning trees.

Definition 19. Suppose T factorizes the complete graph K_{2n} and r, s form a semisurjective pair of vertices in T. Let G be a graph with $2nk - 1$ edges, $n, k > 0$, and the vertex set

$$V(G) = \cup_{i=0}^{2n-1} V_i, \quad \text{where } |V_i| = k \text{ and } V_i \cap V_j = \emptyset \text{ for } i \neq j.$$

Let V_F be a set of fixed vertices in T for the graph G; V_F may be empty.

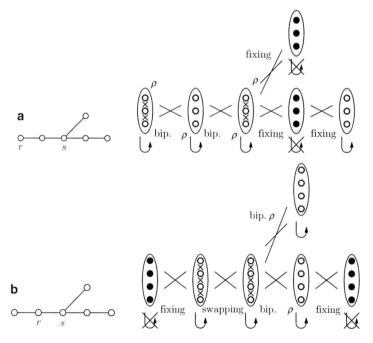

Fig. 7.16 The idea of recursive labeling: (**a**) for k odd; (**b**) for k even

We say G has a *recursive labeling* if there exists an injection $\lambda : V_i \to \{0_i, 1_i, \ldots, (k-1)_i\}$ for all $i = 0, 1, \ldots, 2n-1$ such that the induced labelings have the following property.

- For k odd

  (1) the labeling λ_{ij} induced on H_{ij}, for all $ij \in E(T)$, is either a bipartite ρ labeling or one of the vertices i, j is a fixed vertex in T (for a nonempty V_F),

  (2) the labelings λ_r and λ_s induced on H_r and H_s are ρ-labelings.

- For k even

  (1) the labeling λ_{ij} induced on H_{ij}, for all $ij \in E(T)$, $ij \neq rs$, is either a bipartite ρ labeling or one of the vertices i, j is a fixed vertex in T (for a nonempty V_F),

  (2) the edge $rs \in E(T)$ and the labeling λ_{rs} induced on H_{rs} along with the edges of H_r and H_s is a swapping labeling.

For short we denote the graph H_{rs} along with the edges of H_r and H_s by $H_{rs} \cup H_r \cup H_s$.

7.4.2 The Method of Recursive Labeling

To avoid technical details in the proof of the following theorem we use properties of known labelings. First we prove two lemmas. Why we split the proof into lemmas becomes apparent in the last subsection.

Lemma 2. *Let G be a graph with $2nk - 1$ edges, $n, k > 1$, that allows a recursive labeling. Then there exists a G-decomposition of $U[T; k; r, s]$ into k copies of G.*

Proof. We have to show that we can find k factors $G_1, G_2, \ldots, G_k$ that factorize the graph $U[T; k; r, s]$ and that they are all isomorphic.

Suppose G is a graph with $2nk - 1$ edges that allows a recursive labeling λ. Let V_F be the set of fixed vertices of T for the recursive labeling λ. We can construct the graphs $G_1, G_2, \ldots, G_k$ in the following way. Take $V(G_t) = V(G)$, $t = 1, 2, \ldots, k$, and

$$E(G_t) = \{\pi_i^t(x_i)\pi_j^t(y_j) : x_i \in V_i \text{ and } y_j \in V_j \text{ and } x_i y_j \in E(G)\}$$

where each π_i is a bijection on the set $\{0_i, 1_{,i}, \ldots, (k-1)_i\}$ defined by

$$\pi_i(x_i) = \begin{cases} (x+1 \pmod{k})_i & \text{if } i \notin V_F \\ x_i & \text{if } i \in V_F \end{cases}$$

7 Decompositions and Factorizations of Complete Graphs

for all $i = 0, 1, \ldots, 2n-1$. Notice that if $i \in V_F$ then $\pi_i = \iota$. The power π_i^t denotes the t-th composition $\pi_i \circ \pi_i \circ \ldots \circ \pi_i$. Obviously all graphs G_t for $t = 1, 2, \ldots, k$ are isomorphic to G since $\pi_i^t(x_i)\pi_j^t(y_j) \in E(G_t) \Leftrightarrow x_i y_j \in E(G)$.

To show that $G_1, G_2, \ldots, G_k$ form a G-factorization of $U[T; k; r, s]$ it is enough to show that every edge of $U[T; k; r, s]$ appears in some copy G_t. By the definition of recursive labeling each λ_{ij} (the restriction of λ to a H_{ij} if $ij \in T$) is for k odd either a bipartite ρ labeling if $i, j \notin V_F$ or a fixing labeling of a bipartite graph if $i \in V_F$ and $j \notin V_F$. The restrictions λ_r to H_r and λ_s to H_s give a ρ-labeling. The factors are obtained by π_i, π_j as shown in the proofs of Theorems 3, 2, and 8. For k even the reasoning similar is. All edges of H_{rs}, H_r, and H_s appear in some G_t by Theorem 6. $\qquad \square$

We require $k > 1$ otherwise $G \simeq T$ and the task is trivial and $n > 1$ since there have to be at least two partite sets in $U[T; k; r, s]$.

Lemma 3. *Let T be a tree that factorizes K_{2n} and let r, s be a semisurjective pair of vertices in T. Then $U[T; k; r, s]$ factorizes K_{2nk} into k isomorphic copies of $U[T; k; r, s]$.*

Proof. Let $T_1, T_2, \ldots, T_n$ be a T-factorization of K_{2n}. Obviously since T factorizes K_{2n}, the blown-up graph $T[\overline{K}_k]$ factorizes the complete multipartite graph $K_{2n}[\overline{K}_k] \simeq K_{k,k,\ldots,k}$. Now each $U[T_i; k; r, s]$, $i = 1, 2, \ldots, n$, contains besides the edges of $T_i[\overline{K}_k]$ also $2\binom{k}{2}$ edges of two K_k in the partite sets V_r and V_s. Since vertices r, s are a semisurjective pair, each partite set V_i for $i = 0, 1, \ldots, 2n-1$ will be blown up by K_k in exactly one copy of $U[T; k; r, s]$. This completes the proof. $\quad \square$

Theorem 10. *Let a graph G with $2nk - 1$ edges, $n, k > 1$, allow a recursive labeling. Then there exists a G-decomposition of K_{2nk} into nk copies of G such that if x, y is a semisurjective pair of vertices in T with a fixed set V_F, then all pairs i_x, j_y, $i_x \in V_x$, $j_y \in V_y$ for $i, j = 0, 1, \ldots, k-1$ form a semisurjective pair in G for k odd if $x, y \notin V_F$ and for k even if $x, y \notin (V_F \cup \{r, s\})$.*

Proof. By Lemma 2 the graph G decomposes $U[T; k; r, s]$. Let T be the underlying tree of $U[T; k; r, s]$. Since T factorizes K_{2n} (by the definition of $U[T; k; r, s]$) then $U[T; k; r, s]$ factorizes K_{2nk} into k isomorphic copies as shown in Lemma 3. Thus G decomposes K_{2nk} into nk isomorphic copies.

Now it remains to examine the semisurjective pairs. Since vertices x, y are a semisurjective pair in T, then by definition $V_x = \{\phi_i(x): i = 0, 1, \ldots, n-1\}$, $V_y = \{\phi_i(y): i = 0, 1, \ldots, n-1\}$, $V_x \cap V_y = \emptyset$, and $|V_x| = |V_y| = n$.

Now each vertex x of T, $x \notin V_F$ is taken by the permutation π_x through exactly k vertices $0_x, 1_x, \ldots, (k-1)_x$ in V_x. Thus x is semisurjective and having a semisurjective pair x, y in T, the pairs i_x, j_y are semisurjective for all $i, j = 0, 1, \ldots, k-1$. However, this is not true in general for all vertices in V_r, V_s for k even when λ_{rs} induced on H_{rs} is a swapping labeling. The proof is complete. $\qquad \square$

Notice that for k even some vertices of G such that λ_{rs} is a swapping labeling of $H_{rs} \cup H_r \cup H_s$ may also be a semisurjective pair.

Corollary 1. *If the graph G with $2nk - 1$ edges is a tree, then the G-decomposition becomes a G-factorization of K_{2nk}.*

Now it should be clear why we call such labeling *recursive*. If G is a tree it can become an underlying tree $T' \simeq G$ of a larger graph $U[T'; k'; r', s']$ for some integer k' and a semisurjective pair r', s' in T'. The vertices r', s' exist by Theorem 10.

7.4.3 Examples

Example 13. In [14] it was shown that the caterpillar R_4 in Fig. 7.17 factorizes the complete graph K_8. R_4 has a symmetric ρ-labeling, but it does not have a swapping labeling. A recursive labeling of R_4 is given in Fig. 7.18. Both factors of $U[P_4; 2; r, s]$ are given below. The underlying graph is P_4, which factorizes K_4 so that the two nonleaf vertices form a semisurjective pair; see Fig. 7.14. The vertex labels are implied by the drawing by adopting the convention from Example 11 and choosing a fixed set V_F containing both leaves of P_4.

Example 14. It may seem that the lobster in Fig. 7.19 has a recursive labeling. Indeed the labeling induced on $H_{01} \cup H_0 \cup H_1$ is a swapping labeling, the labeling induced on H_{12} is a bipartite ρ-labeling, and we are fixing the vertices in V_3. Also the underlying graph P_4 factorizes K_4 as in the previous example. But the pair of vertices r (a leaf) and s (nonleaf) of P_4 is *not* a semisurjective pair in P_4. In fact it was shown in [15] that this lobster does not factorize K_8 at all. The proof is rather technical and is based on a careful examination of adjacent nonleaf vertices and counting edges among these vertices and adjacent leaves.

Fig. 7.17 Caterpillar R_4 and the underlying tree P_4

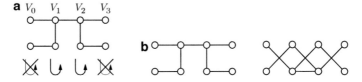

Fig. 7.18 Caterpillar R_4: (**a**) recursive labeling; (**b**) two factors

Fig. 7.19 A lobster that does not factorize K_8 and its underlying tree

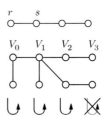

Notice that if T with m edges either decomposes or factorizes K_{2n} then in general it does *not* necessarily decompose K_{2nk}, since there are m edges in T and m divides $n(2n-1)$, but there are $nk(2nk-1)$ edges in K_{2nk} and m does not necessarily divide $nk(2nk-1)$.

7.4.4 Comparison to Previously Defined Labelings

The recursive labeling from Sect. 7.4.1 extends both of the most general methods (fixing and swapping labelings) known so far for spanning tree factorizations of complete graphs. In comparison to fixing labeling the underlying graph T does not have to be symmetric. Taking a symmetric underlying graph and restraining k to odd values the recursive labeling becomes a fixing labeling. Taking $n = 1$ (two partite set V_0 and V_1) the recursive labeling becomes for k even a swapping labeling and for k odd a blended labeling.

Moreover, the pairs of semisurjective vertices are specified. Thus by finding a recursive labeling of some tree G (not a general graph G if we require connectivity) we can easily proceed by blowing it up again and searching for recursive labeling of the blown-up graph as well. Or we can use such a method similarly as in Sect. 7.2 and in certain cases "reduce" a given tree G to smaller trees and search for recursive labeling of the smaller graph. Then G will also have a recursive labeling by the second part of Theorem 10.

7.5 Conclusion

The problem of spanning tree factorization of complete graphs has been studied for more than 10 years. No necessary and sufficient condition is known for a tree to allow such factorization and no such easy condition is expected to exist. There are several methods used to find factorizations; in this chapter we gave a more general approach combining most of the previously introduced concepts. Yet a complete classification of all trees whether they do or do not factorize the corresponding complete graph is far from being solved. On the other hand the presented techniques can be used not only for trees, but also for general graphs.

7.5.1 Further Possible Generalizations

There were several assumptions made during the description of recursive labeling, that in some cases can be lifted and still a decomposition or a factorization of the complete graph K_{2nk} will be obtained. We list them here.

- The permutations π_i used for cyclic decompositions do not have to be cyclic. This would lead to a more general statement in Lemma 2 whereas Lemma 3 and Theorem 10 would remain unchanged.
- The underlying graph T does not have to be a tree. However if T is not spanning, we cannot obtain a G-*factorization* of K_{2nk}. On the other hand, if T is spanning and contains more then $2n - 1$ edges, then more than two vertices of T have to be replaced by K_k in the process of blowing up. This would result in modification of Lemma 3. Hence also the definition of $U[T; k; r, s]$ has to be modified and the notion of a semisurjective *set* has to be introduced. This would lead only to a slight modification of Theorem 10.
- For carefully picked underlying trees and their copies one can use this method not only for decomposition of complete graphs but also of other graphs. On the other hand, one should not expect this method to give general results.

7.5.2 Further Extensions

The firefighters network is usually overdesigned. In idle time we can perform network measuring for each possible connection between two units, and compute and broadcast *reliabilities* for each connection. Based on this a weighted complete graph can be constructed. The zero weight stands for a connection loss between a pair of units. Now the reliability for each of the precomputed factors can be obtained and among all the factors in the list we can instantly pick the "best." We can expect that this network will have the longest operational time.

Another enhancement may lay in permuting the vertex set prior choosing a particular factor. If it is apparent from the measured data that one unit has a weak connection to many units (this unit may have strained from the group), we can reorder the IDs of the vertices so that this particular vertex forms a leaf in the most reliable factors. Hereby we enforce a more stable network.

The proposed method can be used also for large-scale networks, both locally and globally. We can also use various factors on subnets of a larger net, e.g., on a set of relatively close vertices. The main advantage is to control the diameter and a bounded maximal degree. This can be subject to further research.

Acknowledgments Supported by the Ministry of Education of the Czech Republic Grant No. MSM6198910027. The author wants to thank the anonymous referees, whose comments helped in improving the quality of this chapter.

References

1. Alspach B, Gavlas H (2001) Cycle decompositions of K_n and $K_n - I$. J Combin Theory B 81:77–99
2. Bloom GS (1979) A chronology of the Ringel-Kotzig conjecture and the continuing quest to call all trees graceful. In: Topics in graph theory. The New York Academy of Sciences
3. Bosák J (1986) Decompositions of graphs. VEDA, Bratislava
4. Bosák J, Erdös P, Rosa A (1972) Decompositions of complete graphs into factors with diameter two. Mat časopis Slov akad vied 22:14–28
5. Bosák J, Rosa A, Znám Š (1968) On decompositions of complete graphs into factors with given diameters. In: Theory of Graphs (Proc Colloq Tihany 1966), Akademiai Kiadó, Budapest, pp 37–56
6. Colbourn CJ, Dinitz JH (eds) (2007) The CRC handbook of combinatorial designs, 2nd edn. Discrete mathematics and its applications. CRC press, Boca Raton
7. Diestel R (2006) Graph theory, 3rd edn. Springer, Berlin Heidelberg
8. Eldergill P (1997) Decompositions of the complete graph with an even number of vertices. M.Sc. thesis, McMaster University, Hamilton
9. El-Zanati S, Vanden Eynden C (1999) Factorizations of $K_{m,n}$ into spanning trees. Graph Combinator 15:287–293
10. Fronček D (2004) Cyclic decompositions of complete graphs into spanning trees. Discuss Math Graph Theory 24(2):345–353
11. Fronček D (2006) Note on factorizations of complete graphs into caterpillars with small diameters. JCMCC 57:179–186
12. Fronček D (2007) Bi-cyclic decompositions of complete graphs into spanning trees. Discrete Math 307:1317–1322
13. Fronček D, Kovářová T (2007) $2n$-cyclic blended labeling of graphs. Ars Combinatoria 83:129–144
14. Fronček D, Kovář P, Kovářová T, Kubesa M (2010) Factorizations of complete graphs into caterpillars of diameter 5. Discrete Math 310:537–556
15. Fronček D, Kovář P, Kubesa M Factorizations of complete graphs into trees with at most four non-leave vertices, (submitted)
16. Fronček D, Kubesa M (2002) Factorizations of complete graphs into spanning trees. Congr Numer 154:125–134
17. Gallian JA (2003) A dynamic survey of graph labeling. Electron J Combinator DS6
18. Golomb SW (1972) How to number a graph, in graph theory and computing. In: Read RC (ed) Academic, New York, pp 23–37
19. Gronau H-DOF, Grütmüller M, Hartman S, Leck U, Leck V (2002) On orthogonal double covers of graphs. Design Code Cryptogr 27:49–91
20. Kovářová T (2005) Decompositions of complete graphs into isomorphic spanning trees with given diameters. JCMCC 54:67–81
21. Kovářová T (2004) Fixing labelings and factorizations of complete graphs into caterpillars with diameter four. Congr Numer 168:33–48
22. Kovářová T (2004) Spanning tree factorizations of complete graphs. Ph.D. thesis, VŠB – Technical University of Ostrava
23. Kubesa M (2004) Factorizations of complete graphs into [n,r,s,2]-caterpillars of diameter 5 with maximum center. AKCE Int J Graph Combinator 1(2):135–147
24. Kubesa M (2004) Factorizations of complete graphs into caterpillars of diameter four and five. Ph.D. thesis, VŠB – Technical University of Ostrava
25. Kubesa M (2005) Spanning tree factorizations of complete graphs. JCMCC 52:33–49
26. Kubesa M (2005) Factorizations of complete graphs into [r,s,2,2]-caterpillars of diameter 5. JCMCC 54:187–193
27. Kubesa M (2006) Factorizations of complete graphs into [n,r,s,2]-caterpillars of diameter 5 with maximum end. AKCE Int J Graph Combinator 3(2):151–161

28. Kubesa M (2007) Factorizations of complete graphs into [r,s,t,2]-caterpillars of diameter 5. JCMCC 60:181–201
29. Meszka M (2008) Solution to the problem of Kubesa. Discuss Math Graph Theory 28:375–378
30. Palumbíny D (1973) On decompositions of complete graphs into factors with equal diameters. Boll Un Mat Ital 7:420–428
31. Rosa A (1967) On certain valuations of the vertices of a graph. In: Theory of graphs (Intl Symp Rome 1966), Gordon and Breach, Dunod, Paris, pp 349–355
32. Shibata Y, Seki Y (1992) The isomorphic factorization of complete bipartite graphs into trees. Ars Combinatoria 33:3–25
33. Tan ND (2008) On a problem of Fronček and Kubesa. Australas J Combinator 40:237–245
34. Vetrík T (2006) On factorization of complete graphs into isomorphic caterpillars of diameter 6. Magia 22:17–21
35. West DB (2001) Introduction to graph theory, 2nd edn. Prentice-Hall, Upper Saddle River, NJ
36. Yap HP (1988) Packing of graphs – a survey. Discrete Math 72:395–404

Chapter 8
Geodetic Sets in Graphs

Boštjan Brešar, Matjaž Kovše, and Aleksandra Tepeh

Abstract Geodetic sets in graphs are briefly surveyed. After an overview of earlier results, we concentrate on recent studies of the geodetic number and related invariants in graphs. Geodetic sets in Cartesian products of graphs and in median graphs are considered in more detail. Algorithmic issues and relations with several other concepts, arising from various convex and interval structures in graphs, are also presented.

Keywords Geodetic number · Geodetic set · Cartesian product · Median graph · Boundary set

MSC2000: Primary 05C12; Secondary 05C99

8.1 Introduction

In this chapter by the notion of a *graph* we usually mean an undirected graph without loops or multiple edges. The only exception is in Sect. 8.6 where directed graphs are also briefly considered. As usual, $V(G)$ and $E(G)$ denote the vertex and the edge sets of a graph G, respectively. Unless stated differently n and m denote the number of vertices and number of edges of a graph, respectively. We start by formally defining the most basic concepts.

The length of a path is the number of its edges. Let u and v be vertices of a connected graph G. A shortest u, v-path is also called a u, v-*geodesic. The (shortest path) distance* is defined as the length of a u, v-geodesic in G and is denoted by $d_G(u, v)$, or $d(u, v)$ for short if the graph is clear from the context. The corresponding metric space is also called graphic metric space, associated with the graph G; see [40]. *The eccentricity* of a vertex v of a graph G is the maximum distance between v and any other vertex of G. *The diameter* of G, denoted by diam(G), is the maximum

---

A. Tepeh (✉)
University of Maribor, FEECS, Smetanova 17, 2000 Maribor, Slovenia
e-mail: aleksandra.tepeh@uni-mb.si

M. Dehmer (ed.), *Structural Analysis of Complex Networks*,
DOI 10.1007/978-0-8176-4789-6_8, © Springer Science+Business Media, LLC 2011

eccentricity of vertices in G, and *the radius* is the minimum such eccentricity. The *(geodesic) interval* $I(u, v)$ between u and v is the set of all vertices on all shortest u, v-paths. Given a set $S \subseteq V(G)$, its *geodetic closure* $I[S]$ is the set of all vertices lying on some shortest path joining two vertices of S; that is,

$$I[S] = \{v \in V(G) : v \in I(x, y), x, y \in S\} = \bigcup_{x,y \in S} I(x, y).$$

A set $S \subseteq V(G)$ is called *a geodetic set* in G if $I[S] = V(G)$; that is, every vertex in G lies on some geodesic between two vertices from S. *The geodetic number $g(G)$* of a graph G is the minimum cardinality of a geodetic set in G.

See [14] for many aspects of distances in graphs and the Encyclopedia of Distances [40] for a comprehensive collection of different appearances of distances in a variety of areas.

The problem of determining the geodetic number of a graph was initiated by Harary, Loukakis, and Tsouros in 1986 and their result appeared as a published paper in 1993, see [52]. It is shown there that finding the geodetic number of a graph is an NP-hard problem with its decision version being the NP-complete problem. They also determined geodetic numbers in some graph classes. Results about the geodetic number appeared in the literature before the publication of the article [52]; see [14, 15]. Even before that closely related two-person games on graphs, called *achievement* and *avoidance games*, were introduced and examined in [13]; see [50, 53, 75] for more on this topic.

Chartrand et al. [31] proved that for any integer k between 2 and n there exists a graph on n vertices with geodetic number k. Moreover, for positive integers $r, d, k \geq 2$ with $r \leq d \leq 2r$, there exists a connected graph G with $\mathrm{rad}(G) = r$, $\mathrm{diam}(G) = d$, and $g(G) = k$.

Several lower and upper bounds for the geodetic number involving different graph parameters have been found. Obviously $2 \leq g(G) \leq n$ for any connected graph G. Among graphs on n vertices only complete graphs attain the upper bound, while the family of graphs that attain the lower bound is much richer (see Sects. 8.3 and 8.4 for a discussion on median graphs with $g(G) = 2$). In [15] it is shown that $g(G) = 2$ if and only if there exist two vertices u and v with $d(u, v) = \mathrm{diam}(G)$ and every vertex of G lies on a geodesic between vertices u and v. It is also shown there that $g(G) = n-1$ if and only if $G = (K_{n_1} \cup K_{n_2} \cup \cdots \cup K_{n_r}) \oplus K_1$, where $r \geq 2$ and $n_1, n_2, \ldots, n_r$ are positive integers satisfying the equality $n_1 + n_2 + \cdots + n_r = n-1$ and $\oplus$ denotes the join of graphs (the only vertex not needed in a geodetic set is the cut vertex of G). Chartrand et al. [31] obtained the following general upper bound: $g(G) \leq n - \mathrm{diam}(G) + 1$. It is attained, for instance, by complete graphs and paths.

A vertex is *simplicial* (also *extreme* or *complete*) if its neighborhood induces a complete graph. A simplicial vertex obviously belongs to any geodetic set of a graph, hence, the number of simplicial vertices is a lower bound on the geodetic number [31]. A graph G is an *extreme geodesic graph* if all simplicial vertices of G already form a geodetic set (which is therefore also unique). Extreme geodesic graphs are therefore those graphs in which every vertex lies on

8 Geodetic Sets in Graphs

a geodesic between two simplicial vertices; see [29]. Trees and complete graphs provide typical examples of such graphs, and this implies that the following statement is true: for every pair a, n of integers with $2 \leq a \leq n$, there exists a connected extreme geodesic graph G on n vertices with $g(G) = a$. Moreover, in [29] it is also shown that for every pair a, b of integers with $0 \leq a \leq b$ and $b > 2$, there exists a connected graph G with a simplicial vertices and $g(G) = b$. On the other hand, all cut-vertices are clearly excluded from any minimum geodetic set. In addition, for any graph G containing k end-blocks, $g(G) \geq k$, see [28], where an *end-block* of graph G is a block containing only one cut-vertex.

Atici and Vince [2] provided a lower bound for the *edge geodetic number*, denoted by $g_e(G)$. Its definition differs from the definition of the usual geodetic number in that also all edges of a graph G must be included in some geodesic between two vertices from the geodetic set. Clearly $g_e(G) \geq g(G)$. In [2] the following lower bound is shown: $g_e(G) \geq \lceil 3 \log_3 \omega(G) \rceil$, where $\omega(G)$ denotes the clique number of graph G. Moreover, for any k there exists a graph G with $\omega(G) = k$ that contains an edge geodetic set with $\lceil 3 \log_3 k \rceil + \epsilon$ vertices, where $\epsilon \in \{0, 1\}$.

Using the probabilistic method together with a random greedy algorithm for obtaining geodetic sets, Chae et al. [23] determined the asymptotic behavior of the geodetic number of random graphs with fixed edge probability. Starting with an empty set S the so-called random greedy algorithm simply adds two nonadjacent vertices at every step and checks whether S is a geodetic set. In [23] it is shown that given a random graph $G_{n,p}$ on n vertices with a fixed edge probability p, the random greedy geodetic covering algorithm almost surely finds a geodetic set on $O(\log(n))$ vertices. A more detailed analysis shows that almost surely $g(G_{n,p}) = (1 + o(1)) \log_b n$, where $b = \frac{1}{1-p}$.

One can also study geodetic sets that are extremal in some way other than having the smallest cardinality. A geodetic set S is called a *minimal geodetic set* if no proper subset of S is a geodetic set. Of course, every minimum geodetic set is a minimal geodetic set, but the converse is not true as the example in Fig. 8.1 shows. Namely, $\{u, v, w\}$ is a minimal geodetic set of $K_{2,3}$ but not its minimum geodetic set (the unique minimum geodetic set is $\{x, y\}$). *The upper geodetic number* of a graph G, denoted by $g^+(G)$, is defined as the maximum cardinality of a minimal geodetic set of G. It is shown in [30] that for a connected graph G on n vertices and $g^+(G) = n - 1$ it follows that $g^+(G) = g(G)$. Moreover, for every pair a, b of integers with $2 \leq a < b$ there exists a graph G with $g(G) = a$ and $g^+(G) = b$.

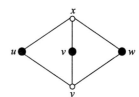

Fig. 8.1 A minimal geodetic set of $K_{2,3}$

A graph G is called a *minimum geodetic subgraph* if there exists a graph H, such that G is an induced subgraph of H and vertices of G form a minimum geodetic set. It is shown in [31] that a nontrivial graph G is a minimum geodetic subgraph if and only if every vertex of G has eccentricity 1 or no vertex of G has eccentricity 1. The subgraph of G induced by the vertices of maximum eccentricity is called the *periphery of a graph* G, denoted $Per(G)$. Together with the result by Bielak and Sysło [7] it follows that a nontrivial graph G is a minimum geodetic subgraph if and only if G is the periphery of some connected graph K.

Concepts related to geodesics also appear in the social sciences [81]. When members of a social group are represented by vertices, and particular relationships between members by edges of a graph, then the geodetic number can be considered as the smallest number n such that every member of a group is contained in a minimal chain of relationships between chosen n members. Beside this rather straightforward application, there are also related studies [8] where different models of (also social) networks and their properties are based on graph distance.

For more results on geodetic number and related concepts that have appeared before 2002 see the survey [32]. The emphasis of this chapter is to present some results about the geodetic number and related concepts that have appeared after [32] was published. They are often connected to graph classes arising in metric graph theory which is not surprising since the geodesic interval is one of the basic notions of this theory. A typical example of an operation suitable for studying graph metrics is the Cartesian product of graphs. The results on geodetic sets concerning this product are presented in Sect. 8.2. In Sect. 8.3 we present relevant results on median graphs which are one of the most important classes in metric graph theory. In Sect. 8.4 we follow with algorithmic issues concerning the geodetic number. In particular we present an efficient algorithm for recognizing median graphs with geodetic number 2. Next, several types of boundary sets have been defined on graphs and were studied with respect to convex properties and geodecity; see Sect. 8.5. In Sect. 8.6 we briefly overview several variations of geodetic sets and geodetic number that were studied intensively in the last 10 years.

8.2 Cartesian Products

The Cartesian product is a widely useful operation on graphs that appears in various settings. It is one of the four standard graph products [61], and the most important one in metric graph theory. Several well-known classes of graphs are Cartesian products: square-grids, hypercubes, Hamming graphs, prisms, etc.

The *Cartesian product* $G \,\square\, H$ of graphs G and H is the graph with the vertex set $V(G) \times V(H)$ in which vertices (g, h) and (g', h') are adjacent whenever $gg' \in E(G)$ and $h = h'$, or $g = g'$ and $hh' \in E(H)$. The most basic metric property of the Cartesian product operation is that for any graphs G and H,

$$d_{G \,\square\, H}((g, h), (g', h')) = d_G(g, g') + d_H(h, h').$$

8 Geodetic Sets in Graphs

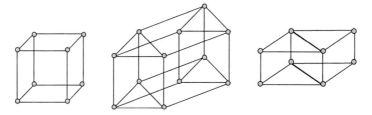

Fig. 8.2 Cartesian products: $K_2 \Box K_2 \Box K_2$, $K_3 \Box K_2 \Box K_2$ and $K_2 \Box (K_4 - e)$

In addition it is also easy to find the following useful observation about intervals in Cartesian products which is already a part of the folklore.

Lemma 1. *Let $X = G \Box H$ be the Cartesian product of (connected) graphs G and H and let (g, h) and (g', h') be vertices of X. Then $I_X[(g, h), (g', h')] = I_G[g, g'] \times I_H[h, h']$. Moreover, $I_X[(g, h), (g', h')] = I_X[(g', h), (g, h')]$.*

The Cartesian product is associative and commutative, and K_1 is the unit. Hence, one can define the Cartesian power in a natural way. The simplest nontrivial examples are powers of edge-graphs $(K_2)^k$ that are known as k-cubes Q_k or hypercubes. Using Lemma 1 and induction on k one quickly finds that $g(Q_k) = 2$; see also [15]. In Fig. 8.2 three examples of Cartesian products are depicted, namely the 3-cube, the Hamming graph $K_3 \Box K_2 \Box K_2$, and the product of K_2 and the diamond $K_4 - e$.

We start with general bounds on the geodetic number of the Cartesian product of two graphs, that are expressed in terms of geodetic numbers of factor graphs.

Theorem 1. *[63] Let G and H be graphs with $g(G) = p \geq g(H) = q \geq 2$. Then*

$$p \leq g(G \Box H) \leq pq - q.$$

These bounds were proved for the first time by Jiang et al. [63], who also constructed infinite families of graphs attaining each of the bounds. We present several classes of graphs that attain the lower bound in Theorem 1. For the upper bound, the following construction was given in [63]. First, graphs D_p^t are defined on vertices $x_1, \ldots, x_p, z$ and $y_1^{(ij)}, \ldots, y_t^{(ij)}$ for all i, j, $1 \leq i < j \leq p$ (altogether $p + 1 + t\binom{p}{2}$ vertices); where each of the vertices $y_1^{(ij)}, \ldots, y_t^{(ij)}$ is adjacent exactly to x_i, x_j and z, z is adjacent to all vertices in $D_p^t - z$, and there are no edges between any x_i and x_j. It is easy to see that $x_1, \ldots, x_p$ is a geodetic set in D_p^t, and in fact $g(D_p^t) = p$. The following result shows that the upper bound in Theorem 1 is attained by products of graphs D_p^t with complete graphs.

Proposition 1. *[63] Let p, q, t be positive integers such that $t > pq - q$. Then $g(D_p^t \Box K_q) = pq - q$.*

The bounds in Theorem 1 are rather far apart and may not be very useful, but they cannot be improved in general. By applying Lemma 1, Brešar et al. [12] found a simple proof of the upper bound in Theorem 1. They also looked for better bounds when a graph has a minimum geodetic set with some special structure.

Let G be a graph and let $S = \{x_1, \ldots, x_k\}$ be a geodetic set of G. Then S is a *linear geodetic set* if for any $x \in V(G)$ there exists an index i, $1 \le i < k$, such that $x \in I[x_i, x_{i+1}]$.

Theorem 2. *[12] Let G and H be graphs on at least two vertices with $g(G) = p$ and $g(H) = q$. Suppose that both G and H contain linear minimum geodetic sets. Then*

$$g(G \,\square\, H) \le \left\lfloor \frac{pq}{2} \right\rfloor.$$

Many graphs admit linear minimum geodetic sets; complete graphs and graphs G with $g(G) = 2$ are obvious instances of such graphs. For another example consider complete bipartite graphs $K_{n,m}$ with $n, m \ge 4$. It is known [31] and easy to see that $g(K_{n,m}) = 4$. Moreover, selecting the first two vertices of a minimum geodetic set from one bipartition set and the last two vertices from the other yields a linear geodetic set.

Combining Theorem 2 with the lower bound from Theorem 1 we infer

Corollary 1. *[12] Let G be a graph on at least two vertices that admits a linear minimum geodetic set and let H be a graph with $g(H) = 2$. Then $g(G \,\square\, H) = g(G)$.*

The above result also implies that the geodetic number of the product of two graphs with geodetic number 2 is also 2. For example, the Cartesian product of an arbitrary finite number of paths has geodetic number 2. In particular, this applies to hypercubes and square-grid graphs.

For another class of graphs attaining the lower bound in Theorem 1 the following property of geodetic sets was introduced in [12]. Let G be a graph. If S is a geodetic set of G such that

$$\forall u \in V(G) \setminus S, \forall v, w \in S : u \in I[v, w], \tag{8.1}$$

we say that S is a *complete geodetic set* of G. (Clearly any complete geodetic set is also a linear geodetic set.)

Proposition 2. *[12] Let G and H be nontrivial graphs both having a complete minimum geodetic set. Then $g(G \,\square\, H) = \max\{g(G), g(H)\}$.*

Examples of graphs having a complete minimum geodetic set include complete graphs, stars, and graphs with geodetic number 2. Odd cycles are examples of graphs that admit a linear minimum geodetic set but not a complete minimum geodetic set.

In several examples that achieve the lower bound in Theorem 1, the class of trees is involved. Note that the unique minimum geodetic set of a tree is the set of its leaves [31]; hence, the geodetic number of a tree is the number of its leaves.

8 Geodetic Sets in Graphs

Proposition 3. *(i) [12] For any trees T_1 and T_2 with ℓ_1 and ℓ_2 leaves,*

$$g(T_1 \,\square\, T_2) = \max\{\ell_1, \ell_2\}.$$

(ii) [84] For any tree T with ℓ leaves and a complete graph K_m,

$$g(T \,\square\, K_m) = \max\{\ell, m\}.$$

Cartesian products of even cycles and complete graphs also attain the best possible lower bound, as follows from Corollary 1. The complete answer for products of complete graphs and cycles was given by Ye et al. [84]: for $m \geq 2$,

$$g(C_n \,\square\, K_m) = \begin{cases} m, & n \text{ even, or } n \text{ odd and } m \geq n \\ m+1, & n \text{ odd and } m < n. \end{cases}$$

The Cartesian product of the form $G \square K_2$ is called the *prism over* G. Geodetic sets of prisms were considered in [31], where the authors observed the inequality $g(G\square K_2) \geq g(G)$, and provided a sufficient condition for the appearance of equality. A necessary and sufficient condition for the equality was obtained by Lu [68], and independently also in [10]. To present it, we need the following notation. For $S_1, S_2 \subseteq V(G)$, and $S_1 \cap S_2 = \emptyset$, let

$$I[S_1, S_2] = \{v \in V(G) : v \in I(x, y), x \in S_1, y \in S_2\},$$

i.e., the set of vertices that lie on some geodesic between a vertex of S_1 and a vertex of S_2.

Theorem 3. *[10, 68] Let G be a nontrivial connected graph. Then $g(G\square K_2) = g(G)$ if and only if G possesses a geodetic set S that can be partitioned into nonempty subsets S_1 and S_2 such that $(I[S_1] \cap I[S_2]) \cup I[S_1, S_2] = V(G)$.*

A characterization of the graph G that enjoys $g(G\square K_3) = g(G)$ was obtained recently in [84]. It is more complex than the above, and involves a partition of a geodetic set of G into three nonempty subsets enjoying several conditions. These conditions are not in an optimal form, and we present a more brief formulation. In addition, our approach can be applied to Cartesian products with arbitrary complete graphs.

Note that the condition in Theorem 3 is equivalent to the conjunction of conditions $I[S_1] \cup I[S_1, S_2] = V(G)$ and $I[S_2] \cup I[S_1, S_2] = V(G)$. We generalize it as follows (for notational convenience $I[S_i, S_i]$ stands for $I[S_i]$).

Theorem 4. *Let G be a nontrivial connected graph. Then $g(G\square K_k) = g(G)$ if and only if G possesses a geodetic set S that can be partitioned into nonempty subsets $S_1, \ldots, S_k$ such that for all $i = 1, \ldots, k$:*

$$\bigcup_{j=1}^{k} I[S_i, S_j] = V(G).$$

Proof. Note that vertices of $G \square K_k$ can be written as follows: for any $x \in V(G)$, and $i \in \{1, \ldots, k\}$, let $x_i = (x, i)$. Then x_i is adjacent to x_j for $j \neq i$, and also to y_i whenever $xy \in E(G)$.

Suppose that G possesses a geodetic set S that can be partitioned into nonempty subsets $S_1, \ldots, S_k$ such that for all $i = 1, \ldots, k$, $\bigcup_{j=1}^{k} I[S_i, S_j] = V(G)$. Then let S' be the set of vertices in $G \square K_k$ defined as

$$\bigcup_{i=1}^{k} S_i \times \{i\}.$$

Let $x_i = (x, i)$ be an arbitrary vertex in $G \square K_k$. Then by the condition of the theorem, x lies in the interval between two vertices y and z from S_i (in which case, x_i clearly lies in the interval between y_i and z_i), or it lies in $I(y, w)$, where $y \in S_i$ and $w \in S_j$ for some $j \neq i$. In the latter case we infer by Lemma 1 that x_i lies in $I(y_i, w_j)$, where clearly $y_i, w_j \in S'$. Hence, $I[S'] = V(G \square K_k)$, and obviously $|S'| = |S| = g(G)$.

For the converse, let S' be a minimum geodetic set of $G \square K_k$ with $|S'| = g(G)$. Let $S_i' = S' \cap (V(G) \times \{i\})$, and note that $S_i' \neq \emptyset$ for all $i = 1, \ldots, k$, since S' is a geodetic set. Fix i and consider an arbitrary vertex x_i. Take any two vertices in S' for which x_i lies in the interval between them. Clearly, at least one of these vertices u_i is in S_i'. Hence, $x_i \in I(u_i, v_j)$, where v_j is the other vertex. Let S_t be the natural projection of S_t' on G for all $t = 1, \ldots, k$. Now, if $v_j \in S_i$, then $x \in I(u, v)$ where both $u, v \in S_i$. Otherwise, $x \in I(u, v)$, where $u \in S_i, v \in S_j$. Hence, $x \in I[S_i]$ or $x \in I[S_i, S_j]$. Since x was arbitrarily chosen, we infer $\bigcup_{j=1}^{k} I[S_i, S_j] = V(G)$, and we have proved this for an arbitrary i. The proof is complete.

8.3 Median Graphs

One of the most important classes of graphs derived from related metric structures is that of *median graphs*. They are defined as the graphs in which for every triple of vertices $u, v, w \in V(G)$ the intersection $I(u, v) \cap I(u, w) \cap I(v, w)$ consists of precisely one vertex (which is called the *median* of the triple u, v, w). Median graphs have been rediscovered several times, and a rich structure theory has been developed, cf. the survey [64] and the book [61]. Applications of median graphs can be found in computer science, phylogenetics, social choice theory, etc.; see, for example, [6, 25, 82]. Geodetic sets and the geodetic number in median graphs were first investigated in [10] and later also in [3, 11].

Brešar and Tepeh [10] studied minimum geodetic sets in median graphs with respect to the procedure that involves so-called peripheral subgraphs of a median graph. For a connected graph and an edge xy of G we denote

$$W_{xy} = \{w \in V(G) \mid d(x, w) < d(y, w)\}.$$

8 Geodetic Sets in Graphs

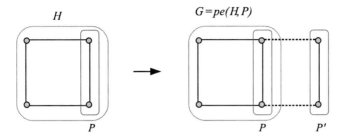

Fig. 8.3 Peripheral expansion G of a graph H

For an edge xy of G let U_{xy} denote the set of vertices u that are in W_{xy} and have a neighbor in W_{yx}. If for some edge xy, $W_{xy} = U_{xy}$, we call the set U_{xy} a *peripheral set* and a subgraph induced by a peripheral set is a *peripheral subgraph* (note that this is different from the concept of periphery as defined in Sect. 8.1).

A subset S of the vertex set of a graph G is *convex* if for any two vertices $u, v \in S$, $I(u, v)$ is a subset of S. Bandelt et al. [5] characterized median graphs as connected bipartite graphs in which for every edge ab of G, the sets U_{ab} and U_{ba} are convex. Another very useful characterization of median graphs due to Mulder is based on a special expansion procedure [70,72]. Let H be a connected graph and P its *convex subgraph*, meaning the subgraph, induced by a convex subset $V(P)$ of $V(H)$. Then the *peripheral expansion* of H along P is the graph G obtained as follows. Take the disjoint union of a copy of H and a copy of P (denoted by P' in Fig. 8.3). Join each vertex u in the copy of P with the vertex that corresponds to u in the copy of H (actually in the subgraph P of H). We say that the resulting graph G is *obtained by a (peripheral) expansion from H along P*, and denote this operation in symbols by $G = pe(H, P)$. Mulder [73] characterized median graphs as graphs that can be obtained from K_1 by a sequence of peripheral expansions; see also [9]. In [10] the following relation between geodetic numbers of a median graph and its peripheral expansion was obtained.

Theorem 5. *[10] Let H be a median graph, P its convex subgraph, and $G = pe(H, P)$. Then $g(H) \leq g(G) \leq g(H) + g(P)$.*

Moreover, it was shown that both bounds are sharp. In fact, by using the structure of minimum geodetic sets in H, situations when $g(G) = g(H)$ were characterized. Let S and T be two disjoint subsets of $V(G)$. Then a u, v-geodesic is called a *geodesic on S* if $u, v \in S$ and a u, v-geodesic is called a *geodesic between S and T* if $u \in S$ and $v \in T$.

Theorem 6. *[10] Let H be a median graph, P its convex subgraph, and $G = pe(H, P)$. Then $g(H) = g(G)$ if and only if there exists a minimum geodetic set S of H and its partition $S = S_1 \cup S_2$ such that $S_1, S_2 \neq \emptyset$, $S_2 \subseteq V(P)$ and for every $v \in V(P)$*

- *v lies on some geodesic between S_1 and S_2, or*
- *v lies on some geodesic on S_1 and on some geodesic on S_2.*

Fig. 8.4 A median graph with geodetic number 4

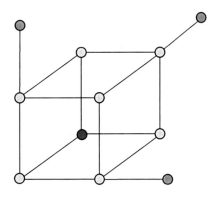

The upper bound in Theorem 5 is attained for an infinite family of median graphs (see [10] for the construction), which also gives the answer to the question of how large the difference can be between $g(pe(H, P))$ and $g(H)$.

Proposition 4. *[10] For every $k \in \mathbb{N}$ there exists a median graph G that can be obtained by the peripheral expansion from a median graph H such that $g(G) = g(H) + k$.*

Trees are median graphs in which peripheral subgraphs are precisely the one-vertex subgraphs induced by each leaf and the minimum geodetic set of a tree consists of all leaves [31]. But unlike in trees, it may happen that in a median graph a minimum geodetic set contains vertices that are not in a peripheral subgraph; see Fig. 8.4. On the other hand, it was shown in [10] that given a geodetic set S of a median graph, every periphery contains a vertex from S. This was the motivation for introducing the concept of the *periphery transversal number* $pt(G)$ as the smallest number of vertices that meet all peripheral subgraphs [3, 11]. Clearly, $pt(G) \leq g(G)$ and it turns out that median graphs with geodetic number 2 and those with periphery transversal number 2 coincide. As already mentioned in Sect. 8.1, general graphs with $g(G) = 2$ were characterized in [15]. Cagaanan et al. [22] proved that the geodetic number of a join of two connected noncomplete graphs is 2 if and only if both the diameter and the geodetic number of one of the graphs is 2. However, much more is known about median graphs with geodetic number 2. Before we give the complete list of thus far known characterizations of such graphs we need some preliminary notions.

A subgraph H of a graph G is *isometrically embeddable* in G if $d_H(u, v) = d_G(u, v)$ for every $u, v \in V(H)$. Edges $e = xy$ and $f = uv$ of a graph G are in the Djoković–Winkler relation Θ if $d_G(x, u) + d_G(y, v) \neq d_G(x, v) + d_G(y, u)$ [41,83]. Relation Θ is reflexive and symmetric. Winkler [83] proved that it is also transitive in *partial cubes* (i.e., isometric subgraphs of hypercubes), and so it is an equivalence relation on the edge set of any median graph, since median graphs are partial cubes [71]. By the result of Eppstein [45] every partial cube G can be isometrically embedded into the n-dimensional grid $\mathbb{Z}^n$ for sufficiently large n. Denote by p_i the projection from $\mathbb{Z}^n$ to a copy of $\mathbb{Z}$, denoted by $\mathbb{Z}_i$. Clearly $p_i(G)$ is a discrete interval

8 Geodetic Sets in Graphs   207

$[a_i, b_i] \subset \mathbb{Z}_i$ of length $d_i = b_i - a_i$. We say that vertices u and v of a graph G which is isometrically embedded in $\mathbb{Z}^n$ are *antipodal* if $d(u, v) = d_1 + d_2 + \cdots + d_n$.

A *profile* $\pi = (x_1, \ldots, x_k)$ on a graph G is a finite sequence of vertices of G. Given a profile π on G and a vertex u of G, the *remoteness* $D(u, \pi)$ see [67] is

$$D(u, \pi) = \sum_{x \in \pi} d(u, x).$$

Theorem 7. *Let G be a median graph. Then the following assertions are equivalent:*

(i) $g(G) = 2$.

(ii) G can be obtained by a sequence of peripheral expansions from K_1 such that in each peripheral expansion step the convex subgraph with respect to which the expansion is performed has a nonempty intersection with some minimum geodetic set [10].

(iii) There exist vertices a, b in $V(G)$ and an a, b-geodesic Π such that every Θ-class in G has an edge on Π [10].

(iv) G can be isometrically embedded in $\mathbb{Z}^n$ in such a way that there exist antipodal vertices a and b [10].

(v) $\mathrm{pt}(G) = 2$ [3].

(vi) $D(x, \pi)$ is constant on G for some profile π [3].

Characterizations (iii) and (iv) hold also in the more general case of partial cubes. A geometric interpretation of characterization (iv) is that a partial cube G has a geodetic set $\{a, b\}$ if and only if G can be isometrically embedded into $\mathbb{Z}^n$ in such a way that $\prod_{i=1}^{n}[a_i, b_i]$ contains $V(G)$. The characterization (vi) answers the following question from location theory: for which median graphs G is their vertex-set the (anti)median set of some profile on G; see [3] for more details. The fifth characterization leads to the question of whether there is a general connection between the geodetic number of a median graph and the structure that is derived from intersecting peripheral subgraphs. For this purpose the *periphery graph* $P(G)$ of a median graph G was introduced in [11] as the graph whose vertices are peripheral subgraphs in G and two vertices are adjacent in $P(G)$ if and only if the peripheral subgraphs intersect. It was proved that there are median graphs whose periphery graph has independence number 2, and have arbitrarily large geodetic number; see also [4].

8.4 Algorithms and Complexity

Harary et al. [52] observed that finding the geodetic number of a graph is an NP-hard problem with its decision variation being NP-complete. Douthat and Kong showed that the decision problem regarding the geodetic number of a graph remains NP-complete when restricted to the class of chordal graphs [43], and when restricted to the class of bipartite graphs [42]. They also noted that the problem becomes polynomially solvable in the class of split graphs. Atici later showed that the following more general decision problem is NP-complete: given a graph G and an integer $k \leq |V(G)|$, is there a set $S \subseteq V(G)$ with $|S| = k$ such that $I[S] = V(G)$ [1].

In [52] an algorithm for finding the geodetic number is also proposed. However, Hansen and van Omme [51] observed that the approach from [52] has a serious mistake and presented an example for which the algorithm fails and does not produce a geodetic set. They proposed another approach by developing a 0–1 integer programming model to find the geodetic number of a graph. We present it next.

For each vertex $v_k \in V(G)$ consider all geodesics passing through v_k. This can be easily done once the distance matrix of G is known, by finding pairs of vertices forming the following set

$$P_k = \{(v_i, v_j) \mid d(v_i, v_k) + d(v_k + v_j) = d(v_i, v_j), \ i < j\}.$$

Let S denote a minimum geodetic set of G. The idea is then to define binary variables of two types: x_k, $1 \leq k \leq |V(G)|$, and y_{ij}, where $(v_i, v_j) \in P_k$ and it holds that

$$x_k = \begin{cases} 1, & \text{if } v_k \in S \\ 0, & \text{if } v_k \notin S \end{cases}$$

and

$$y_{i,j} = \begin{cases} 1, & \text{if } v_i, v_j \in S \\ 0, & \text{otherwise} \end{cases}$$

The following model then constructs a minimum geodetic set:

$$\text{Minimize} \quad \sum_{k=1}^{n} x_k$$

subject to

$$1 - x_k \leq \sum_{(v_i, v_j) \in P_k} y_{ij}, \qquad v_k \in V(G), \tag{8.2}$$

$$y_{ij} \leq x_i, \qquad i = 1, \ldots, n, \tag{8.3}$$

$$y_{ij} \leq x_j, \qquad j = 1, \ldots, n, \tag{8.4}$$

$$x_i + x_j - 1 \leq y_{ij}, \qquad i, j = 1, \ldots, n, \tag{8.5}$$

and all variables are either 0 or 1.

In the objective function we want to minimize the size of a geodetic set. The first inequality ensures that $I[S] = V(G)$, and inequalities (8.3)–(8.5) force all variables

8   Geodetic Sets in Graphs   209

y_{ij} to be 0 or 1, depending on whether v_i and v_j belong to a minimum geodetic set or not. Finally, we note that in [52] results of some computational experiments were also given.

Nonetheless graphs with geodetic number 2 can be recognized in polynomial time. One needs to determine all diametrical vertices and check whether every diametrical vertex has a unique vertex realizing its eccentricity. For all such diametrical pairs, then check whether every other vertex lies on the interval between the pair. Clearly this can be done in polynomial time by using a distance matrix. Median graphs with $g(G) = 2$ can be recognized even more efficiently. The fastest recognition algorithm for median graphs is of complexity $O((m \log n)^{1.41})$ and is due to Imrich and Klavžar [61]. The exponent 1.41 stands for $2\omega/(\omega + 1)$, where ω denotes the exponent of matrix multiplication with its current value 2.376; see [39]. By the result of Imrich et al. [60] the recognition complexity of median graphs is closely related to the recognition complexity of triangle-free graphs. Therefore improving the recognition complexity of median graphs is believed to be very difficult. Some classes of median graphs can be recognized faster. For example, planar median graphs can be recognized in linear time [60] and graphs of acyclic cubical complexes can be recognized in $O(m \log n)$ time; see [59].

We follow with the result from [3] about the recognition of median graphs with $g(G) = 2$ in $O(m \log n)$ time. This is possible because of the upper bound on the maximum degree of a median graph with geodetic number two and the fact that every peripheral subgraph meets any geodetic set; see Theorem 7(ii). The mentioned upper bound reads as follows.

Lemma 2. *[3] If G is a median graph with $g(G) = 2$, then $\Delta(G) \leq 2 \log_2 n$.*

Moreover, if $\Delta(G) \leq 2 \log_2 n$ one can check in $O(m \log n)$ time whether G is a median graph, and determine all Θ-classes and all U_{ab} sets. The next lemma describes all geodetic sets of $G = pe(H; P)$ with regard to H.

Lemma 3. *[3] Let $G = pe(H; P)$ be a median graph and $\{x, y\}$ a geodetic set of H, where $y \in P$. Then the set $\{x, z\}$, where z is the neighbor of y in $G \setminus H$, is a geodetic set in G. Moreover, all minimum geodetic sets of G are of this form.*

If $\{x, y\}$ is a geodetic set in G then this is the only minimum geodetic set containing x, since, by Lemma 3, x is uniquely determined by y and vice versa. Moreover, it also follows that for a median graph $G = pe(H; P)$ with $g(G) = 2$, all minimum geodetic sets of G can be obtained from the minimum geodetic sets of H in $O(|P|)$ time. Therefore if the representation of G as a sequence of peripheral expansions, starting from K_1, is known, then all minimum geodetic sets of G can be obtained in $O(n)$ time. It remains then to find efficiently the representation of G by a sequence of peripheral expansions starting from K_1.

Theorem 8. *[3] Let G be a median graph with $\Delta(G) \leq 2 \log_2 n$. Then a representation of G by a sequence of peripheral expansions can be found in $O(m \log n)$ time.*

In the proof of Theorem 8 the following idea is used: the peripheral U_{ab}-sets are characterized by the fact that ∂U_{ab} consists of $|U_{ab}|$ independent edges that meet every vertex of a U_{ab}-set, where ∂U_{ab} is the set of edges with one endvertex in U_{ab} and the other in $G \setminus U_{ab}$. In other words, U_{ab} is peripheral if

$$\deg_G(v) = \deg_{U_{ab}}(v) + 1,$$

for every $v \in U_{ab}$. Clearly $\deg_{U_{ab}}(v) + 1 \le \deg_G(v)$ for $v \in G$. Thus, setting

$$ex_{U_{ab}}(v) = \deg_G(v) - \deg_{U_{ab}}(v) - 1$$

it is clear that U_{ab} is peripheral if and only if

$$ex(U_{ab}) = \sum_{v \in U_{ab}} ex_{U_{ab}}(v) = 0.$$

If $\{x, y\}$ is a geodetic set then a pair x and y is called a *geodetic pair*. Using the ideas described above an algorithm is presented in [3] that recognizes whether a given graph G is a median graph with $g(G) = 2$, and can be implemented to run within the time complexity $O(m \log n)$. In the case where G is a median graph with $g(G) = 2$ the algorithm also gives the list of all geodetic pairs of G.

The algorithm does the following. First it checks whether the maximum degree of G is bounded by $2 \log_2 n$. If the answer is positive, it checks whether G is also a median graph, and in which case it determines all Θ-classes and all U_{ab} sets. Next it computes the value of $ex(U_{ab})$ for all U_{ab} sets and then determines a peripheral U_{ab} set and removes it. Until one gets K_1, this dismantling procedure is repeated (at each step a peripheral U_{ab} set is determined and removed). Next we construct G back again starting from K_1 by a sequence of peripheral expansions, taking care of all geodetic pairs using Lemma 3. If at any step there are no such pairs, then $g(G) > 2$. Otherwise the algorithm also lists all geodetic pairs. As observed above, it can be implemented to run in $O(m \log n)$ time.

8.5 Boundary and Geodetic Sets

The study of abstract convexity spread in the early 1950s, and resulted in a purely axiomatic definition of convex sets that generalizes the classical concept of convexity in Euclidean spaces (see a survey on this topic [80], or [17] for a brief introduction to the abstract convexity in graphs). In this section we present results in which problems arising in convexity theory relate to geodetic sets.

Recall that a subset W of the vertex set of a graph G is *convex* if for any two vertices $u, v \in S$, $I(u, v)$ is a subset of S. The smallest convex set containing S is called the *convex hull* of S and is denoted by $CH(S)$. The operation of geodetic closure is somewhat similar to the convex hull operation. In fact, the concept of the geodetic set can be regarded as a variation of the following concept introduced

8 Geodetic Sets in Graphs 211

by Everett and Seidman [47]. The set S for which $CH(S) = V(G)$ is called the *hull set* of a graph G, and the size of a minimum hull set of G is called the *hull number* of G, denoted $h(G)$. Given a set S of vertices in a graph G, it is clear that $I[S] \subseteq CH(S)$. Hence, if S is a geodetic set of G then also $CH(S) = V(G)$, and so for any graph G, $h(G) \leq g(G)$. Much work concerning the hull number of graphs was done by Chartrand and Zhang, see [32] for some additional references on these studies. The hull number of median graphs was considered by Mulder [73], showing that $h(G) = pt(G)$ for any median graph G. The hull number of Cartesian products of graphs was studied in Cagaanan and Canoy [20,21], where it was shown that $h(G \square H) = \max\{h(G), h(H)\}$. Canoy et al. [22] considered the hull and the geodetic number of joins of two graphs.

Given a graph G and a convex set $S \subseteq V(G)$, a vertex $v \in S$ is called an *extreme vertex* of S if $S \setminus \{v\}$ is also convex. A graph G is called a *convex geometry* if every convex set of G is the convex hull of its extreme vertices (this property is also called the Minkowski–Krein–Milman property [48, 62, 65]). Graphs with this property are properly contained in the class of distance hereditary graphs [17]. A connected graph G is *distance hereditary* if for every connected induced subgraph H of G and every two vertices $u, v \in H$, $d_H(u, v) = d_G(u, v)$ [57]. A graph is *chordal* if it contains no induced cycle of length greater than 3 and a chordal graph is called *Ptolemaic* if it is distance hereditary. Farber and Jamison [48] proved that a graph G is a convex geometry if and only if G is Ptolemaic. Hence, if S is a convex set in a Ptolemaic graph G, then we can rebuild the set S from its extreme vertices using the convex hull operation. This cannot be done with every graph, however, Cáceres et al. [17] extended the set of extreme vertices of S to the *contour set*, $Ct(G)$, that enables one to rebuild S using the vertices in $Ct(G)$ and the convex hull operation. The contour set $Ct(G)$ of a graph G is defined by

$$\mathrm{Ct}(G) = \{v \in V(G) \mid \mathrm{ecc}(u) \leq \mathrm{ecc}(v), \forall u \in N_G(v)\}.$$

Theorem 9. *[17] Let G be a graph and S a convex subset of vertices. Then $S = CH(Ct(S))$.*

In order to find the convex hull of a set S one can start by taking the union of intervals over all pairs of vertices of S. Hence, one may ask whether the geodetic closure of the contour of S equals S. More specifically, when the contour set of a graph is also its geodetic set, we have the following (partial) result.

Theorem 10. *[17] Let G be a distance hereditary graph. Then $Ct(G)$ is a geodetic set for G.*

The geodeticity of the contour $Ct(G)$ and other related sets was also studied in [18, 19, 33, 36]. The vertex v is said to be a *boundary vertex* of u if no neighbor of v is farther away from u than v. By $\partial(u)$ we denote the set of all boundary vertices of u. Hernando et al. [55] proved that $\{u\} \cup \partial(u)$ is a geodetic set for an arbitrary vertex $u \in V(G)$. The *boundary* $\partial(G)$ of a graph G is the set of all of its boundary vertices [33].

Theorem 11. *[18] The boundary $\partial(G)$ of any connected graph G is a geodetic set.*

In fact this result follows from the stronger result that the so-called *expanded contour* $\Omega(G) = Ct(G) \cup Ecc(Ct(G))$ of every connected graph G is geodetic [18]. (Recall that *eccentricity* $Ecc(G)$ is the set of eccentric vertices of a graph, where a vertex v is called an *eccentric vertex* of G if no vertex in $V(G)$ is farther away from some vertex $u \in V(G)$ than v.)

In [19] the geodeticity of significant subsets of the boundary (periphery, contour set, and eccentricity) was studied for the class of perfect graphs. There are examples of perfect graphs for which neither $Per(G)$, $Ecc(G)$, nor $Ct(G)$ are geodetic sets [16]. However, the contour set of a graph is also geodetic in several perfect graph classes (we have already mentioned that distance hereditary graphs possess this property). In particular, this is true for chordal graphs

Theorem 12. *[19] The contour set of every chordal graph is a geodetic set.*

The state of the art on this topic is presented in [19]. The main open problem is still the case of bipartite graphs (namely, whether the contour set of any bipartite graph is a geodetic set). In particular this was conjectured for median graphs [10].

In [18] the question was posed whether the geodetic closure of the contour set $I[Ct(G)]$ is always a geodetic set. The authors approached this question by trying to prove that for every graph G its expanded contour $\Omega(G)$ is contained in $I[Ct(G)]$ but this still remains an open problem. However, they obtained a number of conditions under which the contour set of a graph is a geodetic set or at least the geodetic closure of the contour set is geodetic. The following partial results were proved in [18].

Theorem 13. *[18] Let G be a connected graph. Then $Ct(G)$ is a geodetic set if one of the following is true:*

(i) $|Ct(G)| = 2$.
(ii) $Ct(G) = Per(G)$.

Theorem 14. *[18] For a connected graph G the geodetic closure of $Ct(G)$ is a geodetic set if one of the following conditions is fulfilled:*

(i) $Ecc(Ct(G)) \subseteq I[Ct(G)]$.
(ii) $Ct(G) \setminus Per(G) = \{y_1, \ldots, y_k\}$ and $ecc(y_i) = ecc(y_j)$, for each $i, j = 1, \ldots, k$.
(iii) $|Ct(G)| = |Per(G)| + 1$.
(iv) $|Ct(G)| = 3$.

8.6 Related Concepts

Several concepts can be found in the literature that in various ways relate to geodetic sets and the geodetic number. We briefly present some of them without an ambition

8 Geodetic Sets in Graphs

to give deeper insights, yet an interested reader is directed to appropriate references. Note that the ideas from convexity theory that are related to geodeticity were presented in Sect. 8.5.

The first group of related concepts pertains to variations of intervals that arise from betweenness structures (for the notion of betweenness we refer to [69, 74]), different from the shortest paths. Typical instances of such intervals come from the so-called *monophonic (i.e., induced) paths* and *detour (i.e., longest) paths* between two vertices, which also yield the corresponding *monophonic convexity* [44, 69] and the *detour distance* in graphs [38]. A set S of vertices in a graph G is called *monophonic* if for every vertex $x \in V(G)$ there exist $u, v \in S$ such that x lies on an induced (i.e., chordless) path between u and v. It was shown in [56] that the contour set in any connected graph is a monophonic set (as we know from Sect. 8.5 it is not always a geodetic set). Similarly, a set S of vertices in a graph G is called a *detour set* if for every vertex $x \in V(G)$ there exist $u, v \in S$ such that x lies on a detour (i.e., longest) path between u and v. The size of a minimum detour set in a graph G is called the *detour number* of G. These concepts were investigated in two papers by Chartrand et al. [35, 37].

Another interval variation (the *Steiner interval*) is of different flavor since it may apply to more than two vertices. Given a graph G and a set of vertices $W \subseteq V(G)$, the *Steiner tree* of W is a minimum (with respect to the number of edges) connected subgraph that contains all vertices of W (it follows easily from the definition that this subgraph is a tree). The *Steiner interval $S[W]$* of a set W consists of all vertices in G that lie on some Steiner tree of W; see [66]. A set $W \subseteq V(G)$ is called a *Steiner set* of a graph G, if $S[W] = V(G)$, and the *Steiner number $sn(G)$* of G is the size of a smallest Steiner set of G. The first general results about the Steiner number in graphs were obtained by Chartrand and Zhang [28]. For instance, they proved that for any positive integers r, d, and $k \geq 2$, where $r \leq d \leq 2r$ there exists a graph G with $\mathrm{diam}(G) = d$, $\mathrm{rad}(G) = r$, and $sn(G) = k$. They also observed that for every pair of integers k and n with $2 \leq k \leq n$ there exists a graph of order n whose Steiner number is k. Unfortunately, as noted by Pelayo [77], it was erroneously stated (and "proved") in [28] that every Steiner set in a connected graph is a geodetic set. (In addition, Pelayo [77] constructed a graph G for which $g(G)$ is greater than $sn(G)$ [77].) This motivated Hernando et al. [54] to address the problem of characterizing the graphs in which every Steiner set is also a geodetic set. They obtained some partial results on this problem, namely that all connected interval graphs have this property, but not all connected chordal graphs. In addition they investigated relations among geodetic, monophonic, Steiner, and hull sets, and proved that every Steiner set in a connected graph is a monophonic set [54]. In particular this implies that in distance hereditary graphs every Steiner set is a geodetic set [54]. The latter result was independently obtained by Oellermann and Puertas [76] who also showed that $g(G)/sn(G)$ can be arbitrarily large if G is not a distance hereditary graph [76]. It was recently proved by Eroh and Oellermann that every Steiner set is a geodetic set in the class of 3-Steiner distance hereditary graphs [46]. However, in general this problem remains open.

Geodetic sets can be studied in directed graphs as well. Chartrand and Zhang [27] introduced the geodetic number of an oriented graph as follows. Let D be an oriented graph, and $S \subseteq V(D)$. Then $I[S]$ is the set of all vertices that lie on some shortest (directed) u, v-path for $u, v \in S$. The minimum size of a set S such that $I[S] = V(D)$ is called the *geodetic number* of an *oriented graph* D. Recall that given a graph G, an *orientation* of G is an oriented graph obtained from G by orienting all of its edges to one of the two directions. Now, if G is a graph, then the *lower orientable geodetic number* $g^-(G)$ of G is the minimum geodetic number among all orientations of G, and the *upper orientable geodetic number* $g^+(G)$ of G is the maximum such geodetic number [27]. It was proved that for every connected graph G of order at least 3, $g^-(G) < g^+(G)$, and for every two integers n and m with $1 \leq n - 1 \leq m \leq \binom{n}{2}$, there exists a connected graph G with n vertices and m edges such that $g^+(G) = n$ [27]. These concepts together with oriented variations of hull numbers were studied in Chartrand et al. [34], developed further by Farrugia [49], and in turn by Hung et al. [58]. Chang et al. [24] considered the concept of a *geodetic spectrum* of a graph G, which is defined as the set of geodetic numbers of all orientations of G, and determined it for several classes of graphs (the concept of a *strong geodetic spectrum* concerning only strongly connected orientations was also studied [24]).

Forcing concepts have been considered for various graph invariants, including the geodetic number. The *forcing geodetic number* was introduced by Chartrand and Zhang [26] as follows. Let G be (an undirected) graph, and let S be a minimum geodetic set of a graph G. A subset T of S is called a *forcing subset* of S, if S is the unique minimum geodetic set that contains T. The minimum size of a forcing subset of a minimum geodetic set in G is called the *forcing geodetic number* $f(G)$ of a graph G. Clearly $f(G) \leq g(G)$ in any graph G, and it was shown that any pair of integers is realizable as the forcing geodetic and the geodetic number of some graph, with only two exceptions (both parameters cannot be 1 (resp. 2) at the same time). An analogous theorem was proved by Zhang [85] for the so-called *upper forcing geodetic number*. Given a minimum geodetic set S a forcing subset T is called *critical* if no subset of T is a forcing subset of S. The maximum size of a critical forcing subset of a minimum geodetic set in G is called the *upper forcing geodetic number* $f^+(G)$. Clearly, $f(G) \leq f^+(G) \leq g(G)$, and, solving a problem of Zhang [85], Tong proved [78] that for any nonnegative integers a, b, and c with $1 \leq a \leq b \leq c - 2$ or $4 \leq a + 2 \leq b \leq c$, there exists a connected graph G with $f(G) = a$, $f^+(G) = b$, and $g(G) = c$. See also [79] for some recent developments in this area.

8.7 Summary and Conclusion

The concept of geodetic number was introduced roughly two decades ago, while most of the corresponding papers were published in the last 10 years. In this chapter we tried to present a concise survey of known results and a state of the art of recent developments. Although to our knowledge we incorporated a great majority of the papers that were published on this subject, emphasis is given to the classes of graphs related to metric graph theory. This in part reveals our subjective preferences, but on the other hand reflects the focus of recent studies. As presented in Sects. 8.5 and 8.6, a lot of attention was given to relations with other concepts from (metric) graph theory, and we believe that this research will continue to develop. An abundance of open problems, some of which we also mentioned in this work, guarantees further interest in these concepts. Natural connections with some other concepts in graph theory (such as Steiner, hull, and several boundary sets) suggest that the geodeticity concepts could be applicable also in other areas.

References

1. Atici M (2002) Computational complexity of geodetic set. Int J Comput Math 79:587–591
2. Atici M, Vince A (2002) Geodesics in graphs, an extremal set problem, and perfect hash families. Graph Combinator 18:403–413
3. Balakrishnan K, Brešar B, Changat M, Imrich W, Klavžar S, Kovše M, Subhamathi AR (2009) On the remoteness function in median graphs. Discrete Appl Math 157(18):3679–3688
4. Bandelt H-J, Chepoi V (1996) Graphs of acyclic cubical complexes. Eur J Combinator 17: 113–120
5. Bandelt H-J, Mulder HM, Wilkeit E (1994) Quasi-median graphs and algebras. J Graph Theory 18:681–703
6. Bandelt H-J, Forster P, Sykes BC, Richards MB (1995) Mitochondrial portraits of human populations using median networks. Genetics 14:743–753
7. Bielak H, Sysło M (1983) Peripheral vertices in graphs. Studia Sci Math Hungar 18:269–275
8. Brandes U (2005) Network analysis. Methodological foundations. Lecture Notes in Computer Science, vol 3418. Springer, Berlin, pp 62–82
9. Brešar B (2003) Arboreal structure and regular graphs of median-like classes. Discuss Math Graph Theory 23:215–225
10. Brešar B, Tepeh Horvat A (2008) On the geodetic number of median graphs. Discrete Math 308:4044–4051
11. Brešar B, Changat M, Subhamathi AR, Tepeh Horvat A (2010) The periphery graph of a median graph. Discuss Math Graph Theory 30:17–32
12. Brešar B, Klavžar S, Tepeh Horvat A (2008) On the geodetic number and related metric sets in Cartesian product graphs. Discrete Math 308:5555–5561
13. Buckley F, Harary F (1986) Geodetic games for graphs. Quaest Math 8:321–334
14. Buckley F, Harary F (1990) Distance in graphs. Addison-Wesley, Redwood City
15. Buckley F, Harary F, Quintas LV (1988) Extremal results on the geodetic number of a graph. Scientia 2A:17–26
16. Cáceres J, Hernando C, Mora M, Pelayo IM, Puertas ML, Seara C (2005) Searching for geodetic boundary vertex sets. Electron Notes Discrete Math 19:25–31
17. Cáceres J, Márquez A, Oellermann OR, Puertas ML (2005) Rebuilding convex sets in graphs. Discrete Math 297:26–37
18. Cáceres J, Hernando C, Mora M, Pelayo IM, Puertas ML, Seara C (2006) On geodetic sets formed by boundary vertices. Discrete Math 306:188–198
19. Cáceres J, Hernando C, Mora M, Pelayo IM, Puertas ML, Seara C (2008) Geodicity of the contour of chordal graphs. Discrete Appl Math 156:1132–1142
20. Cagaanan GB, Canoy SR (2004) On the hull sets and hull number of the Cartesian product of graphs. Discrete Math 287:141–144

21. Cagaanan GB, Canoy SR (2006) On the geodetic covers and geodetic bases of the composition $G[K_m]$. Ars Combinatoria 79:33–45
22. Cagaanan GB, Canoy SR, Gervacio SV (2006) Convexity, geodetic, and hull numbers of the join of graphs. Utilitas Math 71:143–159
23. Chae G, Palmer EM, Siu W (2002) Geodetic number of random graphs of diameter 2. Australas J Combinator 26:11–20
24. Chang GJ, Tong L, Wang H (2004) Geodetic spectra of graphs. Eur J Combinator 25:383–391
25. Chung FRK, Graham RL, Saks ME (1987) Dynamic search in graphs. In: Wilf H (ed) Discrete algorithms and complexity. Perspectives in Computing (Kyoto, 1986), vol 15. Academic, New York, pp 351–387
26. Chartrand G, Zhang P (1999) The forcing geodetic number of a graph. Discuss Math Graph Theory 19:45–58
27. Chartrand G, Zhang P (2000) The geodetic number of an oriented graph. Eur J Combinator 21:181–189
28. Chartrand G, Zhang P (2002) The Steiner number of a graph. Discrete Math 242:41–54
29. Chartrand G, Zhang P (2002) Extreme geodesic graphs. Czech Math J 52:771–780
30. Chartrand G, Harary F, Zhang P (2000) Geodetic sets in graphs. Discuss Math Graph Theory 20:129–138
31. Chartrand G, Harary F, Zhang P (2002) On the geodetic number of a graph. Networks 39:1–6
32. Chartrand G, Palmer EM, Zhang P (2002) The geodetic number of a graph: a survey. Congr Numer 156:37–58
33. Chartrand G, Erwin D, Johns GL, Zhang P (2003) Boundary vertices in graphs. Discrete Math 263:25–34
34. Chartrand G, Fink JF, Zhang P (2003) The hull number of an oriented graph. Int J Math Math Sci 36:2265–2275
35. Chartrand G, Johns G, Zhang P (2003) The detour number of a graph. Utilitas Math 64:97–113
36. Chartrand G, Erwin D, Johns GL, Zhang P (2004) On boundary vertices in graphs. J Combin Math Combin Comput 48:39–53
37. Chartrand G, Johns GL, Zhang P (2004) On the detour number and geodetic number of a graph. Ars Combinator 72:3–15
38. Chartrand G, Escuadro H, Zhang P (2005) Detour distance in graphs. J Combin Math Combin Comput 53:75–94
39. Coppersmith D, Winograd S (1990) Matrix multiplication via arithmetic progressions. J Symbolic Comput 9:251–280
40. Deza M, Deza E (2009) Encyclopedia of distances. Springer, Berlin Heidelberg
41. Djoković D (1973) Distance preserving subgraphs of hypercubes. J Combin Theory B 14: 263–267
42. Douthat AL, Kong MC (1995) Computing the geodetic number of bipartite graphs. Congr Numer 107:113–119
43. Douthat AL, Kong MC (1996) Computing geodetic bases of chordal and split graphs. J Combin Math Combin Comput 22:67–78
44. Duchet P (1988) Convex sets in graphs II. Minimal path convexity. J Combin Theory B 44: 307–316
45. Eppstein D (2005) The lattice dimension of a graph. Eur J Combinator 26:585–592
46. Eroh L, Oellermann OR (2008) Geodetic and Steiner geodetic sets in 3-Steiner distance hereditary graphs. Discrete Math 308:4212–4220
47. Everett MG, Seidman SB (1985) The hull number of a graph. Discrete Math 57:217–223
48. Farber M, Jamison RE (1986) Convexity in graphs and hypergraphs. SIAM J Algebraic Discrete Math 7:433–444
49. Farrugia A (2005) Orientable convexity, geodetic and hull numbers in graphs. Discrete Appl Math 148:256–262
50. Fraenkel AS, Harary F (1989) Geodetic contraction games on graphs. Int J Game Theory 18:327–338
51. Hansen P, van Omme N (2007) On pitfalls in computing the geodetic number of a graph. Opt Lett 1:299–307

8 Geodetic Sets in Graphs

52. Harary F, Loukakis E, Tsouros C (1993) The geodetic number of a graph. Math Comput Model 17:89–95
53. Haynes TW, Henning MA, Tiller C (2003) Geodetic achievement and avoidance games for graphs. Quaest Math 26:389–397
54. Hernando C, Jiang T, Mora M, Pelayo I, Seara C (2005) On the Steiner, geodetic and hull numbers of graphs. Discrete Math 293:139–154
55. Hernando C, Mora M, Pelayo I, Seara C (2006) Some structural, metric and convex properties on the boundary of a graph. Electron Notes Discrete Math 24:203–209
56. Hernando C, Mora M, Pelayo I, Seara C On monophonic sets in graphs. Manuscript
57. Howorka E (1977) A characterization of distance hereditary graphs. Q J Math Oxford, 28: 417–420
58. Hung J, Tong L, Wang H (2009) The hull and geodetic numbers of orientations of graphs. Discrete Math 309:2134–2139
59. Imrich W, Klavžar S (1999) Recognizing graphs of acyclic cubical complexes. Discrete Appl Math 95:321–330
60. Imrich W, Klavžar S, Mulder HM (1999) Median graphs and triangle-free graphs. SIAM J Discrete Math 12:111–118
61. Imrich W, Klavžar S (2000) Product graphs: structure and recognition. Wiley, NY
62. Jamison RE (1980) Copoints in antimatroids. Congr Numer 29:535–544
63. Jiang T, Pelayo I, Pritikin D (2004) Geodesic convexity and Cartesian products in graphs. Manuscript
64. Klavžar S, Mulder HM (1999) Median graphs: characterizations, location theory and related structures. J Combin Math Combin Comp 30:103–127
65. Korte B, Lovász L (1986) Homomorphisms and Ramsey properties of antimatroids. Discrete Appl Math 15:283–290
66. Kubicka E, Kubicki G, Oellermann OR (1998) Steiner intervals in graphs. Discrete Math 81:181–190
67. Leclerc B (2003) The median procedure in the semilattice of orders. Discrete Appl Math 127:285–302
68. Lu C (2007) The geodetic numbers of graphs and digraphs. Sci China Ser A 50:1163–1172
69. Morgana MA, Mulder HM (2002) The induced path convexity, betweenness and svelte graphs. Discrete Math 254:349–370
70. Mulder HM (1978) The structure of median graphs. Discrete Math 24:197–204
71. Mulder HM (1980) n-Cubes and median graphs. J Graph Theory 4:107–110
72. Mulder HM (1980) The interval function of a graph. Mathematical Centre Tracts 132. Mathematisch Centrum, Amsterdam
73. Mulder HM (1990) The expansion procedure for graphs. In: Bodendiek R (ed) Contemporary methods in graph theory. Manhaim/Wien/Zürich: B.I. Wissenschaftsverlag
74. Mulder HM (2008) Transit functions on graphs (and posets). In: Changat M, Klavžar S, Mulder HM, Vijayakumar A (eds) Convexity in discrete structures. Lecture Notes Ser. 5, Ramanujan Math Soc, pp 117–130
75. Necásková M (1988) A note on the achievement geodetic games. Quaest Math 12:115–119
76. Oellermann OR, Puertas ML (2007) Steiner intervals and Steiner geodetic numbers in distance-hereditary graphs. Discrete Math 307:88–96
77. Pelayo I (2004) Comment on: "The Steiner number of a graph" by Chartrand G, Zhang P [Discrete Math 242:1–3, 41–54 (2002)]. Discrete Math 280:259–263
78. Tong L (2009) The (a,b)-forcing geodetic graphs. Discrete Math 309:1623–1628
79. Tong L (2009) The forcing hull and forcing geodetic numbers of graphs. Discrete Appl Math 157:1159–1163
80. van de Vel MLJ (1993) Theory of convex structures. North Holland, Amsterdam
81. Wang F, Wang Y, Chang J (2006) The lower and upper forcing geodetic numbers of block-cactus graphs. Eur J Oper Res 175:238–245
82. William HE, McMorris FR (2003) Axiomatic concensus theory in group choice and bioinformatics. J Soc Ind Appl Math 29:91–94

83. Winkler P (1984) Isometric embeddings in products of complete graphs. Discrete Appl Math 7:221–225
84. Ye Y, Lu C, Liu Q (2007) The geodetic numbers of Cartesian products of graphs. Math Appl (Wuhan) 20:158–163
85. Zhang P (2002) The upper forcing geodetic number of a graph. Ars Combinatoria 62:3–15

Chapter 9
Graph Polynomials and Their Applications I: The Tutte Polynomial

Joanna A. Ellis-Monaghan and Criel Merino

Abstract In this survey of graph polynomials, we emphasize the Tutte polynomial and a selection of closely related graph polynomials such as the chromatic, flow, reliability, and shelling polynomials. We explore some of the Tutte polynomial's many properties and applications and we use the Tutte polynomial to showcase a variety of principles and techniques for graph polynomials in general. These include several ways in which a graph polynomial may be defined and methods for extracting combinatorial information and algebraic properties from a graph polynomial. We also use the Tutte polynomial to demonstrate how graph polynomials may be both specialized and generalized, and how they can encode information relevant to physical applications. We conclude with a brief discussion of computational complexity considerations.

Keywords Tutte polynomial · Graph polynomial · Chromatic polynomial · Flow polynomial · Reliability polynomial · Shelling polynomial · Abelian sandpile model · Spanning tree · Beta invariant

MSC2000: Primary 05-02; Secondary 05C15, 05A15, 05C99

9.1 Introduction

We begin our exploration of graph polynomials and their applications with the Tutte polynomial, a renowned tool for analyzing properties of graphs and networks. This two-variable graph polynomial, due to Tutte [101, 103, 104],

---

J.A. Ellis-Monaghan (✉)
Department of Mathematics, Saint Michael's College, One Winooski Park, Colchester, VT 05439, USA
and
Department of Mathematics and Statistics, University of Vermont, 16 Colchester Avenue, Burlington, VT 05405, USA
e-mail: jellis-monaghan@smcvt.edu

M. Dehmer (ed.), *Structural Analysis of Complex Networks*,
DOI 10.1007/978-0-8176-4789-6_9, © Springer Science+Business Media, LLC 2011

has the important universal property that essentially any multiplicative graph invariant with a deletion/contraction reduction must be an evaluation of it. These deletion/contraction operations are natural reductions for many network models arising from a wide range of problems at the heart of computer science, engineering, optimization, physics, and biology.

In addition to surveying a selection of the Tutte polynomial's many properties and applications, we use the Tutte polynomial to showcase a variety of principles and techniques for graph polynomials in general. These include several ways in which a graph polynomial may be defined and methods for extracting combinatorial information and algebraic properties from a graph polynomial. We also use the Tutte polynomial to demonstrate how graph polynomials may be both specialized and generalized, and how they can encode information relevant to physical applications.

We begin with the Tutte polynomial because it has a rich and well-developed theory, and thus it serves as an ideal model for exploring other graph polynomials in Chap. 10. Furthermore, because of the Tutte polynomial's long history, extensive study, and its universality property, it is often a "point of contact" for research into other graph polynomials in that their study frequently includes exploring their relations to the Tutte polynomial. These interrelationships are a central theme of the Chap. 10.

In this chapter we give both recursive and generating function formulations of the Tutte polynomial, and state its universality in the form of a recipe theorem. We give a number of properties and combinatorial interpretations for various evaluations of the Tutte polynomial. We recover colorings, flows, orientations, network reliability, etc., and related polynomials as specializations of the Tutte polynomial. We discuss the coefficients, zeros, and derivatives of the Tutte polynomial, and conclude with a brief discussion of computational complexity.

9.2 Preliminary Notions

The graph terminology that we use is standard and generally follows Diestel [26]. Graphs may have loops and multiple edges. For a graph G we denote by $V(G)$ its set of vertices and by $E(G)$ its set of edges. An oriented graph, $\vec{G}$, also called a digraph, has a direction assigned to each edge.

9.2.1 Basic Concepts

We first recall some of the notions of graph theory most used in this chapter. Two graphs G_1 and G_2 are *isomorphic*, denoted $G_1 \simeq G_2$, if there exists a bijection $\phi : V(G_1) \rightarrow V(G_2)$ with $xy \in E(G_1)$ if and only if $\phi(x)\phi(y) \in E(G_2)$. We denote by $\kappa(G)$ the number of connected components of a graph G, and by $c(G)$ the number of *nontrivial* connected components, that is, the number of connected

9 The Tutte Polynomial 221

components not counting isolated vertices. A graph is *k-connected* if at least k vertices must be removed to disconnect the graph.

A *cycle* in a graph G is a set of edges $e_1, \ldots, e_k$ such that, if $e_i = v_i w_i$ for $1 \le i \le k$, then $w_i = v_{i+1}$ for $1 \le i \le k - 1$; also $w_k = v_1$ and $v_i \ne v_j$ for $i \ne j$. A *trail* is a path that may revisit a vertex, but not retrace an edge. A *circuit* is a closed trail, and thus a cycle is just a circuit that does not revisit any vertices. In the case of a digraph, the edges of a trail or circuit must be consistently oriented.

The dual notion of a cycle is that of *cut* or *cocycle*. If $\{V_1, V_2\}$ is a partition of the vertex set, and the set C, consisting of those edges with one end in V_1 and one end in V_2, is not empty, then C is called a *cut*. A cycle with one edge is called a *loop* and a cocycle with one edge is called a *cut-edge* or *bridge*. We refer to an edge that is neither a loop nor a bridge as *ordinary*.

A *tree* is a connected graph without cycles. A *forest* is a graph whose connected components are all trees. A subgraph H of a graph G is *spanning* if $V(H) = V(G)$. Spanning trees in connected graphs play a fundamental role in the theory of the Tutte polynomial. Observe that a loop in a connected graph can be characterized as an edge that is in no spanning tree, while a bridge is an edge that is in every spanning tree.

If $V' \subseteq V(G)$, then the *induced subgraph* on V' has vertex set V' and edge set those edges of G with both endpoints in V'. If $E' \subseteq E(G)$, then the *spanning subgraph* induced by E' has vertex set $V(G)$ and edge set E'.

9.2.2 Deletion and Contraction

The operations of deletion and contraction of an edge are essential to the study of the Tutte polynomial. The graph obtained by deleting an edge $e \in E(G)$ is just $G \backslash e := (V, E \backslash e)$. The graph obtained by contracting an edge e in G results from identifying the endpoints of e followed by removing e, and is denoted G/e. When e is a loop, G/e is the same as $G \backslash e$. It is not difficult to check that both deletion and contraction are commutative, and thus, for a subset of edges A, both $G \backslash A$ and G/A are well defined. Also, if $e \ne f$, then $G \backslash e / f$ and $G/f \backslash e$ are isomorphic; thus for disjoint subsets $A, A' \subseteq E(G)$, the graph $G \backslash A / A'$ is well defined. A graph H isomorphic to $G \backslash A / A'$ for some choice of disjoint edge sets A and A' is called a *minor* of G. A class $\mathcal{G}$ of graphs is *minor closed* if whenever G is in $\mathcal{G}$ then any minor of G is also in the class.

A *graph invariant* is a function f on the class of all graphs such that

$$f(G_1) = f(G_2) \text{ whenever } G_1 \simeq G_2.$$

A *graph polynomial* is a graph invariant where the image lies in some polynomial ring.

9.2.3 The Rank and Nullity Functions for Graphs

To simplify notation, we typically identify a subset of edges A of a graph G with the spanning subgraph of G that A induces. Thus, for a fixed graph G we have the following rank and nullity functions on the lattice of subsets of $E(G)$.

Definition 9.2.1. For $A \subseteq E(G)$, the rank and nullity of A, denoted $r(A)$ and $n(A)$ respectively, are defined as

$$r(A) := |V(G)| - \kappa(A) \text{ and } n(A) := |A| - r(A).$$

Three special graphs are important. One is the rank 0 graph L consisting of a single vertex with one loop edge, another is the rank 1 graph B consisting of two vertices with one bridge edge between them, and the third one is the edgeless graph $\mathcal{E}_1$ on one vertex.

9.2.4 Planar Graphs and Duality

A graph is *planar* if it can be drawn in the plane without edges crossing, and it is called a *plane graph* if it is so drawn in the plane. A drawing of a graph in the plane separates the plane into regions called faces. Every plane graph G has a *dual graph*, G^*, formed by assigning a vertex of G^* to each face of G and joining two vertices of G^* by k edges if and only if the corresponding faces of G share k edges in their boundaries. Notice that G^* is always connected. If G is connected, then $(G^*)^* = G$. If G is planar, in principle it may have many plane duals, but when G is 3-connected, all its plane duals are isomorphic. This is not the case when G is only 2-connected.

There is a natural bijection between the edge set of a planar graph G and the edge set of G^*, any one of its plane duals, so we can assume that G and G^* have the same edge set E. It is easy to check that $A \subseteq E$ is a spanning tree of G if and only if $E \setminus A$ is a spanning tree of G^*. Thus, a planar graph and any of its plane duals have the same number of spanning trees. Furthermore, if G is a planar graph with rank function r, and G^* is any of its plane duals, then the rank function of G^*, denoted r^*, can be expressed as

$$r^*(A) = |A| - r(E) + r(E \setminus A). \tag{9.1}$$

These observations reflect a deeper relation between G and G^* that we will see captured by the Tutte polynomial at the end of Sect. 9.3.2.

9.3 Defining the Tutte Polynomial

Here we present several very different, but nevertheless equivalent, definitions of the Tutte polynomial. The interplay among these different formulations is a source of many powerful tools developed to analyze the Tutte polynomial. Furthermore, each formulation lends itself to different proof techniques, for example, induction with the linear recursion form and Möbius inversion with the generating function form. These different formulations also are representative of some of the most common ways of defining any graph polynomial, although we also demonstrate other methods in Chap. 10.

While space prohibits including full proofs of the equivalence of these various expressions for the Tutte polynomial, we note that there are several approaches. One direct way is to specify the linear recursion form as the definition of the Tutte polynomial and then use induction on the number of edges to show that it is equivalent to either the rank-nullity generating function or the spanning trees expansion. Showing that the linear recursion form is equivalent to the rank generating function form also establishes the essential fact that it is well defined, that is, independent of the order in which the edges are deleted and contracted. Another common approach is to establish some definition of the Tutte polynomial, then prove from it that the Tutte polynomial has the universality property discussed in Sect. 9.4. This universality property may then be applied to show that some other function is equivalent to, or an evaluation of, the Tutte polynomial.

The spanning trees expansion formulation in Sect. 9.3.3 was the approach originally used by Tutte to develop versions of this and similar polynomials. See Tutte [101, 103, 104]. A particularly lucid proof of the equivalence between the rank-nullity generating function definition of Definition 9.3.3 and the spanning trees expansion definition of Definition 9.3.5 can be found in [11]. That the Tutte polynomial has a deletion/contraction reduction was shown by Tutte [101, 102, 104] (also see Brylawski [15], and Oxley and Welsh [71]).

9.3.1 Linear Recursion Definition

Broadly speaking, a linear recursion relation is a set of reduction rules together with an evaluation for the terminal forms. The reduction rules rewrite a graph as a weighted (formal) sum of graphs that are in some way "smaller" or "simpler" than the original graph. Furthermore, the reduction rules again apply to the newly generated simpler graphs, hence the recursion. This recursion process eventually terminates in a well-defined set of "most simple" graphs, which are no longer reducible by the reduction rules. These are then each identified with a monomial of independent variables to yield a polynomial. It is essential to show that the reduction rules are independent of the order in which they are applied and that they do in fact terminate. See Yetter [114] for a more formal treatment.

The Tutte polynomial may be defined by a linear recursion relation given by deleting and contracting ordinary edges. The "most simple" terminal graphs are then just forests with loops.

Definition 9.3.1. If $G = (V, E)$ is a graph, and e is an ordinary edge, then

$$T(G; x, y) = T(G \setminus e; x, y) + T(G/e; x, y). \tag{9.2}$$

Otherwise, G consists of i bridges and j loops and

$$T(G; x, y) = x^i y^j. \tag{9.3}$$

In other words, T may be calculated recursively by specifying an ordering of the edges and repeatedly applying (9.2). Remarkably, the Tutte polynomial is well defined in that the polynomial resulting from this recursive process is independent of the order in which the edges are chosen. One way to prove this is by showing that this definition is equivalent to the rank generating form we see in Definition 9.3.3. A proof can be found in [17], for example.

Figure 9.1 gives a small example of computing T using (9.2) and (9.3) for K_4 minus one edge. By adding the monomials at the bottom of Fig. 9.1 we find that $T(G; x, y) = x^3 + 2x^2 + x + 2xy + y + y^2$.

Recall that a *one-point join* $G * H$ of two graphs G and H is formed by identifying a vertex u of G and a vertex v of H into a single vertex w (necessarily a cut vertex) of $G * H$. Also, $G \cup H$ is the disjoint union of G and H.

Proposition 9.3.2. *If G and H are graphs then*

$$T(G \cup H) = T(G) T(H) \text{ and } T(G * H) = T(G) T(H).$$

This follows readily from Definition 9.3.1 by induction on the number of ordinary edges in $G * H$ or $G \cup H$.

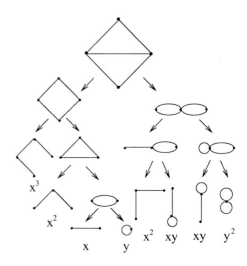

Fig. 9.1 An example of computing the Tutte polynomial recursively

9 The Tutte Polynomial

9.3.2 Rank-Nullity Generating Function Definition

A generating function can often be thought of as a (possible infinite) polynomial whose coefficients count structures that are encoded by the exponents of the variables. Because generating functions count, which is at the very heart of enumerative combinatorics, there is extensive literature on them. Stanley's two volumes [97] and [98] are an excellent resource. In the case of the Tutte polynomial, there are several different generating function formulations, each of which has its advantages. We give one here and another in Sect. 9.3.3, with a variation in Sect. 9.7.2, and refer the readers to Brylawski and Oxley [17] for additional forms.

Definition 9.3.3. If $G = (V, E)$ is a graph, then the Tutte polynomial of G has the following expansion:

$$T(G; x, y) = \sum_{A \subseteq E} (x - 1)^{r(E) - r(A)} (y - 1)^{n(A)}. \tag{9.4}$$

The advantage of a generating function formulation is that it facilitates counting. For example, interpretations for several evaluations of the Tutte polynomial given in Sect. 9.5 follow immediately from Definition 9.3.3.

We can also deduce the following pleasing property of the Tutte polynomial.

Proposition 9.3.4. *If G is a planar graph with dual G^* then*

$$T(G; x, y) = T(G^*; y, x). \tag{9.5}$$

This result follows from routine checking using Definition 9.3.3 and (9.1).

9.3.3 Spanning Trees Expansion Definition

We need to develop a little terminology before presenting the spanning trees definition of the Tutte polynomial. First, given a spanning tree S and an edge $e \notin S$, there is a cycle defined by e, namely the unique cycle in $S \cup e$. Similarly, for an edge $f \in S$, there is a cut defined by f, namely the set of edges C such that if $f' \in C$, then $(S - f) \cup f'$ is a spanning tree.

Assume there is a total ordering $\prec$ on the edges of G, say $E = \{e_1, \ldots, e_m\}$, where $e_i \prec e_j$ if $i < j$. Given a fixed tree S, an edge f is called *internally active* if $f \in S$ and it is the smallest edge in the cut defined by f. Dually, an edge e is *externally active* if $e \notin S$ and it is the smallest edge in the cycle defined by e. The internal activity of S is the number of its internally active edges and its external activity is the number of externally active edges. With this, we have the following definition of the Tutte polynomial.

Definition 9.3.5. If G is a connected graph with a total order on its edge set, then

$$T(G; x, y) = \sum_{i,j} t_{ij} x^i y^j, \tag{9.6}$$

where t_{ij} is the number of spanning trees with internal activity i and external activity j.

Two important observations follow immediately from the equivalence of Definitions 9.3.3 and 9.3.5. One is that the terms t_{ij} in Definition 9.3.5 are independent of the total order used in the edge set, since there is no ordering of the edges in Definition 9.3.3. The other is that the coefficients in Definition 9.3.3 must be nonnegative since the coefficients in Definition 9.3.5 clearly are.

9.4 Universality of the Tutte Polynomial

The universality property discussed here is one of the most powerful aspects of the Tutte polynomial. It says that essentially any graph invariant that is multiplicative on disjoint unions and one-point joins of graphs and that has a deletion/contraction reduction must be an evaluation of the Tutte polynomial. Several applications of this theorem appear throughout the rest of this chapter and in the next chapter as well. Various generalizations of the Tutte polynomial are careful to retain this essential property, and analogous universality properties are sought in the context of other graph polynomials.

Definition 9.4.1. Let $\mathcal{G}$ be a minor closed class of graphs. A graph invariant f from $\mathcal{G}$ to a commutative ring $\mathcal{R}$ with unity is called a *generalized Tutte–Gröthendieck invariant*, or *T–G invariant*, if $f(\mathcal{E}_1)$ is the unity of R, if there exist fixed elements $a, b \in R$ such that for every graph $G \in \mathcal{G}$ and every ordinary edge $e \in G$; then

$$f(G) = af(G \setminus e) + bf(G/e); \tag{9.7}$$

and if for every $G, H \in \mathcal{G}$, whenever $G \cup H$ or $G * H$ is in $\mathcal{G}$, then

$$f(G \cup H) = f(G)f(H) \text{ and } f(G * H) = f(G)f(H). \tag{9.8}$$

Thus, the Tutte polynomial is a T–G invariant, and in fact, since the following two results give both universal and unique extension properties, it is essentially the *only* T–G invariant, in that any other must be an evaluation of it. Theorem 9.4.2 is known as a recipe theorem since it specifies how to recover a T–G invariant as an evaluation of the Tutte polynomial.

Theorem 9.4.2. *Let $\mathcal{G}$ be a minor closed class of graphs, let R be a commutative ring with unity, and let $f : \mathcal{G} \to R$. If there exists $a, b \in R$ such that f is a T–G invariant, then*

$$f(G) = a^{|E(G)|-r(E(G))}b^{r(E(G))}T\left(G; \frac{x_0}{b}, \frac{y_0}{a}\right), \tag{9.9}$$

where $f(B) = x_0$ and $f(L) = y_0$.

Furthermore, we have the following unique extension property, which says that if we specify any four elements $a, b, x_0, y_0 \in R$, then there is a unique well-defined T–G invariant on these four elements.

Theorem 9.4.3. *Let G be a minor closed class of graphs, let R be a commutative ring with unity, and let $a, b, x_0, y_0 \in R$. Then there is a unique T–G invariant f : $G \to R$ satisfying Definition 9.4.1 with $f(B) = x_0$ and $f(L) = y_0$. Furthermore, this function f is given by*

$$f(G) = a^{|E(G)|-r(E(G))}b^{r(E(G))}T\left(G; \frac{x_0}{b}, \frac{y_0}{a}\right). \tag{9.10}$$

If a or b are not units of R, then (9.9) and (9.10) are interpreted to mean using expansion (9.4) of Definition 9.3.3, and cancelling before evaluating.

These results can be proved by induction on the number of ordinary edges from the deletion/contraction definition of the Tutte polynomial. See, for example, Brylawski [15], Oxley and Welsh [71], Brylawski and Oxley [17], Welsh [109], and Bollobás [11] for detailed discussions of these theorems and their consequences.

Examples applying this important theorem may be found throughout Sect. 9.6, where it may be used to show that all of the graph polynomials surveyed there are evaluations of the Tutte polynomial.

9.5 Combinatorial Interpretations of Some Evaluations

A graph polynomial encodes information about a graph. The challenge is in extracting combinatorially useful information from this algebraic object. A number of successful techniques have evolved for meeting this challenge, and we use the Tutte polynomial to showcase some of them while simultaneously demonstrating the richness of the information encoded by the Tutte polynomial.

9.5.1 Spanning Subgraphs

Spanning subgraphs, and in particular spanning trees, play a fundamental role in the theory of Tutte polynomials as we have already seen in Definition 9.3.5. This is also reflected in the most readily attainable interpretations for evaluations of the Tutte polynomial, which enumerate various spanning subgraphs. We begin with these here, writing $\tau(G)$ for the number of spanning trees of a connected graph G.

Theorem 9.5.1. *If $G = (V, E)$ is a connected graph then:*

1. $T(G; 1, 1)$ *equals* $\tau(G)$.
2. $T(G; 2, 1)$ *equals the number of spanning forests of G.*
3. $T(G; 1, 2)$ *equals the number of spanning connected subgraphs of G.*
4. $T(G; 2, 2)$ *equals* $2^{|E|}$.

Proof. To illustrate common proof techniques, we give two short proofs of the first statement. The remaining statements may be proved similarly. When $x = y = 1$, the nonvanishing terms in the rank-nullity expansion (9.4) are $A \subseteq E$ such that $r(E) = r(A)$ and $|A| = r(A)$. That $r(E) = r(A)$ implies that A has the same number of connected components as G, namely one, so (V, A) is connected. Then $|A| = r(A)$ implies that $|A| = |V| - 1$, so A must be a tree, and hence a spanning tree.

Alternatively, we can use Theorem 9.4.2. Let $\tau'(G)$ be the number of maximal spanning forests in a general (not necessarily connected) graph G. We prove that $T(G; 1, 1) = \tau'(G)$. Then, if G is connected, we will have that $T(G; 1, 1) = \tau'(G) = \tau(G)$. Note that the number of maximal spanning forests has a deletion/contraction reduction for ordinary edges; that is, if G is a graph and e is an ordinary edge of G, then $\tau'(G) = \tau'(G \backslash e) + \tau'(G/e)$. This follows because the maximal spanning forests of G can be partitioned into the maximal spanning forests that do not contain e and those that do contain e. The former are the maximal spanning forests of $G \backslash e$ and the latter are in one-to-one correspondence with the maximal spanning forests of G/e.

The result then follows immediately from Theorem 9.4.2 with $a = b = x_0 = y_0 = 1$. $\square$

Computing the number of spanning trees of a graph is easy in that there are polynomial time algorithms to do it. One of these involves a determinant. Recall that the *Laplacian matrix L* of a graph G with vertices $v_1, \ldots, v_n$ is the $n \times n$-matrix defined by

$$
L_{ij} = \begin{cases} \deg(i) & \text{if } i = j \\ -r & \text{if } r \text{ is the number of edges between vertices } i \text{ and } j. \end{cases} \tag{9.11}
$$

Theorem 9.5.2. *If G is a connected graph with Laplacian L, then*

$$
T(G; 1, 1) = \tau(G) = \text{Det}(L'), \tag{9.12}
$$

where L' is any cofactor of L.

A proof of this can be found in [1, 6] using the Binet–Cauchy formula and that $L = DD^t$, where D is the incidence matrix of (an orientation of) G.

This result not only provides an interpretation of the Tutte polynomial at $(1, 1)$ in terms of the incidence matrix of a graph, but also proves that $(1, 1)$ is one of the (very few see Sect. 9.8) points where the Tutte polynomial can be computed in polynomial time.

9.5.2 The Tutte Polynomial when $y = x$

Combinatorial interpretations are known for the Tutte polynomial at all integer values along the line $y = x$. In addition to those for $T(G, 1, 1)$ and $T(G, 2, 2)$ previously given, we have the following interpretation for $T(G; -1, -1)$ due to Read and Rosenstiehl [76]. We also show alternative interpretations for $T(G; -1, -1)$ and $T(G; 3, 3)$ in Sect. 9.5.3.

The incidence matrix D of a graph G defines a vector space over $\mathbb{Z}_2$, called the cycle space $\mathcal{C}$. The bicycle space $\mathcal{B}$ is then just $\mathcal{C} \cap \mathcal{C}^\perp$.

Theorem 9.5.3. $T(G; -1, -1) = (-1)^{|E|}(-2)^{\dim(\mathcal{B})}$.

One method of extracting information from a graph polynomial is via its relation to some other graph invariant. The following interpretations for the Tutte polynomial of a planar graph along the line $x = y$ derive from its relation to the Martin polynomial [61], $m(\vec{G}; x)$, a one-variable graph polynomial for Eulerian digraphs that we discuss further in Chap. 10.

We first recall that the *medial graph* of a connected planar graph G is constructed by placing a vertex on each edge of G and drawing edges around the faces of G. The faces of this medial graph are colored black or white, depending on whether they contain or do not contain, respectively, a vertex of the original graph G. This face two-colors the medial graph. The edges of the medial graph are then directed so that the black face is on the left. We refer to this as the directed medial graph of G and denote it by $\vec{G}_m$. An example is given in Fig. 9.2.

Martin [61] showed that, for a planar graph G, the relation between the Martin polynomial and Tutte polynomial is $m(\vec{G}_m; x) = T(G; x, x)$. Evaluations for the Martin polynomial in [32] then give the following interpretations of the Tutte polynomial.

Let $D_n(\vec{G}_m) = \{(D_1, \ldots, D_n)\}$, where $(D_1, \ldots, D_n)$ is an ordered partition of $E(\vec{G}_m)$ into n subsets such that G restricted to D_i is 2-regular and consistently oriented for all i.

Theorem 9.5.4. *Let G be a planar graph with oriented medial graph $\vec{G}_m$. Then, for n a positive integer,*

$$(-n)^{c(G)} T(G; 1-n, 1-n) = \sum_{D_n(\vec{G}_m)} (-1)^{\sum_{i=1}^n c(D_i)}.$$

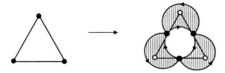

Fig. 9.2 On the left-hand side we have a planar graph G. On the right-hand side we have $\vec{G}_m$ with the vertex faces colored black, oriented so that black faces are to the left of each edge

230　　　　　　　　　　　　　J.A. Ellis-Monaghan and C. Merino

Theorem 9.5.5. *Let G be a planar graph with oriented medial graph $\overrightarrow{G_m}$. Then, for n a positive integer,*

$$n^{c(G)} T(G; 1+n, 1+n) = \sum 2^{\mu(\phi)},$$

where the sum is over all edge colorings ϕ of $\overrightarrow{G_m}$ with n colors so that each (possibly empty) set of monochromatic edges forms an Eulerian digraph, and where $\mu(\phi)$ is the number of monochromatic vertices in the coloring ϕ.

9.5.3　Orientations and Score Vectors

The combinatorial interpretations of the Tutte polynomial in Theorem 9.5.1 are given in terms of the number of certain subgraphs of the graph G. However, they can also be given in terms of orientations of the graph and its score vectors. Given a graph $G = (V, E)$, an orientation of G may be obtained by directing every edge from one of its endpoints to the other. From this follows that $T(G; 2, 2)$ equals the number of possible orientations of G.

The *score vector* of an orientation $\vec{G}$ is the vector $(s_1, s_2, \ldots, s_n)$ such that vertex i has outdegree s_i in the orientation. In the following theorem we gather several similar results about the Tutte polynomial and orientations of a graph.

Theorem 9.5.6. *Let $G = (V, E)$ is a connected graph with Tutte polynomial $T(G; x, y)$. Then*

1. *$T(G; 2, 0)$ equals the number of acyclic orientations of G, that is, orientations without oriented cycles.*
2. *$T(G; 0, 2)$ equals the number of totally cyclic orientations of G, that is, orientations in which every arc is in a directed cycle.*
3. *$T(G; 1, 0)$ equals the number of acyclic orientations with exactly one predefined source v.*
4. *$T(G; 2, 1)$ equals the number of score vectors of orientations of G.*
5. *$T(G; 0, 1)$ equals the number of score vectors of strongly connected orientations, that is, orientations where there is a directed path between each pair of vertices.*
6. *$T(G; 1, 1)$ equals the number of score vectors of all orientations with a given vertex reachable from any other vertex.*

Item 1 was proved by Stanley in [94] by using a deletion/contraction reduction; a proof using the universality of the Tutte polynomial is given by Brylawski and Oxley in [17]. For Items 2 and 3, see Green and Zaslavsky [43] and Las Vergnas [54]. The former also gives the number of strongly connected orientations of G, and note that the latter is independent of the choice of source vertex v. Item 4 was first proved by Stanley in [95], with a bijective proof given by Kleitman and Winston in [52], and a

9 The Tutte Polynomial

proof using Theorem 9.4.2 by Brylawski and Oxley in [17]. Comparing Item 4 with Theorem 9.5.1, Item 2, shows that the number of score vectors equals the number of spanning forests of G. For Items 5 and 6, see Gioan [39].

Some other evaluations of the Tutte polynomial can also be interpreted in terms of orientations. Recall that an anticircuit in a digraph is a closed trail so that the directions of the edges alternate as the trail passes through any vertex of degree greater than 2. Note that in a 4-regular Eulerian digraph such as $\overrightarrow{G_m}$, the set of anticircuits can be found by pairing the two incoming edges and the two outgoing edges at each vertex.

Two surprising results from Las Vergnas [56] and Martin [62] are the following.

Theorem 9.5.7. *Let G is a connected planar graph. Then*

$$T(G; -1, -1) = (-1)^{|E(G)|}(-2)^{a(\overrightarrow{G_m})-1} \tag{9.13}$$

and

$$T(G; 3, 3) = K2^{a(\overrightarrow{G_m})-1}, \tag{9.14}$$

where $a(\overrightarrow{G_m})$ is the number of anticircuits in the directed medial graph of G and K is some odd integer.

Comparing (9.13) to Theorem 9.5.3 gives the following corollary.

Corollary 9.5.8. *If G is a connected planar graph, then the dimension of the bicycle space is $a(\overrightarrow{G_m}) - 1$.*

9.6 Some Specializations

Here we illustrate the wide range of applicability of the Tutte polynomial while demonstrating some proof techniques for showing that a graph invariant is related to the Tutte polynomial. The advantage of recognizing an application-driven function as a specialization of the Tutte polynomial is that the large body of knowledge about the Tutte polynomial is then available to inform the desired application. We say a graph polynomial is a specialization of the Tutte polynomial if it may be recovered from the Tutte polynomial by some substitution for x and y, with possibly some prefactor.

For various substitutions along different algebraic curves in x and y, the Tutte polynomial has interpretations as the generating function of combinatorial quantities or numerical invariants associated with a graph. Some of these were considered long before the development of the Tutte polynomial, and others were discovered to be unexpectedly related to the Tutte polynomial. We survey six of the more well known of these application-driven generating functions.

9.6.1 The Chromatic Polynomial

The chromatic polynomial, introduced by Birkhoff [8], and see also Whitney [112], because of its theoretical and applied importance, has generated a large body of work. Chia [21] provides an extensive bibliography on the chromatic polynomial, and Dong et al. [29] give a comprehensive treatment.

For positive integer λ, a λ-coloring of a graph G is a mapping of $V(G)$ into the set $[\lambda] = \{1, 2, \ldots, \lambda\}$. Thus there are exactly λ^n colorings for a graph on n vertices. If ϕ is a λ-coloring such that $\phi(i) \neq \phi(j)$ for all $ij \in E$, then ϕ is called a *proper* (or *admissible*) coloring.

We wish to find the number, $\chi(G; \lambda)$, of admissible λ-colorings of a graph G. As noted by Whitney [112], the four-color theorem can be formulated in this general setting as follows: If G is a planar graph, then $\chi(G; 4) > 0$.

The following theorem is due to Birkhoff in [8] and independently by Whitney in [112]. We sketch the latter's proof.

Theorem 9.6.1. *If $G = (V, E)$ is a graph, then*

$$\chi(G; \lambda) = \sum_{A \subseteq E} (-1)^{|A|} \lambda^{\kappa(A)}. \tag{9.15}$$

Proof. Let P_{ij} be the set of λ-colorings such that vertices i and j receive the same color. Let $\bar{P}_{ij}$ be the complement of P_{ij} in the set of λ-colorings. Then, the value $\chi(G; \lambda)$ can be computed using the inclusion–exclusion principle.

$$\chi(G; \lambda) = \left| \bigcap_{ij \in E} \bar{P}_{ij} \right|$$

$$= \lambda^n - \sum_{ij \in E} |P_{ij}| + \sum_{\substack{ij, kl \in E \\ ij \neq kl}} |P_{ij} \cap P_{kl}| -$$

$$\cdots + (-1)^{|E|} \left| \bigcap_{ij \in E} P_{ij} \right|. \tag{9.16}$$

Every term is of the form $\left| \bigcap_{ij \in A} P_{ij} \right|$ for some $A \subseteq E$, and hence corresponds to the subgraph (V, A), where A is the set of edges given by the indices of the P_{ij}'s. Thus, the cardinality of this set is the number of λ-colorings that have a constant value on each of the connected components of (V, A), that is, $\lambda^{\kappa(A)}$. The sum on the right-hand side of (9.15) is then precisely (9.16). $\qquad \square$

Thus, $\chi(G; \lambda)$ is a polynomial on λ and it is called the *chromatic polynomial* of G. Some easily seen properties of $\chi(G; \lambda)$, which can be found in Read's seminal work [75] on the chromatic polynomial, are the following.

9   The Tutte Polynomial   233

Proposition 9.6.2. *If G is a graph with chromatic polynomial $\chi(G;\lambda)$, then:*

1. *If G has no edges, then $\chi(G;\lambda) = \lambda^n$.*
2. *If G has a loop, then $\chi(G;\lambda) = 0$, for all λ.*
3. *$\chi(K_n;\lambda) = \lambda(\lambda - 1)\cdots(\lambda - n + 1)$.*
4. *If e is any edge of G, then*

$$\chi(G;\lambda) = \chi(G \setminus e;\lambda) - \chi(G/e;\lambda).$$

Note that Items 1 and 4 together give a recursive alternative definition of the chromatic polynomial.

Also in Read [75] is the following not so trivial, but not difficult to prove, property of the chromatic polynomial.

Theorem 9.6.3. *If G is the union of two vertex set induced subgraphs H_1 and H_2 such that the intersection $H_1 \cap H_2$ is a vertex set induced subgraph isomorphic to K_p, then*

$$\chi(G;\lambda) = \frac{\chi(H_1;\lambda)\chi(H_2;\lambda)}{\chi(K_p;\lambda)}.$$

Thus, although Proposition 9.6.2, Item 4 suggests that the chromatic polynomial might be a T–G invariant, by Theorem 9.6.3, it is *not* multiplicative on the one point join of two graphs. However, as is frequently the case, this can be addressed by a simple multiplier; it is easy to check that $\lambda^{-\kappa(G)}\chi(G;\lambda)$ is a T–G invariant. The relation between the Tutte and chromatic polynomials may then be found by applying Theorem 9.4.2 with the help of Proposition 9.6.2, Item 4. We give an alternative proof of this relationship deriving from Theorem 9.6.1.

Theorem 9.6.4. *If $G = (V, E)$ is a graph, then*

$$\chi(G;\lambda) = (-1)^{r(E)}\lambda^{\kappa(G)}T(G;1 - \lambda, 0).$$

Proof. Since $r(E) - r(A) = \kappa(A) - \kappa(G)$ we have that

$$\chi(G;\lambda) = \sum_{A \subseteq E} (-1)^{|A|}\lambda^{\kappa(A)}$$

$$= (-1)^{r(E)}\lambda^{\kappa(G)} \sum_{A \subseteq E} (-1)^{|A|-r(A)}(-\lambda)^{r(E)-r(A)}$$

$$= (-1)^{r(E)}\lambda^{\kappa(G)}T(G;1 - \lambda, 0),$$

with the last equality following from Definition 9.3.3.   □

9.6.2 The Bad Coloring Polynomial

One way to generalize the chromatic polynomial is to count *all* possible colorings of the graph G, not just proper colorings. In order to differentiate between proper

234 J.A. Ellis-Monaghan and C. Merino

and improper colorings, we keep track of the edges between vertices of the same color, calling them *bad edges*. This leads to the *bad coloring polynomial*.

Definition 9.6.5. The *bad coloring polynomial* is the generating function

$$B(G; \lambda, t) = \sum_j b_j(G; \lambda)t^j,$$

where $b_j(G; \lambda)$ is the number of λ-colorings of G with exactly j bad edges.

Now consider $B(G; \lambda, t + 1)$, which can be written as

$$B(G; \lambda, t + 1) = \sum_{\phi: V \to [\lambda]} (1 + t)^{|b(\phi)|}, \tag{9.17}$$

where $b(\phi)$ is the set of bad edges in the λ-coloring ϕ. With this last expression it is again easy to get the relation to the Tutte polynomial using the following derivation of Noble (private communication).

Theorem 9.6.6. *For a graph $G = (V, E)$ we have that*

$$B(G; \lambda, t + 1) = t^{r(E)} \lambda^{\kappa(G)} T\left(G; \frac{\lambda + t}{t}, 1 + t\right).$$

Proof.

$$\begin{aligned}
B(G; \lambda, t + 1) &= \sum_{\phi: V \to [\lambda]} (1 + t)^{|b(\phi)|} \\
&= \sum_{\phi: V \to [\lambda]} \sum_{A \subseteq b(\phi)} t^{|A|} \\
&= \sum_{A \subseteq E} \sum_{\substack{\phi: V \to [\lambda] \\ A \subseteq b(\phi)}} t^{|A|} \\
&= \sum_{A \subseteq E} t^{|A|} \lambda^{\kappa(A)}.
\end{aligned}$$

Thus,

$$\begin{aligned}
B(G; \lambda, t + 1) &= \sum_{A \subseteq E} t^{|A|} \lambda^{\kappa(A)} \\
&= t^{r(E)} \lambda^{\kappa(G)} \sum_{A \subseteq E} t^{|A| - r(A)} \left(\frac{\lambda}{t}\right)^{r(E) - r(A)} \\
&= t^{r(E)} \lambda^{\kappa(G)} T\left(G; 1 + \frac{\lambda}{t}, t + 1\right).
\end{aligned}$$

$\square$

9 The Tutte Polynomial

Again, the above result could also be obtained from the universal property of the Tutte polynomial given in Theorem 9.4.2 by applying it to $\bar{B}(G; \lambda, t) = \lambda^{-\kappa(G)} B(G; \lambda, t)$ and verifying that:

1. $\bar{B}(G; \lambda, t) = \bar{B}(G \setminus e; \lambda, t) + (t - 1)\bar{B}(G/e; \lambda, t)$, if e is an ordinary edge.
2. $\bar{B}(G; \lambda, t) = t\bar{B}(G \setminus e; \lambda, t)$, if e is a loop.
3. $\bar{B}(G; \lambda, t) = (t + \lambda - 1)\bar{B}(G/e; \lambda, t)$, if e is a bridge.

9.6.3 The Flow Polynomial

The dual notion to a proper λ-coloring is a nowhere zero λ-flow. A standard resource for the material in this section is Zhang [116], and Jaeger [48] gives a good survey.

Let G be a graph with an arbitrary but fixed orientation, and let H be an Abelian group with 0 as its identity element. An H-flow is a mapping ϕ of the oriented edges $\vec{E}(G)$ into the elements of the group H such that Kirchhoff's law is satisfied at each vertex of G; that is,

$$\sum_{\vec{e}=u \to v} \phi(\vec{e}) + \sum_{\vec{e}=u \leftarrow v} \phi(\vec{e}) = 0,$$

for every vertex v, and where the first sum is taken over all arcs towards v and the second sum is over all arcs leaving v. An H-flow is *nowhere zero* if ϕ never takes the value 0.

By replacing the group element on an edge e by its inverse, it is clear that two orientations that differ only in the direction of exactly one arc $\vec{e}$ have the same number of nowhere zero H-flows for any H. Thus, this number does not depend on the choice of orientation of G. In fact, when H is finite, it does not depend on the structure of the group, but rather only on its cardinality. The following, due to Tutte [103], relates the number of nowhere zero flows of G over a finite group and Tutte polynomial of G. The reason for the notation χ^* becomes clear with Corollary 9.6.8.

Theorem 9.6.7. *Let $G = (V, E)$ be a graph and H a finite Abelian group. If $\chi^*(G; H)$ denotes the number of nowhere zero H-flows then*

$$\chi^*(G; H) = (-1)^{|E|-r(E)} T(G; 0, 1 - |H|).$$

Proof (sketch). Here we use the universality of the Tutte polynomial. If e is an ordinary edge of G, then the number of nowhere zero H-flows in G/e can be partitioned into two sets P_1 and P_2. We let P_1 consist of those that are also nowhere zero H-flows in $G \setminus e$, and P_2 be the complement of P_1. Clearly then $|P_1| = \chi^*(G \setminus e; H)$. Furthermore, there is a bijection between the elements in P_2 and the nowhere zero H-flows in G, and thus $|P_2| = \chi^*(G; H)$. It follows that

$$\chi^*(G; H) = \chi^*(G \setminus e; H) - \chi^*(G/e; H),$$

and hence $\chi^*(G; H)$ satisfies (9.7). It is also easy to check that $\chi^*(G; H)$ satisfies (9.8). Since $\chi^*(L; H) = 0$ and $\chi^*(B; H) = |H| - 1$, the result follows from Theorem 9.4.2. □

Consequently, $\chi^*(G; \lambda)$ is a polynomial called the *flow polynomial* which for λ an integer at least 1 gives the number of nowhere zero flows of G in a group of order λ. We call any nowhere zero H-flow simply a λ-flow if $|H| = \lambda$.

If the Abelian group is $\mathbb{Z}_3$, and the graph is 4-regular, then the Tutte polynomial at $(0, -2)$ counts the number of nowhere zero $\mathbb{Z}_3$-flows on G. But these flows are in one-to-one correspondence with orientations such that at each vertex exactly two edges are directed in and two out. Such an orientation is called an *ice configuration* of G (see Lieb [58] and Pauling [72] for this important model of ice and its physical properties). Thus, we have the following corollary.

Corollary 9.6.8. *If G is a 4-regular graph, then $T(G; 0, -2)$ equals the number of ice configurations of G.*

We mentioned previously that proper colorings are the dual concept of nowhere zero flows, and now with Theorem 9.6.3 and (9.5) we observe that

$$\chi(G; \lambda) = \lambda \chi^*(G^*; \lambda),$$

for G a connected planar graph and G^*, any of its plane duals. Thus, to each λ-proper coloring in G corresponds λ nowhere zero $\mathbb{Z}_\lambda$-flows of G^*. A bijective proof can be found in Diestel [26].

The four-color theorem and the duality relation between colorings and nowhere zero H-flows then mean that every bridgeless planar graph has a 4-flow. For a cubic graph, having a nowhere zero $\mathbb{Z}_2 \times \mathbb{Z}_2$-flow is equivalent to being 3-edge-colorable. Therefore, as the Petersen graph is not 3-edge-colorable, it has no 4-flow. However, the Petersen graph does have a 5-flow. In fact, the famous 5-flow conjecture of Tutte [103] postulates that every bridgeless graph has a 5-flow.

The 5-flow conjecture is clearly difficult as it is not even apparent that every graph will have a λ-flow for some λ. However, Jaeger [47] proved that every bridgeless graph has an $\mathbb{Z}_2 \times \mathbb{Z}_2 \times \mathbb{Z}_2$-flow, thus every bridgeless graph has an 8-flow. Subsequently Seymour [85] proved that every bridgeless graph has a $\mathbb{Z}_2 \times \mathbb{Z}_3$-flow, thus every graph has a 6-flow.

Not much is currently known about properties of the flow polynomial apart from those that can be deduced from its duality with the chromatic polynomial and efforts to solve the 5-flow conjecture. However, for some recent work in this direction, see Dong and Koh [28] and Jackson [46].

9.6.4 Abelian Sandpile Models

Self-organized criticality is a concept widely considered in various domains since Bak et al. [4] introduced it. One of the paradigms in this framework is the Abelian sandpile model, introduced by Dhar [25].

9   The Tutte Polynomial   237

We begin by recalling the definition of the general Abelian sandpile model on a set of N sites labeled $1, 2, \ldots, N$, that we refer to as the system. A sandpile at each site i has height given by an integer h_i. The set $\vec{h} = \{h_i\}$ is called a *configuration* of the system. For every site i, a threshold H_i is defined; configurations with $h_i < H_i$ for all i are called *stable*. For every stable configuration, the height h_i increases in time at a constant rate; this is called the *loading* of the system. This loading continues until $h_i \geq H_i$ for some i. The site i then "topples" and all the values h_j, for $1 \leq j \leq N$, are updated according to the rule:

$$h_j = h_j - M_{ij}, \quad \text{for all } j, \tag{9.18}$$

where M is a given fixed integer matrix satisfying

$$M_{ii} > 0, \quad M_{ij} \leq 0 \quad \text{and} \quad s_i = \sum_j M_{ij} \geq 0.$$

If, after this redistribution, the height at some vertex exceeds its threshold, we again apply the toppling rule (9.18), and so on, until we arrive at a stable configuration and the loading resumes. The sequence of topplings is called an *avalanche*. We assume that an avalanche is "instantaneous", so that no loading occurs during an avalanche.

The value s_i is called the *dissipation* at site i. We say that s_i is *dissipative* if $s_i > 0$ and *nondissipative* if $s_i = 0$. It may happen that an avalanche continues without end. We can avoid this possibility by requiring that from every non-dissipative site i, there exists a path to a dissipative site j. In other words, there is a sequence $i_0, \ldots, i_n$, with $i_0 = i$, $i_n = j$ and $M_{i_{k-1}, i_k} < 0$, for $k = 1, \ldots, n$. In this case, following Gabrielov [37], we say that the system is *weakly dissipative*, and we assume that a system is always weakly dissipative. In a weakly dissipative system, any configuration $\vec{h}$ will eventually arrive at a stable configuration. But the process is infinite, and the stable configurations are clearly finite. Thus, some stable configurations recur, and these are called *critical configurations*.

The sandpile process has an Abelian property, in that if at some stage, two sites can topple, the resulting stable configuration after the avalanche is independent of the order in which the sites toppled. Thus, for any configuration $\vec{h}$, the process eventually arrives at a unique critical configuration $\vec{c}$.

Let G be a connected graph, $q \in V(G)$, and L' be the minor of the Laplacian of G resulting from deleting the row and column corresponding to q. When the matrix M is L' for some vertex q, the Abelian sandpile model coincides with the chip-firing game or dollar game on a graph that was defined by Biggs [7]. For the rest of this section we assume M is given in this way.

For a configuration $\vec{h}$, we define its *weight* to be $w(\vec{h}) = \sum_{i=0}^N h_i$. If $\vec{c}$ is a critical configuration, we define its *level* as

$$\text{level}(\vec{c}) = w(\vec{c}) - |E(G)| + \deg(q).$$

This definition may seem a little unnatural, but it is justified by the following theorem of Biggs [7], which tells us that it is actually the right quantity to consider if we want to grade the critical configurations.

Theorem 9.6.9. *If $G = (V, E)$ is a connected graph and $\vec{h}$ a critical configuration of G, then*

$$0 \leq \text{level}(\vec{h}) \leq |E| - |V| + 1.$$

The rightmost quantity is called the *cyclomatic number* of G. We now consider the generating function of these critical configurations.

Definition 9.6.10. Let $G = (V, E)$ be a graph and for nonnegative integers i let c_i be the number of critical configurations with level i. Then the *critical configuration polynomial* is

$$P_q(G; y) = \sum_{i=0}^{|E|-|V|+1} c_i \, y^i.$$

Theorem 9.6.11. *For a connected graph G and any vertex q, the generating function of the critical configurations equals the Tutte polynomial of G along the line $x = 1$, that is,*

$$P_q(G; y) = T(G; 1, y),$$

and thus $P_q(G; y)$ is independent of the choice of q.

A proof using deletion and contraction of an edge incident with the special vertex q can be found in [64].

New combinatorial identities frequently arise when a new generating function can be shown to be related to the Tutte polynomial, as in the following corollary.

Corollary 9.6.12. *If G is a connected graph, then the number of critical configurations of G is equal to the number of spanning trees, and the number of critical configurations with level 0 is equal to the number of acyclic orientations with a unique source.*

This follows from comparing Theorem 9.6.11 with Theorems 9.5.1 and 9.5.6.

9.6.5 The Reliability Polynomial

Many of the invariants reviewed thus far have various applications in the sciences, engineering, and computer science. However, when a graph models some kind of network (e.g., electrical, communication, etc.), then the applicability of the reliability polynomial we discuss next is particularly apparent.

Definition 9.6.13. Let G be a connected graph or network with n vertices and m edges, and suppose that each edge is independently chosen to be active with probability p. Then the (all terminal) *reliability polynomial* is

9 The Tutte Polynomial

$$R(G; p) = \sum_{\substack{A \text{ spanning} \\ \text{connected}}} p^{|A|}(1-p)^{|E-A|}$$

$$= \sum_{k=0}^{m-n+1} g_k \, p^{k+n-1}(1-p)^{m-k-n+1}, \tag{9.19}$$

where g_k is the number of spanning connected subgraphs with $k + n - 1$ edges.

Thus the reliability polynomial, $R(G; p)$, is the probability that in this random model there is a path of active edges between each pair of vertices of G. In other words, it gives the probability that the overall network is functioning.

Theorem 9.6.14. *If G is a connected graph with m edges and n vertices, then*

$$R(G; p) = p^{n-1}(1-p)^{m-n+1} T\left(G; 1, \frac{1}{1-p}\right).$$

Proof. We first note from the rank generating expansion of Definition 9.3.3 that

$$T(G; 1, y+1) = \sum_{k=0}^{m-n+1} g_k y^k,$$

since the only nonvanishing terms are those corresponding to $A \subseteq E$ with $r(E) = r(A)$, that is, spanning connected subgraphs.

We then observe that

$$R(G; p) = \sum_{k=0}^{m-n+1} g_k \, p^{k+n-1}(1-p)^{m-k-n+1}$$

$$= p^{n-1}(1-p)^{m-n+1} \sum_{k=0}^{m-n+1} g_k \left(\frac{p}{1-p}\right)^k$$

$$= p^{n-1}(1-p)^{m-n+1} T\left(G; 1, 1 + \frac{p}{1-p}\right)$$

$$= p^{n-1}(1-p)^{m-n+1} T\left(G; 1, \frac{1}{1-p}\right).$$

$\square$

If we extend the reliability polynomial to graphs with more than one component by defining $R(G \cup H; p) := R(G, p)R(H, p)$, then this result may also be proved using the universality property of the Tutte polynomial. Observe that if an ordinary edge is not active (this happens with probability $1 - p$), then the reliability of the network is the same as if the edge were deleted. Similarly, if an edge is active (which happens with probability p), then the reliability is the same as it would

240 J.A. Ellis-Monaghan and C. Merino

be if the edge were contracted. Thus, the reliability polynomial has the following deletion/contraction reduction:

$$R(G; p) = (1 - p)R(G \setminus e) + pR(G/e).$$

With this, and noting that $R(G * H; p) = R(G; p)R(H; p)$ with $R(L, p) = 1$ and $R(B; p) = p$, Theorem 9.6.14 also follows immediately from Theorem 9.4.2.

There is a vast literature about reliability and the reliability polynomial; for a good survey, including a wealth of open problems, we refer the reader to Chari and Colbourn [20].

9.6.6 The Shelling Polynomial

A *simplicial complex* Δ is a collection of subsets of a set of vertices V such that if $v \in V$, then $\{v\} \in \Delta$, and also if $F \in \Delta$ and $H \subseteq F$, then $H \in \Delta$. The elements of Δ are called *faces*. Maximal faces are called *facets*, and if all the facets have the same cardinality, Δ is called *pure*. The dimension of a face is its cardinality minus one and the dimension of a pure simplicial complex is the dimension of any of its facets.

If f_k is the number of faces of cardinality k in a simplicial complex Δ, then the vector $(f_0, f_1, \ldots, f_d)$ is called the *face vector* or f-*vector* of Δ, and

$$f_\Delta(x) = \sum_{k=0}^{d} f_k x^{d-k}, \tag{9.20}$$

is the generating function of the faces of Δ, or *face enumerator*.

The collection of spanning forests of a connected graph G forms a pure $(d - 1)$-dimensional simplicial complex $\Delta(G)$. The points of $\Delta(G)$ are the nonloop edges of G and its facets are the spanning trees, so $d = r(E)$. The collection of complements of spanning connected subgraphs of G also forms a pure $(d^* - 1)$-dimensional simplicial complex $\Delta^*(G)$. Here the elements are the nonbridge edges, while the facets are complements of spanning trees, when viewed as subsets of E; in general, if A is the edge set of a spanning connected subgraph of G of cardinality $k + n - 1$, then $E \setminus A$ is a face of size $m - n + 1 - k$ in $\Delta^*(G)$. Thus, $d^* = m - n + 1$ and if, as before, g_k is the number of spanning connected subgraphs with $k + n - 1$ edges, the f-vector of $\Delta^*(G)$ is $(f_0^*, \ldots, f_{d^*}^*)$, where $f_i^* = g_{d^*-i}$.

Theorem 9.6.15. *The Tutte polynomial gives the face enumerators for both* $\Delta(G)$ *and* $\Delta^*(G)$:

$$T(G; x + 1, 1) = \sum_{k=0}^{d} f_k x^{d-k} = f_{\Delta(G)}(x),$$

9 The Tutte Polynomial

and

$$T(G; 1, y+1) = \sum_{i=0}^{d^*} f_i^* y^{d^*-i} = f_{\Delta^*(G)}(x).$$

Proof. This follows readily by comparing (9.20) with Definition 9.3.3. $\square$

For a pure simplicial complex Δ, a *shelling* is a linear ordering of the facets $F_1, F_2, \ldots, F_t$ such that, if $1 \le k \le t$, then F_k meets the complex generated by its predecessors, denoted Δ_{k-1}, in a nonempty union of maximal proper faces. A complex is said to be *shellable* if it is pure and admits a shelling. A good exposition of the following results can be found in Björner [9].

For $1 \le k \le t$, define $\mathcal{R}(F_k) = \{x \in F_k \mid F_k \setminus x \in \Delta_{k-1}\}$, where here $\Delta_0 = \emptyset$. The number of facets such that $|F_k - \mathcal{R}(F_k)| = i$ is denoted by h_i and it does not depend on the particular shelling (this follows, for example, from (9.21) below). The vector $(h_0, h_1, \ldots, h_d)$ is called the *h-vector* of Δ. The *shelling polynomial* is the generating function of the h-vector, and is given by

$$h_\Delta(x) = \sum_{i=0}^{d} h_i x^{d-i}.$$

The face enumerator and shelling polynomial are related in a somewhat surprising way, namely

$$h_\Delta(x+1) = f_\Delta(x). \tag{9.21}$$

Both $\Delta(G)$ and $\Delta^*(G)$ are known to be shellable (see, for example, Provan and Billera [74]), and thus (9.21) gives the following corollary to Theorem 9.6.15, relating the two shelling polynomials to the Tutte polynomial (see Björner [9]).

Corollary 9.6.16. *Let G be a graph. Then*

$$T(G; x, 1) = h_{\Delta(M)}(x) = \sum_{i=0}^{d} h_i x^{d-i}$$

and

$$T(G; 1, y) = h_{\Delta^*(G)}(y) = \sum_{i=0}^{d^*} h_i^* y^{d^*-i}.$$

The reader may have noticed that the reliability polynomial as well as the face enumerator and shelling polynomial of $\Delta^*(G)$ are all specializations of the Tutte polynomial along the line $x = 1$. There is an important open conjecture in algebraic combinatorics about the h-vectors (and hence the shelling polynomials), of the two complexes coming from a graph (or, more generally, a matroid), namely that they are "pure O-sequences". For more details see Stanley [96] or [65]. The latter also relates the shelling polynomial and the chip firing game. Let G be a graph with n vertices and m edges. From Corollary 9.6.16 and Theorem 9.6.11, we get that

$c_i = h^*_{m-n+1-i}$, where c_i is the number of critical configurations of level i of G and $(h^*_0, \ldots, h^*_{m-n+1})$ is the h-vector of $\Delta^*(G)$. In [65] it is proved that $(c_{m-n+1}, \ldots, c_0)$ is a pure O-sequence. Thus, the conjecture is true for the simplicial complex $\Delta^*(G)$ but is still open for $\Delta(G)$.

It is also clear from Theorem 9.6.14 and Corollary 9.6.16 that the reliability and shelling polynomials are related. This connection is explored, and open questions related to it presented, by Chari and Colbourn [20].

9.7 Some Properties of the Tutte Polynomial

There is a large and ever-growing body of information about properties of the Tutte polynomial. Here, we present some of them, again with an emphasis on illustrating general techniques for extracting information from a graph polynomial.

9.7.1 The Beta Invariant

Even a single coefficient of a graph polynomial can encode a remarkable amount of information. It may characterize entire classes of graphs and have a number of combinatorial interpretations. A noteworthy example is the β invariant, introduced (in the context of matroids) by Crapo in [23].

Definition 9.7.1. Let $G = (V, E)$ be a graph with at least two edges. The β invariant of G is

$$\beta(G) = (-1)^{r(G)} \sum_{A \subseteq E} (-1)^{|A|} r(A).$$

The beta invariant is a deletion/contraction invariant; that is, it satisfies (9.7). However, the β invariant is zero if and only if G either has loops or is not 2-connected. Thus, the β invariant is not a Tutte–Gröthendieck invariant in the sense of Sect. 9.4. While the β invariant may be defined to be 1 for a single edge or a single loop, it still will not satisfy (9.8), and it is not multiplicative with respect to disjoint unions and one-point joins. Nevertheless, the β invariant derives from the Tutte polynomial.

Theorem 9.7.2. *If G has at least two edges, and we write $T(G; x, y)$ in the form $\sum t_{ij} x^i y^j$, then $t_{0,1} = t_{1,0}$, and this common value is equal to the β invariant.*

Proof. This can easily be proved by induction, using deletion/contraction for an ordinary edge, and otherwise noting that the β invariant is zero if the graph has loops or is not 2-connected. $\square$

The β invariant does not change with the insertion of parallel edges or edges in series. Thus, homeomorphic graphs have the same β invariant. The β invariant is

9 The Tutte Polynomial

also occasionally called the chromatic invariant, because the derivative $\chi'(G;1) = (-1)^{r(G)} \beta(G)$, where $\chi(G;x)$ is the chromatic polynomial.

Definition 9.7.3. A series–parallel graph is a graph constructed from a digon (two vertices joined by two edges in parallel) by repeatedly adding an edge in parallel to an existing edge, or adding an edge in series with an existing edge by subdividing the edge. Series–parallel graphs are loopless multigraphs, and are planar.

Brylawski [14] and also [16], in the context of matroids, showed that the β invariant completely characterizes series–parallel graphs.

Theorem 9.7.4. *G is a series–parallel graph if and only if $\beta(G) = 1$.*

Using the deletion/contraction definition of the Tutte polynomial, it is quite easy to show that the β invariant is unchanged by adding an edge in series or in parallel to another edge in the graph. This, combined with the β invariant of a digon being one, suffices for one direction of the proof. The difficulty is in the reverse direction, and the proof is provided in [15] by a set of equivalent characterizations for series–parallel graphs, one by excluded minors and another that the β invariant is 1 for series–parallel graphs. For graphs, the excluded minor is K_4 (cf. Duffin [30] and Oxley [70]). Succinct proofs may also be found in Zaslavsky [115]. The fundamental observation, which may be applied to other situations, is that there is a graphical element, here an edge which is in series or parallel with another edge, which behaves in a tractable way with respect to the computation methods of the polynomial.

The β invariant has been explored further, for example, by Oxley in [70] and by Benashshki et al. in [5]. Oxley characterized 3-connected matroids with $\beta \leq 4$, and a complete list of all simple 3-connected graphs with $\beta \leq 9$ is given in [5].

A wide variety of combinatorial interpretations have also been found for the β invariant. Most interpretations involve objects other than graphs, but we give two graphical interpretations below. The first is due to Las Vergnas [55].

Theorem 9.7.5. *Let G be a connected graph. Then $2\beta(G)$ gives the number of orientations of G that have a unique source and sink, independent of their relative locations.*

This result is actually a consequence of a more general theorem giving an alternative formulation of the Tutte polynomial, which we discuss further in Sect. 9.7.2. We also have the following result from [32].

Theorem 9.7.6. *Let $G = (V, E)$ be a connected planar graph with at least two edges. Then*

$$\beta = \frac{1}{2} \sum (-1)^{c(E \backslash P)+1},$$

where the sum is over all closed trails P in $\vec{G}_m$ which visit each vertex at least once.

Like the interpretations for $T(G;x,x)$ given in Sect. 9.5.2, this result follows from the Tutte polynomial's relation to the Martin polynomial.

Graphs in a given class may have β invariants of a particular form. McKee [63] provides an example of this in dual-chordal graphs. A *dual-chordal graph* is 2-connected, 3-edge-connected, such that every cut of size at least four creates a bridge. A θ *graph* has two vertices with three edges in parallel between them. A dual-chordal graph has the property that it may be reduced to a θ graph by repeatedly contracting induced subgraphs of the following forms: digons, triangles, and $K_{2,3}$'s, where in all cases each vertex has degree 3 in G.

Theorem 9.7.7. *If G is a dual-chordal graph, then $\beta(G) = 2^a 5^b$. Here, a is the number of triangles in G, where each vertex has degree 3 in G, that are contracted in reducing G to a θ graph. Similarly, b is the number of induced $K_{2,3}$'s in G, again where each vertex has degree 3 in G, that are contracted in reducing G to a θ graph.*

The proof follows from considering the acyclic orientations of G with unique source and sink and applying the results of Green and Zaslavsky [43].

9.7.2 Coefficient Relations

After observing that $t_{1,0} = t_{0,1}$ in the development of the β invariant, it is natural to ask if there are similar relations among the coefficients t_{ij} of the Tutte polynomial $T(G; x, y) = \sum t_{ij} x^i y^j$ and whether there are combinatorial interpretations for these coefficients as well. The answer is yes, although less is known. The most basic fact, and one which is not obvious from the rank-nullity formulation of Definition 9.3.3, is that all the coefficients of the Tutte polynomial are nonnegative.

That $t_{1,0} = t_{0,1}$ is one of an infinite family of relations among the coefficients of the Tutte polynomial. Brylawski [16] has shown the following.

Theorem 9.7.8. *If G is a graph with at least m edges, then*

$$\sum_{i=0}^{k} \sum_{j=0}^{k-i} (-1)^j \binom{k-i}{j} t_{ij} = 0,$$

for $k = 0, 1 \ldots, m - 1$.

Additionally, Las Vergnas in [55] found combinatorial interpretations in the context of oriented matroids for these coefficients by determining yet another generating function formulation for the Tutte polynomial. Las Vergnas [55] gives the following specialization to orientations of graphs, and see also Gioan and Las Vergnas [40] for a bijective interpretation of this result.

Theorem 9.7.9. *Let G be a graph with a linear ordering of its edges. Let $o_{i,j}$ be the number of orientations of G such that the number of edges that are smallest on some consistently directed cocycle is i and the number of edges that are smallest on a consistently directed cycle is j. Then*

9 The Tutte Polynomial

Fig. 9.3 A counterexample to the unimodularity conjecture

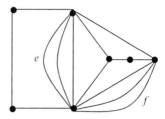

$$T(G;x,y) = \sum_{i,j} o_{i,j} 2^{-(i+j)} x^i y^j,$$

and thus $t_{ij} = o_{i,j}/(2^{i+j})$.

The proof is modeled on Tutte's proof that the t_{ij}'s are independent of the ordering of the edges by using deletion/contraction on the greatest edge in the ordering.

Another natural question is to ask if these coefficients are unimodular or perhaps log concave, for example in either x or y. While this was originally conjectured to be true (see Seymour and Welsh [86], Tutte [105]), then Schwärzler [83] found a contradiction in the graph in Fig. 9.3. This counterexample can be extended to an infinite family of counterexamples by increasing the number of edges parallel to e or f.

The unimodularity question for the chromatic polynomial, raised by Read in [75], is still unresolved.

9.7.3 Zeros of the Tutte Polynomial

Because the Tutte polynomial is after all a polynomial, it is natural to ask about its zeros and factorizations. The importance of its zeros is magnified by their interpretations. For example, since $T(G;0,y)$ is essentially the flow polynomial, a root of the form $(0, 1 - \lambda)$, for λ a positive integer, means that G does not have a nowhere zero flow for any Abelian group of order λ. Similarly, since $T(G;x,0)$ is essentially the chromatic polynomial, a root of the form $(1 - \lambda, 0)$ with λ a positive integer, means that G cannot be properly colored with λ colors. In particular, a direct proof the four-color theorem would follow if it could be shown that the Tutte polynomial has no zero of the form $(-3, 0)$ on the class of planar graphs. Of course, because of the duality between the flow and chromatic polynomials, results for the zeros of the one informs the other, and vice versa. Jackson [45] surveys zeros of both chromatic and the flow polynomials.

As we discuss in Chap. 10, the chromatic polynomial has an additional interpretation as the zero-temperature antiferromagnetic Potts model of statistical mechanics. In this context, its zeros correspond to numbers of spins for which the ground-state degeneracy function may be nonanalytic. This has led to research into its zeros by theoretical physicists as well as mathematicians. Traditionally, the focus from a graph theory perspective was on positive integer roots of the chromatic polynomial, corresponding to graphs not being properly colorable with q colors. In statistical mechanics, however, the relevant quantity involves the limit of an increasing family of graphs as the number, n, of vertices goes to infinity. This shifted the focus to the complex roots of the chromatic polynomial, since the sequence of complex roots as $n \to \infty$ may have an accumulation point on the real axis.

Because of this, a significant body of work has emerged in recent years devoted to clearing regions of the complex plane (in particular, regions containing intervals of the real axis) of roots of the chromatic polynomial. Results showing that certain intervals of the real axis and certain complex regions are free of zeros of chromatic polynomials include those of Woodall [113], Jackson [44], Shrock and Tsai [87, 88], Thomassen [99], Sokal [92], Procacci et al. [73], Choe et al. [22], Borgs [12], and Fernandez and Procacci [36]. One particular question concerns the maximum magnitude of a zero of a chromatic polynomial and of zeros comprising region boundaries in the complex plane as the number of vertices $n \to \infty$. An upper bound is given by Sokal in [92], depending on the maximal vertex degree. There are, however, families of graphs where both of these magnitudes are unbounded (see Read and Royle [77], Shrock and Tsai [87, 89], Brown et al. [13], and Sokal [93]). For recent discussions of some relevant research directions concerning zeros of chromatic polynomials and properties of their accumulation sets in the complex plane, as well as approximation methods, see, e.g., Shrock and Tsai [88], Shrock [90], Sokal [91, 92], Chang and Shrock [18], Chang et al. [19], Choe et al. [22], Dong and Koh [27], and more recently Royle [81, 82].

If G is a graph with chromatic number $k + 1$, then $\chi(G; x)$ has integer roots at $0, 1, \ldots, k$. Thus, the chromatic polynomial of G can be written as

$$\chi(G; x) = x^{a_0}(x - 1)^{a_1} \cdots (x - k)^{a_k} q(x),$$

where $a_0, \ldots, a_k$ are integers and $q(x)$ is a polynomial with no integer roots in the interval $[0, k]$. In contrast to this we have the following result of Merino et al. [66].

Theorem 9.7.10. *If G is a 2-connected graph, then $T(G; x, y)$ is irreducible in $\mathbb{Z}[x, y]$.*

The proof is quite technical and it heavily relies on Theorem 9.7.8 and that $\beta(G) \neq 0$ if and only if G has no loops and it is 2-connected.

If G is not 2-connected, then $T(G; x, y)$ can be factored. From Proposition 9.3.2 we get that if G is a disconnected graph with connected components $G_1, \ldots, G_\kappa$, then $T(G; x, y) = \prod_{i=1}^{\kappa} T(G_i; x, y)$. So we assume G is connected but not 2-connected.

One of the basic properties mentioned in [17] is that $y^s | T(G; x, y)$ if and only if G has s loops. Thus, we assume G is loopless and connected but not 2-connected. It is well known that such graphs have a decomposition into their blocks; see, for example, [11]. A *block of* a graph G is either a bridge or a maximal 2-connected

9   The Tutte Polynomial   247

subgraph. If two blocks of G intersect, they do so in a cut vertex. By Theorem 9.7.10 and Proposition 9.3.2 we get the following.

Corollary 9.7.11. *If G is a loopless connected graph that is not 2-connected with blocks $H_1, \ldots, H_p$, then the factorization of $T(G; x, y)$ in $\mathbb{Z}[x, y]$ is exactly*

$$T(G; x, y) = T(H_1; x, y) \cdots T(H_p; x, y).$$

9.7.4   Derivatives of the Tutte Polynomial

It is also most natural to differentiate the Tutte polynomial and to ask for combinatorial interpretations of its derivatives. For example, Las Vergnas [57] has found the following combinatorial interpretation of the derivatives of the Tutte polynomial. It first requires a slight generalization of the notions of internal and external activities given in Sect. 9.3.3.

Definition 9.7.12. Let $G = (V, E)$ be a graph with a linear order on its edges, and let $A \subseteq E$. An edge $e \in A$ and a cut C are internally active with respect to A if $e \in C \subseteq (E \backslash A) \cup \{e\}$ and e is the smallest element in C. Similarly, an edge $e \in E \backslash A$ and a cycle C are externally active with respect to A if $e \in C \subseteq A \cup \{e\}$.

In the case that A is a spanning tree, this reduces to the previous definitions of internally and externally active.

Theorem 9.7.13. *Let G be a graph with a linear ordering on its edges. Then*

$$\frac{\partial^{p+q}}{\partial x^p \partial y^q} T(G; x, y) = p! \, q! \sum x^{\text{in}(A)} y^{\text{ex}(A)},$$

where the sum is over all subsets A of the edge set of G such that $r(G) - r(A) = p$ and $|A| - r(A) = q$, and where $\text{in}(A)$ is the number of internally active edges with respect to A, and $\text{ex}(A)$ is the number of externally active edges with respect to A.

The proof begins by differentiating the spanning tree definition of the Tutte polynomial, Definition 9.3.5, which gives a sum over i and j restricted by p and q. This is followed by showing that the coefficients of $x^{i-p} y^{j-q}$ enumerate the edge sets described in the theorem statement. The enumeration comes from examining, for each subset A of E, the set of $e \in E \backslash A$ such that there is a cut-set of G contained in $E \backslash A$ with e as the smallest element (and dually for cycles).

The Tutte polynomial along the line $x = y$ is a polynomial in one variable that, for planar graphs, is related to the Martin polynomial via a medial graph construction. From this relationship, [31] derives an interpretation for the n-th derivatives of this one variable polynomial evaluated at 2 in terms of edge disjoint closed trails in the oriented medial graph.

Definition 9.7.14. For an oriented graph $\vec{G}$, let P_n be the set of ordered n-tuples $\bar{p} := (p_1, \ldots, p_n)$, where the p_i's are consistently oriented edge-disjoint closed trails in $\vec{G}$.

Theorem 9.7.15. *If G is a connected planar graph with oriented medial graph $\overrightarrow{G_m}$, then, for all nonnegative integers n,*

$$\frac{\partial^n}{\partial x} T(G; x, x)\Big|_{x=2} = \sum_{k=0}^{n} (-1)^{n-k} \frac{n!}{k!} \sum_{\bar{p} \in P_k\left(\overrightarrow{G_m}\right)} 2^{m(\bar{p})},$$

where $m(\bar{p})$ is the number of vertices of $\overrightarrow{G_m}$ not belonging to any of the trails in $\bar{p}$.

9.7.5 Convolution and the Tutte Polynomial

Since the Tutte polynomial can also be formulated as a generating function, the tools of generating functions, such as Möbius inversion and convolution, are available to analyze it. A comprehensive treatment of convolution and Möbius inversion can be found in Stanley [97]. Convolution identities are valuable because they write a graph polynomial in terms of the polynomials of its substructures, thus facilitating induction techniques. We have the following result from Kook et al. [53] using this approach (see also [33]).

Theorem 9.7.16. *The Tutte polynomial can be expressed as*

$$T(G; x, y) = \sum T(G/A; x, 0) T(G|_A; 0, y),$$

where the sum is over all subsets A of the edge set of G, and where $G|_A$ is the restriction of G to the edges of A; i.e., $G|_A = G \setminus (E \setminus A)$.

This result is particularly interesting in that it essentially writes the Tutte polynomial of a graph in terms of the chromatic and flow polynomials of its minors. It may be proved in several ways, for example, by induction using the deletion/contraction relation, or from the spanning trees expansion of the Tutte polynomial. However, we present the first proof from [53], which is dependent on results of Crapo [24], to illustrate an application of convolution.

Proof (sketch). We begin with a convolution product of two functions on graphs into the ring $\mathbb{Z}[x, y]$ given by $f * g = \sum_{A \subseteq E(G)} f(G|_A) g(G/A)$. The identity for convolution is $\delta(G)$ which is 1 if and only if G is edgeless and 0 otherwise. From Crapo [24], we have that

$$T(G; x + 1, y + 1) = (\zeta(1, y) * \zeta(x, 1))(G),$$

where $\zeta(x, y)(G) = x^{r(G)} y^{r(G^*)}$. Kook et al. [53] then show that $\zeta(x, y)^{-1} = \zeta(-x, -y)$. From this it follows that $T(G; x + 1, 0) = (\zeta(1, -1) * \zeta(x, 1))(G)$ and

9   The Tutte Polynomial   249

$T(G; 0, y + 1) = (\zeta(1, y) * \zeta(-1, 1))(G)$. Thus, $\sum T(G|_A; 0, y + 1) T(G/A; x + 1, 0) = (\zeta(1, y) * \zeta(-1, 1)) * (\zeta(1, -1) * \zeta(x, 1))(G)$. By associativity, the last expression is the same as $(\zeta(1, y) * (\zeta(-1, 1) * \zeta(1, -1)) * \zeta(x, 1))(G) = (\zeta(1, y) * \zeta(x, 1))(G) = T(G; x + 1, y + 1)$.   □

A formula, known as Tutte's identity for the chromatic polynomial, with a similar form, exists for the chromatic polynomial.

Theorem 9.7.17. *The chromatic polynomial can be expressed as*

$$\chi(G; x + y) = \sum \chi(G|_A; x) \chi(G|_{A^c}; y),$$

where the sum is over all subsets A of the set of vertices of G, and where $G|_A$ is the restriction of G to the vertices of A.

Proof. Consider an $(m + n)$-coloring of G, and let A be the vertices colored by the first m colors. Then an $(m + n)$-coloring of G decomposes into an m coloring of $G|_A$ using the first m colors and an n coloring of $G|_{A^c}$ using the remaining colors. Thus, for any two nonnegative integers m and n, it follows that $\chi(G; m + n) = \sum \chi(G|_A; m) \chi(G|_{A^c}; n)$. Since the expressions involve finite polynomials, this establishes the result for indeterminates x and y.   □

9.8   The Complexity of the Tutte Polynomial

We assume the reader is familiar with the basic notions of computational complexity, but for formal definitions in the present context, see, for example, Garey and Johnson [38] or Welsh [109].

We have seen that along different algebraic curves in the x–y plane, the Tutte polynomial evaluates to many diverse quantities. Some of these, such as $T(G; 2, 2) = 2^{|E|}$ are very easy to compute, and others such as $T(G; 1, 1)$ may also be computed efficiently, as in Sect. 9.5.1. In general though, the Tutte polynomial is intractable, as shown in the following theorem of Jaeger et al. [49].

Theorem 9.8.1. *The problem of evaluating the Tutte polynomial of a graph at a point (a, b) is $\#P$-hard except when (a, b) is on the special hyperbola*

$$H_1 \equiv (x - 1)(y - 1) = 1$$

or when (a, b) is one of the special points $(1, 1), (-1, -1), (0, -1), (-1, 0), (i, -i), (-i, i), (j, j^2),$ and $(j^2, j),$ where $j = e^{2\pi i/3}$.

In each of the exceptional cases the evaluation can be done in polynomial time; see Vertigan [107] and Gioan and Las Vergnas [41].

For planar graphs there is a significant difference. The technique developed using the Pfaffian to solve the Ising problem for the plane square lattice by Kasteleyn [51] can be extended to give a polynomial time algorithm for the evaluation of the Tutte polynomial of any planar graph along the special hyperbola

$$H_2 \equiv (x - 1)(y - 1) = 2.$$

However, even restricting a class of graphs to its planar members, or further restricting colouring enumeration on the square lattice, does not necessarily yield any additional tractability, as shown by the following results, the first due to Vertigan and Welsh [108], and the second to Farr [34].

Theorem 9.8.2. *The evaluation of the Tutte polynomial of bipartite planar graphs at a point (a, b) is #P-hard except when*

$$(a, b) \in H_1 \cup H_2 \cup \{(1, 1), (-1, -1), (j, j^2), (j^2, j)\}$$

where again $j = e^{2\pi i/3}$.

Theorem 9.8.3. *For $\lambda \geq 3$, computing the number of λ-colorings of induced subgraphs of the square lattice is #P-complete.*

A natural question then arises as to how well an evaluation of the Tutte polynomial might be approximated. That is, if there is a fully polynomial randomized approximation scheme, or FPRAS, for T at a point (x, y) for a well-defined family of graphs. Here, FPRAS refers to a probabilistic algorithm that takes the input s and the degree of accuracy ϵ to produce, in polynomial time on $|s|$ and ϵ^{-1}, a random variable which approximates $T(G; x, y)$ within a ratio of $1 + \epsilon$ with probability greater than or equal to 3/4. For example, Jerrum and Sinclair [50] show that there exists an FPRAS for T along the positive branch of the hyperbola H_2.

However, in general approximating is provably difficult as well. Recently, Goldberg and Jerrum [42] have extended the region of the x–y plane for which the Tutte polynomial does not have an FPRAS, to essentially all but the first quadrant (under the assumption that $RP \neq NP$). A consequence of this is that there is no FPRAS for counting nowhere zero λ-flows for $\lambda > 2$. They also provide a good overview of prior results. For a somewhat more optimistic prognosis in the case of dense graphs, we refer the reader to [111], and to Alon et al. [2].

There has been an increasing body of work since the seminal results of Robertson and Seymour [78–80] impacting computational complexity questions for graphs with bounded tree-width (see Bodlaender's accessible introduction to tree-width in [10]). A powerful aspect of this work is that many NP-hard problems become tractable for graphs of bounded tree-width. For example, Noble [67] and Andrzejak [3] have shown that the Tutte polynomial may be computed in polynomial time (in fact it requires only a linear number of multiplications and additions) for rational points on graphs with bounded tree width. Makowsky et al. [60] provide similar results for bounded clique-width (a notion with significant computational complexity consequences analogous to those for bounded tree-width; see Oum and Seymour [69]). Noble [68] gives a recent survey of complexity results for

9   The Tutte Polynomial   251

this area, including new monadic second-order logic methods and extensions to the multivariable generalizations of the Tutte polynomial discussed in Chap. 10 (see also [59, 100]).

Although the Tutte polynomial is not in general computationally tractable, there are some resources for reasonably sized graphs (up to about 100 edges). These include Sekine et al. [84], which provides an algorithm to implement the recursive definition. Common computer algebra systems such as Maple and Mathematica will compute the Tutte polynomial for smallish graphs, and there are also some implementations freely available on the Web, such as http://ada.fciencias.unam.mx/~rconde/tulic/ by R. Conde or http://homepages.mcs.vuw.ac.nz/~djp/tutte/ by G. Haggard and D. Pearce.

9.9   Conclusion

For further exploration of the Tutte polynomial and its properties, we refer the reader to the relevant chapters of Welsh [109] and Bollobás [11] for excellent introductions, and to Brylawki [16], Brylawski and Oxley [17], and Welsh [110] for an in-depth treatment of the Tutte polynomial, including generalizations to matroids. Although we focused on graphs here to broaden accessibility, matroids, rather than graphs, are the natural domain of the Tutte polynomial, and Crapo [24] gives a compelling justification for this viewpoint. Farr [35] gives a recent treatment and engaging history of the Tutte polynomial. Finally, we especially recommend Tutte's own account of how he "became acquainted with the Tutte polynomial" in [106].

Acknowledgments We thank all the friends and colleagues who offered many helpful comments and suggestions during the writing of this chapter.

The first author was supported by the National Security Agency and by the Vermont Genetics Network through Grant Number P20 RR16462 from the INBRE Program of the National Center for Research Resources (NCRR), a component of the National Institutes of Health (NIH).

The second author was supported by CONACYT of Mexico, Grant 83977.

References

1. Aigner M, Ziegler GM (2001) Proofs from the book. Springer, Berlin, Heidelberg
2. Alon N, Frieze AM, Welsh DJA (1995) Polynomial time randomized approximation schemes for Tutte–Gröthendieck invariants: the dense case. Random Struct Algorithm 6:459–478
3. Andrzejak A (1998) An algorithm for the Tutte polynomials of graphs of bounded treewidth. Discrete Math 190:39–54
4. Bak P, Tang C, Wiesenfeld K (1988) Self-organized criticality. Phys Rev A 38:364–374
5. Benashski J, Martin R, Moore J, Traldi L (1995) On the β-invariant for graphs. Congr Numer 109:211–221
6. Biggs N (1996) Algebraic graph theory, 2nd edn. Cambridge University Press, Cambridge
7. Biggs N (1996) Chip firing and the critical group of a graph. Research report, London school of economics, London

8. Birkhoff GD (1912) A determinant formula for the number of ways of coloring a map. Ann Math 14:42–46
9. Björner A (1992) Homology and shellability of matroids and geometric lattices. In: White N (ed) Matroid applications, encyclopedia of mathematics and its applications. Cambridge University Press, Cambridge
10. Bodlaender HL (1993) A tourist guide through treewidth. Acta Cybernet 11:1–21
11. Bollobás B (1998) Modern graph theory. Graduate texts in mathematics. Springer, New York
12. Borgs C (2006) Absence of zeros for the chromatic polynomial on bounded degree graphs. Combinator Probab Comput 15:63–74
13. Brown JI, Hickman CA, Sokal AD, Wagner DG (2001) On the chromatic roots of generalized theta graphs. J Combin Theory B 83:272–297
14. Brylawski T (1971) A combinatorial model for series–parallel networks. Trans Am Math Soc 154:1–22
15. Brylawski T (1972) A decomposition for combinatorial geometries. Trans Am Math Soc 171:235–282
16. Brylawski T (1982) The Tutte polynomial, Part 1: general theory. In: Barlotti A (ed) Matroid theory and its applications. Proceedings of the 3rd international mathematical summer center (C.I.M.E. 1980)
17. Brylawski T, Oxley J (1992) The Tutte polynomial and its applications. In: White N (ed) Matroid applications, encyclopedia of mathematics and its applications. Cambridge University Press, Cambridge
18. Chang SC, Shrock R (2001) Exact Potts model partition functions on wider arbitrary-length strips of the square lattice. Physica A 296:234–288
19. Chang SC, Jacobsen J, Salas J, Shrock R (2004) Exact Potts model partition functions for strips of the triangular lattice. J Stat Phys 114:768–823
20. Chari MK, Colbourn CJ (1997) Reliability polynomials: a survey. J Combin Inf System Sci 22:177–193
21. Chia GL (1997) A bibliography on chromatic polynomials. Discrete Math 172:175–191
22. Choe YB, Oxley JG, Sokal AD, Wagner DG (2004) Homogeneous multivariate polynomials with the half-plane property. Adv Appl Math 32:88–187
23. Crapo HH (1967) A higher invariant for matroids. J Combin Theory 2:406–417
24. Crapo HH (1969) The Tutte polynomial. Aeq Math 3:211–229
25. Dhar D (1990) Self-organized critical state of sandpile automaton models. Phys Rev Lett 64:1613–1616
26. Diestel R (2000) Graph theory. Graduate texts in mathematics. Springer, New York
27. Dong FM, Koh KM (2004) On upper bounds for real roots of chromatic polynomials. Discrete Math 282:95–101
28. Dong FM, Koh KM (2007) Bounds for the coefficients of flow polynomials. J Combin Theory B 97:413–420
29. Dong FM, Koh KM, Teo KL (2005) Chromatic polynomials and chromaticity of graphs. World Scientific, Hackensack, NJ
30. Duffin RJ (1965) Topology of series–parallel networks. J Math Anal Appl 10:303–318
31. Ellis-Monaghan J (2004) Exploring the Tutte–Martin connection. Discrete Math 281:173–187
32. Ellis-Monaghan J (2004) Identities for the circuit partition polynomials, with applications to the diagonal Tutte polynomial. Adv Appl Math 32:188–197
33. Etienne G, Las Vergnas M (1998) External and internal elements of a matroid basis. Discrete Math 179:111–119
34. Farr GE (2006) The complexity of counting colourings of subgraphs of the grid. Combinator Probab Comput 15:377–383
35. Farr GE (2007) Tutte–Whitney polynomials: some history and generalizations. In: Grimmett GR, McDiarmid CJH (eds) Combinatorics, complexity, and chance: a tribute to Dominic Welsh. Oxford University Press, Oxford
36. Fernandez R, Procacci A (2008) Regions without complex zeros for chromatic polynomials on graphs with bounded degree. Combinator Probab Comput 17:225–238

9 The Tutte Polynomial

37. Gabrielov A (1993) Abelian avalanches and the Tutte polynomials. Physica A 195:253–274
38. Garey MR, Johnson DS (1979) Computers and intractability – a guide to the theory of *NP*-completeness. W.H. Freeman, San Francisco
39. Gioan E (2007) Enumerating degree sequences in digraphs and a cycle–cocycle reversing system. Eur J Combinator 28:1351–1366
40. Gioan E, Las Vergnas M (2005) Activity preserving bijections between spanning trees and orientations in graphs. Discrete Math 298:169–188
41. Gioan E, Las Vergnas M (2007) On the evaluation at (j, j^2) of the Tutte polynomial of a ternary matroid. J Algebr Combinator 25:1–6
42. Goldberg LA, Jerrum MR (2007) Inapproximability of the Tutte polynomial. In STOC '07: Proceedings of the 39th annual ACM symposium on theory of computing. ACM Press, New York
43. Green C, Zaslavsky T (1983) On the interpretation of whitney numbers through arrangements of hyperplanes, zonotopes, non-radon partitions and orientations of graphs. Trans Am Math Soc 280:97–126
44. Jackson B (1993) A zero-free interval for chromatic polynomials of graphs. Combinator Probab Comput 2:325–336
45. Jackson B (2003) Zeros of chromatic and flow polynomials of graphs. J Geom. 76:95–109
46. Jackson B (2007) Zero-free intervals for flow polynomials of near-cubic graphs. Combinator Probab Comput 16:85–108
47. Jaeger F (1976) On nowhere-zero flows in multigraphs. In: Nash-Williams Crispin St JA, Sheehan J (eds) Proceedings of the 5th British combinatorial conference, Winnipeg
48. Jaeger F (1988) Nowhere-zero flow problems. In: Beineke LW, Wilson RJ (eds) Selected topics in graph theory, vol 3. Academic, New York
49. Jaeger F, Vertigan DL, Welsh DJA (1990) On the computational complexity of the Jones and Tutte polynomials. Math Proc Cambridge Philos Soc 108:35–53
50. Jerrum MR, Sinclair A (1993) Polynomial time approximation algorithms for the Ising model. SIAM J Comput 22:1087–1116
51. Kasteleyn PW (1961) The statistics of dimers on a lattice. Physica 27:1209–1225
52. Kleitman DJ, Winston KJ (1981) Forests and score vectors. Combinatorica 1:49–54
53. Kook W, Reiner V, Stanton D (1999) A convolution formula for the Tutte polynomial. J Combin Theory B 76:297–300
54. Las Vergnas M (1977) Acyclic and totally cyclic orientations of combinatorial geometries. Discrete Math 20:51–61
55. Las Vergnas M (1984) The Tutte polynomial of a morphism of matroids II. Activities of orientations. In: Bondy JA, Murty USR (eds) Progress in graph theory, Proceedings of Waterloo Silver Jubilee Combinatorial Conference 1982. Academic, Toronto
56. Las Vergnas M (1988) On the evaluation at (3,3) of the Tutte polynomial of a graph. J Combin Theory B 44:367–372
57. Las Vergnas M. The Tutte polynomial of a morphism of matroids V. Derivatives as generating functions, Preprint
58. Lieb EH (1967) Residual entropy of square ice. Phys Rev 162:162–172
59. Makowsky JA (2005) Colored Tutte polynomials and Kauffman brackets for graphs of bounded tree width. Discrete Appl Math 145:276–290
60. Makowsky JA, Rotics U, Averbouch I, Godlin B (2006) Computing graph polynomials on graphs of bounded clique-width. In: Lecture Notes in Computer Science 4271. Springer, New York
61. Martin P (1977) Enumérations eulériennes dans le multigraphs et invariants de Tutte–Gröthendieck. PhD thesis, Grenoble
62. Martin P (1978) Remarkable valuation of the dichromatic polynomial of planar multigraphs. J Combin Theory B 24:318–324
63. McKee TA (2001) Recognizing dual-chordal graphs. Congr Numer 150:97–103
64. Merino C (1997) Chip firing and the Tutte polynomial. Ann Combin 1:253–259
65. Merino C (2001) The chip firing game and matroid complex. Discrete Mathematics and Theoretical Computer Science, Proceedings, vol AA, pp 245–256

66. Merino C, de Mier A, Noy M (2001) Irreducibility of the Tutte polynomial of a connected matroid. J Combin Theory B 83:298–304
67. Noble SD (1998) Evaluating the Tutte polynomial for graphs of bounded tree-width. Combinator Probab Comput 7:307–321
68. Noble SD (2007) The complexity of graph polynomials. In: Grimmett GR, McDiarmid CJH (eds) Combinatorics, complexity, and chance: a tribute to Dominic Welsh. Oxford University Press, Oxford
69. Oum S, Seymour PD (2006) Approximating clique-width and branch-width. J Combin Theory B 96:514–528
70. Oxley J (1982) On Crapo's beta invariant for matroids. Stud Appl Math 66:267–277
71. Oxley J, Welsh DJA (1979) The Tutte polynomial and percolation. In: Bondy JA, Murty USR (eds) Graph theory and related topics. Academic, London
72. Pauling L (1935) The structure and entropy of ice and of other crystals with some randomness of atomic arrangement. J Am Chem Soc 57:2680–2684
73. Procacci A, Scoppola B, Gerasimov V (2003) Potts model on infinite graphs and the limit of chromatic polynomials. Commun Math Phys 235:215–231
74. Provan JS, Billera LJ (1980) Decompositions of simplicial complexes related to diameters of convex polyhedra. Math Oper Res 5:576–594
75. Read RC (1968) An introduction to chromatic polynomials. J Combin Theory B 4:52–71
76. Read RC, Rosenstiehl P (1978) On the principal edge tripartition of a graph. Ann Discrete Math 3:195–226
77. Read RC, Royle G (1991) Chromatic roots of families of graphs. In: Alavi Y et al (eds) Graph theory, combinatorics, and applications. Wiley, New York
78. Robertson N, Seymour PD (1984) Graph minors. I. Excluding a forest. J Combin Theory B 35:39–61
79. Robertson N, Seymour PD (1984) Graph minors. III. Planar tree-width. J Combin Theory B 36:49–64
80. Robertson N, Seymour PD (1986) Graph minors. II. Algorithmic aspects of tree-width. J Algorithm 7:309–322
81. Royle G (2007) Graphs with chromatic roots in the interval (1,2). Electron J Combinator 14(1):N18
82. Royle G (2008) Planar triangulations with real chromatic roots arbitrarily close to four. Ann Combinator 12:195–210
83. Schwärzler W (1993) The coefficients of the Tutte polynomial are not unimodal. J Combin Theory B 58:240–242
84. Sekine K, Imai H, Tani S (1995) Computing the Tutte polynomial of a graph of moderate size. In: Lecture Notes in Computer Science. Springer, Berlin
85. Seymour PD (1981) Nowhere-zero 6-flows. J Combin Theory B, 30, 130–135
86. Seymour PD, Welsh DJA (1975) Combinatorial applications of an inequality of statistical mechanics. Math Proc Cambridge Philos Soc 77:485–495
87. Shrock R, Tsai SH (1997) Asymptotic limits and zeros of chromatic polynomials and ground state entropy of Potts antiferromagnets. Phys Rev E 55:5165–5179
88. Shrock R, Tsai SH (1997) Families of graphs with $W_r(G, q)$ functions that are nonanalytic at $1/q = 0$. Phys Rev E 56:3935–3943
89. Shrock R, Tsai SH (1998) Ground state entropy of Potts antiferromagnets: cases with noncompact W boundaries having multiple points at $1/q = 0$. Physica A 259:315–348
90. Shrock R (2001) Chromatic polynomials and their zeros and asymptotic limits for families of graphs. Discrete Math 231:421–446
91. Sokal AD (2001) A personal list of unsolved problems concerning lattice gases and antiferromagnetic Potts models. Markov Process Relat Fields 7:21–38
92. Sokal AD (2001) Bounds on the complex zeros of (di)chromatic polynomials and Potts-model partition functions. Combinator Probab Comput 10:41–77
93. Sokal AD (2004) Chromatic roots are dense in the whole complex plane. Combinator Probab Comput 13:221–261

9 The Tutte Polynomial 255

94. Stanley R (1973) Acyclic orientations of graphs. Discrete Math 5:171–178
95. Stanley R (1980) Decomposition of rational polytopes. Ann Discrete Math 6:333–342
96. Stanley R (1996) Combinatorics and commutative algebra, 2nd edn. Progress in Mathematics, vol 41. Birkhäuser, Boston
97. Stanley R (1996) Enumerative combinatorics, vol 1. Cambridge University Press, Cambridge
98. Stanley R (1999) Enumerative combinatorics, vol 2. Cambridge University Press, Cambridge
99. Thomassen C (1997) The zero-free intervals for chromatic polynomials of graphs. Combinator Probab Comput 6:497–506
100. Traldi L (2006) On the colored Tutte polynomial of a graph of bounded treewidth. Discrete Appl Math 154:1032–1036
101. Tutte WT (1947) A ring in graph theory. Proc Cambridge Philos Soc 43:26–40
102. Tutte WT (1948) An algebraic theory of graphs, PhD thesis, University of Cambridge
103. Tutte WT (1954) A contribution to the theory of chromatic polynomials. Can J Math 6:80–91
104. Tutte WT (1967) On dichromatic polynomials. J Combin Theory 2:301–320
105. Tutte WT (1984) Graph theory. Cambridge University Press, Cambridge
106. Tutte WT (2004) Graph-polynomials. Special issue on the Tutte polynomial. Adv Appl Math 32:5–9
107. Vertigan D (1998) Bicycle dimension and special points of the Tutte polynomial. J Combin Theory B 74:378–396
108. Vertigan DL, Welsh DJA (1992) The computational complexity of the Tutte plane: the bipartite case. Combinator Probab Comput 1:181–187
109. Welsh DJA (1993) Complexity: knots, colorings and counting. Cambridge University Press, Cambridge
110. Welsh DJA (1999) The Tutte polynomial, in statistical physics methods in discrete probability, combinatorics, and theoretical computer science. Random Struct Algorithm 15:210–228
111. Welsh DJA, Merino C (2000) The Potts model and the Tutte polynomial. J Math Phys 41:1127–1152
112. Whitney H (1932) A logical expansion in mathematics. Bull Am Math Soc 38:572–579
113. Woodall D (1992) A zero-free interval for chromatic polynomials. Discrete Math 101: 333–341
114. Yetter D (1990) On graph invariants given by linear recurrence relations. J Combin Theory B 48:6–18
115. Zaslavsky T (1987) The Möbius function and the characteristic polynomial. In: White N (ed) Combinatorial geometries, encyclopedia of mathematics and its applications. Cambridge University Press, Cambridge
116. Zhang CQ (1997) Integer flows and cycle covers of graphs. Marcel Dekker Inc., New York

Chapter 10
Graph Polynomials and Their Applications II: Interrelations and Interpretations

Joanna A. Ellis-Monaghan and Criel Merino

Abstract We survey a variety of graph polynomials, giving a brief overview of techniques for defining a graph polynomial and then for decoding the combinatorial information it contains. These polynomials are not generally specializations of the Tutte polynomial, but they are each in some way related to the Tutte polynomial, and often to one another. We emphasize these interrelations and explore how an understanding of one polynomial can guide research into others. We also discuss multivariable generalizations of some of these polynomials and the theory facilitated by this. We conclude with two examples, the interlace polynomial in biology and the Tutte polynomial and Potts model in physics, that illustrate the applicability of graph polynomials in other fields.

Keywords Tutte polynomial · Characteristic polynomial · Matching polynomial · Penrose polynomial · Martin polynomial · Circuit partition polynomial · Ehrhart polynomial · Interlace polynomial · U-polynomial · W-polynomial · Bollobás–Riordan polynomial · Ribbon graph polynomial · Topological Tutte polynomial · Multivariable Tutte polynomial · Parametrized Tutte polynomial · Transition polynomial · Polychromate · Symmetric function · DNA sequencing · Potts model

MSC2000: Primary 05-02; Secondary 05C90, 05C45, 05C70, 05E05

10.1 Introduction

A graph polynomial is an algebraic object associated with a graph that is usually invariant at least under graph isomorphism. As such, it encodes information about the graph, and enables algebraic methods for extracting this information.

J.A. Ellis-Monaghan (✉)
Department of Mathematics, Saint Michaels College, One Winooski Park, Colchester, VT 05439, USA
and
Department of Mathematics and Statistics, University of Vermont, 16 Colchester Avenue, Burlington, VT 05405, USA
e-mail: jellis-monaghan@smcvt.edu

M. Dehmer (ed.), *Structural Analysis of Complex Networks*,
DOI 10.1007/978-0-8176-4789-6_10, © Springer Science+Business Media, LLC 2011

This chapter surveys a comprehensive, although not exhaustive, sample of graph polynomials. It concludes Chap. 9 by continuing the goal of providing a brief overview of a variety of techniques for defining a graph polynomial and then for decoding the combinatorial information it contains.

The polynomials we discuss here are not evaluations of the Tutte polynomial. However, they are each in some way related to the Tutte polynomial, and often to one another. We emphasize these interrelations and explore how an understanding of one polynomial can guide research into others. We also present multivariable generalizations of some of these polynomials and discuss the theory facilitated by this. We conclude with two examples, one from biology and one from physics, that illustrate the applicability of graph polynomials in other fields.

10.2 Formulating Graph Polynomials

We have seen two methods for formulating a graph polynomial with the linear recursion (deletion/contraction) and generating function definitions of the Tutte polynomial in Chap. 9. Here we show several more. We begin with one of the earliest graph polynomials, the edge-difference polynomial, a multivariable polynomial defined as a product and originally studied by Sylvester [129] and Peterson [113] in the late 1800s. More recently, it has been used to address list coloring questions (see Alon and Tarsi [8] and Ellingham and Goddyn [53]), where a list coloring of a graph is a proper coloring of the vertices of a graph with the color of each vertex selected from a predetermined list of colors assigned to that vertex.

Definition 10.2.1. The edge-difference polynomial. Let $(v_1, \ldots, v_n)$ be an ordering of the vertices of a graph G. Then $D\,(G; x_1, \ldots, x_n) = \prod_{i<j} (x_i - x_j)$, where the product is over all edges (v_i, v_j) of G.

Note that a proper coloring of G corresponds to finding positive integer values N_i (not necessarily distinct) for each of the x_i's so that $D\,(G; N_1, \ldots, N_n) \neq 0$.

There are also several polynomials based on various determinants (or even permanents; see Pathasarthy [111] for a survey) involving the adjacency matrix of a graph. Recall that $A(G)$, the adjacency matrix of a graph, has entries $a_{ij} = 1$ if (i, j) is an edge of the graph and 0 if it is not. The characteristic polynomial is the classic example of a such a graph polynomial, and is discussed further in Sect. 10.3.

Definition 10.2.2. The characteristic polynomial. Let $A(G)$ be the adjacency matrix of a graph G. Then $f(G; x) = |xI - A(G)|$.

Other examples of such polynomials are the *idiosyncratic polynomial* introduced by Tutte [133], that is defined by $v(G; x, y) = |A(G) + y(J - I - A) - xI|$, where J is the matrix having all entries equal to 1. Also $\mu(G; x, y) = |xI - D(G) + A(G)|$, where $D(G)$ is the *degree matrix* of G, that is, the diagonal matrix with $\deg(i)$ in the position (i, i), introduced by Kel'mans [93]. Note that $J - I - A$ is the adjacency matrix of the complement of G and when G is a simple graph, $D(G) - A(G)$ is just the Laplacian matrix $L(G)$ of the graph.

10 Interrelations and Interpretations

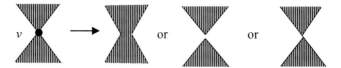

Fig. 10.1 The three possible local reconfigurations at a vertex v, identified, from left to right, as white, black, and crossing. Which strand passes over which in the crossing configuration does not affect the computation

A number of important graph polynomials may be defined by state model formulas. Loosely speaking, a state of a graph is some configuration resulting from making local assignments for substructures (e.g., the edges or vertices) of the graph. These assignments may be, for example, associating an element of a given set to each vertex, or even the result of reconfiguring the edges incident with a vertex. A graph polynomial is formed by associating an expression, often a weighted monomial, to each state of the graph, and then summing over all possible graph states. The language comes from physics, and is also found in knot theory. We show several state model graph polynomials among those surveyed below, as well as an application of this method in the Potts model of statistical mechanics in Sect. 10.5.

An early example of a graph polynomial given by a state model formulation is $P(G;x)$, the *Penrose polynomial*. This polynomial graph invariant for planar graphs was defined implicitly by Penrose [112] in the context of tensor diagrams in physics, but an excellent graph theoretical exposition can be found in Aigner [1]. To compute $P(G;x)$, let G be a plane graph, and let G_m be its medial graph, face two-colored with the unbounded face colored white. At each vertex, we consider three possible local reconfigurations, as in Fig. 10.1. A state S of G_m results from choosing one of these three reconfigurations at each vertex of G_m and consists of a set of disjoint closed curves (like a knot diagram). Furthermore, to each local reconfiguration at a vertex v, we assign a weight $\omega(S,v)$ that is $+1$, 0, or -1 for a white, black, or crossing configuration, respectively.

Definition 10.2.3. The Penrose polynomial. Let G be a planar graph with medial graph G_m, let $St(G_m)$ be the set of states of G_m, and let $St'(G_m)$ be the set of states with no black configurations. Then,

$$P(G;x) = \sum_{St(G_m)} \left(\left(\prod_{v \in G_m} \omega(S,v) \right) x^{k(S)} \right) = \sum_{St'(G_m)} \left((-1)^{cr(S)} x^{c(S)} \right),$$

where $c(S)$ is the number of components in the graph state S, and $cr(S)$ is the number of crossing vertex configurations chosen in the state S.

For example, if G is the θ-graph consisting of two vertices joined by three edges in parallel, then $P(G;x) = x^3 - 3x^2 + 2x$, as in Fig. 10.2. The Penrose polynomial may also be computed via a linear recursion relation (see Jaeger [89], for example).

The Penrose polynomial has some surprising properties, particularly with respect to graph coloring. The four-color theorem is equivalent to showing that every planar,

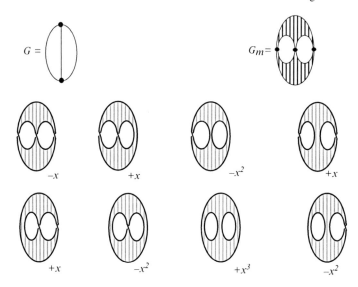

Fig. 10.2 Computing the Penrose polynomial of a graph G from the states of its medial graph

cubic, connected graph can be properly edge-colored with three colors. The Penrose polynomial, when applied to planar, cubic, connected graphs, encodes exactly this information (see Penrose [112]):

$$P(G;3) = \left(\frac{-1}{4}\right)^{\frac{|V|}{2}} P(G;-2) = \text{the number of 3-edge-colorings of } G.$$

10.3 Some Interrelated Polynomial Invariants

We present a further sampling of graph polynomials here. They are each related in some way to the Tutte polynomial, and have additional relations among themselves. These relations lead to combinatorial insights as results for any one polynomial then inform those related to it.

10.3.1 Characteristic and Matching Polynomials

The characteristic and matching polynomials are particularly interrelated, so we treat them together here, beginning with the characteristic polynomial $f(G;x)$ already introduced in Definition 10.2.2. Note that $f(G;x)$ is a monic polynomial of degree n. Furthermore, since the adjacency matrix A is real and symmetric, all its eigenvalues are real, and thus all the zeros of $f(G;x)$ are real.

10 Interrelations and Interpretations 261

By using properties of determinants we can find interpretations of the coefficients of $f(G; x)$ in terms of the *principal minors* of A. A principal minor of order r is the determinant of an $r \times r$ submatrix of A obtained by choosing r rows and columns with the same set of indices.

Proposition 10.3.1. *Suppose that* $f(G; x) = \sum_{i=0}^{n} a_i x^{n-i}$. *Then* $(-1)^i a_i$ *is equal to the sum of the principal minors of A with order i.*

This property of the characteristic polynomial can be found, for example, in Horn and Johnson [85].

Since the diagonal elements of A are all zero, we have that $a_1 = 0$. The principal minors of order two and three that are not zero are of the form $|J - I|$, where J is the matrix having all entries $+1$ and $J - I$ has order 2 or 3. The 2×2 submatrices $J - I$ of $A(G)$ correspond naturally to the edges of G and the 3×3 submatrices $J - I$ correspond to the K_3 subgraphs of G. Thus $a_2 = -|E(G)|$ and $-a_3$ is twice the number of K_3 subgraphs of G.

A *linear subgraph* of G is a subgraph whose components are edges or cycles. An expression for the coefficients of $f(G; x)$ in terms of linear subgraphs is given in the following result.

Proposition 10.3.2. *The coefficients of the characteristic polynomial may be expressed as*

$$(-1)^i a_i = \sum_{\Lambda} (-1)^{r(\Lambda)} 2^{r^*(\Lambda)},$$

where r is the rank function and the sum is over all linear subgraphs Λ of G having i vertices.

Note that because Λ is a linear subgraph, $r^*(\Lambda)$ is simply the number of components in Λ that are cycles. The proof of Proposition 10.3.2 uses Proposition 10.3.1 and can be found in Harary [82], while a detailed history of this result is given by Cvetković et al. [45].

As with the Tutte polynomial we also have here some reduction formulas and an expression for the derivative of the characteristic polynomial.

Theorem 10.3.3. *The characteristic polynomial of a graph satisfies the following identities:*

1. $f(G \cup H; x) = f(G; x) f(H; x),$
2. $f(G; x) = f(G \setminus e, x) - f(G - u - v; x)$ *if $e = uv$ is a cut-edge of G,*
3. $\frac{\partial}{\partial x} f(G; x) = \sum_{v \in V(G)} f(G - v; x).$

A proof of these properties can be found in Godsil [75]. Item 1 is an easy exercise in matrix theory, as in Horn and Johnson [85]. Item 2 can be proved by using Proposition 10.3.2 and considering the linear subgraphs of G that use the edge e and the ones that do not use it. The result follows because e is in no cycle of G if and only if it is a cut-edge. Item 3 can be proved by using Proposition 10.3.1 since any principal minor of order i is counted $n - i$ times in the right-hand side of the formula in Item 3.

Given a graph G, the collection of (unlabeled) subgraphs $G - v$, for $v \in V(G)$, is called the *deck* of G, and the individual subgraphs are called *cards*. Thus, the deck for a graph on n vertices consists of n graphs, each of which has $n - 1$ vertices. Ulam's reconstruction conjecture in [135] asserts that any finite graph G with more than two vertices is uniquely determined by its deck (see [135]), and we call any graph that satisfies the conjecture *reconstructible*. Similarly, an invariant of G which can be deduced from the deck is called reconstructible.

Clearly, the number n of vertices of a graph is reconstructible. Also, as every edge is present in exactly $n - 2$ cards, the number of edges is also reconstructible. The following useful example of reconstructibility is due to Kelly [92].

Lemma 10.3.4 (Kelly's lemma). *Let G and H be graphs, and let $v(H, G)$ denote the number of subgraphs of G isomorphic to H. Then*

$$(|V(G)| - |V(H)|)v(H, G) = \sum_{v \in V(G)} v(H, G - v).$$

The proof is by a double counting argument, and it follows that $v(H, G)$ is reconstructible whenever $|V(H)| < |V(G)|$.

Tutte proved in [132] that the Tutte polynomial is reconstructible, and thus the chromatic polynomial, the flow polynomial, the number of spanning trees and any invariant mentioned in Chap. 9 are also reconstructible. Tutte also proved that the characteristic polynomial is reconstructible in [133].

Theorem 10.3.5. *The characteristic polynomial of a graph is reconstructible.*

For the proof, note that we have immediately from Theorem 10.3.3 that $f'(G; x)$ is reconstructible. It then remains to prove that the constant term of $f(G; x)$ is reconstructible. But by Proposition 10.3.1, this is the same as proving that $|A(G)|$ is reconstructible. Then, using Theorem 10.3.2 and an extension of Kelly's lemma (see [94]), the problem is reduced to proving that the number of Hamiltonian cycles is reconstructible. A complete proof of Theorem 10.3.5 based on the proof in Kocay [94], can be found in [75].

Let us turn now to the matching polynomial. An i-*matching* in a graph G is a set of i edges, no two of which have a vertex in common. Let $\Phi_i(G)$ denote the number of i-matchings, setting $\Phi_0(G) = 1$. Thus $\Phi_1(G) = m$ is the number of edges of G, and if n, the number of vertices, is even, then $\Phi_{n/2}(G)$ is the number of perfect matchings of G.

Definition 10.3.6. Let G be a graph. Then the matching polynomial of G is

$$\mu(G; x) = \sum_{i \geq 0} (-1)^i \Phi_i(G) x^{n-2i}.$$

10 Interrelations and Interpretations 263

A more natural polynomial might be the *matching generating polynomial*, given as the generating function of i-matchings by

$$g(G;x) = \sum_{i \geq 0} \Phi_i(G)x^i.$$

However, the two polynomials are related by the identity

$$\mu(G;x) = x^n g(G;(-x)^2),$$

so there is no essential difference between them.

The matching polynomial is also known as the *acyclic polynomial* in Gutman and Trinajstić [81], *matching defect polynomial* in Lovász and Plummer [101] and *reference polynomial* in Aihara [6]. It has appeared independently in several different contexts. In combinatorics, it was probably introduced by Farrell in [65], but since the matching polynomial is essentially the same as the rook polynomial for bipartite graphs (see Farrell [67]), then its origin can be traced back at least to Riordan [117]. In statistical physics it appears because of the monomer-dimer problem and was introduced by Heilmann and Lieb in [83] and independently by Kunz in [95]. Finally, in theoretical chemistry it was introduced by Hosaya in [86] and later in connection with the so-called topological resonance energy by Gutman et al. in [79–81] and independently by Aihara in [6]. For a full account of the history of the matching polynomial see Gutman [77].

As with the Tutte and characteristic polynomials, we have some reduction formulas for the matching polynomial. The proof of the following theorem can be found in Godsil [75].

Theorem 10.3.7. *The matching polynomial satisfies the following identities:*

1. $\mu(G \cup H; x) = \mu(G;x)\mu(H;x)$,
2. $\mu(G;x) = \mu(G \setminus e;x) - \mu(G - u - v;x)$ *if $e = uv$ is an edge of G,*
3. $\mu(G;x) = x\mu(G - u;x) - \sum_{uv \in E(G)} \mu(G - v - u;x)$ *if $u \in V(G)$,*
4. $\frac{\partial}{\partial x}\mu(G;x) = \sum_{v \in V(G)} \mu(G - v;x)$.

An i-matching in $G \cup H$ corresponds to an s-matching in G and a t-matching in H such that $s + t = i$. Item 1 then follows by the fundamental counting principle. For Item 2, notice that the set of i-matchings can be partitioned into those i-matchings that use the edge e and those that do not use it. Item 3 follows similarly. Finally, every i-matching of G with $i < n/2$ is counted $n - 2i$ times in the right-hand side of the formula in Item 4, so the result follows.

When the graph G is a forest, a linear subgraph of G with j vertices corresponds to a matching covering j vertices, with j even. Thus, Proposition 10.3.2 has the following corollary, observed by Hosaya [86] and by Heilmann and Lieb [84].

Corollary 10.3.8. *If G is a forest then $f(G;x) = \mu(G;x)$.*

An unexpected property of the matching polynomial, proved by Heilmann and Lieb [84], is that all its zeros are real, and furthermore the zeros for any graph G interlace with the zeros of any of the cards in its deck. The same paper also gives bounds for the zeros.

Theorem 10.3.9. *For any graph G, the matching polynomial $\mu(G;x)$ has only real zeros. Furthermore, if u is any vertex in G and if $a_1, a_2, \ldots, a_n$ are the zeros of $\mu(G;x)$ while the zeros of $\mu(G-u;x)$ are $a'_1, a'_2, \ldots, a'_{n-1}$, then*

$$a_1 \le a'_1 \le a_2 \le a'_2 \le \cdots \le a'_{n-1} \le a_n;$$

that is, the zeros of $\mu(G;x)$ and $\mu(G-u;x)$ interlace.

Theorem 10.3.10. *The (real) zeros, $a_1, a_2, \ldots, a_n$, of $\mu(G;x)$, satisfy*

$$|a_i| < 2\sqrt{maxdeg(G) - 1}.$$

We outline a proof of Theorem 10.3.9 from Godsil [73] that uses some of the results already mentioned for $\mu(G;x)$ and $f(G;x)$. First, given a graph G and a vertex u in G, the *path tree* $T(G,u)$ is the tree that has as its vertices the paths in G which start at u, and where two such vertices are joined by an edge if one represents a maximal proper subpath (i.e., all but the last edge) of the other. We then have the following proposition from [73] that leads to a proof of Theorem 10.3.9.

Proposition 10.3.11. *Let u be a vertex in a graph G, let $T = T(G,u)$ be the path tree of G with respect to u, and let u' be the vertex of T corresponding to the path of length 0 beginning at u. Then*

$$\frac{\mu(G-u;x)}{\mu(G;x)} = \frac{\mu(T-u';x)}{\mu(T;x)} = \frac{f(T-u';x)}{f(T;x)}.$$

The last equality follows from Corollary 10.3.8. Because all the roots of the characteristic polynomial are real, we conclude that all zeros and poles of the rational function $\mu(G-u;x)/\mu(G;x)$ are real. An induction argument on the number of vertices in G then yields the conclusion that all the zeros of $\mu(G;x)$ are real.

An interesting combinatorial consequence of Theorem 10.3.9 is the following result of Heilmann and Lieb [84], which gives a stark contrast with how little is known about the coefficients of the chromatic polynomial.

Theorem 10.3.12. *For any graph G, the sequence $\Phi_0(G)$, $\Phi_1(G)$, $\ldots$ of coefficients of $g(G;x)$ is log-concave, that is $\Phi_i^2 \ge \Phi_{i-1}\Phi_{i+1}$.*

The characteristic polynomial has been well studied, particularly with respect to graphs with the same characteristic polynomial. Godsil [75, 76] gives a thorough treatment of both the characteristic and the matching polynomials. Another good reference for the characteristic polynomial is Biggs [20]. Just as the matching polynomial is a way to study the matchings of a graph, the characteristic polynomial is a way to study the spectra of the adjacency matrix of a graph. Cvetković et al. have written a book [45] dedicated to the spectra of the adjacency matrix, and Lovász and Plummer [101] have a book devoted to the theory of matchings. Furthermore, although the characteristic polynomial is not a complete invariant of graphs, it is conjectured that the characteristic polynomial of a graph G is reconstructible from its polynomial deck, i.e., from the set of characteristic polynomials of the cards of G. See Gutman and Cvetković [78] for the conjecture, and then Cvetković and Lepović [46], where it is proved in the case of trees.

10.3.2 Ehrhart Polynomial

A *convex polytope* P is the convex hull of a finite set of points in $\mathbb{R}^m$. We denote the interior of P (in the usual topological sense) by P^o. A convex polytope P is said to be a *rational*, or *integral*, polytope if all its vertices have rational, or integral, coordinates, respectively. We write $d = \dim P$ and call P a d-polytope.

For $P \subset \mathbb{R}^m$ a rational d-polytope and t a nonnegative integer we define the functions $i(P;t) = |tP \cap \mathbb{Z}^m|$ and $\bar{i}(P;t) = |tP^o \cap \mathbb{Z}^m|$, where $tP = \{ta | a \in P\}$ is the t-fold dilatation of P. Ehrhart proved in [50–52] that these functions are *quasi-polynomials*; that is, they are of the form

$$c_d(t)t^d + c_{t-1}(t)t^{d-1} + \cdots + c_0(t),$$

where each $c_i(t)$ is a periodic function with integer period. Since $i(P;t)$ is a quasi-polynomial, it can be defined for all $t \in \mathbb{Z}$. In fact, we have the following reciprocity law due to Ehrhart [49]:

$$i(P;-t) = (-1)^d \bar{i}(P;t).$$

From Ehrhart [50–52] (also see Stanley [127]), we have that when $P \subset \mathbb{R}^m$ is an integral d-polytope, then $i(P;t)$ and $\bar{i}(P;t)$ are polynomials, which leads to the following definition of the Ehrhart polynomial.

Definition 10.3.13. Let P be an integral convex d-polytope. Then the Ehrhart polynomial of P is

$$i(P;t) = c_0 + c_1 t + \cdots + c_{d-1}t^{d-1} + c_d t^d.$$

From the early works of Ehrhart [50] and Macdonald [102] it is known that $c_0 = 1$ and $c_d = \text{vol}(P)$, and that c_{d-1} is half of the surface area of P, normalized with respect to the sub-lattice on each face of P. Specifically, $c_{d-1} = 1/2 \sum_F \text{vol}_{d-1}(F)$, where F ranges over all facets of P and the volume of a facet is measured intrinsically with respect to the lattice $\mathbb{Z}^m \cap L_F$, where L_F is the affine hull of F. The other coefficients were not well understood, until the later work of Betke and Kneser [19], Pommersheim [116], Kantor and Khovanskii [90], and

Diaz and Robins [47], but such interpretations go beyond the scope of this chapter. For the complexity of computing these coefficients see Barvinok [16]. A reference for integer point enumeration in polytopes is Beck and Robins [18].

In the special case that P is a *zonotope* there is a combinatorial interpretation for the coefficients of the Ehrhart polynomial. First recall that if A is an $r \times m$ real matrix written in the form $A = [a_1, ..., a_r]$, then it defines a *zonotope* $Z(A)$ which consists of those points p of $\mathbb{R}^m$ that can be expressed in the form

$$p = \sum_{i=1}^{m} \lambda_i a_i, \quad 0 \le \lambda_i \le 1.$$

In other words, $Z(A)$ is the *Minkowski sum* of the line segments $[0, a_i]$, for $1 \le i \le n$. For more on zonotopes, see McMullen [105].

When A has integer entries, Stanley [125], using techniques from Shephard [121], proved that $i(P; t) = \sum_X f(X) t^{|X|}$, where X ranges over all linearly independent subsets of columns of A and where $f(X)$ denotes the greatest common divisor of all minors of sizes $|X|$ of the matrix A.

When A is a *totally unimodular matrix*, that is, the determinant of every square submatrix is 0 or ± 1, then $Z(A)$ is described as a *unimodular zonotope*. For these polytopes the previous result shows that

$$i(Z(A); t) = \sum_{k=0}^{r} f_k t^k,$$

where f_k is the number of subsets of columns of the matrix A which are linearly independent and have cardinality k. In other words, the Ehrhart polynomial $i(Z(A); t)$ is the generating function of the number of independent sets in the regular matroid $M(A)$.

The incidence matrix $D(G)$ of a graph G is totally unimodular, a long-standing result due to Poincaré [115] with a modern treatment given by Biggs [20]. A linearly independent subset of columns in D corresponds to a subset of edges with no cycle. Thus, the coefficient f_k in this case is the number of spanning forests of G with exactly k edges. From the previous chapter we know that $T(G; x + 1, 1) = \sum_{k=0}^{r} f_k x^{r-k}$, where r is the rank of the graph G. With these ingredients we get the following relation to the Tutte polynomial from Welsh [138].

Theorem 10.3.14. *If G is a graph and D is its incident matrix then the Ehrhart polynomial of the unimodular zonotope $Z(D)$ is given by*

$$i(Z(D); t) = t^r T\left(G; 1 + \frac{1}{t}, 1\right),$$

where r is the rank of G.

10 Interrelations and Interpretations 267

In this case, the zonotope $Z(D)$ is a r-polytope in $\mathbb{R}^n$, where n is the number of vertices of G.

The reciprocity law of Theorem 10.3.14 leads to the following geometric result, also from Welsh [138].

Corollary 10.3.15. *If D is the incidence matrix of a graph G with rank r and n vertices then for any positive integer λ the number of lattice points of $\mathbb{R}^n$ lying strictly inside the zonotope $t Z(D)$ is given by*

$$\bar{i}(Z(D);t) = (-t)^r T\left(G; 1 - \frac{1}{t}, 1\right).$$

In particular we have that the number of lattice points strictly inside $Z(D)$ is $(-1)^r T(G;0,1)$.

10.3.3 The Topological Tutte Polynomial of Bollobás and Riordan

The classical Tutte polynomial discussed in the previous chapter is an invariant of abstract graphs, so it encodes no information specific to graphs embedded in surfaces. In [24, 25], Bollobás and Riordan generalize the classical Tutte polynomial to topological graphs, that is, graphs embedded in surfaces. In [24], Bollobás and Riordan define the *cyclic graph polynomial*, a three-variable deletion/contraction invariant for graphs embedded in oriented surfaces. They extend this work in [25], using a different approach, with the four-variable *ribbon graph polynomial*. Both of these polynomials extend the classical Tutte polynomial, but in such a way that topological information about the embedding is encoded. The version for oriented surfaces is subsumed by the version for arbitrary surfaces, so we focus on the latter here. The ribbon graph polynomial is also sometimes called the *Bollobás–Riordan polynomial* after the authors or the *topological Tutte polynomial* to emphasize that it simultaneously encodes topological information while generalizing the classical Tutte polynomial.

First recall that a cellular embedding of a graph in an orientable or unorientable surface can be specified by providing a sign for each edge and a rotation scheme for the set of half edges at each vertex, where a rotation scheme is simply a cyclic ordering of the half edges about a vertex. This is equivalent to a *ribbon* (or *fat*) *graph*, which is a surface with boundary where the vertices are represented by a set of disks and the edges by ribbons, with the ribbon of an edge with a negative sign having a half-twist. This can also be thought of as taking a slight "fattening" of the edges of the graph as it is embedded in the surface, or equivalently as "cutting out" the graph together with a small neighborhood of it from the surface. Figure 10.3 shows a graph with two vertices and two parallel edges, one positive and one negative. It is embedded on a Klein bottle, and the ribbon graph is a Möbius band with boundary.

In addition to the usual graphic characteristics such as number of vertices, connected components, rank, and nullity for a ribbon graph G, we also consider $bc\,(G)$,

Fig. 10.3 A ribbon graph which is a Möbius band with boundary

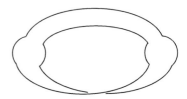

the number of boundary components of the surface, and $t(G)$, an index of the orientability of the surface. The value of $t(G)$ is 0 if the surface is orientable, and 1 if it is not. Thus, $t(G)$ is 1 if and only if for some cycle in G, the product of the signs of the edges is negative.

Definition 10.3.16. Let G be a ribbon graph, that is, a graph embedded in a surface. Then the topological Tutte polynomial of Bollobás and Riordan is given by

$$R(G; x, y, z, w) = \sum_{A \subseteq E(G)} (x-1)^{r(G)-r(A)} y^{n(A)} z^{\kappa(A)-bc(A)+n(A)} w^{t(A)}$$

as an element of $Z[x, y, z, w]/\langle w^2 - w \rangle$.

As previously, $r(A), \kappa(A), n(A)$, and now also $bc(A)$ and $t(A)$, refer to the spanning subgraph of G with edge set A, here with its embedding inherited from G.

Clearly, by comparing with the rank-nullity generating function definition of the classical Tutte polynomial given in Chap. 9, this generalizes the classical Tutte polynomial. Like the classical Tutte polynomial, $R(G; x, y, z, w)$ is multiplicative on disjoint unions and one-point joins of ribbon graphs. More importantly, it retains the essential properly of obeying a deletion/contraction reduction relation.

To see this, we must first define deletion and contraction in the context of embedded graphs. The ribbon graph resulting from deleting an edge is clear, but contraction requires some care. Let e be a nonloop edge. First assume the sign of e is positive, by flipping one endpoint if necessary to remove the half-twist (this reverses the cyclic order of the half edges at that vertex and toggles their signs). Then G/e is formed by deleting e and identifying its endpoints into a single vertex v. The cyclic order of edges at v comes from the original cyclic order at one endpoint, beginning where e had been, and continuing with the cyclic order at the other endpoint, again beginning where e had been.

Theorem 10.3.17. *If G is a ribbon graph, then*

$$R(G; x, y, z, w) = R(G/e; x, y, z, w) + R(G - e; x, y, z, w)$$

if e is an ordinary edge and $R(G; x, y, z, w) = xR(G/e; x, y, z, w)$ if e is a bridge.

The proof depends on a careful analysis of how each of the relevant parameters $r(A), \kappa(A), n(A), bc(A)$, and $t(A)$ changes with the deletion or contraction of an edge.

Repeated application of this theorem reduces a ribbon graph to a disjoint union of embedded *bouquet graphs*, that is, graphs each consisting of a single vertex with some number of loops. Because of the embedding, the loops are signed, and there is a rotation system of the half-edges about the single vertex. Not surprisingly, the topological information is distilled into these minors of the original graph, and to complete a deletion/contraction linear recursion computation, it is necessary to specify an evaluation of these terminal forms.

Signed chord diagrams provide a useful device for determining the relevant parameters of an embedded bouquet graph. Recall that a *chord diagram* consists of a circle with n symbols on its perimeter, with each symbol appearing twice and a chord drawn between each pair of like symbols. A *signed chord diagram* simply has a sign on each chord. A signed chord diagram D corresponds to an embedded bouquet graph G by assigning a symbol to each loop and arranging them on the perimeter of the circle in the chord diagram in the same order as the cyclic order of the half-edges about the vertex. A chord receives the same sign as the loop it represents. If we "fatten" the chords as in Fig. 10.4, with a negative chord receiving a half-twist, then $bc(G)$ is equal to the number of components in the resulting diagram, which is denoted $bc(D)$. Similarly, since G has only one vertex, $n(G)$ is the number of edges of G, which is the number of chords of D, so we denote this by $n(D)$. We also set $t(D) = t(G)$, and note that $t(D) = 0$ if all chords of D have a positive sign, and $t(D) = 1$ otherwise. This, combined with the definition of $R(G; x, y, z, w)$ above, gives the following evaluation for these terminal forms.

Theorem 10.3.18. *If G is an embedded bouquet graph with corresponding signed chord diagram D, then*

$$R(G; x, y, z, w) = \sum_{D' \subseteq D} y^{n(D')} z^{1-bc(D')+n(D')} w^{t(D')},$$

where the sum is over all subdiagrams D' of D.

Theorems 10.3.17 and 10.3.18 taken together give a linear recursion definition for $R(G; x, y, z, w)$. There are a number of technical considerations, similar to the care that must be taken in contracting edges, but nevertheless many other properties analogous to those of the classical Tutte polynomial hold. For example, $R(G; x, y, z, w)$ has a spanning tree expansion, a universality property, and duality relation (in addition to Bollobás and Riordan's work in [24, 25], see also

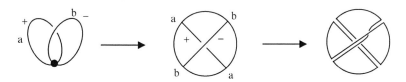

Fig. 10.4 A signed bouquet graph, its signed chord diagram, and the boundary components of the signed chord diagram

Las Vergnas' early exploration [97], and recent work by Chmutov [39], Moffatt [108], and [58, 59]). Furthermore, Chmutov and Pak [40], and Moffatt [107, 108] have shown that $R(G; x, y, z, w)$ also extends the relation between the classical Tutte polynomial and the Kauffman bracket and the Jones polynomial of knot theory due to Thistlethwaite [130] and Kauffman [91].

10.3.4 Martin, or Circuit Partition, Polynomials

In his 1977 thesis, Martin [103] recursively defined polynomials $M(G, x)$ and $m(\vec{G}; x)$ that encode, respectively, information about the families of circuits in 4-regular Eulerian graphs and digraphs. Las Vergnas subsequently found a state model expression for these polynomials, extended their properties to general Eulerian graphs and digraphs, and further developed their theory (see [96, 98, 99]). Both Martin [104] and Las Vergnas [99] found combinatorial interpretations for some small integer evaluations of the polynomials, while combinatorial interpretations for all integer values as well as some derivatives were given in [55–57], and by Bollobás [22].

Transforms of the Martin polynomials, $J(G; x)$ and $j(\vec{G}; x)$, given in [54], and then aptly named circuit partition polynomials in [11], facilitate these computations, and for this reason we give the definitions below in terms of J and j. As do many of the polynomials surveyed here, the circuit partition polynomials have several definitions, including linear recursion formulations, generating function formulations, and state model formulations. We give the state model definition, and refer the reader to [56, 57] for the others.

As with other state model formulations, we must first specify what we mean by a state of a graph (or digraph) in this context. Here an Eulerian graph must have vertices all of even degree, but it need not be connected. An Eulerian digraph must have the indegree equal to the outdegree at each vertex, and again need not be connected.

Definition 10.3.19. An Eulerian graph state of an Eulerian graph G is the result of replacing each $2n$-valent vertex v of G with n 2-valent vertices joining pairs of half edges originally adjacent to v. An Eulerian graph state of an Eulerian digraph $\vec{G}$ is defined similarly, except here each incoming half edge must be paired with an outgoing half edge.

Note that a Eulerian graph state is a disjoint union of cycles, each consistently oriented in the case of a digraph.

Definition 10.3.20. The circuit partition polynomial. Let G be an Eulerian graph, let $St(G)$ be the set of states of G, and let $c(S)$ be the number of components in a state $S \in St(G)$. Then the circuit partition polynomial has a state model formulation given by

$$J(G; x) = \sum_{S \in St(G)} x^{c(S)}.$$

10 Interrelations and Interpretations

The circuit partition polynomial is defined similarly for Eulerian digraphs as

$$j(\vec{G};x) = \sum_{S \in St(\vec{G})} x^{c(S)}.$$

The transforms between the circuit partition polynomials and the original Martin polynomial, as extended to general Eulerian graphs and digraphs by Las Vergnas, are:

$$J(G;x) = xM\,(G;x+2)\,,\text{ for } G \text{ an Eulerian graph, and} \qquad (10.1)$$

$$j(\vec{G};x) = xm\left(\vec{G};x+1\right)\text{ for } \vec{G} \text{ an Eulerian digraph.} \qquad (10.2)$$

The circuit partition polynomials have "splitting" formulas, analogous to Tutte's identity for the chromatic polynomial given in Chap. 9, proofs for which may be found in [54, 57]. These formulas derive from the Hopf algebra structures of the generalized transition polynomial we discuss in Sect. 10.4.3, but may also be proved combinatorially, as in [57] and by Bollobás [22].

Theorem 10.3.21. *Let G be an Eulerian graph and $\vec{G}$ be an Eulerian digraph. Then*

$$J(G;x+y) = \sum J\,(A;x)\,J\,(A^c;y),$$

where the sum is over all subsets $A \subseteq E(G)$ such that G restricted to both A and $A^c = E(G) - A$ is Eulerian. Also,

$$j(\vec{G};x+y) = \sum j\left(\vec{A};x\right) j\left(\vec{A}^c;y\right),$$

where the sum is over all subsets $\vec{A} \subseteq E(\vec{G})$ such that $\vec{G}$ restricted to both $\vec{A}$ and $\vec{A}^c$ is an Eulerian digraph.

The connection between the circuit partition polynomial of a digraph and the Tutte polynomial of a planar graph G is through the oriented medial graph $\overrightarrow{G_m}$ described in the previous chapter. Martin [103] proved the following, which we extend to the circuit partition polynomial via (10.2).

Theorem 10.3.22. *Let G be a connected planar graph, and let $\overrightarrow{G_m}$ be its oriented medial graph. Then relationships among the Martin polynomial, circuit partition polynomial, and Tutte polynomial are:*

$$j(\overrightarrow{G_m};x) = xm(\overrightarrow{G_m};x+1) = xt(G;x+1,x+1).$$

The proof of this theorem depends on a fundamental observation relating deletion/contraction in G with choices of configurations at a vertex in an Eulerian graph state of $\overrightarrow{G_m}$, as illustrated in Fig. 10.5. Theorems 10.3.21 and 10.3.22 combine to

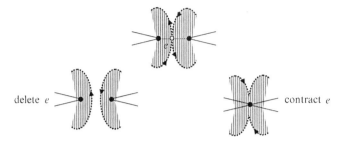

Fig. 10.5 An edge e in a planar graph G, with the corresponding vertex v in the oriented medial graph $\vec{G}_m$ (*dotted edges*). Deleting e corresponds to one possible configuration at v in an Eulerian graph state of $\vec{G}_m$, while contracting e corresponds to the other

give the basis for many of the combinatorial interpretation of the Tutte polynomial along the line $y = x$ described in the previous chapter. For more details, see Martin [103, 104], Las Vergnas [96, 98, 99], Bollobás [22], and also [54–57].

Evolving from the relation between the Tutte and Martin polynomials is the theory of isotropic systems, which unifies essential properties of 4-regular graphs and pairs of dual binary matroids. A series of papers throughout the 1980s and 1990s, including work by Bouchet [26–31], as well as Bouchet and Ghier [37], and Jackson [87], significantly extends the relationship between the Tutte polynomial of a planar graph and the Martin polynomial of its medial graph via the theory of isotropic systems.

10.3.5 Interlace Polynomial

In [11], Arratia, Bollobás, and Sorkin defined a one-variable graph polynomial motivated by questions arising from DNA sequencing by hybridization addressed by Arratia, Bollobás, Coppersmith, and Sorkin in [10], an application we will return to in Sect. 10.5. In [12], Arratia, Bollobás, and Sorkin defined a two-variable interlace polynomial, and showed that the original polynomial of [11] is a specialization of it, renaming the original one-variable polynomial as the vertex-nullity interlace polynomial due to its relationship with the two-variable generalization.

Remarkably, despite very different terminologies, motivations, and approaches, the original vertex-nullity interlace polynomial of a graph may be realized as the Tutte–Martin polynomial of an associated isotropic system (see Bouchet [36]). For exploration of this relationship, see the works mentioned in Sect. 10.3.4, as well as Aigner [3], Aigner and Mielke [4], Aigner and van der Holst [5], Allys [7], and also Bouchet's series on multimatroids [32–35].

Both the vertex-nullity interlace polynomial of a graph and the two-variable interlace polynomial may be defined recursively via a pivot operation. This pivot is defined as follows. Let vw be an edge of a graph G, and let A_v, A_w, and A_{vw} be

10 Interrelations and Interpretations

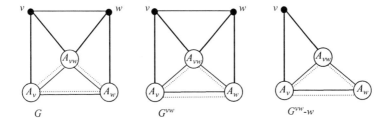

Fig. 10.6 Pivoting on the edge vw. A_v, A_w, and A_{vw} are the sets of vertices of G adjacent to v only, w only, and to both v and w, respectively. These sets are constant in all the diagrams. Vertices of G adjacent to neither v nor w are omitted. Heavy lines indicate that all edges are present, and dotted lines represent nonedges. Note interchange of edges and nonedges among A_v, A_w, and A_{vw}

the sets of vertices in $V(G) \setminus \{v, w\}$ adjacent to v only, w only, and to both v and w, respectively. The pivot operation "toggles" the edges among A_v, A_w and A_{vw}, by deleting existing edges and inserting edges between previously nonadjacent vertices. The result of this operation is denoted G^{vw}. More formally, G^{vw} has the same vertex set as G, and edge set equal to the symmetric difference $E(G) \triangle S$, where S is the complete tripartite graph with vertex classes A_v, A_w and A_{vw}. See Fig. 10.6.

Also, G^a is the *local complementation* of G, defined as follows. Let $N(a)$ be the neighbors of a, that is, the set $\{w \in V : a \text{ and } w \text{ are joined by an edge}\}$. The graph G^a is equal to G except that we "toggle" the edges among the neighbors of a, switching edges to nonedges and vice versa.

Definition 10.3.23. Let G be a graph of order n, which may have loops, but no multiple loops or multiple edges. The two-variable interlace polynomial may be given recursively by $q(\mathcal{E}_n) = y^n$ for $\mathcal{E}_n$, the edgeless graph on $n \geq 0$ vertices, with

$$q(G) = q(G - a) + q(G^{ab} - b) + ((x-1)^2 - 1)q(G^{ab} - a - b),$$

for any edge ab where neither a nor b has a loop, and

$$q(G) = q(G - a) + (x - 1)q(G^a - a),$$

for any looped vertex a.

Alternatively, the interlace polynomial has the following generating function formulation.

Definition 10.3.24. Let G be a graph of order n, which may have loops, but no multiple loops or multiple edges. Then the two-variable interlace polynomial may be given by

$$q(G; x, y) = \sum_{S \subseteq V(G)} (x-1)^{r(G|_S)} (y-1)^{n(G|_S)},$$

where $r(G|_S)$ and $n(G|_S) = |S| - r(G|_S)$ are, respectively, the $\mathbb{F}_2$-rank and nullity of the adjacency matrix of $G|_S$, the subgraph of G restricted to S.

Definition 10.3.25. The vertex-nullity interlace polynomial is defined recursively as:

$$q_N(G;x) = \begin{cases} x^n \text{ if } G = \mathcal{E}_n, \text{ the edgeless graph on } n \text{ vertices} \\ q_N(G-v;x) + q_N(G^{vw}-w;x) \text{ if } vw \in E(G). \end{cases}$$

This polynomial was shown to be well defined by Arratia, Bollobás, and Sorkin for all simple graphs in [11], and then was shown in [12] to be a specialization of the two-variable interlace polynomial as follows:

$$q_N(G;y) = q(G;2,y) = \sum_{W \subseteq V(G)} (y-1)^{n(G|w)}.$$

An equivalent formulation for $q_N(G;x)$ for simple graphs is given by Aigner and van der Holst in [5].

A somewhat circuitous route through the circuit partition polynomial relates the vertex-nullity interlace polynomial to the Tutte polynomial. First recall that a *circle graph* on n vertices is a graph G derived from a chord diagram.

Two vertices v and w in G share an edge if and only if their corresponding chords intersect in the chord diagram. Note that G is necessarily simple.

For circle graphs, the vertex-nullity interlace polynomial and the circuit partition polynomial are related by the following theorem, noting that although $\vec{G}$ may be a multigraph, H is necessarily simple.

Theorem 10.3.26. *([11], Theorem 6.1). If $\vec{G}$ is a 4-regular Eulerian digraph, C is any Eulerian circuit of $\vec{G}$, and H is the circle graph of the chord diagram determined by C, then $j(\vec{G};x) = xq_N(H;x+1)$.*

This now allows us to relate the vertex-nullity interlace polynomial to the Tutte polynomial, a relation proved in [62] and also observed by Arratia et al. at the end of Sect. 7 in [12].

Theorem 10.3.27. *If G is a planar graph, and H is the circle graph of some Eulerian circuit of $\overrightarrow{G_m}$, then $q_N(H;x) = t(G;x,x)$.*

Proof. By Theorem 10.3.26, $j(\overrightarrow{G_m};x) = xq_N(H;x+1)$, but recalling that the circuit partition and Martin polynomials are simple translations of each other, we have from Theorem 10.3.22 that $j(\overrightarrow{G_m};x) = xm(\overrightarrow{G_m};x+1)$, and hence $q_N(H;x) = m(\overrightarrow{G_m};x) = t(G;x,x)$. □

The interlace polynomial has generated further interest and other applications in Balister et al. [14, 15], Glantz and Pelillo [72], and Ellis-Monaghan and Sarmiento [62].

10 Interrelations and Interpretations

10.4 Multivariable Extensions

Multivariable extensions have proved valuable theoretical tools for many of the polynomials we have seen since they capture information not encoded by the original polynomial. More critically, powerful algebraic tools not applicable to the original polynomial may be available to the multivariable version, providing new means of extracting combinatorial information from the polynomial. Although the multivariable indexing may make the defining notation somewhat bulky, these generalizations are natural extensions of classical versions, computed in exactly the same ways, only now also keeping track of some additional parameters in the computation processes.

10.4.1 Generalized Coloring Polynomials and the U-Polynomial

The evaluation of the chromatic polynomial at λ can be written as

$$\chi_G(\lambda) = \sum_{\substack{\phi:V \to \{1,\dots,\lambda\} \\ \text{proper}}} 1. \tag{10.3}$$

This was generalized to a symmetric function over (commuting) indeterminates $x_1, x_2, \dots$ by Stanley [126] in the following way.

Definition 10.4.1. Let $G = (V, G)$ be a graph, let $\phi : V \to \mathbb{P} = \{1, 2, \dots\}$, and denote the product $\prod_{v \in V} x_{\phi(v)}$ by x^{ϕ}. Then the symmetric function generalization of the chromatic polynomial is

$$X_G(x) := X(G; x_1, x_2, \dots) = \sum_{\substack{\phi:V \to \mathbb{P} \\ \text{proper}}} x^{\phi}.$$

That this is a generalization of the chromatic polynomial can be seen by setting $x_i = 1$ for $1 \leq i \leq \lambda$ and $x_j = 0$ for $j > \lambda$ and noting that the expression in (10.3) for the chromatic polynomial evaluated at λ results.

Generalizing polynomial graph invariants is not a theoretical exercise. The original invariant encodes combinatorial information, and the multivariable generalization will encode not only the same information but also more refined information. For example, the chromatic polynomial of any tree with n vertices has chromatic polynomial $x(x-1)^{n-1}$. But not all trees have the same $X_G(x)$. For example, if $K_{1,3}$ is the 4-star graph and P^4 is the path of order 4, then $X_{K_{1,3}}(x)$ has a term $x_i x_j^3$ for all $i \neq j$, but such a term is not present in $X_{P4}(x)$. In fact, it is still an open question if X distinguishes trees, that is, if $X_{T_1}(x) \neq X_{T_2}(x)$, whenever T_1 and T_2 are not isomorphic trees.

A similar multivariable extension of the bad coloring polynomial is also natural, especially given the importance of the latter because of its being equivalent to the Tutte polynomial. The following generalization of the bad coloring polynomial is also due to Stanley [128].

Definition 10.4.2. Let $G = (V, E)$ be a graph, let $\phi : V \to \mathbb{P} = \{1, 2, \ldots\}$, and let $b(\phi)$ be the set of monochromatic edges in the coloring given by ϕ. Then the symmetric function generalization of the bad coloring polynomial over indeterminates $x_1, x_2, \ldots$ and t is

$$X_G(\boldsymbol{x}, t) = \sum_{\phi:V \to \mathbb{P}} (1 + t)^{|b(\phi)|} x^\phi,$$

where the sum is over all possible colorings ϕ of the graph G.

Again, by setting $x_i = 1$ for $1 \leq i \leq \lambda$ and $x_j = 0$ for $j > \lambda$ we get the bad-coloring polynomial, and hence the Tutte polynomial. Therefore, $X_G(\boldsymbol{x}, t)$ is a multivariable generalization of the Tutte polynomial.

There is another multivariable generalization of the Tutte polynomial that was developed independently and for very different reasons. This generalization is called the U-polynomial and is due to Noble and Welsh in [109].

Definition 10.4.3. Let $G = (V, E)$ be a graph. Then the U-polynomial of G is

$$U_G(\boldsymbol{x}, y) = \sum_{A \subseteq E} x_{n_1} \cdots x_{n_k} (y - 1)^{|A| - r(A)},$$

where $n_1, \ldots, n_k$ are the numbers of vertices in the k different components of G restricted to A.

Clearly, this is a generalization of the Tutte polynomial, as by setting $x_i = (x-1)$ for all i in $U_G(\boldsymbol{x}, y)$ we get $(x - 1)^{\kappa(G)} T_G(x, y)$. Note that the factor $x_{n_1} \cdots x_{n_k}$ in every term keeps track of the number of vertices in the different components in A. Thus, this is a refinement of the rank-nullity generating-function definition of the Tutte polynomial where the factors $x^{r(G)-r(A)} = x^{\kappa(A)-\kappa(G)}$ in each term keep track of the number of components in A.

That U_G captures more combinatorial information from G than the Tutte polynomial can be seen by noting that U_G contains the matching generating polynomial, and thus the matching polynomial, as a specialization as well.

Theorem 10.4.4. *For any graph G,*

$$g(G; x) = U_G(1, t, 0, \ldots, 0, \ldots, y = 1).$$

The U-polynomial has a deletion/contraction reduction relationship not in the class of graphs but in the class of weighted graphs. To see this, we turn to the W-polynomial also due to Noble and Welsh in [109]. A *weighted graph* consists of a graph $G = (V, E)$, together with a weight function $\omega : V \to \mathbb{Z}^+$.

10 Interrelations and Interpretations 277

If e is an edge of (G, ω) then $(G \setminus e, \omega)$ is the weighted graph obtained from (G, ω) by deleting e and leaving ω unchanged. If e is not a loop, $(G/e, \omega/e)$ is the weighted graph obtained from (G, ω) by contracting e, and the weight function ω/e is defined as $\omega/e(u) = \omega(u)$ for all $u \in V \setminus \{v, v'\}$ and $\omega/e(v'') = \omega(v) + \omega(v')$.

Definition 10.4.5. Let (G, ω) be a weighted graph. The W-polynomial may be given recursively by the following rules. If e is an ordinary edge or a bridge, then

$$W_{(G,\omega)}(x, y) = W_{(G \setminus e, \omega)}(x, y) + W_{(G/e, \omega/e)}(x, y).$$

If e is a loop, then $W_{(G,\omega)}(x, y) = y W_{(G \setminus e, \omega)}(x, y)$. Finally, if (G, ω) is $\mathcal{E}_n$, the edgeless graph on $n \geq 0$ vertices, with weights $a_1, \ldots, a_n$, then $W_{(\mathcal{E}_n, \omega)}(x, y) = x_{a_1} \cdots x_{a_n}$.

That the resulting multivariate polynomial W is independent of the order in which the edges are deleted and contracted is proved in [109]. This can easily be done by induction on the number of edges once it is proved that the order in which you contract or delete edges in (G, ω) does not affect the weighted graph that you obtain.

The U-polynomial is obtained from the W-polynomial by setting all weights equal to 1 and a proof that this definition is equivalent to Definition 10.4.3 can be found in [109]. Actually in [109] it is proved that W has a representation of the form

$$W_{(G,\omega)}(x, y) = \sum_{A \subseteq E} x_{c_1} \cdots x_{c_k} (y - 1)^{|A| - r(A)},$$

where c_i, for $1 \leq i \leq k$, is the total weight of the ith component of the weighted subgraph (A, ω).

Noble and Welsh [109] show that the symmetric function generalization of the bad coloring polynomial and the U-polynomial are equivalent in the following sense.

Theorem 10.4.6. *For any graph G, the polynomials U_G and X_G determine each other in that if $p_0 = 1$ and $p_r = \sum_i x_i^r$, then*

$$X_G(x, t) = t^{|V|} U_G \left(x_j = \frac{p_j}{t}, y = t + 1 \right).$$

There is yet another polynomial, the *polychromate*, introduced originally by Brylawski in [38], that is as general as U_G or X_G. Given a graph G and a partition π of its vertices into nonempty blocks, let $e(\pi)$ be the number of edges with both ends in the same block of the partition. If $\tau(\pi) = (n_1, \ldots, n_k)$ is the type of the partition π, we denote by $x_{\tau(\pi)}$ the monomial $\prod_{i=1}^k x_i^{n_i}$.

Definition 10.4.7. Let G be a graph. Then the polychromate $\chi_G(x, y)$ is

$$\chi_G(x, y) = \sum_{\pi} y^{e(\pi)} x_{\tau(\pi)},$$

where the sum is over all partitions of $V(G)$.

We have the following theorem due to Sarmiento in [119], with an alternative proof given by Merino and Noble [106].

Theorem 10.4.8. *The polynomials $U_G(x, y)$ and $\chi_G(x, y)$ are equivalent.*

The story does not end here. All three polynomials $U_G(x, y)$, $X_G(x, t)$ and $\chi_G(x, y)$ have natural extensions. For example, the extension of the $X_G(x, t)$ replaces the t variable by countably infinitely many variables $t_1, t_2, \ldots$, thus enumerating not just the total number of monochromatic edges but the number of monochromatic edges of each color. It is defined as follows:

$$X_G(x, t) = \sum_{\phi:V\to\mathbb{P}} \left(\prod_{i=1}^{\infty} (1 + t_i)^{|b_i(\phi)|} \right) x^{\phi},$$

where the sum is over all colorings ϕ of G and $b_i(\phi)$ is the set of monochromatic edges for which both end points have color i. By setting $t_i = t$ for all $i \geq 1$ we regain $X_G(x, t)$.

For the other extensions the reader is referred to Merino and Noble [106], where it is also proved that all of these extensions are equivalent.

10.4.2 The Parametrized Tutte Polynomial

The basic idea of a *parametrized Tutte polynomial* is to allow each edge of a graph to have four parameters (four ring values specific to that edge), which apply as the Tutte polynomial is computed via a deletion/contraction recursion. Which parameter is applied in a linear recursion reduction depends on whether the edge is deleted or contracted as an ordinary edge, or whether it is contracted as an isthmus or deleted as a loop. The difficulty lies in ensuring that a well-defined function, that is, one independent of the order of deletion/contraction, results. This requires a set of relations, coming from three very small graphs, to be satisfied. Interestingly, additional constraints are necessary for there to be a corank-nullity expansion or even for the function to be multiplicative or a graph invariant, that is, equal on isomorphic graphs.

The motivation for allowing edge-specific values for the deletion/contraction recursion comes from a number of applications where it is natural. This includes graphs with signed edges coming from knot theory, graphs with edge-specific failure probabilities in network reliability, and graphs whose edges represent various interaction energies within a molecular lattice in statistical mechanics. Although there is compelling motivation for allowing various edge parameters, the technical details of a general theory are challenging. The two major works in this area are Zaslavsky [142] and Bollobás and Riordan [23]. However, these two works take different approaches, which were subsequently reconciled with a mild generalization in [63], and for this reason we adopt the formalism of [63]. Bollobás and Riordan [23] also give a succinct historical overview of the development of these multivariable extensions.

10 Interrelations and Interpretations 279

For the purposes of the following, we consider a class of graphs minor-closed if it is closed under the deletion of loops, the contraction of bridges, and the contraction and deletion of ordinary edges; however, we do not require closure under the deletion of bridges. Some formalism is necessary to handle the parameters.

Definition 10.4.9. Let U be a class, and let R be a commutative ring. Then an R-parametrization of U consists of four parameter functions $x, y, X, Y : U \rightarrow R$, denoted $e \rightarrow x_e, y_e, X_e, Y_e$.

Definition 10.4.10. Let U be an R-parametrized class, and let Γ be a minor-closed class of graphs with $E(G) \subseteq U$ for all $G \in \Gamma$. Then a parametrized Tutte polynomial on Γ is a function $T : \Gamma \rightarrow R$ that satisfies the following: $T(G) = X_e T(G/e)$ for any bridge e of $G \in \Gamma$, and $T(G) = Y_e T(G - e)$ for any loop e of $G \in \Gamma$, and $T(G) = y_e T(G - e) + x_e T(G/e)$ for any ordinary edge e.

The following theorem gives the central result. The identity in Item 1 comes from requiring to be equal the two ways of carrying out deletion/contraction reductions on a graph on two vertices with two parallel edges e_1 and e_2 having parameters $\{x_{e_i}, y_{e_i}, X_{e_i}, Y_{e_i}\}$. Similarly, the identities in Items 2 and 3 come from considering the θ-graph and K_3. Here again $\mathcal{E}_n$ is the edgeless graph on n vertices.

Theorem 10.4.11 (The generalized Zaslavsky–Bollobás–Riordan theorem for graphs). *Let R be a commutative ring, let Γ be a minor-closed class of graphs whose edge-sets are contained in an R-parametrized class U, and let a_1, $a_2, \ldots \in R$. Then there is a parametrized Tutte polynomial T on Γ with $T(\mathcal{E}_n) = a_n$ for all n with $\mathcal{E}_n \in \Gamma$ if and only if the following identities are satisfied.*

1. *Whenever e_1 and e_2 appear together in a circuit of a k-component graph $G \in \Gamma$, then $a_k(x_{e_1} Y_{e_2} + y_{e_1} X_{e_2}) = a_k(x_{e_2} Y_{e_1} + y_{e_2} X_{e_1})$.*
2. *Whenever e_1, e_2 and e_3 appear together in a circuit of a k-component graph $G \in \Gamma$, then $a_k X_{e_3}(x_{e_1} Y_{e_2} + y_{e_1} x_{e_2}) = a_k X_{e_3}(Y_{e_1} x_{e_2} + x_{e_1} y_{e_2})$.*
3. *Whenever e_1, e_2 and e_3 are parallel to one another in a k-component graph $G \in \Gamma$, then $a_k Y_{e_3}(x_{e_1} Y_{e_2} + y_{e_1} x_{e_2}) = a_k Y_{e_3}(Y_{e_1} x_{e_2} + x_{e_1} y_{e_2})$.*

A most general parametrized Tutte polynomial, which possibly could be called *the* parametrized Tutte polynomial, might begin with the polynomial ring on independent variables $\{x_e, y_e, X_e, Y_e : e \in U\} \cup \{a_i : i \geq 1\}$. However, the resulting function is not technically a polynomial, in that it must take its values not in the polynomial ring, but has as R the polynomial ring modulo the ideal generated by the identities in Theorem 10.4.11.

The question also arises as to whether "the most general" parametrized Tutte polynomial should be multiplicative on disjoint unions and the one-point joint of graphs, as this introduces additional relations among the a_i's. This is because a parametrized Tutte polynomial is not necessarily multiplicative. A sufficient condition is the following.

Proposition 10.4.12. *Suppose T is a parametrized Tutte polynomial on a minor-closed class of graphs that contains at least one graph with k components for every k and that is closed under one-point unions and the removal of isolated vertices. Then T is multiplicative with respect to both disjoint unions and one-point joins if and only if the $a_i = T(\mathcal{E}_i$ is idempotent, and $a_k = a_1$ for all $k \geq 1$.*

Bollobás and Riordan [23] emphasize graph invariants, and hence require that the parametrization be a coloring of the graph. That is, graphs are edge-colored (not necessarily properly), with edges of the same color having the same parameter sets. This enables consideration of parametrized Tutte polynomials that are invariants of colored graphs, but requires the following additional constraints. For every $e_1 \in U$, there are $e_2, e_3 \in U$ with $e_1 \neq e_2 \neq e_3 \neq e_1$ such that $x_{e_1} = x_{e_2} = x_{e_3}$, $y_{e_1} = y_{e_2} = y_{e_3}$, $X_{e_1} = X_{e_2} = X_{e_3}$, and $Y_{e_1} = Y_{e_2} = Y_{e_3}$.

Proofs of the above results and further details may be found in [23, 63, 142]. We note that any relation between this Tutte polynomial generalization with its edge parameters, and the W- and U-polynomials of Sect. 10.4.1 with their vertex weights, has not yet been studied.

Interestingly, although the parametrized Tutte polynomial has an activities expansion analogous to that of the classical Tutte polynomial, it does not necessarily have an analog of the rank-nullity formulation. However, under modest assumptions involving nonzero parameters and some inverses, the parametrized Tutte polynomial may be expressed in a rank-nullity form. This is fortunate, because significant results for the zeros of the chromatic and Tutte polynomial have arisen from such a multivariable realization. Examples may be found in Sokal [124], Royle and Sokal [118], and Choe et al. [41].

10.4.3 The Generalized Transition Polynomial

A number of state model polynomials, for example the circuit partition polynomials, Penrose polynomial, the Kauffman bracket for knots and links, and the transition polynomials of Jaeger [89], which are not specializations of the Tutte polynomial, are specializations of the multivariable generalized transition polynomial of [61] which we describe here. This multivariable extension is a Hopf algebra map, which leads to structural identities that then inform its various specializations. The medial graph construction that relates the circuit partition polynomial and the classical Tutte polynomial extends to similarly relate the generalized transition polynomial and the parametrized Tutte polynomial when it has a rank-nullity formulation.

The graphs here are Eulerian, although not necessarily connected, with loops and multiple edges allowed. A vertex state, or transition, is a choice of local reconfiguration of a graph at a vertex by pairing the half edges incident with that vertex. A graph state, or transition system, $S(G)$, is the result of choosing a vertex state at each vertex of degree greater than 2, and hence is a union of disjoint cycles. We write $St(G)$ for the set of graph states of G, and throughout we assume weights have values in R, a commutative ring with unity.

10 Interrelations and Interpretations

A *skein relation* for graphs is a formal sum of weighted vertex states, together with an evaluation of the terminal forms (the graph states). See [54,61] for a detailed discussion of these concepts, which are appropriated from knot theory, in their most general form, and Yetter [141] for a general theory of invariants given by linear recursion relations. A *skein type* (or state model, or transition) *polynomial* is one which is computed by repeated applications of skein relations. See Jaeger [89] for a comprehensive treatment of these in the case of 4-regular graphs.

For brevity, we elide technical details such as free loops and isomorphism classes of graphs with weight systems which may be found in [61].

Definition 10.4.13. Pair, vertex, and state weights.

1. A *pair weight* is an association of a value $p\left(e_v, e_v'\right)$ in a unitary ring R to a pair of half edges incident with a vertex v in G. A *weight system*, $W(G)$, of an Eulerian graph G is an assignment of a pair weight to every possible pair of adjacent half edges of G.
2. The *vertex state weight* of a vertex state is $\prod p(e_v, e_v')$ where the product is over the pairs of half edges comprising the vertex state.
3. The *state weight* of a graph state S of a graph G with weight system W is $\omega(S) = \prod \omega(v, S)$, where $\omega(v, S)$ is the vertex state weight of the vertex state at v in the graph state S, and where the product is over all vertices of G.

When A is an Eulerian subgraph of an Eulerian graph G with weight system $W(G)$, then A inherits its weight system $W(A)$ from G in the obvious way, with each pair of adjacent edges in A having the same pair weight as it has in G. When A is a graph resulting from locally replacing the vertex v by one of its $\prod_{i=0}^{n-1} (2n - (2i + 1))$ vertex states, then all the pair weights are the same as they are in G, except that all the pairs of half edges adjacent to the newly formed vertices of degree 2 in A have pair weight equal to 1, the identity in R.

The generalized transition polynomial $N(G; W, x)$ has several formulations, and we give two of them, a linear recursion formula and a state model formula, here.

Definition 10.4.14. The generalized transition polynomial, $N(G; W, x)$, is defined recursively by repeatedly applying the skein relation

$$N(G; W, x) = \sum \beta_i N(G_i; W(G_i), x)$$

at any vertex v of degree greater than 2. Here the G_i's are the graphs that result from locally replacing a vertex v of degree $2n$ in G by one of its vertex states. The β_i's are the vertex state weights. Repeated application of this relation reduces G to a weighted (formal) sum of disjoint unions of cycles (the graph states). These terminal forms are evaluated by identifying each cycle with the variable x, weighted by the product of the pair weights over all pairs of half edges in the cycle.

Definition 10.4.15. The state model definition of the generalized transition polynomial is:

$$N(G; W, x) = \sum_{St(G)} \left(\left(\prod \omega(v, S) \right) x^{k(S)} \right) = \sum_{St(G)} \omega(S) x^{k(S)}.$$

Note that vertex states commute; that is, if G_{uv} results from choosing a vertex state at u, and then at v, we have $G_{uv} = G_{vu}$. Thus, Definition 10.4.14 gives a well-defined function, and Definitions 10.4.14 and 10.4.15 are equivalent.

Several of the polynomials we have already seen are specializations of this generalized transition polynomial. For example, if all the pair weights are 1, then the circuit partition polynomial for an unoriented Eulerian graph results. If $\vec{G}$ is an Eulerian digraph, and G is the underlying undirected graph with pair weights of 1 for pairs half edges corresponding to one inward and one outward oriented half edge of $\vec{G}$ and 0 otherwise, then the oriented version of the circuit partition polynomial results.

In the special case that G is 4-regular, the polynomial $N(G; W, x)$ is essentially the same as the transition polynomial $Q(G, A, \tau)$ of Jaeger [89], where G is a 4-regular graph and A is a system of vertex state weights (rather than pair weights). If the vertex state weight in (G, A) is w, then define $W(G)$ by letting the pair weights for each of the two pairs of edges determined by the state be $\sqrt{w}$. The two polynomials then just differ by a factor of x, so $N(G; W, x) = xQ(G, A, x)$, and here we retain vertices of degree 2 in the recursion while they are elided in [89]. Thus $N(G; W, x)$ gives a generalization of Jaeger's transition polynomials to all Eulerian graphs.

Because $Q(G, A, \tau)$ assimilates the original Martin polynomial for 4-regular graphs and digraphs, the Penrose polynomial, and the Kauffman bracket of knot theory (see [89]), and $N(G; W, x)$ assimilates $Q(G, A, \tau)$, we have that the Penrose polynomial and Kauffman bracket are also specializations of $N(G; W, x)$. Specifically, if G is a planar graph with face two-colored medial graph G_m, and we give a weight system to G_m by assigning a value of 1 to pairs of edges that either cross at a vertex or bound the same black face and 0 otherwise, then $N(G_m; W, x) = P(G; x)$. Similarly, if L is a link, and G_L is the signed, face two-colored universe of L, then a weight system can be assigned to G_L so that $N(G_L; W, a^2 + a^{-2}) = (a^2 + a^{-2})K[L]$ where $K[L]$ is the Kauffman bracket of the link.

Because of these specializations, the algebraic properties of the generalized transition polynomial are available to inform these other polynomials as well. In particular, $N(G; W, x)$ is a Hopf algebra map from the freely generated (commutative) hereditary Hopf algebra of Eulerian graphs with weight systems to the binomial bialgebra $R[x]$ (details may be found in [61]). This leads to two structural identities, the first from the comultiplication in the Hopf algebra, the second from the antipode.

10 Interrelations and Interpretations

Theorem 10.4.16. *Let G be an Eulerian graph. Then*

$$N(G;W,x+y) = \sum N(A_1;W(A_1),x) N(A_2;W(A_2),y)$$

where the sum is over all ordered partitions of G into two edge-disjoint Eulerian subgraphs A_1 and A_2, and

$$N(G;W,-x) = N(\zeta(G;W),x),$$

where ζ is the antipode $\zeta(G,W) = \sum(-1)^{|P|}(A_1 \cdots A_{|P|})$, with the sum over all ordered partitions P of G into $|P|$ edge-disjoint Eulerian subgraphs each with inherited weight system. Here $N(G;W,x)$ is extended linearly over such formal sums.

This type of Hopf algebraic structure has already been used to considerably extend the known combinatorial interpretations for evaluations of the Martin, Penrose, and Tutte polynomials implicitly by Bollobás [22], and explicitly by Ellis-Monaghan and Sarmiento [54, 56, 57, 60, 120]. The first identity has been used to find combinatorial interpretations for the Martin polynomials for all integers, where this was previously only known for 2, -1, 0, 1 in the oriented case, and 2, 0, 2 in the unoriented case. This then led to combinatorial interpretations for the Tutte polynomial (and its derivatives) of a planar graph for all integers along the line $x = y$, where previously 1, 3 were the only known nontrivial values. These results for the Tutte polynomial were mentioned in Chap. 9 and for the circuit partition polynomial in Sect. 10.3.4. The second identity has been used to determine combinatorial interpretations for the Penrose polynomial for all negative integers, where this was previously only known for positive integers.

10.5 Two Applications

Graph polynomials have a wide range of applications throughout many fields. We have already seen some examples of this with various applications of the classical Tutte polynomial in the previous chapter. Here we present two representative important applications (out of many possible) and show how they may be modeled by graph polynomials.

10.5.1 DNA Sequencing

We begin with string reconstruction, a problem that may be modeled by the interlace and circuit partition polynomials (and hence indirectly in special cases by the Tutte polynomial). String reconstruction is the process of reassembling a long string of symbols from a set of its subsequences together with some (possibly incomplete, redundant, or corrupt) sequencing information. Although we focus on DNA sequencing, which was the original motivation for the development of the interlace polynomial, the methods here apply to any string reconstruction problem. For example, fragmenting and reassembling messages is a common network protocol, and reconstruction techniques might be applied when the network protocol has been disrupted, yet the original message must be reassembled from the fragments.

DNA sequences are typically too long to read at once with current laboratory techniques, so researchers probe for shorter fragments (reads) of the strand. They then are faced with the difficulty of recovering the original long sequence from the resulting set of subsequences. DNA sequencing by hybridization is a method of reconstructing the nucleotide sequence from a set of short substrings (see Waterman [136] for an overview). The problem of determining the number of possible reconstructions may be modeled using Eulerian digraphs, with a correct sequencing of the original strand corresponding to exactly one of the possible Eulerian circuits in the graph. The probability of correctly sequencing the original strand is thus the reciprocal of the total number of Euler circuits in the graph.

The most basic (two-way repeats only) combinatorial model for DNA sequencing by hybridization uses an Eulerian digraph with two incoming and two outgoing edges at each vertex (see Pevzner [114] and Arratia et al. [9]). The raw data consist of all subsequences of the DNA strand of a fixed length L, called the L-spectrum of the sequence. As L increases, the statistical probability that the beginning and end of the DNA strand are the same approaches zero, as does the likelihood of three or more repeats of the same pattern of length L or more in the strand (see Dyer et al. [48]). Thus, this model assumes that the only consideration in reconstructing the original sequence is the appearance of interlaced two-way repeats, that is, alternating patterns of length L or greater, for example, $\ldots ACTG \ldots CTCT \ldots ACTG \ldots CTCT \ldots$.

From the multiset (duplicates are allowed) of subsequences of length L, create a single vertex of the de Bruijn graph for each subsequence of length L-1 that appears in one of the subsequences. For example, if L = 4 and ACTG appears as a subsequence, create two vertices, one labeled ACT and one labeled CTG. Edges are directed from head to tail of a subsequence; e.g., there would be a directed edge labeled ACTG from the vertex labeled ACT to the vertex labeled CTG. If there is another subsequence ACTT, we do not create another vertex ACT, but rather draw an edge labeled ACTT from the vertex ACT to a new vertex labeled CTT. If, in the multiset of subsequences, ACTG appears twice, then we draw two edges from ACT to CTG.

The beginning and end of the strand are identified to be represented by the same vertex, and, since by assumption no subsequence appears more than twice, the result is an Eulerian digraph of maximum degree 4. Tracing the original DNA sequence in this graph corresponds to an Eulerian circuit that starts at the vertex representing the beginning and end of the strand. All other possible sequences that could be (mis)reconstructed from the multiset of subsequences correspond to other Eulerian circuits in this graph. Thus (up to minor reductions for long repeats and

10 Interrelations and Interpretations 285

forced subsequences), finding the number of DNA sequences possible from a given multiset of subsequences corresponds to enumerating the Eulerian circuits in this directed graph.

The generalized transition polynomial models this problem directly: when the pair weights are identically 1, it reduces to the circuit partition polynomial. This is a generating function for families of circuits in a graph, so the coefficient of x is the number of Eulerian circuits. The interlace polynomial informs the problem as follows. Consider an Eulerian circuit through the de Bruijn graph, which gives a sequence of the vertices visited in order. Now construct the interlace graph by placing a vertex for each symbol and an edge between symbols that are interlaced (occur in alternation) in the sequence. The interlace polynomial of the interlace graph is then a translation of the circuit partition polynomial of the original de Bruijn graph, as in Theorem 10.3.26, where again the coefficient of x is the number of Eulerian circuits (see Arratia et al. [10, 11, 13]).

One of the original motivating goals of Arratia et al. [10] was classifying Eulerian digraphs with a given number of Eulerian circuits. The BEST theorem, a formula for the number of the circuits of an Eulerian graph in terms of its Kirchhoff matrix (see Fleischner [70] for good exposition) gives only a tautological classification: the Eulerian digraphs with m Eulerian circuits are those where the BEST theorem formula gives m circuits. Critically, all of the above graph polynomials encode much more information than is available from the BEST theorem, and all of them are embedded in broader algebraic structures that provide tools for extracting information from them. Thus, they better serve the goal of seeking structural characterizations of graph classes with specified Eulerian circuit properties.

10.5.2 The Potts Model of Statistical Mechanics

Here we have an important physics model that remarkably was found to be exactly equivalent to the Tutte polynomial.

Complex systems are networks in which very simple interactions at the microscale level determine the macroscale behavior of the system. The Potts model of statistical mechanics models complex systems whose behaviors depend on nearest neighbor energy interactions. This model plays an important role in the theory of phase transitions and critical phenomena, and has applications as widely varied as magnetism, adsorption of gases on substrates, foam behaviors, and social demographics, with important biological examples including disease transmission, cell migration, tissue engulfment, diffusion across a membrane, and cell sorting.

Central to the Potts model is the Hamiltonian,

$$ h(\omega) = -J \sum_{\{i,j\} \in E(G)} \delta(\sigma_i, \sigma_j), $$

a measure of the energy of the system. Here a spin, σ_i, at a vertex i, is a choice of condition (for example, healthy, infected, or necrotic for a cell represented by the vertex). J is a measure of the interaction energy between neighboring vertices, ω is a state of a graph G (here a state is a fixed choice of spin at each vertex), and δ is the Kronecker delta function.

The Potts model partition function is the normalization factor for the Boltzmann probability distribution. Systems such as the Potts model, following Boltzmann distribution laws, will have the number of states with a given energy (Hamiltonian value) exponentially distributed. Thus, the probability of the system being in a particular state ω at temperature t is:

$$\Pr(\omega, \beta) = \frac{\exp(-\beta h(\omega))}{\sum \exp(-\beta h(\varpi))}.$$

Here, the sum is over all possible states ϖ of G, and $\beta = \frac{1}{\kappa t}$, where $\kappa = 1.38 \times 10^{-23}$ J/K is the Boltzmann constant. The parameter t is an important variable in the model, although it may not represent physical temperature, but some other measure of volatility relevant to the particular application (for example, ease of disease transmission/reinfection). The denominator of this expression, $P(G; q, \beta) = \sum \exp(-\beta h(\varpi))$, called the Potts model partition function, is the most critical, and difficult, part of the model.

Remarkably, the Potts model partition function is equivalent to the Tutte polynomial:

$$P(G; q, \beta) = q^{k(G)} v^{|v(G)| - k(G)} T\left(G; \frac{q+v}{v}, v+1\right),$$

where q is the number of possible spins, and $v = \exp(J\beta) - 1$. See Fortuin and Kasteleyn [71] for the nascent stages of this theory, later exposition in Tutte [134], Biggs [20], Bollobás [21], Welsh [137, 139], and surveys by Welsh and Merino [140], and Beaudin et al. [17].

One common extension of the Potts model involves allowing interaction energies to depend on individual edges. With this, the Hamiltonian becomes $h(\omega) = \sum_{e \in E(G)} J_e \delta(\sigma_i, \sigma_j)$, where J_e is the interaction energy on the edge e. The partition function is then

$$P(G) = \sum_{A \subseteq E(G)} q^{k(A)} \prod_{e \in A} v_e,$$

where $v_e = \exp(\beta J_e) - 1$. Again see Fortuin and Kasteleyn [71], and more recently Sokal [122,123]. As we have seen in Sect. 10.4.2, the Tutte polynomial has also been extended to parametrized Tutte functions that incorporate edge weights. The generalized partition function given above satisfies the relations of Theorem 10.4.11, however, and thus is a special case of a parametrized Tutte function.

This relationship between the Potts model partition function and the Tutte polynomial has led to a remarkable synergy between the fields, particularly, for example, in the areas of computational complexity and the zeros of the Tutte and chromatic polynomials. For overviews, see Welsh and Merino [140], and Beaudin et al. [17].

10.6 Conclusion

There are a great many other graph polynomials equally interesting to those surveyed here, including, for example, the F-polynomials of Farrell, the Hosaya or Wiener polynomial, the clique/independence and adjoint polynomials, etc. In particular, Farrell [66] has a circuit cover polynomial (different from the circuit partition polynomial of Sect. 10.3.4) with noteworthy interrelations with the characteristic polynomial. There are rich connections between graph theory and knot theory via the Tutte polynomial, including a relation to the HOMFLY polynomial given by Jaeger [88], with applications in biology such as Emmert-Streib [64]. Also, Chung and Graham developed a "Tutte-like" polynomial for directed graphs in [43]. The resultant cover polynomial is extended to a symmetric function generalization, like those in Sect. 10.4.1, by Chow [42]. Similarly, Courcelle [44] and Traldi [131] have also very recently developed multivariable extensions of the interlace polynomial. Some surveys of graph polynomials with complementary coverage to this one include Pathasarthy [111], Jaeger [89], Farrell [68], Fiol [69], Godsil [74], Aigner [2], Noy [110], and Levit and Mandrescu [100].

Acknowledgments We thank all the friends and colleagues who offered many helpful comments and suggestions during the writing of this chapter.

The first author was supported by the National Security Agency and by the Vermont Genetics Network through Grant Number P20 RR16462 from the INBRE Program of the National Center for Research Resources (NCRR), a component of the National Institutes of Health (NIH).

The second author was supported by Conacyt of Mexico, Grant 83977.

References

1. Aigner M (1997) The Penrose polynomial of a plane graph. Ann Math 307:173–189
2. Aigner M (1997) The Penrose polynomial of graphs and matroids. In: Hirschfeld JWP (ed) Surveys in combinatorics, 2001. Cambridge University Press, Cambridge
3. Aigner M (2000) Die Ideen von Penrose zum 4-Farbenproblem. Jahresber Deutsch Math-Verein 102:43–68
4. Aigner M, Mielke H (2000) The Penrose polynomial of binary matroids. Monatsh Math 131:1–13
5. Aigner M, van der Holst H (2004) Interlace polynomials. Lin Algebra Appl 377:11–30
6. Aihara J (1976) A new definition of Dewar-type resonance energies. J Am Chem Soc 98:2750–2758
7. Allys L (1994) Minimally 3-connected isotropic systems. Combinatorica 14:247–262
8. Alon N, Tarsi M (1992) Colorings and orientations of graphs. Combinatorica 12:125–134
9. Arratia R, Martin D, Reinert G, Waterman M (1996) Poisson process approximation for sequence by hybridization. J Comput Biol 3:425–463
10. Arratia R, Bollobás B, Coppersmith D, Sorkin G (2000) Euler circuits and DNA sequencing by hybridization, combinatorial molecular biology. Discrete Appl Math 104:63–96
11. Arratia R, Bollobás B, Sorkin G (2000) The interlace polynomial: a new graph polynomial. In: Proceedings of the 11th annual ACM-SIAM symposium on discrete algorithms. San Francisco, CA
12. Arratia R, Bollobás B, Sorkin G (2004) A two-variable interlace polynomial. Combinatorica 24:567–584

13. Arratia R, Bollobás B, Sorkin G (2004) The interlace polynomial of a graph. J Combin Theory B 92:199–233
14. Balister PN, Bollobás B, Riordan OM, Scott AD (2001) Alternating knot diagrams, Euler circuits and the interlace polynomial. Eur J Combinator 22:1–4
15. Balister PN, Bollobás B, Cutler J, Pebody L (2002) The interlace polynomial of graphs at −1. Eur J Combinator 23:761–767
16. Barvinok AI (1994) Computing the Ehrhart polynomial of a convex lattice polytope. Discrete Comput Geom 12:35–48
17. Beaudin L, Ellis-Monaghan JA, Pangborn G, Shrock R (2010) A little statistical mechanics for the graph theorist. Discrete Math 310:2037–2053
18. Beck M, Robins S (2007) Computing the continuous discretely: integer-point enumeration in polyhedra. Springer, New York
19. Betke U, Kneser M (1985) Zerlegungen und Bewertungen von Gitterpolytopen. J Reine Angew Math 358:202–208
20. Biggs N (1996) Algebraic graph theory, 2nd edn. Cambridge University Press, Cambridge
21. Bollobás B (1998) Modern graph theory. Graduate text in mathematics. Springer, Berlin, New York
22. Bollobás B (2002) Evaluations of the circuit partition polynomial. J Combin Theory B 85:261–268
23. Bollobás B, Riordan O (1999) A Tutte polynomial for coloured graphs. Combinator Probab Comput 8:45–93
24. Bollobás B, Riordan O (2001) A polynomial invariant of graphs on orientable surfaces. Proc Lond Math Soc 83:513–531
25. Bollobás B, Riordan O (2002) A polynomial of graphs on surfaces. Ann Math 323:81–96
26. Bouchet A (1987) Isotropic systems. Eur J Combinator 8:231–244
27. Bouchet A (1987) Reducing prime graphs and recognizing circle graphs. Combinatorica 7:243–254
28. Bouchet A (1988) Graphic presentations of isotropic systems. J Combin Theory B 45:58–76
29. Bouchet A (1989) Connectivity of isotropic systems. (J Combin Math, Proceedings of the 3rd international conference, New York, 1985) Ann N Y Acad Sci 555:81–93
30. Bouchet A (1991) Tutte–Martin polynomials and orienting vectors of isotropic systems. Graph Combinator 7:235–252
31. Bouchet A (1993) Compatible Euler tours and supplementary Eulerian vectors. Eur J Combinator 14:513–520
32. Bouchet A (1997) Multimatroids. I. Coverings by independent sets. SIAM J Discrete Math 10:626–646
33. Bouchet A (1998) Multimatroids. II. Orthogonality, minors and connectivity. Electron J Combinator 5, R8
34. Bouchet A (1998) Multimatroids. IV. Chain-group representations. Lin Algebra Appl 277:271–289
35. Bouchet A (2001) Multimatroids. III. Tightness and fundamental graphs. Eur J Combinator 22:657–677
36. Bouchet A (2005) Graph polynomials derived from Tutte–Martin polynomials. Discrete Math 302:32–38
37. Bouchet A, Ghier L (1996) Connectivity and β invariants of isotropic systems and 4-regular graphs. Discrete Math 161:25–44
38. Brylawski TH (1981) Intersection theory for graphs. J Combin Theory B 30:233–246
39. Chmutov S (2009) Generalized duality for graphs on surfaces and the signed Bollobás-Riordan polynomial. J Combin Theory B 99:617–638
40. Chmutov S, Pak I (2007) The Kauffman bracket of virtual links and the Bollobás-Riordan polynomial. Mosc Math J 7:409–418, 573
41. Choe Y-B, Oxley J, Sokal A, Wagner D (2004) Homogeneous multivariate polynomials with the half-plane property. Adv Appl Math 32:88–187
42. Chow T (1996) The path-cycle symmetric function of a digraph. Adv Math 118:71–98

10 Interrelations and Interpretations 289

43. Chung FRK, Graham RL (1995) On the cover polynomial of a digraph. J Combin Theory B 65:273–290
44. Courcelle B (2008) A multivariate interlace polynomial and its computation for graphs of bounded clique-width. Electron J Combinator 15(1):R69
45. Cvetković DM, Doob M, Sachs H (1980) Spectra of graphs: theory and applications. Academic, New York
46. Cvetković DM, Lepović M (1998) Seeking counterexamples to the reconstruction conjecture for characteristic polynomials of graphs and a positive result. Bull Acad Serbe Sci Arts, Cl Sci Math Natur, Sci Math 116:91–100
47. Diaz R, Robins S (1997) The Ehrhart polynomial of a lattice polytope. Ann Math 145: 503–518
48. Dyer M, Frieze A, Suen S (1994) The probability of unique solutions of sequencing by hybridization. J Comput Biol 1:105–110
49. Ehrhart E (1967) Démonstration de la loi de réciprocité du polyèdre rationnel. C R Acad Sci Paris Sér A–B 265:A91–A94
50. Ehrhart E (1967) Sur un problème de géometrie diophantienne linéaire I. J Reine Angew Math 226:1–29
51. Ehrhart E (1967) Sur un problème de géometrie diophantienne linéaire II. J Reine Angew Math 227:25–49
52. Ehrhart E (1977) Polynômes Arithmétiques et Méthode des Polyèdres en Combinatoire. International Series of Numerical Mathematics vol 35. Birkhäuser, Basel-Stuttgart
53. Ellingham MN, Goddyn L (1996) List edge colourings of some 1-factorable multigraphs. Combinatorica 16:343–352
54. Ellis-Monaghan JA (1998) New results for the Martin polynomial. J Combin Theory B 74:326–352
55. Ellis-Monaghan JA (2000) Differentiating the Martin polynomial. Cong Numer 142:173–183
56. Ellis-Monaghan JA (2004) Exploring the Tutte–Martin connection. Discrete Math 281: 173–187
57. Ellis-Monaghan JA (2004) Identities for circuit partition polynomials, with applications to the Tutte polynomial. Adv Appl Math 32:188–197
58. Ellis-Monaghan JA, Moffatt I. Twisted duality and polynomials of embedded graphs, preprint arXiv:0906.5557
59. Ellis-Monaghan JA, Sarmiento I. A recipe theorem for the topological Tutte polynomial of Bollobás and Riordan, preprint arXiv:0903.2643
60. Ellis-Monaghan JA, Sarmiento I (2001) Medial graphs and the Penrose polynomial. Congr Numer 150:211–222
61. Ellis-Monaghan JA, Sarmiento I (2002) Generalized transition polynomials. Congr Numer 155:57–69
62. Ellis-Monaghan JA, Sarmiento I (2007) Distance hereditary graphs and the interlace polynomial. Combinator Probab Comput 16:947–973
63. Ellis-Monaghan JA, Traldi L (2006) Parametrized Tutte polynomials of graphs and matroids. Combinator Probab Comput 15:835–834
64. Emmert-Streib F (2006) Algorithmic computation of knot polynomials of secondary structure elements of proteins. J Comput Biol 13(8):1503–1512
65. Farrell EJ (1979) An introduction to matching polynomials. J Combin Theory B 27:75–86
66. Farrell EJ (1979) On a class of polynomials obtained from the circuits in a graph and its application to characteristic polynomials of graphs. Discrete Math 25:121–133
67. Farrell EJ (1988) On the matching polynomial and its relation to the rook polynomial. J Franklin Inst 325:527–543
68. Farrell EJ (1993) The impact of F-polynomials in graph theory, Quo vadis, graph theory? Ann Discrete Math, North-Holland, Amsterdam, 55:173–178
69. Fiol MA (1997) Some applications of the proper and adjacency polynomials in the theory of graph spectra. Electron J Combinator 4(1):R21
70. Fleischner H (1991) Eulerian graphs and related topics, part 1, vol 2, p 50. Ann Discrete Math, North-Holland Publishing Co., Amsterdam

71. Fortuin CM, Kasteleyn PW (1972) On the random cluster model. Physica 57:536–564
72. Glantz R, Pelillo M (2006) Graph polynomials from principal pivoting. Discrete Math 306:3253–3266
73. Godsil CD (1981) Matchings and walks in graphs. J Graph Theory 5:285–297
74. Godsil CD (1984) Real graph polynomials. In: Bondy JA, Murty USR (eds) Progress in graph theory. Academic, Toronto
75. Godsil CD (1993) Algebraic combinatorics. Chapman & Hall, New York
76. Godsil CD (1995) Tools from linear algebra. In: Graham RL, Grötschel M, Lovász L (eds) Handbook of combinatorics, vol 2. Elsevier, Amsterdam
77. Gutman I (1991) Polynomials in graph theory. In: Bonchev D, Rouvray DH (eds) Chemical graph theory: introduction and fundamentals. Abacus Press, New York
78. Gutman I, Cvetković DM (1975) The reconstruction problem for the characteristic polynomial of graphs. Publ Electrotehn Fac Ser Fiz 488–541:45–48
79. Gutman I, Milun M, Trinajstić N (1976) Graph theory and molecular orbitals. XVIII. On topological resonance energy. Croat Chem Acta 48:87–95
80. Gutman I, Milun M, Trinajstić N (1977) Graph theory and molecular orbitals. 19. Nonparametric resonance energies of arbitrary conjugated systems. J Am Chem Soc 99:1692–1704
81. Gutman I, Trinajstić N (1976) Graph theory and molecular orbitals, XIV. On topological definition of resonance energy. Acta Chim Acad Sci Hung 91:203–209
82. Harary F (1962) The determinant of the adjacency matrix of a graph. SIAM Rev 4:202–210
83. Heilmann OJ, Lieb EH (1970) Monomers and dimers. Phys Rev Lett 24:1412–1414
84. Heilmann OJ, Lieb EH (1972) Theory of monomer-dimer systems. Commun Math Phys 25:190–232
85. Horn RA, Johnson CR (1990) Matrix analysis. Cambridge University Press, Cambridge
86. Hosaya H (1971) Topological Index. A newly proposed quantity characterizing the topological nature of structural isomers of saturated hydrocarbons. Bull Chem Soc Jpn 44:2332–2339
87. Jackson B (1991) Supplementary Eulerian vectors in isotropic systems. J Combin Theory B 53:93–105
88. Jaeger F (1988) Tutte polynomials and link polynomials. Proc Am Math Soc 103(2):647–654
89. Jaeger F (1990) On Transition polynomials of 4-regular graphs, cycles and rays. NATO Adv Sci Inst Ser C: Math Phys Sci 301. Kluwer, Dordrecht
90. Kantor JM, Khovanskii A (1993) Une application du Théorème de Riemann-Roch combinatoire au polynôme d'Ehrhart des polytopes entier de $\mathbb{R}^d$. C R Acad Sci Paris Ser I 317:501–507
91. Kauffman LH (1989) A Tutte polynomial for signed graphs. Discrete Appl Math 25:105–127
92. Kelly PJ (1957) A congruence theorem for trees. Pac J Math 7:961–968
93. Kel'mans AK (1965) The number of trees in a graph. I. Automat i Telemeh 26:2194–2204 (1965) (in Russian); transl. Autom Rem Contr 26:2118–2129
94. Kocay WL (1981) An extension of Kelly's lemma to spanning subgraphs. Congr Numer 31:109–120
95. Kunz H (1970) Location of the zeros of the partition function for some classical lattice systems. Phys Lett A 32:311–312
96. Las Vergnas M (1979) On Eulerian partitions of graphs. In: Wilson RJ (ed) Graph theory and combinatorics. Pitman, Boston London
97. Las Vergnas M (1981) Eulerian circuits of 4-valent graphs imbedded in surfaces. Algebric methods in graph theory, Vol. I, II (Szeged, 1978), pp 451–477, Colloq math soc Jnos Bolyai, 25, North-Holland, Amsterdam, New York
98. Las Vergnas M (1983) Le polynôme de Martin d'un graphe eulérien. In: Berge C, Bresson D, Camion P, Maurras J-F, Sterboul F (eds) Combinatorial mathematics. North-Holland, Amsterdam
99. Las Vergnas M (1988) On the evaluation at (3,3) of the Tutte polynomial of a graph. J Combin Theory B 44:367–372
100. Levit VE, Mandrescu E (2005) The independence polynomial of a graph – a survey. In: Bozapalidis S, Kalampakas A, Rahonis G (eds) Proceedings of the 1st international conference on algebraic informatics. Aristotle University of Thessaloniki, Thessaloniki

10 Interrelations and Interpretations 291

101. Lovász L, Plummer MD (1986) Matching theory. Ann Discrete Math 29, Amsterdam
102. Macdonald IG (1971) Polynomials associated with finite cell complexes. J Lond Math Soc 4:181–192
103. Martin P (1977) Enumérations eulériennes dans le multigraphs et invariants de Tutte–Grothendieck. Thesis, Grenoble
104. Martin P (1978) Remarkable valuation of the dichromatic polynomial of planar multigraphs. J Combin Theory B 24:318–324
105. McMullen P (1971) On zonotopes. Trans Am Math Soc 159:91–109
106. Merino C, Noble SD (2009) The equivalence of two graph polynomials and a symmetric function. Combinator Probab Comput 18:601–615
107. Moffatt I (2010) Unsigned state models for the Jones polynomial. Ann Combinator (to appear).
108. Moffatt I (2008) Knot invariants and the Bollobás-Riordan Polynomial of embedded graphs. Eur J Combinator 29:95–107
109. Noble SD, Welsh DJA (1999) A weighted graph polynomial from chromatic invariants of knots. Annales de l'institute Fourier 49:1057–1087
110. Noy M (2003) Graphs determined by polynomial invariants. Theor Comput Sci 307:365–384
111. Pathasarthy KR (1989) Graph polynomials. In: Kulli VR (ed) Recent studies in graph theory. Vishwa International Publications, Gulbarga
112. Penrose R (1971) Applications of negative dimensional tensors. In: Welsh DJA (ed) Combinatorial mathematics and its applications: Proceedings of a conference held at the mathematical institute, Oxford, 1969. Academic, London/New York
113. Peterson J (1891) Die theorie der regularen graphs. Acta Math 15:193–220
114. Pevzner PA (1989) l-tuple DNA sequencing: computer analysis. J Biomol Struct Dynam 7:63–73
115. Poincaré H (1901) Second complément à l'analysis situs. Proc Lond Math Soc 65:23–45
116. Pommersheim J (1993) Toric varieties, lattice points, and Dedekind sums. Ann Math 295: 1–24
117. Riordan J (1958) An introduction to combinatorial analysis. Wiley, New York
118. Royle G, Sokal A (2004) The Brown–Colbourn conjecture on zeros of reliability polynomials is false. J Combin Theory B 91:345–360
119. Sarmiento I (2000) The polychromate and a chord diagram polynomial. Ann Combinator 4:227–236
120. Sarmiento I (2001) Hopf algebras and the Penrose polynomial. Eur J Combinator 22: 1149–1158
121. Shephard GC (1974) Combinatorial properties of associated zonotopes. Can J Math 26: 302–321
122. Sokal AD (2000) Chromatic polynomials, Potts models and all that. Physica A 279:324–332
123. Sokal AD (2001) A personal list of unsolved problems concerning lattice gases and antiferromagnetic Potts models. Markov Process Relat Fields 7:21–38
124. Sokal AD (2001) Bounds on the complex zeros of (di)chromatic polynomials and Potts-model partition functions. Combinator Probab Comput 10:41–77
125. Stanley RP (1980) Decompositions of rational convex polytopes. Ann Discrete Math 6:333–342
126. Stanley RP (1995) A symmetric function generalization of the chromatic polynomial of a graph. Adv Math 111:166–194
127. Stanley RP (1996) Enumerative combinatorics, vol 1. Cambridge University Press, Cambridge
128. Stanley RP (1998) Graph colourings and related symmetric functions: ideas and applications. A description of results, interesting applications, and notable open problems. Discrete Math 193:267–286
129. Sylvester JJ (1878) On an application of the new atomic theory to the graphical representation of the invariants and covariants of binary quantics, with three appendices. Am J Math 1:64–125

130. Thistlethwaite MB (1987) A spanning tree expansion of the Jones polynomial. Topology 26:297–309
131. Traldi L (2010) Weighted interlace polynomials. Combinator Probab Comput 19:133–157
132. Tutte WT (1967) On dichromatic polynomials. J Combin Theory 2:301–320
133. Tutte WT (1979) All the kings horses. In: Bondy JA, Murty USR (eds) Graph theory and related topics. Academic, London
134. Tutte WT (1984) Graph theory. Addison-Wesley, New York
135. Ulam S (1960) A collection of mathematical problems. Wiley, New York
136. Waterman MS (1995) Introduction to computational biology: maps, sequences and genomes. Chapman & Hall, New York
137. Welsh DJA (1993) Complexity: knots, colorings and counting. Cambridge University Press, Cambridge
138. Welsh DJA (1997) Approximate counting. In: Bailey R (ed) Surveys in Combinatorics, 1997. Cambridge University Press, Cambridge
139. Welsh DJA (1999) The Tutte polynomial. Statistical physics methods in discrete probability, combinatorics, and theoretical computer science. Random Struct Algorithm 15:210–228
140. Welsh DJA, Merino C (2000) The Potts model and the Tutte polynomial. J Math Phys 41:1127–1152
141. Yetter DN (1990) On graph invariants given by linear recurrence relations. J Combin Theory B 48:6–18
142. Zaslavsky T (1992) Strong Tutte functions of matroids and graphs. Trans Am Math Soc 334:317–347

Chapter 11
Reconstruction Problems for Graphs, Krawtchouk Polynomials, and Diophantine Equations

Thomas Stoll

Abstract We give an overview about some reconstruction problems in graph theory, which are intimately related to integer roots of Krawtchouk polynomials. In this context, Tichy and the author recently showed that a binary Diophantine equation for Krawtchouk polynomials only has finitely many integral solution. Here, this result is extended. By using a method of Krasikov, we decide the general finiteness problem for binary Krawtchouk polynomials within certain ranges of the parameters.

Keywords Krawtchouk polynomials · Graph reconstruction · Diophantine equations · Discrete orthogonal polynomials · Laguerre inequality

MSC2000: Primary 11D45; Secondary 33C05, 33C45, 39B72

11.1 Introduction

11.1.1 The Reconstruction Conjecture

A famous conjecture in graph theory states that graphs are determined (up to isomorphism) by their subgraphs. This conjecture is known as the *(Kelly-Ulam-) Reconstruction Conjecture* and the literature on solving the conjecture for special graphs is vast (see [2] for a survey). Also, negative results are known, for example, digraphs and hypergraphs are in general not *reconstructible*. On the other hand, there is much freedom in formulating reconstructions problems, namely, one may remove edges, vertices, or specific sets of vertices for the subgraphs under question. The aim of the present chapter is to give a short overview on how these reconstruction problems relate to the investigation of integral zeroes of so-called *Krawtchouk*

T. Stoll (✉)
Faculty of Mathematics, School of Computer Science,
University of Waterloo, Waterloo, ON, Canada
e-mail: tstoll@cs.uwaterloo.ca

M. Dehmer (ed.), *Structural Analysis of Complex Networks*,
DOI 10.1007/978-0-8176-4789-6_11, © Springer Science+Business Media, LLC 2011

polynomials as well as to report on known results on this connection. Indeed, reconstruction can be put in terms of a one-variable Diophantine problem for Krawtchouk polynomials. It is a great challenge to study this Diophantine problem in the most general setting, hereby making a substantial attempt to unify several of the dispersed results in the area of graph reconstruction.

11.1.2 Reconstruction Problems and Zeroes of Krawtchouk Polynomials

Given a finite, simple graph G with $|V(G)| = n \geq 3$. For $U \subset V(G)$, the switching G_U of G at U is the graph obtained from G by replacing all edges between U and $V(G) \backslash U$ by the nonedges. The multiset of unlabeled graphs $D_s(G) = \{G_U : |U| = s\}$ is called the *s*-switching deck of G. The *vertex-switching reconstruction problem* asks whether G is uniquely defined up to isomorphism by $D_s(G)$. Stanley [17] pointed out that the vertex-switching reconstruction problem has a negative answer in general, as illustrated by the following simple example. Let G be the totally disconnected graph $4K_1$ on four vertices, respectively, the cycle of length four, C_4. Then, in both cases, $D_1(G)$ consists of the star $K_{1,3}$ only (see Fig. 11.1, where we switched at the left-upper vertex of the graph).

On the other hand, it is natural to ask which conditions have to be imposed on the underlying graphs in order to solve the reconstruction problem. Many special graphs have been investigated and several bounds on the degree of reconstructible graphs have been shown (cf. [4–7, 12, 13]). A major result in this area has been obtained by Krasikov and Roditty [13, Remark 2]. They proved an analogue of Kelly's lemma to reconstruct the number of subgraphs in a graph. To state the result, some more notation is needed. Given graphs G and H, let $X_s(G \to H)$ denote the number of sets $U \subset V(G)$, $|U| = s$, such that G_U is isomorphic to H. Furthermore, let A_s^n denote the matrix with rows and columns indexed by the unlabeled graphs on n vertices, with the (G, H) entry being $X_s(G \to H)$. Denote by

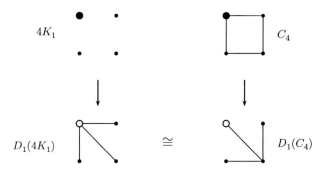

Fig. 11.1 Vertex-reconstruction for $4K_1$ and C_4

11 Reconstruction Problems

$$P_k^n(x) = \sum_{j=0}^{k} (-1)^j \binom{x}{j} \binom{n-x}{k-j} \tag{11.1}$$

the binary Krawtchouk polynomial of degree k (for more details see Sect. 11.2).

Theorem 1 ([13]). *The s-switching deck $D_s(G)$ of G determines the number of induced subgraphs of G isomorphic to a given m-vertex graph provided no eigenvalue of A_1^m is a root y of*

$$R_s^m(y) = \sum_{k=\max(0,s+m-n)}^{\min(m,s)} \binom{n-m}{s-k} P_k^n((m-y)/2).$$

Ellingham [4] used an idea about m-cubes to simplify the result, thus directly relating the reconstruction to the existence of integral roots of Krawtchouk polynomials. Recall that G has n vertices.

Theorem 2 ([4]). *The s-switching deck $D_s(G)$ of G determines the number of induced subgraphs of G isomorphic to a given m-vertex graph provided $P_s^n(x)$ has no even root in $[0, m]$.*

Several other reconstruction problems relate to integer roots of Krawtchouk polynomials [11]. Mention, for example, the *reorientation reconstruction problem*, which refers to a reconstruction problem for directed graphs. We are given a directed graph Γ with $E(\Gamma) = m$. For any $A \subset E$ denote by Γ_A the graph obtained by flipping the orientation of all arcs in A. Similarly as before, define the s-reorientation deck $D_s(\Gamma) = \{\Gamma_A : |A| = s\}$. The reorientation reconstruction problem asks whether Γ is uniquely defined up to isomorphism by $D_s(\Gamma)$. The following connection holds [11]:

Theorem 3 ([10]). *If $P_s^m(x)$ has no integer root then Γ can be reconstructed.*

A similar connection holds for the *sign reconstruction problem*.

11.1.3 Outline of Chapter

In the present chapter we study integral roots of Krawtchouk polynomials from a Diophantine point of view and prove our main result (see Sect. 11.1.4). The chapter is organized as follows. In Sect. 11.2 we recall several well-known facts on Krawtchouk polynomials, which we use in the sequel. Section 11.3 is devoted to a short account on Diophantine equations of the type $f(x) = g(y)$, where $f, g \in \mathbb{Q}[x]$. Most important, we present the algorithmic criterion for finiteness of solutions of Bilu and Tichy [1]. In Sect. 11.4 we recall the discrete Laguerre inequality, a striking result by Krasikov [9]. After presenting the connection of monotonicity of stationary points and indecomposability of polynomials (Sect. 11.5), we use

Krasikov's result for the stationary points of Krawtchouk polynomials (Sect. 11.6) to decompose these polynomials in Sect. 11.7. In the final section, Sect. 11.8, we treat the remaining possibilities for decomposing the polynomials with the standard pairs. The exposition ends with a short summary and perspectives for future work.

11.1.4 Main Result

Our main result is the following (for the exact notion we refer to Sect. 11.3).

Theorem 4. *Let $g(x) \in \mathbb{Q}[x]$ with $\deg g \geq 3$ and assume that $n, k \in \mathbb{Z}^+$ with*

$$16 \leq n \leq 100, \qquad \theta(n) \leq k \leq \theta(n) + 10,$$

where

$$\theta(n) = \max\left(7, \left\lceil \frac{17}{40} n - \frac{19}{2} \right\rceil\right).$$

Suppose that the Diophantine equation

$$P_k^n(x) = g(y) \tag{11.2}$$

with Krawtchouk polynomials $P_k^n(x)$ has infinitely many rational solutions (x, y) with a bounded denominator. Then we are in one of the following cases.

(i) $g(x) = P_k^n(\tilde{g}(x))$ for some polynomial $\tilde{g} \in \mathbb{Q}[x]$.

(ii) $k = 2k'$, $k' \geq 2$ and $g(x) = \phi(\tilde{g}(x))$, where $\tilde{g}$ is a polynomial over $\mathbb{Q}$, whose square-free part has at most two zeroes, such that $\tilde{g}$ takes infinitely many square values in $\mathbb{Z}$.

11.2 Krawtchouk Polynomials

11.2.1 Basic Facts

The *Krawtchouk polynomials* $P_k^{(p,n)}(x)$ resp. $P_k^n(x)$ are often found while studying combinatorial problems where some sort of involution on the underlying structure takes place. This is well explained by the generating function,

$$\sum_{k=0}^{\infty} P_k^{(p,n)}(x) z^k = \left(1 - \frac{1-p}{p} z\right)^x (1+z)^{n-x}. \tag{11.3}$$

According to the Askey-scheme [14] (see also [22, pp. 35/36]), the (general) Krawtchouk polynomials $P_k^{(p,n)}(x)$ form a family of polynomials which are orthogonal with respect to the discrete measure μ defined by $\mu(i) = \binom{n}{i} p^i (1-p)^{n-i}$,

11 Reconstruction Problems   297

$i = 0, \ldots, n$ with $0 < p < 1$. The special case $p = 1/2$ yields the standard *binary* Krawtchouk polynomials, which – for the sake of brevity – we denote by $P_k^n(x) = P_k^{(1/2,n)}(x)$. From (11.1) it is easy to derive that

$$P_k^n(x) = \sum_{j=0}^{k}(-2)^j \binom{x}{j}\binom{n-j}{k-j},\tag{11.4}$$

from which again the uppermost coefficients of

$$P_k^n(x) = c_k x^k + c_{k-1} x^{k-1} + c_{k-2} x^{k-2} + \cdots + c_0$$

follow at once,

$$c_k = \frac{(-2)^k}{k!}, \quad c_{k-1} = \frac{(-2)^{k-1}n}{(k-1)!}, \quad c_{k-2} = \frac{(-2)^{k-2}}{6(k-2)!}(3n^2 - 3n + 2k - 4),$$

$$\tag{11.5}$$

$$c_{k-3} = \frac{(-2)^{k-3}n}{6(k-3)!}(n^2 - 3n + 2k - 4),$$

$$c_{k-4} = \frac{(-2)^{k-2}}{360(k-4)!}(20k^2 - 108k - 60kn + 60kn^2 + 150n$$
$$- 90n^3 + 15n^4 - 75n^2 + 112),$$

$$c_{k-5} = \frac{(-2)^{k-2}n}{360(k-5)!}(20k^2 - 108k - 60kn + 20kn^2 + 150n$$
$$+ 5n^2 + 112 + 3n^4 - 30n^3),\ \text{etc.}$$

We also recall the three-term recurrence relation

$$(k+1)P_{k+1}^n(x) = (n-2x)P_k^n(x) - (n-k+1)P_{k-1}^n(x), \qquad k \geq 1,\quad(11.6)$$

and the difference equation

$$(n-x)P_k^n(x+1) = (n-2k)P_k^n(x) - xP_k^n(x-1), \qquad k \geq 0,\qquad(11.7)$$

which is especially important for the method presented in this chapter. Another useful recurrence relation is [10, relation (7)],

$$(n-k+1)P_k^{n+1}(x) = (3n-2k-2x+1)P_k^n(x) - 2(n-x)P_k^{n-1}(x).\ (11.8)$$

11.2.2 Zeroes and Upper Bounds

As for a detailed study of the zeroes of $P_k^n(x)$ (such as interlacing properties, bounds, etc.), we refer to [10, 11]. Here we briefly recall some well-known facts, which are crucial for our discussion. One easily notes that

$$P_k^n(n/2) = \begin{cases} (-1)^{k/2} \binom{n/2}{k/2}, & n \text{ even}; \\ 0, & n \text{ odd}. \end{cases} \tag{11.9}$$

The Krawtchouk polynomial $P_k^n(x)$ has k simple roots

$$0 < r_{1,n}(k) < r_{2,n}(k) < \cdots < r_{k,n}(k) < n. \tag{11.10}$$

Since

$$P_k^n(x) = (-1)^k P_k^n(n - x), \tag{11.11}$$

they lie symmetric around the point $x = n/2$. Moreover, for $k < n/2$ the distance between consecutive zeroes decreases towards $n/2$. Also, recall that for $1 \le k < n/2$ we have

$$r_{i+1,n}(k) - r_{i,n}(k) > 2, \tag{11.12}$$

and for $k < n$ we have $r_{i+1,n}(k) - r_{i,n}(k) > 1$. Levenshtein [15] proved the following explicit formula for the smallest root,

$$r_{1,n}(k) = n/2 - \max \left(\sum_{i=0}^{k-2} x_i x_{i+1} \sqrt{(i + 1)(n - i)} \right), \tag{11.13}$$

where the maximum is taken over all $(x_1, \dots, x_n)$ with $\sum_{i=0}^{k-1} x_i^2 = 1$. It is not difficult to see that

$$r_{1,n}(k) > 1. \tag{11.14}$$

It is well known that the zeroes of Krawtchouk polynomials for small k can be approximated by the corresponding roots of the Hermite polynomials. If $(n - k) \to \infty$ then the zeroes of $P_k^n(x)$ indeed approach

$$\frac{n}{2} + \frac{\sqrt{n - k - 1}}{2} h_i(x), \tag{11.15}$$

where $h_1(k) < \cdots < h_k(k)$ are the roots of the Hermite polynomial $H_k(x)$. Finally, we also mention a result due to Krasikov [8] which gives a bound of $P_k^n(x)$ at integer values provided $k \le n/2$. Let $q = 2\sqrt{k(n - k)}$; then it holds that

$$(P_k^n(x))^2 \le \frac{x!(n - x)!}{\lfloor \frac{k}{2} \rfloor!^2 \lfloor \frac{n-k}{2} \rfloor!^2} \tau(n, k, x), \qquad x = 0, 1, \dots, \lfloor n/2 \rfloor, \tag{11.16}$$

11 Reconstruction Problems 299

where $\tau(n,k,x)$ is

$$\frac{q^2+2n}{4(n-x)}, \quad n,k \text{ even}; \qquad \frac{4}{n-x}, \quad n \text{ even}, k \text{ odd};$$

$$\frac{2k+1}{n-x}, \quad n \text{ odd}, k \text{ even}; \qquad \frac{2n-2k+1}{n-x}, \quad n,k \text{ odd}.$$

In the vicinity of $n/2$ there are better estimates available [8].

11.3 The Diophantine Equation $P_k^n(x) = g(y)$

11.3.1 Introduction

The integrality of zeroes of Krawtchouk polynomials relates to the study of the solution set of the one-variable Diophantine equation

$$P_k^n(x) = 0 \tag{11.17}$$

in rational integers x. Much interest has been focused on classifying the zeroes for certain values of k and n (see [11]). For instance, the zeroes are completely classified for $k \le 7$, for $k = (n-t)/2$ with $t \le 6$ and $t = 8$ when the root is odd. It is conjectured that for any choice of the pair (k,n) the number of integral zeroes does not exceed 4. On the other hand, there are also results of a typical Diophantine nature. For example, for every $k \ge 4$, the polynomial $P_k^n(x)$ can have nontrivial integer roots only for finitely many values n.

An interesting generalization is to allow an arbitrary rational polynomial $g(y)$ on the right-hand side of (11.17),

$$P_k^n(x) = g(y), \tag{11.18}$$

which makes up the hub of the present chapter. How many integral solutions (x,y) does (11.18) have? Is it possible to find an infinite set of solutions which can be constructed via a suitable integer-valued parametrization?

The study of Diophantine equations of the shape $f(x) = g(y)$ has a long history. In order to settle the problem of finiteness of integral solutions (x,y) for a specific equation (i.e., without parameters involved), one can resort to Siegel's theorem on integral points on algebraic curves [16]. The procedure is as follows: First, one computes the genus of the algebraic curve under question, and in the case of zero genus one calculates the number of points at infinity to conclude. If the polynomials f and g themselves depend on several parameters (e.g., on k and n in (11.18)), such a direct calculation is not possible. In 2000, Bilu and Tichy [1], while extending work of Davenport, Ehrenfeucht, Fried, Lewis, MacRae, Ritt, Schinzel, Siegel, and others, proved an algorithmic criterion which makes it possible to apply Siegel's theorem also in the multiparametric case.

11.3.2 The Criterion of Bilu and Tichy

In order to formulate the criterion we need the definition of the five so-called *standard pairs* (over $\mathbb{Q}$). In what follows, let $\gamma, \delta \in \mathbb{Q} \setminus \{0\}$, $q, s, t \in \mathbb{Z}_{>0}$, $r \in \mathbb{Z}_{\geq 0}$ and $v(x) \in \mathbb{Q}[x]$ a nonzero polynomial (which may be constant). We also make use of the *Dickson polynomials* which can be defined by

$$D_s(x, \gamma) = \sum_{i=0}^{\lfloor s/2 \rfloor} d_{s,i} x^{s-2i} \quad \text{with} \quad d_{s,i} = \frac{s}{s-i} \binom{s-i}{i} (-\gamma)^i. \tag{11.19}$$

We say that the equation $f(x) = g(y)$ has *infinitely many rational solutions with a bounded denominator*, if there is $v \in \mathbb{Z}^+$ such that $f(x) = g(y)$ has infinitely many rational solutions (x, y) with $vx, vy \in \mathbb{Z}$. If an equation has only finitely many rational solutions with a bounded denominator then, in particular, it has only finitely many solutions in integers.

The list of *standard pairs* (over $\mathbb{Q}$), which is referred to in Theorem 5, includes five different pairs of polynomials (f_1, g_1).

A standard pair of the *first* kind is of the type

$$(x^q, \gamma x^r v(x)^q) \tag{11.20}$$

(or switched), where $0 \leq r < q$, $\gcd(r, q) = 1$, and $r + \deg v > 0$.

A standard pair of the *second* kind is given by

$$(x^2, (\gamma x^2 + \delta) v(x)^2) \tag{11.21}$$

(or switched).

A standard pair of the *third* kind is

$$(D_s(x, \gamma^t), D_t(x, \gamma^s)) \tag{11.22}$$

with $s, t \geq 1$ and $\gcd(s, t) = 1$.

A standard pair of the *fourth* kind is

$$(\gamma^{-s/2} D_s(x, \gamma), -\delta^{-t/2} D_t(x, \delta)) \tag{11.23}$$

(or switched) with $s, t \geq 1$ and $\gcd(s, t) = 2$.

11 Reconstruction Problems

A standard pair of the *fifth* kind is of the form

$$((\gamma x^2 - 1)^3, 3x^4 - 4x^3) \tag{11.24}$$

(or switched).

We are now ready to state the criterion of Bilu and Tichy [1].

Theorem 5 ([1]). *Let* $f(x), g(x) \in \mathbb{Q}[x]$ *be nonconstant polynomials. Then the following two assertions are equivalent.*

 (i) *The equation* $f(x) = g(y)$ *has infinitely many rational solutions with a bounded denominator.*
 (ii) *We can express* $f \circ \kappa_1 = \phi \circ f_1$ *and* $g \circ \kappa_2 = \phi \circ g_1$ *where* $\kappa_1, \kappa_2 \in \mathbb{Q}[x]$ *are linear,* $\phi(x) \in \mathbb{Q}[x]$*, and* (f_1, g_1) *is a standard pair over* $\mathbb{Q}$*.*

Observe that if we were able to get a contradiction for decompositions of f and g as demanded in (i) of Theorem 5, then finiteness of the number of integral solutions (x, y) of the original Diophantine equation $f(x) = g(y)$ is guaranteed.

The proof of Theorem 5 relies on the celebrated ineffective theorem of Siegel [16] from 1929 on the finiteness of the number of integer solutions of the equation $F(x, y) = 0$, where $F(x, y)$ is absolutely irreducible. In fact, this number is finite except when the projective completion of the curve has genus 0 and at most 2 points at infinity. Thus, in principle, one splits $f(x) - g(y)$ into irreducible factors in $\mathbb{Q}[x, y]$, and for each factor that is irreducible over $\bar{\mathbb{Q}}$ one determines the genus and the number of points of infinity. A number of people have studied the irreducibility of $f(x) - g(y)$, for instance, Ehrenfeucht, Fried, Lewis, MacRae, Runge, and Schinzel. The main contribution of Bilu and Tichy was to drop a condition on the gcd of f and g, so as to obtain the full general result.

Tichy and Stoll [21] used the special form of the leading coefficient c_k in (11.5) and Theorem 5 to prove

Theorem 6 ([21]). *Let* n *and* m *be distinct integers satisfying* $m, n \geq 3$. *Further, let* $N \geq \max(m, n)$ *and* $p_1, p_2 \in \mathbb{Q} \setminus \{0, 1\}$*. Then the equation*

$$\binom{N}{m} P_n^{(p_1, N)}(x) = \binom{N}{n} P_m^{(p_2, N)}(y) \tag{11.25}$$

has only finitely many solutions in integers (x, y)*.*

Despite the generality of Theorem 6, which addresses general Krawtchouk polynomials $P_k^{(p,n)}(x)$ (recall (11.3)), it is not possible to extend the proof to remove the binomial coefficient factors in (11.25). The aim of the present chapter is to outline a method which uses an ingenious tool from the geometry of polynomials to get a finiteness result of the same shape for (11.18).

11.4 The Discrete Laguerre Inequality

11.4.1 Introduction and Statement

In order to apply Theorem 5 in the most general form for the binary Krawtchouk polynomials one has to prove a general decomposition theorem for $P_k^n(x)$ and to exclude possible decompositions involving the standard pairs. Although this is rather straightforward for the classical continuous orthogonal polynomials (Laguerre, Hermite, Jacobi) [20], it has not even been proved for a single family of discrete classical orthogonal polynomials (Krawtchouk, Meixner, Meixner-Pollaczek, Hahn, Wilson, Charlier, etc.). At least, due to the similarity to Hermite polynomials (11.15), one may strongly expect an analogous result for Krawtchouk polynomials. We use here a method due to Krasikov [9] to get a first result in this direction. We do not aim to optimize our argument; indeed, in the end, we use concrete numerical data in place of the general parameters k and n. However, with more technical effort it is possible to enlarge the parameter sets in our main theorem and to get a statement for polynomials $P_k^n(x)$ with $k = k(n)$ as well.

The classical Laguerre inequality states that for any polynomial $f \in \mathbb{R}[x]$ with only real zeroes there holds $f'^2 - ff'' \geq 0$. A higher-degree generalization has been obtained by Jensen and used by Patrick (see [9] for the references), namely,

$$L_m(f) = \sum_{j=-m}^{m} (-1)^{m+j} \frac{f^{(m-j)}(x) f^{(m+j)}(x)}{(m-j)!(m+j)!} \geq 0. \tag{11.26}$$

In 2003, Krasikov [9] showed a surprising difference analogue of (11.26). Let $x_1 < x_2 < \cdots < x_n$ be the zeroes of $f(x)$ and denote by $M(f)$ the mesh defined by $M(f) = \min_{2 \leq i \leq n}(x_i - x_{i-1})$.

Theorem 7 ([9]). Let $M(f) \geq \sqrt{4 - \frac{6}{m+2}}$, then

$$V_m(f) = \sum_{j=-m}^{m} (-1)^j \frac{f(x-j)f(x+j)}{(m-j)!(m+j)!} \geq 0. \tag{11.27}$$

The proof of Theorem 7 is elementary. The crucial step is to show that $V_m(f)$ satisfies some recursion relation and is a quadratic polynomial in x with nonpositive discriminant provided f satisfies the mesh condition given in the statement.

Relation (11.27) can be used to get explicit inequalities on the size of polynomials, respectively, to bound the extreme zeroes.

11.4.2 Krasikov's Application to Krawtchouk Polynomials

A nice application to Krawtchouk polynomials has been outlined in [9]. Therein, Theorem 7 is used with $m = 2$ to get very sharp envelopes for $P_k^n(x)$ with $k < n/2$.

11 Reconstruction Problems

As we need these numerical data in our investigations, we recall the method and the calculations from [9] (we also fix a misprint in (11.29)).

By the difference relation (11.7) it is possible to write $V_2(P_k^n)$ only in terms of $P_k^n(x)$ and $P_k^n(x-1)$. Moreover, by (11.12) the mesh condition of Theorem 7 is satisfied. This gives

$$V_2(P_k^n) = \frac{A(x)t^2 + B(x)t + C(x)}{12(n-x)(n-x-1)(x-1)} (P_k^n(x))^2 \geq 0, \tag{11.28}$$

where $t = t(x) = P_k^n(x-1)/P_k^n(x)$ and

$$\begin{aligned}
A(x) &= -x(4x^2 - 4nx + 4n + m^2 - 4), \\
B(x) &= m(4x^2 - 4nx + 2x + 3n + m^2 - 4), \\
C(x) &= 4x^3 - 8nx^2 + (4n^2 + 2n + m^2 - 4)x - 2n^2 - m^2n + 4n - m^2,
\end{aligned}$$

with $m = n - 2k$. Note that by (11.10) and (11.14) the denominator in (11.28) is positive. Having at hand (11.28), it is possible to derive bounds on $P_k^n(x)$ and $P_k^n(x-1)$ inside the oscillatory region. This is obtained by looking at the ellipse described by V_2. Define

$$\begin{aligned}
W(x) &= \frac{V_2(P_k^n(x+1)) - zV_2(P_k^n(x))}{(P_k^n(x))^2} \\
&= \frac{x\alpha(x)t^2 - m\beta(x)t - (n-x-2)\gamma(x)}{12(n-x)^2(n-x-2)(n-x-1)(x-1)}, \tag{11.29}
\end{aligned}$$

where

$$\begin{aligned}
\alpha(x) &= (n-x-2)(n-x)(4x^2 - 4xn + m^2 + 4n - 4)z \\
&\quad + (x-1)(4x^3 + (12-8n)x^2 + (4n^2 - 14n + m^2 + 8)x \\
&\quad - n(m^2 - 2n + 2)), \\
\beta(x) &= (n-x-2)(n-x)(m^2 + 3n - 4xn + 2x + 4x^2 - 4)z \\
&\quad + (x-1)(4x^3 + (14-8n)x^2 + (m^2 - 17n + 4n^2 + 14)x \\
&\quad - n(m^2 + 2 - 3n)), \\
\gamma(x) &= (n-x)(4x^3 - 8x^2n + (m^2 + 4n^2 + 2n - 4)x + 4n \\
&\quad - nm^2 - m^2 - 2n^2)z - (x-1)(4x^3 + (4-8n)x^2 \\
&\quad + (m^2 - 8n + 4n^2)x + n^2 - 12k^2 - nm^2 + 12nk).
\end{aligned}$$

Note that the discriminant $m^2\beta(x)^2 + 4x(n-x-2)\alpha(x)\gamma(x)$ can be interpreted as a quadratic polynomial in z. We choose z from setting the discriminant equal to zero, in which case the signs of $W(x)$ and $\alpha(x)$ coincide. This yields,

$$z_{1,2}(x) = \frac{(x-1)(\triangle(x+\frac{1}{2}) - 3S \pm 6\sqrt{R})}{(n-x-2)\triangle(x)},$$

where (with the abbreviation $y = n - 2x$),

$$\begin{aligned}
\triangle(x) &= (y^2 - (n-1)^2 + m^2 - 1)^3 - 2(y^2 + m^2 - 1)^2 + m^2 y^2 \\
&\quad + 2(n-1)^2(n^2 - 2n + 5), \\
S &= y(y-2)(y-1)^2 - (n^2 - 2n - m^2 + 2)^2 + 7(n-1)^2, \\
R &= (n^2 - y^2 - 2n + 2y)(n^2 - m^2)((n-2)^2 - m^2)((n-1)^2 \\
&\quad - m^2 - (y-1)^2).
\end{aligned}$$

Recall that $\triangle(x) < 0$ in the oscillatory region [8], provided that

$$2 \leq k < \frac{n}{2} - 2 \cdot 3^{-3/4} \sqrt{n}. \tag{11.30}$$

Within this range we therefore have

$$z_1(x) \leq \frac{V_2(P_k^n(x+1))}{V_2(P_k^n(x))} \leq z_2(x). \tag{11.31}$$

As Krasikov points out, one can use $V_2(P_k^n(n/2))$ as an initial value in (11.31) to obtain upper bounds for $P_k^n(n/2 + i)$, $i \geq 1$, consecutively. For our purpose, we need explicit *upper and lower bounds for the maximum of* $P_k^n(x)$ between consecutive zeroes, i.e., for *real* x in the interval $[x_{i-1}, x_i]$. This is motivated by the connection of monotonicity of stationary points to decomposability of polynomials, which is the subject of the next section.

11.5 Monotonicity of Stationary Points and Indecomposability

11.5.1 Definitions

Polynomial decomposition theory is aimed at a characterization of all representations of a given polynomial $f = \phi \circ h \in \mathbb{R}[x]$, where $\phi, h \in \mathbb{R}[x]$, $\min(\deg \phi, \deg h) \geq 2$ and "$\circ$" denotes the functional composition applied for polynomials.[1] The left term ϕ is called the *left* and the right term h the *right component* of the decomposition. Two decompositions $f = \phi_1 \circ h_1 = \phi_2 \circ h_2$ are called

---

[1] More precisely, such a decomposition is called a *nontrivial decomposition*.

11 Reconstruction Problems

equivalent (and thus regarded as basically the same), if there is a linear polynomial κ such that $\phi_2 = \phi_1 \circ \kappa$ and $h_2 = \kappa^{-1} \circ h_1$. A polynomial f is called *decomposable* (over $\mathbb{R}$) if it has at least one nontrivial decomposition with real components.

11.5.2 Decomposition and Orthogonal Polynomials

Orthogonal polynomials – besides having simple real zeroes – have simple stationary points. A main theme, for instance in approximation theory, is to prove a monotonicity result for the extremal points of the polynomials under question. Denote by

$$\delta(f;\gamma) = \deg \gcd(f - \gamma, f'), \quad \gamma \in \mathbb{R},$$

which counts the number of stationary points of $f(x)$ with equal ordinate value. An important connection to polynomial decomposition theory is given by the following fact [3].

Lemma 1 ([3]). *Let $f = \phi \circ h$, where $\phi, h \in \mathbb{R}[x]$. If $\deg \phi \geq 2$, then there exists $\gamma \in \mathbb{R}$ with $\delta(f;\gamma) \geq \deg h$. In particular, if $\delta(f;\gamma) \leq s$ for all $\gamma \in \mathbb{R}$ then $\deg h \leq s$.*

According to Lemma 1 we have $\deg h \leq s \in \mathbb{Z}_{>0}$ provided that there are at most s intervals for which the stationary points of $f(x)$ are monotone increasing/decreasing on the respective intervals. In that context we recall a result due to Tichy and the author [20].

Theorem 8 ([20]). *Let $f(x) \in \mathbb{R}[x]$ with only real zeroes satisfy*

$$\sigma(x) f''(x) + \tau(x) f'(x) - \lambda(x) f(x) = 0, \tag{11.32}$$

with $\sigma(x) = ax^2 + bx + c$, $\tau(x) = dx + e$ and $a, b, c, d, e \in \mathbb{R}$, $ad \neq 0$. Furthermore, suppose that $\sigma'(x) - 2\tau(x)$ does not vanish identically. Then $\delta(f;\gamma) \leq 2$ for all $\gamma \in \mathbb{R}$.

The general continuous classical orthogonal polynomials (Laguerre, Jacobi, Hermite) satisfy (11.32), whereas the Chebyshev polynomials exactly make up the exceptional case of Theorem 8.[2] However, for Krawtchouk polynomials there is no differential equation of Sturm–Liouville type available, such that one has to use another method.

---

[2] In fact, it is well known that the standard (*nonmonic*) Chebyshev polynomials of the first kind $T_k(x)$ have all stationary points of equal ordinate value. Moreover, they are decomposable for any nonprime k by the relation $T_m(T_n(x)) = T_n(T_m(x)) = T_{mn}(x)$.

11.6 Stationary Points of Krawtchouk Polynomials

11.6.1 Iteration of Krasikov's Bound

The present section is devoted to a detailed study of relation (11.31) which delivers the needed information to bound $\delta(P_k^n; \gamma)$ for all $\gamma \in \mathbb{R}$. To start with, iterating (11.31) yields

$$\rho_1(x) := V_2 \left(P_k^n \left(\left\{ x - \frac{n}{2} \right\} + \frac{n}{2} \right) \right) \prod_{i=0}^{\lfloor x \rfloor - \frac{n}{2} - 1} z_1 \left(\left\{ x - \frac{n}{2} \right\} + \frac{n}{2} + i \right) \qquad (11.33)$$

$$\leq V_2(P_k^n(x))$$

$$\leq V_2 \left(P_k^n \left(\left\{ x - \frac{n}{2} \right\} + \frac{n}{2} \right) \right) \prod_{i=0}^{\lfloor x \rfloor - \frac{n}{2} - 1} z_2 \left(\left\{ x - \frac{n}{2} \right\} + \frac{n}{2} + i \right) =: \rho_2(x).$$

Herein, $\{x\} = x - \lfloor x \rfloor$ denotes the fractional part of x. It may be possible to relax (11.33) in order to improve on our results, but only at the cost of extensive computational work. In fact, although z_1, z_2 are monotone increasing functions in the oscillatory region, this behaviour changes near the extreme zeroes and one has to use more tricky arguments (see [7]). Moreover, it is a rather (computationally) complex task to prove that $V_2(P_k^n(x))$ takes its minimal, resp. maximal, value on $[n/2, n/2+1]$ at the left, resp. right, point of the interval (one may use (11.8), (11.9), and (11.28)).

Now, consider (11.28) and the ellipse with

$$A(x)t^2 + B(x)t + C(x) = \text{const.} \qquad (11.34)$$

The upper bound for $\max_{x_i < x < x_{i+1}} P_k^n(x)$ follows by calculating the major axis of (11.34). This gives (we omit the details)

$$\lambda(x) = \frac{A(x) + C(x)}{2} - \frac{1}{2} \sqrt{A(x)^2 - 2A(x)C(x) + C(x)^2 + B(x)^2}$$

and

$$P_k^n(x) \leq \sqrt{\frac{\rho_2(x)}{\lambda(x)}} =: u(x), \qquad n/2 \leq x \leq x_n. \qquad (11.35)$$

On the other hand, considering the minor axis yields that for all $i = 1, \ldots, n$ there exists $x \in [x_{i-1}, x_i]$ such that

$$P_k^n(x) \geq \sqrt{\frac{\rho_1(x)}{C(x)}} := l(x). \qquad (11.36)$$

11   Reconstruction Problems

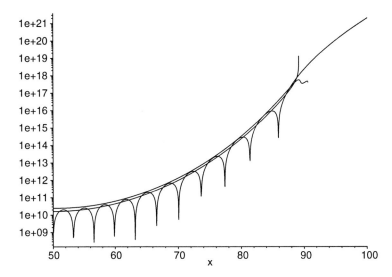

Fig. 11.2 $|P_{21}^{100}(x)|$ on a logarithmic scale

We illustrate these two bounds in Fig. 11.2 for the case $n = 100, k = 21$.

Obviously, comparing the upper and lower bound it is possible to get a bound for the number of stationary points of equal ordinate value.

11.6.2 Admissible Parameter Ranges

We use a very rough criterion to conclude, namely (motivated by (11.12)), if

$$\min\{1 \leq j \leq n/2 : \quad l(n/2 + 2j) > u(n/2 + j)\} = s \qquad (11.37)$$

then $\delta(P_k^n; \gamma) \leq 2s$. For every $n \leq 100$ we have calculated the values for k subject to (11.30) which satisfy (11.37) with $s \leq 3$. The data are illustrated in Fig. 11.3.[3] From the plot we see that the bounds are most helpful in the vicinity of the bound in (11.30), which is the upper envelope of the represented points.

According to Lemma 1, for the values (n, k) given in Fig. 11.3 (which we call *admissible* in the sequel) we have that $P_n^k(x) = \phi(h(x))$ with $\phi, h \in \mathbb{R}[x]$ implies $\deg h \leq 6$. In the next section we deal with these possible decompositions by a recent method proposed by the author [19]. Observe that the set referred to in Theorem 4, i.e.,

---

[3] One may considerably improve these estimates for k odd, however, we aim for a more uniform result.

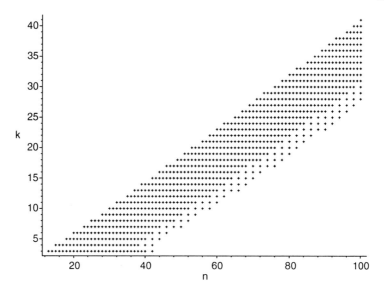

Fig. 11.3 Values of (n, k) with $P_n^k(x)$ having at most six stationary points of equal value

$$16 \leq n \leq 100, \qquad \theta(n) \leq k \leq \theta(n) + 10,$$

with

$$\theta(n) = \max\left(7, \left\lceil \frac{17}{40} n - \frac{19}{2} \right\rceil\right),$$

is a subset of the *admissible pairs* (n, k).

11.7 Decomposition of Krawtchouk Polynomials

11.7.1 An Indecomposability Criterion

Given a polynomial $f(x) \in \mathbb{R}[x]$, suppose that there is a decomposition of the form

$$f = \phi \circ h \tag{11.38}$$

with $\deg h = s$ being a small number (in our case ≤ 6). One way to disprove that there cannot exist such a decomposition consists in comparing coefficients on both sides of the decomposition equation (11.38). Since the uppermost coefficients of f (cf. 11.5) are given, one may try to come to a contradiction while equating with the parametric coefficients on the right-hand side of (11.38). An algorithmic, well-organized way of performing this task has recently been given by the author [18, 19]. We recall the main ingredients. First, a polynomial $\hat{h}$ of degree s is

11 Reconstruction Problems

computed which is the *only* (normed) candidate of degree s which could make up a right decomposition factor for f (see [19, Algorithm 1]). Using $\hat{h}$ we have at hand a convenient algorithmic criterion for impossibility of polynomial decomposition.

Lemma 2 (Stoll [18]). *Let f be monic and $s \geq 2$ a positive integer. Furthermore, let*

$$f(x) = \hat{h}(x)^k + \beta_1 \hat{h}(x)^{k-1} + \cdots + \beta_l \hat{h}(x)^{k-l} + \mathcal{R}(x), \qquad (11.39)$$

for some constants $\beta_j \in \mathbb{R}, 0 \leq l \leq k$ with $\deg \mathcal{R} \leq sk - s$ and $m \nmid \deg R$. Then f is indecomposable with right components of degree s.

From a practical point of view, Lemma 2 fits the problem best, when the degree of $\hat{h}$ is small. In fact, given $f(x)$, one expands $f(x)$ regarding $\hat{h}(x)$ up to sufficiently large order (indicated by l) such that the remainder polynomial $\mathcal{R}(x)$ has the wanted properties. Regarding the Krawtchouk polynomials with parameter constrictions given in Fig. 11.3, we have to come to a contradiction when considering right decomposition factors with $\deg h \leq 6$. In the sequel, we give an outline of these calculations. For more details on the computational aspects (addressing both the Gröbner bases and the implementation issues) we refer the interested reader to the article [19].

11.7.2 Application to Krawtchouk Polynomials

The main result which is proved in the remaining part of this section is the following.

Theorem 9. *Suppose $P_k^n(x) = \phi(h(x))$ with $\phi(x), h(x) \in \mathbb{R}[x]$ and $2 \leq \deg h \leq 6$. Then $\deg h = 2$ and the decomposition is equivalent to*

$$P_k^n(x) = \hat{\phi}(x^2 - nx) \qquad (11.40)$$

for some unique polynomial $\hat{\phi}(x) \in \mathbb{Q}[x]$.

To start with the proof, let $\deg h = 2$. By (11.11) we see that $P_{2k}^n(x) = P_{2k}^n(n - x)$ from which easily follows that there are unique polynomials $\phi_1(x)$, $\phi_2(x) \in \mathbb{Q}[x]$ with

$$P_{2k}^n(x) = \phi_1\left((x - n/2)^2\right) = \phi_2\left(x^2 - nx\right),$$

which is (11.40).

The only possible candidate of degree 3 (we use the first algorithm given in [19]) is

$$\hat{h}(x) = x^3 - \frac{3}{2}nx^2 + \left(\frac{9}{4}k^2 - \frac{9}{8}kn - \frac{9}{4}k + \frac{3}{4}n^2 + \frac{3}{8}n + \frac{1}{2}\right)x.$$

Taylor expansion with respect to this polynomial (use the second algorithm of [19]) leads to

$$P_{3k}^n(x) = \hat{h}(x)^k - \frac{1}{16}nk(18k^2 - 18k - 9nk + 4 + 2n^2 + 3n)\,\hat{h}(x)^{k-1}$$
$$- \frac{9}{540}k(3k-1)(k-1)(48k^2 - 56k - 30kn + 16 + 20n + 5n^2)\,x^{3k-4}$$
$$+ O(x^{3k-5}).$$

Lemma 2 implies that in the equation above we have $[x^{3k-4}] = 0$, giving

$$n = k - \frac{1}{2} + \frac{1}{6}\sqrt{-60k^2 + 108k - 39}.$$

The expression under the square root symbol is ≥ 0 if and only if

$$\frac{1}{2} \leq k \leq \frac{13}{10} < 2,$$

a contradiction. Thus, there cannot exist a decomposition of a Krawtchouk polynomial $P_k^n(x)$ with right component of degree three.

The calculations for $\deg h = 4$ are much more involved. First suppose that $k \geq 3$. We start with the expansion

$$P_{4k}^n(x) = \hat{h}(x)^k + \beta_1\hat{h}(x)^{k-1} + r_1 x^{4k-6} + r_2 x^{4k-7}$$
$$+ \beta_2\hat{h}(x)^{k-2} + r_3 x^{4k-9} + r_4 x^{4k-10} + O(x^{4k-11}),$$

which yields $r_1 = r_2 = r_3 = r_4 = 0$. The equations for r_1, r_2, r_3 are basically the same (see below), so that we need one further (independent) equation to conclude. The four equations are

$$k(k-1)(2k-1)(4k-1)(-63488k^3 + 102592k^2 + 46368k^2n - 52528k$$
$$- 51408kn - 11340kn^2 + 8192 + 13734n + 6615n^2 + 945n^3) = 0,$$

$$nk(k-1)(4k-1)(2k-1)(2k-3)(-63488k^3 + 102592k^2 + 46368k^2n - 52528k$$
$$- 51408kn - 11340kn^2 + 8192 + 13734n + 6615n^2 + 945n^3) = 0,$$

$$n^3k(k-1)(k-2)(4k-1)(2k-1)(2k-3)(4k-7)(-63488k^3 + 102592k^2$$
$$+ 46368k^2n - 52528k - 51408kn - 11340kn^2 + 8192 + 13734n$$
$$+ 6615n^2 + 945n^3) = 0,$$

$$k(k-1)(k-2)(4k-1)(2k-1)(2k-3)(-380764160 + 3731295872k - 855679440n$$
$$- 763247100n^2 - 436104900n^3 + 226600605n^4 + 656537805n^5$$
$$+ 337265775n^6 + 49116375n^7 + 4552320960kn^2 - 1791912540kn^4$$
$$- 8016988320k^2n^3 + 6200551104kn - 3051556200kn^5 + 2789111160kn^3$$

11 Reconstruction Problems

$$
\begin{aligned}
&- 908730900k\,n^6 - 49896000k\,n^7 - 10817991360k^2n^2 + 671101200k^2n^6 \\
&+ 12474000k^2n^7 - 16177793280k^2n - 9804833280k^4n^3 + 4726600560k^2n^4 \\
&+ 612057600k^4n^5 + 12445614720k^3n^3 - 149688000k^3n^6 - 12401326080k^4n^2 \\
&- 5779583040k^3n^4 + 4749267600k^2n^5 + 16998812160k^3n + 14100514560k^3n^2 \\
&- 2907273600k^3n^5 - 2681733120k^6n^2 + 2964234240k^5n^3 - 838041600k^5n^4 \\
&- 9931874304k^5n - 1421629440k^4n + 3492910080k^4n^4 + 7937740800k^5n^2 \\
&+ 8102150144k^6 - 3575644160k^7 + 24003993600k^3 - 18583019520k^4 \\
&+ 402751488k^5 - 13733508864k^2 + 5293178880k^6n) = 0.
\end{aligned}
$$

With the aid of the Gröbner-package in MAPLE we get the *complete* solution set

$$
\begin{aligned}
&\{\{k = -1/4, n = -2\}, \{k = 1/4, n = -3\}, \{k = -1, n = -5\}, \{k = 1/4, n = -6\} \\
&\{k = -1/2, n = -3\}, \{k = 1/4, n = n\}, \{k = 5/4, n = 2\}, \{k = 1/2, n = -4\}, \\
&\{k = 1/2, n = n\}, \{k = 0, n = n,\}, \{k = 1/4, n = 2/3\}, \{k = 1, n = n\}, \\
&\{k = 1/2, n = 2/3\}, \{k = 1, n = 2/3\}, \{k = 1/2, n = -8\}, \{k = 1, n = -6\}, \\
&\{k = 1/4, n = -8\}, \{k = 1/2, n = -2\}, \{k = 1/2, n = 2\}, \{k = -1/2, n = -6\}, \\
&\{k = 1/2, n = -3\}, \{k = 1/2, n = -6\}, \{k = 3/4, n = 2/3\}, \{k = -1, n = -8\}, \\
&\{k = -1, n = -6\}, \{k = 13/36, n = -50/27\}, \{k = 1/2, n = -5\}, \{k = -1/2, n = -4\}, \\
&\{k = 105057/2998036Z_2^2 + 1848969/2998036Z_2 + 1696531/2998036, n = 2\}, \\
&\{k = 3/2, n = Z_3/3\}, \{k = 2, n = Z_1\},\}.
\end{aligned}
$$

Herein, Z_1, Z_2, Z_3, respectively, satisfy the equations

$$
7Z_1^3 - 119Z_1^2 + 714Z_1 - 1440 = 0,
$$
$$
105057Z_2^3 + 316794Z_2^2 - 759285Z_2 + 193378 = 0,
$$
$$
Z_3^3 - 33Z_3^2 + 390Z_3 - 1544 = 0.
$$

No member in the solution set satisfies the integrality constraints for k and n. One easily comes to a contradiction also for $k = 2$ by inspecting the single equation $r_1 = 0$.

Next, assume $\deg h = 5$. We here get the expansion

$$
P_{5k}^n(x) = \hat{h}(x)^k + \beta_1 \hat{h}(x)^{k-1} + r_1 x^{5k-6} + r_2 x^{5k-7} + r_3 x^{5k-8} + O(x^{5k-9}),
$$

where

$$
\begin{aligned}
\beta_1 = -\frac{1}{2304}nk(&-5000k^4 - 3750nk^3 + 15000k^3 + 4125n^2k^2 - 1500nk^2 \\
&- 11000k^2 - 900n^3k - 1800n^2k + 1950nk + 3000k \\
&+ 180n^3 + 195n^2 - 300n - 272 + 72n^4) = 0.
\end{aligned}
$$

Obviously $r_1 = 0$. The equation $r_2 = 0$ does not yield any new information on the parameters k and n with respect to the first equation. We therefore also need $r_3 = 0$. More explicitly,

$$nk(k-1)(5k-1)(5k-6)(-740000k^4 + 409500k^3n + 1254000k^3$$
$$- 535500k^2n - 75600k^2n^2 - 774800k^2 + 229320kn + 68040kn^2$$
$$+ 4725kn^3 + 205440k - 19520 - 32256n - 15120n^2 - 2079n^3) = 0,$$

$$k(5k-1)(k-1)(11101440 - 164212480k + 22925952n + 6119568n^2$$
$$- 11716488n^3 - 7169715n^4 - 1047816n^5 + 377751000k^3n^2 - 452655000k^4n^3$$
$$- 2641350000k^5n - 3186633600k^3 - 22680000k^4n^4 - 850672500k^4n^2$$
$$+ 4046100000k^4n - 3086382000k^3n + 77962500k^3n^4 + 604894500k^3n^3$$
$$+ 1417500k^3n^5 - 373577400k^2n^3 + 1266640800k^2n + 5907912000k^4$$
$$- 267948480kn - 4309200k^2n^5 - 22427700k^2n^2 - 91868175k^2n^4$$
$$- 6278640000k^5 - 29378400kn^2 + 4003020kn^5 + 107764020kn^3$$
$$+ 43630650kn^4 + 3478800000k^6 - 740000000k^7 + 631500000k^6n$$
$$- 222000000k^6n^2 + 754950000k^5n^2 + 122850000k^5n^3 + 990888000k^2) = 0.$$

This system of equations has no admissible solution.[4]

Finally, consider $\deg h = 6$. We use the three coefficient equations $[x^{6k-7}] = [x^{6k-8}] = [x^{6k-9}] = 0$ to conclude that there is no admissible solution pair (n, k). For the sake of completeness, we append the three relevant equations,

$$k(k-1)(6k-1)(3k-1)(2k-1)(2598912k^4 - 4133376k^3 - 1626480k^3n$$
$$+ 2393664k^2 + 381780k^2n^2 + 1980000k^2n - 39900kn^3 - 317520kn^2$$
$$- 585696k - 781860kn + 49152 + 1575n^4 + 97960n + 64575n^2 + 17150n^3) = 0,$$

$$n(1077737253527224432k - 67223682337996800 - 75805048910774400k^2$$
$$- 8055435775759680k^4 + 30850177972149600k^3 + 1034250n^3k^2$$
$$+ 80325n^4k^2 - 175659840nk^9 - 1495000800nk^7 - 6300n^4k + 799372800nk^8$$
$$- 4309200n^3k^7 + 41232240n^2k^8 + 1505800800nk^6 - 171732960n^2k^7$$
$$+ 285064920n^2k^6 + 170100n^4k^6 - 567000n^4k^5 - 244981800n^2k^5$$
$$+ 16216200n^3k^6 + 1411897888929696k^5 + 13511581833600k^7$$
$$- 696037950720k^8 + 19669174272k^9 - 168416458660800k^6 - 23291100n^3k^5$$

---

[4] Again, we used MAPLE-V11 to perform the computations.

11 Reconstruction Problems

$$- 897154020nk^5 + 324647400nk^4 + 16294950n^3k^4 + 118297935n^2k^4$$

$$+ 675675n^4k^4 - 5876500n^3k^3 - 32185440n^2k^3 - 352800n^4k^3$$

$$- 69737900k^3n + 8123400k^2n + 4563405k^2n^2 - 258300kn^2 - 68600kn^3$$

$$- 391840kn) = 0,$$

$$k(k-1)(6k-1)(3k-1)(2k-1)(-19660800 + 314930688k - 45724800n$$

$$- 19203888n^2 + 23587740n^3 + 22753500n^4 + 6592740n^5 + 623700n^6$$

$$- 1651985280k^5n^2 + 1648896480k^4n^2 + 1197302040k^4n^3 - 10152980640k^4n$$

$$+ 75592440k^4n^4 - 537604848k^3n^2 + 7281972720k^3n - 7900200k^3n^5$$

$$- 267899940k^3n^4 - 1584615780k^3n^3 + 311850k^2n^6 + 945784290k^2n^3$$

$$+ 113532012kn^2 + 568854176kn - 147192045kn^4 - 2831232624k^2n$$

$$+ 318468150k^2n^4 - 119138184k^2n^2 + 25467750k^2n^5 - 883575kn^6$$

$$- 24759735kn^5 - 253449185kn^3 - 1802756736k^6n - 322043040k^5n^3$$

$$+ 7092131904k^5n + 514584576k^6n^2 + 6836900736k^3 - 13430568960k^4$$

$$- 9060470784k^6 + 15235057152k^5 - 2014594176k^2 + 2058338304k^7) = 0.$$

Observe that in principle the first and second equations are sufficient to conclude. However, the Gröbner calculations become much more efficient (and faster) if one includes an additional polynomial equation.

11.8 Decompositions with Standard Pairs

11.8.1 Introduction

Regarding Theorem 5, we have to treat decompositions of $P_k^n(x)$ involving the standard pairs given by (11.20)–(11.24). Recall that by Theorem 9 the only nontrivial decomposition of $P_k^n(x)$ is equivalent to $P_k^n(x) = \hat{\phi}(x^2 - nx)$ with $k \in 2\mathbb{Z}^+$, provided we assume the parameter restrictions for n and k given in Theorem 4.[5] To begin with, suppose that the Diophantine equation

$$P_k^n(x) = g(y)$$

has infinitely many rational solutions (x, y) with a bounded denominator. Then by Theorem 5,

$$P_k^n = \phi \circ f_1 \circ \kappa_1 \quad \text{and} \quad g = \phi \circ g_1 \circ \kappa_2,$$

---

[5] Therein, we assume $k \geq 7$. It is possible to consider the smaller values of k also, however, at the cost of some more case distinctions.

where κ_1, κ_2 are some linear polynomials, $\phi \in \mathbb{Q}[x]$, and (f_1, g_1) is a standard pair as given by the list in Sect. 11.3. By Theorem 9, we have one of the three cases:

(i) $\deg \phi = k$,
(ii) $\deg \phi = k'$ with $k = 2k'$ and $P_k^n(x) = \hat{\phi}(x^2 - nx)$,
(iii) $\deg \phi = 1$.

11.8.2 Case $\deg \phi = n$

By comparison of degrees, it holds that $P_k^n = \phi \circ \kappa$ for some linear polynomial $\kappa(x)$ and thus

$$g = P_k^n \circ (\kappa^{-1} \circ g_1 \circ \kappa_2) = P_k^n \circ \tilde{g}$$

for some nonconstant polynomial $\tilde{g} \in \mathbb{Q}[x]$. Obviously, there are infinitely many solutions with a bounded denominator of $P_k^n(x) = P_k^n(\tilde{g}(y))$. This gives Case (i) in Theorem 4.

11.8.3 Case $\deg \phi = k$ with $k = 2k'$ and $P_k^n = \hat{\phi}(x^2 - nx)$

Let $P_k^n = \phi \circ f_1 \circ \kappa_1$ and κ be the unique linear polynomial such that $\phi \circ \kappa = \hat{\phi}$. Then $P_k^n = (\phi \circ \kappa) \circ (\kappa^{-1} \circ f_1 \circ \kappa_1) = \hat{\phi} \circ l_1$ and Theorem 9 yields $l_1 = x^2 - nx$. On the other hand,

$$g = \phi \circ g_1 \circ \kappa_2 = (\phi \circ \kappa) \circ (\kappa^{-1} \circ g_1 \circ \kappa_2) = \hat{\phi} \circ l_2,$$

where $l_2 = \kappa^{-1} \circ g_1 \circ \kappa_2$. If the equation $(x - n/2)^2 = l_2(y) + n^2/4$ has infinitely many solutions with a bounded denominator, then by Siegel's theorem l_2 has at most two zeroes of odd multiplicity. This yields Case (ii).

11.8.4 Case $\deg \phi = 1$

In this case $\phi(x) = \phi_1 x + \phi_0$ with $\phi_1, \phi_0 \in \mathbb{Q}$. Since ϕ is a linear polynomial we have to treat $P_k^n = \phi \circ f_1 \circ \kappa_1$ and $g = \phi \circ g_1 \circ \kappa_2$, where (f_1, g_1) is a standard pair with $\deg f_1 = k$. We now have to analyze all decompositions with the special polynomials of the standard pairs.

First, recall the standard pair of the *second* kind $(x^2, (\gamma x^2 + \delta) v(x)^2)$ given in (11.21). Since both $k \geq 3$ and $\deg g \geq 3$, there cannot exist a decomposition involving (f_1, g_1) of the second kind. In the same manner we can exclude the standard pair of the *fifth* kind (11.24).

11 Reconstruction Problems   315

Next we want to exclude decompositions with the Dickson polynomials, namely, the standard pairs of the *third* (11.22) and *fourth* kind (11.23),

$$(f_1, g_1) = (D_s(x, \gamma^t), D_t(x, \gamma^s)).$$

Assume that $P_k^n \circ \kappa = \phi \circ D_s(x, \gamma^t)$ with a linear polynomial κ, or in other words,

$$P_k^n(x) = \phi_1 D_s(\alpha x + \beta, \gamma^t) + \phi_0. \tag{11.41}$$

In view of (11.19) here we have to cope with the six variables $k, n, \phi_1, \alpha, \beta$, and γ^t. It is again a straightforward (but involved) computation to come to a contradiction. In fact, for $k \geq 6$ we may write down six coefficient equations from (11.41) and conclude. Here we omit the details.

Finally, consider the standard pair of the *first* kind given by (11.20), namely $(x^q, \gamma x^r v(x)^q)$. The polynomial $(P_k^n(x))'$ has zeroes of multiplicity one. Hence, for $k \geq 7$, there cannot be a representation with $P_k^n(\alpha x + \beta) = \phi_1 x^q + \phi_0$. On the other hand, suppose that

$$P_k^n(x) = \hat{\phi}_1 (\beta_1 x + \beta_0)^r \hat{v}(x)^q + \phi_0, \tag{11.42}$$

where $\hat{\phi}_1 = \phi_1 \gamma$, $\hat{v}(x) = v(\beta_1 x + \beta_0)$ with $\beta_0, \beta_1 \in \mathbb{Q}$ and $0 \leq r < q$, $\gcd(r, q) = 1, r + \deg \hat{v} > 0$ as demanded in (11.20). Since $q \geq 3$ by $\deg g \geq 3$, we here again come to a contradiction by arguing in the same way as above.

This concludes the investigation with linear polynomials $\phi(x)$ and finishes the proof of Theorem 4.

11.9 Summary and Conclusion

In the present chapter we have outlined an analytic method to study the Diophantine equation

$$P_k^n(x) = g(y) \tag{11.43}$$

in integral variables x, y, where $P_k^n(x)$ denotes a binary Krawtchouk polynomial of degree $k \geq 7$ and $g \in \mathbb{Q}[x]$ is an arbitrary polynomial of degree ≥ 3. Within certain parameter ranges (informally speaking, k growing like $n/2$) we have shown that the Diophantine equation (11.43) only has finitely many integral solutions x, y (Theorem 4). This Diophantine equation is motivated by the close relationship between integrality of zeroes of Krawtchouk polynomials and the resolution of reconstruction problems in graphs (Sect. 11.3).

Our machinery ranges from a recent indecomposability criterion due to the author (Lemma 2) to the discrete Laguerre inequality (Theorem 7) applied to Krawtchouk polynomials, as obtained and outlined by Krasikov. The method used in this chapter describes a new approach in the theory of polynomial decomposition, and well fits the decomposition of discrete orthogonal polynomials. Also, the longstanding question, whether the stationary points of discrete orthogonal polynomials – or at least, a special family like the Krawtchouk polynomials – are convex, could be treated by this method. On the other hand, convexity results are well known for the continuous orthogonal polynomial families (Laguerre, Hermite, Jacobi), but it would be a major breakthrough to show such a result for the instance of a discrete family of polynomials.

The present chapter makes this attempt for certain ranges of the degree k and the parameter $n \leq 100$ in (11.43). With more computational work it seems possible to get a general parametric result, i.e., where the result holds uniformly for all $n \geq n_0$ and $k \in I_n$, where I_n denotes a set of consecutive integers depending on n.

Acknowledgments The author is a recipient of an APART-fellowship of the Austrian Academy of Sciences at the University of Waterloo, Canada. He also wishes to express his gratitude to I. Krasikov for several helpful discussions.

References

1. Bilu Y, Tichy RF (2000) The Diophantine equation $f(x) = g(y)$. Acta Arith 95:261–288
2. Bondy JA (1991) A graph reconstruction manual. In: Keedwell AD (ed) Surveys in combinatorics. LMS-Lecture Note Series, vol 166. Cambridge University Press, Cambridge, pp 221–252
3. Dujella A, Tichy RF (2001) Diophantine equations for second-order recursive sequences of polynomials. Q J Math 52:161–169
4. Ellingham MN (1996) Vertex-switching reconstruction and folded cubes. J Combin Theory B 66:361–364
5. Ellingham MN, Royle GF (1992) Vertex-switching reconstruction of subgraph numbers and triangle-free graphs. J Combin Theory B 54:167–177
6. Krasikov I (1994) Applications of balance equations to vertex switching reconstruction. J Graph Theory 18:217–225.
7. Krasikov I (1996) Degree conditions for vertex switching reconstruction. Discrete Math 160:273–278
8. Krasikov I (2001) Nonnegative quadratic forms and bounds on orthogonal polynomials. J Approx Theory 111:31–49
9. Krasikov I (2003) Discrete analogues of the Laguerre inequality. Anal Appl (Singap) 1:189–197
10. Krasikov I, Litsyn S (1996) On integral zeros of Krawtchouk polynomials. J Combin Theory A 74:71–99
11. Krasikov I, Litsyn S (2001) Survey of binary Krawtchouk polynomials, Codes and association schemes. (Piscataway, NJ, 1999), DIMACS Ser Discrete Math Theor Comput Sci 56:199–211. American Mathematical Society, Providence, RI
12. Krasikov I, Roditty Y (1992) Switching reconstruction and Diophantine equations. J Combin Theory B 54:189–195
13. Krasikov I, Roditty Y (1994) More on vertex-switching reconstruction. J Combin Theory B 60:40–55
14. Koekoek R, Swarttouw RF (1998) The Askey-Scheme of Hypergeometric Orthogonal Polynomials and its q-Analogue. Report 98-17, Delft, Netherlands
15. Levenshtein V (1995) Krawtchouk polynomials and universal bounds for codes and designs in Hamming spaces. IEEE Trans Inf Theory 41:1303–1321

11 Reconstruction Problems

16. Siegel CL (1929) Über einige Anwendungen Diophantischer Approximationen. Abh Preuss Akad Wiss Math Phys Kl 1:209–266
17. Stanley RP (1985) Reconstruction from vertex-switching. J Combin Theory B 38:132–138
18. Stoll T (2008) Complete decomposition of Dickson-type recursive polynomials and related Diophantine equations. J Number Theory 128:1157–1181
19. Stoll T (2008) Decomposition of perturbed Chebyshev polynomials. J Comput Appl Math 214:356–370
20. Stoll T, Tichy RF (2003) Diophantine equations for continuous classical orthogonal polynomials. Indagat Math 14:263–274
21. Stoll T, Tichy RF (2005) Diophantine equations involving general Meixner and Krawtchouk polynomials. Quaest Math 28:105–115
22. Szegő G (1975) Orthogonal polynomials, vol. 23, 4th edn. American Mathematical Society Colloquium Publications, Providence, RI

Chapter 12
Subgraphs as a Measure of Similarity

Josef Lauri

Abstract How similar can two graphs be? The ultimate positive answer to this question is, of course, when the two graphs are isomorphic. However, how much internal structure can two nonisomorphic graphs share? We show what the answer can look like if the measure of similarity between the two graphs is taken to be the number of isomorphic subgraphs which they share. We see how this notion is related to the internal symmetries of a graph and that therefore, for most graphs, their internal structure forces them to be very dissimilar to other graphs. We also indicate some attempts to find nonisomorphic graphs which are very similar in terms of the common subgraphs which they share. We also point out some issues of computational complexity and some possible applications associated with this measure of graph similarity.

Keywords Graph similarity · Isomorphic subgraphs · Graph reconstruction · Reconstruction numbers

MSC2000: Primary 05C60; Secondary 05C25

12.1 Introduction

Defining the similarity, or distance, between mathematical objects in some class is generally always an important undertaking, and this is no exception for graphs. Ideally we would like to define the similarity between two graphs G, H as a parameter which is easy to compute, achieves some maximum value if and only if G and H are isomorphic, and in some sense captures how different G and H are when they are not isomorphic. In a sense, all graphical parameters can be considered candidates for such a similarity measure, but no measure which satisfies all these conditions

---

J. Lauri (✉)
Department of Mathematics, University of Malta, Tal-Qroqq, Malta
e-mail: josef.lauri@um.edu.mt

M. Dehmer (ed.), *Structural Analysis of Complex Networks*,
DOI 10.1007/978-0-8176-4789-6_12, © Springer Science+Business Media, LLC 2011

is known. An easily computable parameter which determines when two graphs are isomorphic would solve the Graph Isomorphism (GI) problem, one of graph theory's diseases [24]. Easily computable parameters such as the degree sequence and the spectrum do not always distinguish between nonisomorphic graphs. But devising measures which are efficiently computable although not always able to distinguish between nonisomorphic graphs is still an important realm of investigation, especially in applications. A recent example of work in this field (sometimes called *inexact graph matching* [8]) is [9], where the authors derive a hierarchy of similarity measures related to the degree sequence parameter and which can be computed efficiently. In this paper the authors give experimental results obtained by applying their similarity measures to more than 400 directed graphs representing web-based hypertext structures.

In this chapter we focus on measuring the similarity of two graphs in terms of their subgraphs. Complexity considerations and practical use are only discussed briefly in the last section. The first paper to study this way of measuring similarity or distance between graphs was probably [25]. In this paper, motivated by a question of Vizing, Zelinka defines the distance $\delta(G, H)$ between two graphs on n vertices as the minimum k such that G and H are both induced subgraphs of a graph on $n + k$ vertices and he shows that δ is a metric on the set of graphs with n vertices. He also proves the simple result that G and H are induced subgraphs of a graph on at most $n + k$ vertices if and only if they have a common induced subgraph on at least $n - k$ vertices. We consider a similarity measure which takes into consideration all induced subgraphs and which is also related to another well-known graph theory disease.

In the following all graphs are simple and undirected. Let G be a graph and v, e a vertex and an edge, respectively, of G. Then $G - v$ denotes the graph obtained by deleting from G the vertex v and all the edges incident to v; this is called a *vertex-deleted subgraph* of G. More generally, if X is a set of vertices of G then $G - X$ denotes the graph obtained by deleting from G all vertices in X and all edges incident to at least one vertex in X. The resulting graph $G - X$ is said to be *induced* by the vertices $V(G) - X$.

Similarly, $G - e$ denotes the graph obtained by deleting the edge e; it is called an *edge-deleted subgraph* of G. We are mostly concerned with vertex-deleted subgraphs, but we often indicate how the results and questions we present relate to the edge-deletion case.

The measure of similarity between two graphs which we discuss is the number of vertex-deleted subgraphs that they possess in common. We define the *subgraph similarity* sim(G, H) between two graphs G and H with the same number of vertices n as follows. Let $\mathcal{D}(G)$, called the *deck* of G, be the list of vertex-deleted subgraphs of G, where isomorphic subgraphs appear with the appropriate multiplicity. Similarly let $\mathcal{D}(H)$ be the deck of H. Then sim(G, H) is equal to the number of vertex-deleted subgraphs in $\mathcal{D}(G)$ which are also in $\mathcal{D}(H)$, where a subgraph that appears more than once in $\mathcal{D}(G)$ is counted as many times as it appears in $\mathcal{D}(H)$. (Therefore Zelinka's result quoted above for $k = 1$ states that G and H are both in

Fig. 12.1 Graphs G, H with $\text{sim}(G, H) = 5$

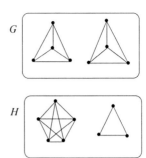

the deck of some graph if and only if $\text{sim}(G, H) \geq 1$.) To make these definitions clear note that the two graphs G and H in Fig. 12.1 have $\text{sim}(G, H) = 5$.

All of this is, of course, related to the reconstruction conjecture (RC) which can now be stated as

Reconstruction Conjecture

If G and H are two graphs on $n \geq 3$ vertices with $\text{sim}(G, H) = n$ then G and H are isomorphic.

Most results in graph reconstruction can now be stated in this fashion, for example, if $\text{sim}(G, H) = n$ then G and H have the same degree sequence, and the same characteristic and chromatic polynomials [16, 17]; if $\text{sim}(G, H) = n$ and G is in one of these classes of graphs, then $G \simeq H$: regular, disconnected [17], trees [12] and maximal planar [15]. (For a survey on the RC the reader is referred to [16].)

But the new insight which this point of view brings is that now, new and perhaps more amenable structural questions about graphs arise. Basically, even if we assume that the RC is true, we can still ask questions such as how large can $\text{sim}(G, H)$ be when G and H are not isomorphic. This enables us to revisit classes of graphs for which the question of reconstruction is easily settled but for which the issue of similarity in terms of subgraphs is still a very interesting unresolved question.

The notion of $\text{sim}(G, H)$ is at the heart of two important parameters which have been studied in the literature on the RC. Both of these parameters indicate how similar or dissimilar a given graph is to all others, and therefore how easy or difficult it is to determine it from its deck. The *universal reconstruction number* $\forall \text{rn}(G)$ of a graph G is defined to be

$$\forall \text{rn}(G) = 1 + \max_{H \not\simeq G} \{\text{sim}(G, H)\}.$$

This means that, given *any* $\forall \text{rn}(G)$ vertex-deleted subgraphs from $\mathcal{D}(G)$, these subgraphs determine G uniquely because no other nonisomorphic graph can have all of them in its deck. This interpretation, which tacitly assumes that RC is true, explains the name given to this parameter and its notation. This parameter is also often called the *adversary reconstruction number* of G [4].

The other reconstruction number $\exists rn(G)$, called the *existential reconstruction number* of G, is defined a little differently. Again tacitly assuming the truth of the RC, $\exists rn(G)$ is defined to be the smallest number of vertex-deleted subgraphs of G which are not found in the deck of any other graph. This means that there exist $\exists rn(G)$, and no less, vertex-deleted subgraphs of G which alone determine G uniquely, and this again explains the name of this parameter and the notation used. This parameter is also often called the *ally reconstruction number* or simply the *reconstruction number* of G [4]. For reasons which will become clear in the next section, in this chapter we mostly discuss these reconstruction numbers from the point of view of finding graphs with a high value for these parameters, that is, graphs which are in some sense very similar to other graphs. For more results about these two reconstruction numbers the reader is referred to the survey [4] and the book [17].

When we discuss the analogous situation with edge-deleted subgraphs we denote these parameters by the suffix e: sim_e, $\forall rn_e$, and $\exists rn_e$.

It is clear that $\exists rn(G) \leq \forall rn(G)$ but sometimes the two can be equal. For the two graphs in Fig. 12.1 one can check that $\exists rn(G) = \forall rn(G) = 6$, $\exists rn(H) = 3$, and $\forall rn(H) = 6$. It is also clear that $\exists rn(G) > 2$ because suppose we claim that $\exists rn(G) = 2$ for some graph G. Let $G - u$ and $G - v$ be the two vertex-deleted subgraphs which alone determine G. Construct H as follows. If u and v are adjacent in G then remove the edge uv, if they are not adjacent then add the new edge uv. Then, H is not isomorphic to G but it contains the two graphs $G - u$ and $G - v$ in its deck.

So the question becomes: how large can $\exists rn(G)$ and $\forall rn(G)$ be? We have seen that for the graphs in Fig. 12.1 $\exists rn(G)$ is as small as it can be and $\forall rn(G), \exists rn(H)$ and $\forall rn(H)$ are almost as large as the truth of the RC would allow. We show in the next section that such large reconstruction numbers are very rare.

12.2 Most Graphs Are Dissimilar

It turns out that most graphs are so dissimilar that their universal reconstruction number is three; that is, any three vertex-deleted subgraphs of most graphs will determine the graph uniquely. We shall make this statement more precise and give the proof in full because it illustrates very well how the concept of subgraph similarity which we are using depends heavily on the internal structure of graphs. The proof is based on [2] (Chap. 10).

It is well-known that almost every graph has a trivial automorphism group. However, a stronger result is possible which will tell us a lot about $sim(G, H)$, but we need first to explain what we mean when we say that almost every graph has some property. So, let $\mathcal{P}$ be a graph-theoretic property such as 'planar' or 'vertex-transitive'. Let r_n denote the proportion of labelled graphs on n vertices that have property $\mathcal{P}$. If $\lim_{n \to \infty} r_n = 1$, then we say that *almost every (a.e.) graph has property* $\mathcal{P}$. To show that a.e. graph has our desired property we use the simplest

12 Subgraphs as a Measure of Similarity

probability space which is set up when studying random graphs. Let $\mathcal{G}(n, \frac{1}{2})$ be the set of all labelled graphs on the set of vertices $\{1, 2, \ldots, n\}$ where, for each pair i, j,

$$P(ij \text{ is an edge}) = P(ij \text{ is not an edge}) = \frac{1}{2}$$

independently. Therefore each graph G in $\mathcal{G}(n, \frac{1}{2})$ has probability $(\frac{1}{2})^{\binom{n}{2}}$, which is, of course, equal to the probability of choosing G randomly from amongst all $2^{\binom{n}{2}}$ labelled graphs on n vertices when all are equally likely to be chosen. So, in order to show that a.e. graph has a particular property $\mathcal{P}$ one has to show that the probability that $G \in \mathcal{G}(n, \frac{1}{2})$ has property $\mathcal{P}$ tends to 1 as n tends to infinity.

The property we are interested in is the following. Let k be fixed. We say that a graph G has *property A_k* if all induced subgraphs of G on $n - k$ vertices are mutually nonisomorphic. In other words, G has property A_k means that, if X, Y are two distinct k-subsets of $V(G)$, then $G - X \not\cong G - Y$. It is easy to see that if G has property A_{k+1}, then it also has property A_k and that if it has property A_1, then it is asymmetric. Therefore having property A_k is stronger than just being asymmetric. We show that, for any fixed k, a.e. graph has property A_k.

Lemma 1. *Let $W \subseteq V$, $|W| = t$, $|V| = n$, and let $\rho : W \to V$ be an injective function that is not the identity. Let $g = g(\rho)$ be the number of elements $w \in W$ such that $\rho(w) \neq w$. Then there is a set I_ρ of pairs of (distinct) elements of W, containing at least $2g(t - 2)/6$ pairs, such that $I_\rho \cap \rho(I_\rho) = \emptyset$.*

Proof. Consider those pairs $v, w \in W$ such that at least one is moved. (All pairs are taken to contain distinct elements.) There are $g(t - g) + \binom{g}{2}$ such pairs. For all but at most $g/2$ of these pairs, $\{v, w\} \neq \{\rho(v), \rho(w)\}$ (the exceptions are when $\rho(v) = w$ and $\rho(w) = v$). Let E_ρ be the set of all such pairs. Then

$$|E_\rho| \geq g(t - g) + \binom{g}{2} - \frac{g}{2} = g \left(t - \frac{g}{2} - 1 \right) \geq g \left(\frac{t}{2} - 1 \right).$$

Define a graph H_ρ with vertex-set the pairs in E_ρ and such that each pair $\{v, w\}$ is adjacent to the pair $\{\rho(v), \rho(w)\}$. In H_ρ, all degrees are at most 2. Degrees equal to 1 could arise because $\{\rho(v), \rho(w)\}$ could contain an element not in W, and so the pair would not be in E_ρ. Degrees equal to 2 could arise because $\{v, w\}$ could be adjacent to both $\{\rho(v), \rho(w)\}$ and $\{\rho^{-1}(v), \rho^{-1}(w)\}$.

Therefore the components of H_ρ are isolated vertices, paths, or cycles. Let I_ρ be a set of independent (that is, not adjacent) vertices in H_ρ. Therefore, for any pair $\{v, w\} \in I_\rho$, $\{\rho(v), \rho(w)\}$ is not in I_ρ.

Now, all isolated vertices in H_ρ are independent, at least half of the vertices on a path are independent, and at least one third of the vertices on a cycle are independent, the extreme case here being a triangle. Therefore

$$|I_\rho| \geq |E_\rho|/3 \geq \frac{2g(t - 2)}{6},$$

as required. $\qquad\square$

Corollary 1. *Let $G \in \mathcal{G}(n, \frac{1}{2})$, $W \subset V = V(G)$, and $|W| = t$. Let $\rho : W \to V$ be an injective function that is not the identity. Let $g = g(\rho)$ be the number of elements $w \in W$ such that $\rho(w) \neq w$. Let S_ρ be the event*

> "*ρ gives an isomorphism from $G[W]$ to $G[\rho(W)]$*".

Then

$$P(S_\rho) \leq \left(\frac{1}{2}\right)^{2g\,(t-2)/6}.$$

Proof. Let I_ρ be the set constructed in the previous lemma. Now, for a given pair $\{v, w\} \in I_\rho$, the event

> "$\{v, w\}$ and $\{\rho(v), \rho(w)\}$ are both edges or nonedges"

has probability $1/2$. These events, as they range over all pairs $\{v, w\} \in I_\rho$, are mutually independent, because they involve distinct pairs. But S_ρ requires all these events simultaneously. Therefore, by independence,

$$P(S_\rho) \leq \left(\frac{1}{2}\right)^{|I_\rho|} \leq \left(\frac{1}{2}\right)^{2g\,(t-2)/6},$$

as required. $\qquad\square$

The result of this corollary is the crux of the matter. There are too many independent correct 'hits' required for ρ to be an isomorphism, and the probability therefore becomes small as n increases.

Theorem 2 ([6, 14, 21]). *Let k be a fixed nonnegative integer and let $G \in \mathcal{G}(n, \frac{1}{2})$. Let p_n denote the probability that*

$$\exists W \subseteq V(G) = V = \{1, 2, \ldots, n\},$$

with $|W| = n - k$ and such that

$$\exists \rho : W \to V, \rho \neq id, \rho \text{ is an isomorphism from } G[W] \text{ to } G[\rho(W)].$$

Then, $\lim_{n \to \infty} p_n = 0$.

Hence, a.e. graph has property A_k.

Proof. Pick a particular $W \subset V$ with $|W| = n - k$. This can be done in $\binom{n}{n-k}$ ways, and

$$\binom{n}{n-k} = \frac{n(n-1)\cdots(n-k+1)}{k!} < n^k.$$

Let $t = n - k$. Let $\rho : W \to V$ be injective and not the identity, and let $g = g(\rho)$ be the number of vertices of W that are moved by ρ. Let S_ρ be the event defined in the previous corollary.

Now, for a given value of g between 1 and t, how many functions ρ are there such that $g(\rho) = g$? Such a function is determined by the set $\{w : \rho(w) \neq w\}$ and by the values it takes on this set. Therefore, there are less than n^{2g} such ρ. Therefore, for a given fixed W, the probability of a nontrivial isomorphism is given by

$$
\begin{aligned}
\sum_{\rho \neq \mathrm{id}} P(S_\rho) &= \sum_{g=1}^{t} \sum_{\rho: g(\rho)=g} P(S_\rho) \\
&\leq \sum_{g=1}^{t} n^{2g} \left(\frac{1}{2}\right)^{2g(t-2)/6} \\
&= \sum_{g=1}^{t} \left[n^2 2^{(2-t)/3}\right]^g \\
&< \sum_{g=1}^{t} \left[4^{1/3} n^2 2^{-t/3}\right]^g.
\end{aligned}
$$

Now $t = n - k > 12(k+1) \lg n$ for sufficiently large n. Therefore

$$
\begin{aligned}
4^{1/3} n^2 2^{-t/3} &< 4^{1/3} n^2 2^{-4(k+1)\lg n} \\
&= \frac{4^{1/3} n^2}{n^{4(k+1)}} \\
&\leq \frac{4^{1/3}}{n^{2(k+1)}} \\
&< \frac{1}{n^{k+1}}
\end{aligned}
$$

where the last inequality follows if $4^{1/3} < n^{k+1}$.

Therefore

$$
\begin{aligned}
\sum_{\rho \neq \mathrm{id}} P(S_\rho) &< \sum_{g=1}^{t} \left(\frac{1}{n^{k+1}}\right)^g \\
&< \sum_{g=1}^{n} \left(\frac{1}{n^{k+1}}\right)^g \\
&= \frac{n^{n(k+1)} - 1}{n^{n(k+1)}(n^{k+1} - 1)}.
\end{aligned}
$$

But all this is for fixed W. Therefore the required probability is

$$
p_n < n^k \frac{n^{n(k+1)} - 1}{n^{n(k+1)}(n^{k+1} - 1)},
$$

and this tends to 0 as n tends to infinity. $\qquad \square$

Now the following theorem explains the relationship between property A_k and the subgraph similarity between graphs which we have been discussing.

Theorem 3 ([6, 21, 22]). *Let G have property A_3. Then G can be uniquely determined from any three vertex-deleted subgraphs in its deck. That is, $sim(G, H) \leq 2$ for any graph H not isomorphic to G and $\forall rn(G) = 3$.*

Proof. Let $u, v, w \in V(G)$. We show that G is uniquely determined from just $G - u$, $G - v$, and $G - w$.

Note first that v is identifiable in $G - u$ and u is identifiable in $G - v$; because G has property A_3 (and hence A_2), the only pair of vertices $x \in V(G - u)$, $y \in V(G - v)$ such that $G - u - x \simeq G - v - y$ are $x = v$ and $y = u$. Let $X = G - u - x$ and $Y = G - v - y$. There can only be one isomorphism from X to Y. For suppose α and β are two such isomorphisms. Let $z \in V(X)$ such that $\alpha(z) \neq \beta(z)$. Then $X - z \simeq Y - \alpha(z) \simeq Y - \beta(z)$, contradicting property A_3. Therefore we can label X and Y uniquely, and, from $X = G - u$, we can determine uniquely all of the neighbours of v in G, except possibly u. All we need to know is whether u and v are adjacent. To determine this we repeat the above procedure with $G - w$ instead of $G - u$. $\square$

From Theorem 2 and this lemma the following surprising result is immediate.

Theorem 4. *Almost every graph G has $sim(G, H) \leq 2$ for any graph $H \not\simeq G$ and therefore $\forall rn(G) = 3$.*

In an analogous manner one prove this result on edge-deleted subgraphs.

Theorem 5. *Almost every graph G has the property that any two edge-deleted subgraphs from its edge-deck determine it uniquely, that is, $sim_e(G, H) \leq 1$ for any graph $H \not\simeq G$, and $\forall rn_e(G) = 2$.*

12.2.1 Empirical Evidence

The data in Table 12.1, obtained by McMullen and Radziszowski [18], give a very good idea of how strong Theorem 4 really is. Out of more than 12,000,000 graphs on ten vertices, only 12 have $\exists rn$ greater than the minimum possible value of 3. This

Table 12.1 Number of graphs with given order and given $\exists rn$

$\exists rn$	Order							
	3	4	5	6	7	8	9	10
3	4	8	34	150	1044	12,334	274,666	12,005, 156
4		3		4		8		6
5				2		2	2	4
6						2		
7								2

situation sets the scene for the search of graphs with large values of ∃rn and ∀rn, and sometimes even a value of four can be considered large and graphs with this value could be difficult to find. In the next section we look at some results which have been obtained in this vein.

12.3 Graphs with Large Subgraph Similarity

We look at the problem of finding graphs with large subgraph similarity from two angles, that of the existential reconstruction number ∃rn and the universal reconstruction number ∀rn.

12.3.1 Large Values of ∃rn

The first graphs for which ∃rn were studied were disconnected graphs. Myrvold [23] and Molina [20] showed the following.

Theorem 6. *A disconnected graph with nonisomorphic components has ∃rn equal to 3. A disconnected graph with all components isomorphic each having c vertices has ∃rn $\leq c + 2$.*

So here it seems that we have a rich supply of graphs with large ∃rn. The example which Myrvold gave of disconnected graphs with ∃rn $= c + 2$ was the graph G consisting of disjoint copies of the complete graph K_c. The graph G in Fig. 12.1 is the special case $K_4 \cup K_4$. However, Asciak and Lauri [5] showed that in fact these are the only examples of disconnected graphs with ∃rn $= c + 2$ and that there are no disconnected graphs with ∃rn $= c + 1$. The computer searches of McMullen and Radziszowski [18] amongst all graphs on at most ten vertices unearthed only two examples of disconnected graphs with ∃rn > 3. These are the graph made up of two disjoint copies of the cycle on four vertices and the graph made up of two disjoint copies of the path on four vertices. Both have ∃rn $= 4$ and no other disconnected graphs with ∃rn > 3 are known. The big gap between ∃rn $= 4$ and ∃rn $= c$ is waiting to be explored.

The situation with regular graphs is somewhat similar. Myrvold [22] has shown that r-regular graphs have ∃rn at most $r + 3$ but Asciak [3] has shown that again the disconnected graph consisting of disjoint copies of K_{r+1} is the only r-regular graph with ∃rn $= r + 3$. Here too, knowledge about the gap between ∃rn $= 4$ and ∃rn $= r + 2$ is very scant. The computer searches of McMullen and Radziszowski led them to this construction. The graph $RCC_{n,j}$ is obtained as follows (RCC stands for redundantly connected cycles). Take $n \geq 2$ disjoint copies of cycles each of length $j \geq 3$. Let $v_{c,i}, i \in \{0, 1, \ldots, j-1\}$ denote the i-th vertex of the c-th cycle. For each $c \neq d$ join the vertices $v_{c,i}$ and $v_{d,i+1}$, where addition is modulo j. The resulting graph $RCC_{n,j}$ is regular and McMullen and Radziszowski prove the following.

Theorem 7. $\exists rn(RCC_{n,j}) > n + 1$, for all $n \geq 2$ and $j \geq 3$.

However, in their computer searches, McMullen and Radziszowski found no other regular graphs with $\exists rn > 3$, apart from pK_n and their complements. And as they say, there seems to be no clear idea on how to establish in general the exact value of $\exists rn(RCC_{n,j})$ for all n, j.

These cases illustrate the new set of problems which the notion of reconstruction numbers creates. The classical reconstruction of regular graphs is trivial and that of disconnected graphs is an easy exercise [17]. But even finding examples with $\exists rn > 3$ is a difficult task. The reader who is interested in finding out more about graphs with large $\exists rn$ is invited to read [18].

12.3.2 Large Values of $\forall rn$

The definition of $\forall rn$ is more closely related to that of $sim(G, H)$, and it seems more difficult to tackle. It certainly seems easier to find disconnected graphs with large $\forall rn$ than ones with large $\exists rn$. For example, Hemaspaandra et al. [11] observe that since

$$sim(K_{t+1} \cup K_{t-1}, 2K_t) = t + 1$$

then $\forall rn(K_{t+1} \cup K_{t-1})$ and $\forall rn(2K_t)$ are both at least $t + 2$ and therefore greater than the corresponding $\exists rn$ numbers which are both three. However, the proof in [11] that these two $\forall rn$ numbers are actually $t + 2$ is not simple, even for such straightforward graphs; determining $\forall rn$ seems to be quite difficult in general. Also, it is not clear that these and the other two examples given in [11] are not exceptional cases similar to the usual suspects: the graphs pK_n with large $\exists rn$. Therefore the question of finding disconnected graphs with large $\forall rn$ might be as open as it is for finding disconnected or regular graphs with large $\exists rn$.

Until recently, most of the results obtained about $\forall rn$ and $sim(G, H)$ were found in [10, 22]. An early result was the following.

Theorem 8 ([22]). *Let G and H be two graphs on n vertices and with $sim(G, H)$ $= n - 1$. Then G and H have the same degree sequence.*

Again we see that what is an easy exercise in reconstruction [17] becomes a difficult result when seen in terms of the subgraph similarity between graphs. The obvious, and difficult, question here is: for given n, what is the largest value of k such that there exist graphs G and H on n vertices with $sim(G, H) = k$ but with different degree sequences?

Of course, the most general problem here is to determine the largest value of $sim(G, H)$ for nonisomorphic graphs on n vertices. But since this would solve the RC, all authors have attempted this question by restricting G and H to particular classes and generally trying to determine the maximum possible value of $sim(G, H)$.

12 Subgraphs as a Measure of Similarity

Significant advances in this direction have recently been reported by Bowler et al. in [7]. For example, they show the following.

Theorem 9. *Let G be a tree and H a unicyclic graph on n vertices ($n \geq 19$). Then*

$$sim(G, H) \leq \left\lfloor \frac{2}{5}(n + 1) \right\rfloor.$$

Moreover, this bound is attained.

From this result and other work in [22] the following holds.

Theorem 10. *Let G and H be two graphs on n vertices ($n \geq 19$) and such that*

$$sim(G, H) \geq \left\lfloor \frac{n}{2} \right\rfloor + 1.$$

Then if G is a tree H must also be a tree.

Francalanza [10] also considered the number of edge-deleted subgraphs in common between a tree and a unicyclic graph plus an isolated vertex. She proved the following.

Theorem 11. *Let G be a tree and H a unicyclic graph with an isolated vertex, both on n vertices. Then*

$$sim_e(G, H) \leq \frac{n}{2} + 1.$$

Bowler et al. make a conjecture that if G is a tree and H is a unicyclic graph plus an isolated vertex, both on n vertices, then in fact

$$sim_e(G, H) \leq \frac{n}{2}.$$

The structures of the trees and the unicyclic graphs which attain large subgraph similarity between them are very particular. The trees are *caterpillars*, that is, trees the deletion of whose endvertices gives a path, and the unicyclic graphs are what Myrvold and Francalanza call *sunshine graphs*, that is, unicyclic graphs the deletion of whose endvertices gives a cycle.

The main question which these researchers would like to answer here is certainly the following: What is the largest possible value of $sim(G, H)$ when G and H are two nonisomorphic trees on n vertices?

This construction from [7] gives a family of pairs of nonisomorphic trees with large subgraph similarity. Let

$$G^* = K_{1,p-1} \cup K_{1,p+1} \cup K_{1,p+1}$$
$$H^* = K_{1,p} \cup K_{1,p} \cup K_{1,p+1}.$$

Let G be the tree obtained from G^* by adding a new central vertex and three new edges joining the new vertex to the three cutvertices of G^*. Similarly, construct H from H^*. These two trees are nonisomorphic, have $n = 3p + 5$ vertices, and

$$\text{sim}(G, H) = 2p = \frac{2}{3}(n - 5).$$

This family of tree pairs has the highest known subgraph similarity between non-isomorphic trees. A similar construction in [7] gives examples of pairs G, H of nonisomorphic trees on n vertices with the same degree sequence and

$$\text{sim}(G, H) = \frac{2}{3}(n + 1 - 2\sqrt{3n - 6}).$$

The best result known to date regarding the highest possible value of $\text{sim}(G, H)$ for general graphs is again found in [7]. First we require a definition. A *2UC graph pair* is a pair of nonisomorphic graphs, G and H, on n vertices, at least one of which is disconnected, such that in G or in H there are at least two components which cannot be matched with the components of the other graph by isomorphism. A particular example is when G is connected and H is disconnected. (2UC stands for Two Unmatched Components.) The motivation behind this definition is that if A and B are two nonisomorphic connected graphs with the same deck (hence counterexamples to the RC) and on $n - 1$ vertices, then $\text{sim}(A \cup K_1, B \cup K_1) = n - 1$. Bowler et al. prove the following theorem.

Theorem 12. *Let G and H be two 2UC graphs. Then*

$$\text{sim}(G, H) \leq 2 \left\lfloor \frac{1}{3}(n - 1) \right\rfloor.$$

For $n \geq 22$ and $n \equiv 1 \pmod 3$, they also give the following infinite family of pairs of 2UC graphs attaining this bound:

$$G = K_{p-1} \cup K_{p+1} \cup K_{p+1}$$
$$H = K_p \cup K_p \cup K_{p+1}.$$

They also show that this pair is unique for the given values of the parameter n. Note that although G and H are disconnected, their complements are connected and also have the same subgraph similarity.

More examples are given in [7] including uniqueness of some families of pairs attaining the upper bound in Theorem 12. Their work also gives an example of pairs G, H of 2UC graphs with $n = 3p^2 - 2, (p \geq 3)$, having the same degree sequence, and

$$\text{sim}(G, H) = \frac{2}{3}(n + 5 - 2\sqrt{3n + 6}).$$

12 Subgraphs as a Measure of Similarity

This number is smaller than the upper bound in Theorem 12. Therefore it seems natural to ask what is the maximum possible value of $\text{sim}(G, H)$ when G, H are two nonisomorphic 2UC graphs on n vertices with the same degree sequence.

Motivated by Theorem 12, Bowler et al. make the following conjecture which, of course, is a considerable strengthening of the RC.

Strong Reconstruction Conjecture

Let G and H be nonisomorphic graphs on n vertices. For large enough n,

$$\text{sim}(G, H) \leq 2 \left\lfloor \frac{1}{3}(n - 1) \right\rfloor.$$

Therefore for any graph G on n vertices and sufficiently large n,

$$\forall\text{rn}(G) \leq 2 \left\lfloor \frac{1}{3}(n - 1) \right\rfloor + 1.$$

Finally, what about $\forall\text{rn}_e(G)$, the universal edge-reconstruction number? In classical graph reconstruction, determining G from edge-deleted subgraphs is always easier than determining it from vertex-deleted subgraphs. However, the relationship between the vertex and the edge versions of the parameters which we have been discussing in this chapter does not seem to be so straightforward (see [4] for more on this). Sometimes the edge parameter is larger than the corresponding vertex parameter, and often determining the former is at least as difficult as finding the latter. Certainly, very little work, if any, has been done on $\forall\text{rn}_e(G)$, especially the search for graphs with large $\forall\text{rn}_e$, so this is a field wide open for investigation.

12.4 Algorithmic and Other Issues

The RC is not an algorithm question. The issue is not whether there is an efficient way of obtaining G from its deck but it is a question of uniqueness: is there more than one graph with the given deck? However, a few variants of the RC have been adapted into questions of algorithmic complexity. Subgraph similarity and reconstruction numbers, being so closely related to the GI problem and the subgraph isomorphism problem which is known to be NP-complete [13] are perhaps the most natural variants of the reconstruction problem to be treated algorithmically.

In [11], the authors define these four decision problems:

1. EXIST-VRN $= \{\langle G, k \rangle | \exists\text{rn}(G) \leq k\}$.
2. UNIV-VRN $= \{\langle G, k \rangle | \forall\text{rn}(G) \leq k\}$.
3. EXIST-ERN $= \{\langle G, k \rangle | \exists\text{rn}_e(G) \leq k\}$.
4. UNIV-ERN $= \{\langle G, k \rangle | \forall\text{rn}_e(G) \leq k\}$.

They remark that it is easy to see that EXIST-VRN $\in \Sigma_2^p$ (since GI is low for Σ_2^p), UNIV-RN $\in$ coNP^{GI}, EXIST-ERN $\in$ NP^{GI}, and UNIV-ERN $\in$ coNP^{GI} and they suggest that obtaining tight, or tighter, bounds on the complexity of these problems should be interesting. (For explanations of the above complexity terms the reader is referred to [13].)

And finally, what about possible applications? Any measure of similarity between mathematical objects is bound to have some relevance in situations modeled by the objects, and graphs are certainly amongst the mathematical structures most often used as models. One in which notions that relate to the concept of subgraph similarity seem to be useful is in systems biology. The way a cell processes information from its environment in order to determine the rate of production of the proteins it requires is often modeled by what are called *transcription networks*, which are basically directed graphs [1]. Biologists try to identify particular subgraphs of transcription networks in order to explain their functionality. These *network motifs* are often identified as those subgraphs in the transcription network which appear significantly more often than they do in a random graph of the same size. This seems quite reminiscent of the notion we have been discussing of comparing two graphs by counting the number of subgraphs they have in common. Here, the comparison is usually between the given transcription network and the general random graph. In this comparison, the number of symmetries of the network motif (the size of its automorphism group) often plays an important part. The notion of how a subgraph embeds in a graph, a notion which involves the number of appearances of the subgraph and the size of its automorphism group, seems to be the central issue in the reconstruction problem (see, for example, Chap. 10 and especially Chap. 11 in [17]). An investigation of how these ideas from subgraph similarity and graph reconstruction might apply to the study of network motifs in transcription networks could therefore be very useful.

Similar ideas have cropped up in the unlikely area of counterterrorism. Interactions between agents in a society (conversations, emails, telephone calls, etc.) can be modeled by a graph. Within this "transactional noise" one would like to detect the emergence of unlikely configurations (subgraphs) which could signify the existence of networks of terrorist activities [19]. Again, this is done by comparing the transactional network with some appropriate random graph model to detect subgraphs which appear more frequently than expected by the model. The similarities with the previous application and what we have been discussing is clear.

It is, after all, not surprising that such applications should exist. With the ability to collect and handle ever larger amounts of data in various fields from biology to sociology comes the need of modeling situations with large graphs. And very often a natural way to investigate certain aspects of the internal structure of such graphs is through smaller subgraphs which are more manageable. And these applications are often closely related to issues of algorithmic complexity. When tackling empirically questions about subgraphs in common between two graphs one cannot escape from the Graph Isomorphism problem in some guise or another.

12.5 Conclusion

We have discussed a way of measuring similarity between graphs in terms of subgraphs which does not simply give an alternative framework for wording the RC. It raises simple questions which are difficult to solve about graphs which are very easily reconstructible, and it gives some new twists to old ideas, such as the relationship between vertex-reconstruction and edge-reconstruction. Independently of the status of RC, finding classes of graphs with large subgraph similarity or reconstruction number is an interesting nontrivial problem. And the notion of comparing graphs in terms of the number of common subgraphs of some type or another that they share seems to be a promising area of modern applied graph theory, which is closely connected to algorithmic complexity issues related to reconstruction numbers, which are in turn of important theoretical interest. It seems that subgraph similarity has a lot to offer to graph theorists with different interests and tastes.

Note added in proof: Bowler, Brown, Fenner and Myrvold have recently shown that if G is a connected graph and H a disconnected graph, then $\text{sim}(G, H) \leq \lfloor n/2 \rfloor + 1$ and they have also characterised those pairs of graphs that achieve this bound.

References

1. Alon U (2007) An introduction to systems biology: design principles of biological circuits. Chapman & Hall, London
2. Alon N, Spencer JH (1992) The probabilistic method. Wiley, New York
3. Asciak KJ (1998) On certain classes of graphs with large reconstruction number. Master's thesis, University of Malta
4. Asciak KJ, Francalanza MA, Lauri J, Myrvold W (2010) A survey of some open questions in reconstruction numbers. Ars Combinatoria (to appear)
5. Asciak KJ, Lauri J (2002) On disconnected graphs with large reconstruction number. Ars Combinatoria 62:173–181
6. Bollobás B (1990) Almost every graph has reconstruction number 3. J Graph Theory 14:1–4
7. Bowler A, Brown P, Fenner T. Families of pairs of graphs with a large number of common cards. Preprint
8. Bunke H (2000) Recent developments in graph matching. In Proceedings of the 15th international conference on pattern recognition, Spain, vol 2
9. Dehmer M, Mehler A (2007) A new method of measuring similarity for a special class of directed graphs. Tatra Mt Math Publ 36:39–59
10. Francalanza MA (1999) The adversary reconstruction of trees: the case of caterpillars and sunshine graphs. Master's thesis, University of Malta
11. Hemaspaandra E, Hemaspaandra LA, Radziszowski SP, Tripathi R (2007) Complexity results in graph reconstruction. Discrete Appl Math 155:103–118
12. Kelly PJ (1957) A congruence theorem for trees. Pac J Math 7:961–968
13. Köbler J, Schöning U, Torán J (1993) The graph isomorphism problem: its structural complexity. Birkhäuser, Switzerland
14. Korshunov AD (1971) Number of nonisomorphic graphs in an n-point graph. Math Notes Acad Sci USSR 9:155–160
15. Lauri J (1981) The reconstruction of maximal planar graphs, II: reconstruction. J Combin Theory B 30:196–214

16. Lauri J (2004) The reconstruction problem. In Gross JL, Yellen J (eds) Handbook of graph theory. CRC Press, Boca Raton, FL, pp 79–98
17. Lauri J, Scapellato R (2003) Topics in graph automorphisms and reconstruction. Cambridge University Press, Cambridge
18. McMullen B, Radziszowski SP (2005) Graph reconstruction numbers. J Combin Math Combin Comput 62:85–96
19. Mifflin T, Boner C, Godfrey G, Greenblatt M (2006) Detecting terrorist activities in the twenty-first century: a theory of detection for transactional networks. In Popp RL, Yen J (eds) Emergent information technologies and enabling policies for counter-terrorism. Series on computational intelligence. IEEE Press, New York, pp 349–365
20. Molina R (1995) Correction of a proof on the ally-reconstruction number of a disconnected graph. Ars Combinatoria 40:59–64
21. Müller V (1977) The edge reconstruction hypothesis is true for graphs with more than $n \log_2 n$ edges. J Combin Theory B 22:281–283
22. Myrvold W (1988) Ally and adversary reconstruction problems. PhD thesis, University of Waterloo, ON, Canada
23. Myrvold W (1989) The ally-reconstruction number of a disconnected graph. Ars Combinatoria 28:123–127
24. Read RC, Corneil DG (1977) The graph ismorphism disease. J Graph Theory 1:339–363
25. Zelinka B (1975) On a certain distance between isomorphism classes of graphs. Časopis pro p̌est Mathematiky 100:371–373

Chapter 13
A Chromatic Metric on Graphs

Gerhard Benadé

Abstract In this chapter, we introduce the concept of relatedness of graphs, based upon the generalized chromatic number. This allows the definition of a graph metric. It is proved that the distance between any two graphs is at most three.

Keywords Graphs · Chromatic number · Generalized graph coloring · Graph distance · Metric

MSC2000: Primary 05C12; Secondary 05C15

13.1 Introduction

Ever since the four-color conjecture was formulated over a century ago, the study of graph coloring played an active and vital part in this graph theory. The ideas and techniques developed there, and the problems and questions generated by the research provided a stimulus to much of nonchromatic graph theory. An abundance of parameters associated with various ways of partitioning the vertex set of a graph has been studied over the years.

In 1977, Harary [16] gave a very broad definition of a vertex coloring which subsumed most of the sometimes quite disparate concepts defined previously. If P is any class of graphs, satisfying only the condition that it must contain the one-vertex graph, then a P-coloring of a graph is a partition of the vertex set of the graph for which the subgraph induced by each partition class is an element of P. The corresponding parameter is the P-chromatic number, defined as the smallest number of partition classes in any P-coloring of a graph.

G. Benadé (✉)
School of Computer Science, Statistics and Mathematics, North-West University,
Potchefstroom, South Africa
e-mail: gerhard.benade@nwu.ac.za

M. Dehmer (ed.), *Structural Analysis of Complex Networks*,
DOI 10.1007/978-0-8176-4789-6_13, © Springer Science+Business Media, LLC 2011

This definition opened up the theory of generalized graph coloring, which has experienced a high level of activity in subsequent years. In this field, one endeavors to uncover the essential structure underlying chromatic theory. This has proved to be a quite difficult task; the very generality of the setting implies a concomitant dearth of available knowledge. For example, some very easily stated questions, such as determining the $(P \bigcup Q)$-chromatic number of a given graph, are still unsolved (and perhaps unsolvable!). Nevertheless, some remarkable results have been proved, for instance, Folkman's theorem [13], asserting that, for every positive integer k and for some classes P of graphs, there exists a graph G_k such that the P-chromatic number of G_k is k, and the theorem of Bollobás and Thomason [5] about the existence of uniquely P-colorable graphs for some general classes P of graphs.

A concept that seems to strike a good balance between generality and tractability is the notion of F-free coloring. This special case of a P-coloring is defined as follows. For any graph F which is not the one-vertex graph, an F-free coloring of some graph is a partition of the vertex set of that graph in such a way that F does not occur as an induced subgraph in the subgraph induced by any of the partition classes – in other words, the partition classes induce F-free subgraphs. The parameter corresponding to the P-chromatic number defined above is defined in the same way and is called the F-free chromatic number. Many well-known vertex partitioning parameters are in fact F-free chromatic numbers, the most well-known certainly being the usual chromatic number χ, which corresponds to $F = K_2$, the complete graph on two vertices. Notable exceptions are the cochromatic number and the vertex arboricity; they are accommodated only under the general definition of the P-chromatic number.

Using the notion of an F-free coloring, we introduce a relation on the class of graphs that ultimately leads to a metric. Two graphs F and G are called related, denoted by $F \asymp G$, if, for every positive integer k, there exists a graph for which both the F-free and the G-free chromatic numbers are equal to k. They are called n-distantly related, for some integer n, if there exist graphs $R_0, R_1, \ldots, R_n$ such that $F \cong R_0 \asymp R_1 \asymp \cdots \asymp R_n \cong G$, and they are called distantly related if there is some integer n for which they are n-distantly related. Making essential use of Folkman's theorem, it is proved that any two graphs are distantly related. This means that the metric, defined for two graphs F and G as the minimum number n for which F and G are n-distantly related, assumes some finite value for any pair of graphs. It is a rather remarkable fact that, indeed, this value does not exceed three.

In Sect. 13.2, we introduce the notion of relatedness for general classes of graphs, then confine our attention to graph relatedness in Sect. 13.3. In Sect. 13.4, various conjectures giving sufficient conditions for relatedness to hold between two graphs are discussed. Relatedness between specific graphs types are given in Sect. 13.5. In Sect. 13.6, a transitive relation is defined, leading to the definition of the chromatic metric in Sect. 13.7.

13.2 Relatedness of Classes

Although the main results on relatedness are about relatedness of graphs rather than of classes, the concept can equally naturally be defined for classes and some interesting general results may be deduced. Recall that the meaning of **class** is restricted to exclude the class of all graphs and to always include the empty graph and the one-vertex graph.

Definition 1. Two classes P and Q are called n-**chromatically related** for some positive integer n, denoted by $P \times_n Q$, if there exists a graph M_n such that

$$\chi(M_n : P) = \chi(M_n : Q) = n.$$

If $P \times_n Q$ for every positive integer n, then P and Q are called **chromatically related**, denoted by $P \times Q$. Furthermore, should P and Q not be chromatically related ($P \not\times Q$), they are called **chromatically alien**.

Often, the word "chromatically" in the above definition is dropped when it can unambiguously be inferred from the context, and we simply refer to n-related, related, and alien classes.

Two classes are thus called related if, for each possible value of the chromatic numbers defined by them, a graph exists for which these two chromatic numbers both have this value. The chromatic numbers defined by two alien classes, on the other hand, may well coincide for some values in the spectrum of possible values, but there will be at least one integer n for which, when applied to any graph, they will not both be n.

Example 1. The concepts of n-relatedness and relatedness are different, at least for classes of graphs. An example is furnished by the classes Q_l, consisting of all graphs with order less than l, for some $l \geq 3$. It follows that $\chi(G : Q_l) = \lceil \frac{v(G)}{l-1} \rceil$. Now, for some integer $l \geq 3$, $Q_l \times_n Q_{l+1}$ if and only if $n < l$.

Note that $Q_l \times_l Q_{l+1}$ if and only if there is a graph G of order p such that $\lceil \frac{p}{l} \rceil = n = \lceil \frac{p}{l} \rceil$, that is, $n - 1 < \frac{p}{l}$ and $\frac{p}{l-1} \leq n$. Such a p exists if and only if $nl - l < p \leq nl - n$; that is, $n < l$.

This means that, for all integers $l \geq 3$, the classes Q_l and Q_{l+1} are not n-related for all $n \geq l$ and hence that they are alien.

Example 2. The relations $\times_n$ and $\times$ are not equivalence relations, since, although they are symmetric and reflexive, they are not transitive. Again, the classes Q_l supply an example. Suppose we take a class of graphs P such that, for any pair of positive integers n and k with $k \geq n$, there is a graph G of order k and with P-chromatic number n. (An example of such a class is the class of null graphs.) Then $Q_l \times P$ and $Q_{l+1} \times P$: For each integer n, a P n-chromatic graph of order $n(l - 1)$ will have Q_l-chromatic number equal to n too, whereas a P n-chromatic graph of order nl will have Q_{l+1}-chromatic number n too. Thus Q_l and Q_{l+1} are both related to P, but by the previous example they are alien.

Example 3. Another example showing that the relation $\asymp$ is not transitive is the following. It is later proved that the classes $-K_2$ and $-P_3$ are related, and also $-P_3$ and $-K_3$, but that $-K_2$ and $-K_3$ are alien.

A few elementary properties of these relations are now listed. In the first of these, 1-relatedness is established between any two classes. Thus, showing that two classes are n-chromatically related for $n \geq 2$ is sufficient to prove that they are chromatically related.

Remark 1. Any two classes P and Q are 1-chromatically related.

Proof. Since it was assumed that K_1 is an element of any class, the one-vertex graph K_1 can be both P 1-colored and Q 1-colored for any two classes P and Q. $\square$

Remark 2. Let P and Q be classes. Then, if for some positive integer n, P and Q are n-chromatically related, so are $\overline{P}$ and $\overline{Q}$. Also, if P and Q are related, then so are $\overline{P}$ and $\overline{Q}$.

Proof. Recall that the complement of a class P consists of the class of the complements of the graphs in P. Let $P \asymp_n Q$ for some n. Let M_n be a graph establishing this; in other words, $\chi(M_n : P) = \chi(M_n : Q) = n$. Then, using Remark 3.2.7 of [1], it follows that $\chi(\overline{M_n} : \overline{P}) = \chi(\overline{M_n} : \overline{Q}) = n$ and hence the graph $\overline{M_n}$ establishes n-relatedness between $\overline{P}$ and $\overline{Q}$. Should P and Q be n-related for all $n \geq 2$, this proves that their complements will also be, hence $P \asymp Q$ implies $\overline{P} \asymp \overline{Q}$. $\square$

The next result is a corollary to Theorem 2.3.3 of [1], and shows that there are in fact many pairs of chromatically alien classes.

Remark 3. Let P and Q be classes of graphs and let Q^2 denote the class of all graphs having Q-chromatic number at most 2. Then $Q^2 \subseteq P$ implies that $P \not\asymp Q$.

Proof. Recall the condition in the above-mentioned theorem, holding for two classes P and Q and an integer k: Every graph I, with $\chi(I : Q) \leq k$, is in P. This condition, with $k = 2$, is equivalent with $Q^2 \subseteq P$, thus the theorem may be applied and hence the following inequality holds between the two chromatic numbers, defined by P and Q, of any graph H:

$$\chi(H : P) \leq \left\lceil \frac{\chi(H : Q)}{2} \right\rceil.$$

This implies that P and Q are alien. $\square$

Note that $Q \subset Q^2$ for any class of graphs Q. This result thus asserts that two classes of graphs are alien whenever their sizes differ too much. This provides an intuitive way to think about the relations "is alien to" and "is related to" in set-theoretic terms.

The converse of this result does not hold. It is possible for two classes P and Q that the condition $Q^2 \subseteq P$ fails, but that they still are alien. This happens, for

13 A Chromatic Metric on Graphs

instance, in the case of the two classes Q_l and Q_{l+1} discussed above. They are alien, but there are Q_l 2-chromatic graphs that do not occur in Q_{l+1}; any graph with order at least $l + 1$ and at most $2l - 2$ falls into this category. Thus $(Q_l)^2 \not\subseteq Q_{l+1}$.

Remark 4. Let P and Q be classes with $P \subseteq Q$ and let $P \asymp_n Q$ for some $n \geq 2$. Let $P \asymp_n R \asymp_n Q$. Also, if $P \asymp Q$ and R is any class with $P \subseteq R \subseteq Q$, then $P \asymp R \asymp Q$.

Proof. Let P and Q be n-related for some $n \geq 2$, and let M_n be a graph such that $\chi(M_n : P) = \chi(M_n : Q) = n$. Then, by Remark 2.3.2 of [1], $n = \chi(M_n : P) \geq \chi(M_n : R) \geq \chi(M_n : Q) = n$. This implies that all three of these classes are mutually n-related. This proof holds for any integer $n \geq 2$, thus the corresponding statement for chromatic relatedness is also true. □

The following result gives sufficient conditions for n-relatedness to imply m-relatedness, for some positive integers m and n.

Remark 5. Let P and Q be classes of graphs and let m and n be positive integers. If $P \subseteq Q$, then $P \asymp_n Q$ implies that $P \asymp_m Q$ for all $m \leq n$.

Proof. Consider any P-coloring in n classes of a graph M_n establishing the given n-relatedness between P and Q. Let M_{n-1} be a subgraph of M_n induced by any $n - 1$ of the color classes; say V is the omitted color class. Then the P-chromatic number of M_{n-1} is $n - 1$. Since $P \subseteq Q$,

$$\chi(M_{n-1} : Q) \leq \chi(M_{n-1} : P) = n - 1.$$

Furthermore, equality must hold, since M_{n-1} is $n - 1$-chromatic: Suppose that $\chi(M_{n-1} : Q) = n - 2$. Then a Q-coloring for M_n in $n - 1$ colors can be obtained by using any Q $n - 2$-coloring for M_{n-1} and adding the P-color class V, which is also a Q-color class by the assumption on P and Q. This contradicts the fact that M_n is Q n-chromatic.

Thus there exists a graph for which the Q- and the P-chromatic numbers are both equal to $n - 1$, proving that $P \asymp_{n-1} Q$. By iterating this procedure another $n - 2$ times, the result is obtained. □

13.3 Relatedness of Graphs

A special case of relatedness of classes is obtained if one considers only classes $-F$ for some graph F. In this case, the relatedness will be said to hold between the *graphs*.

Definition 2. Two graphs F and G are called n-**chromatically related** for some positive integer n, denoted by $F \asymp_n G$, if $-F \asymp_n -G$, and they are called

chromatically related, denoted by $F \asymp G$, if $-F \asymp -G$. Furthermore, should F and G be not chromatically related, they are called **chromatically alien**, denoted by $F \not\asymp G$.

Thus, F and G are n-chromatically related if there exists a graph M_n such that $\chi(M_n : -F) = \chi(M_n : -G) = n$, and they are chromatically related if $F \asymp_n G$ for all integers $n \geq 1$. Also, for two chromatically alien graphs F and G there exists an integer $n \geq 2$ such that for no graph H, the F-free chromatic number of H and the G-free chromatic number of H are both equal to n.

As before, when referring to these concepts further on, the word "chromatically" might sometimes be omitted if the meaning is sufficiently clear from the context. Note also that whenever a graph appears in any of these relations, it has been used to define a chromatic number and thus the convention requiring such graphs to have at least two vertices applies.

The following few remarks reinterpret the previous section's results for graphs.

Remark 6. Any two graphs are 1-chromatically related.

Remark 7. Let F and G be graphs. If, for some positive integer n, the graphs F and G are n-chromatically related, then so are $\overline{F}$ and $\overline{G}$. Likewise, if F and G are chromatically related, $\overline{F}$ and $\overline{G}$ are also chromatically related.

The next result gives a necessary condition for n-relatedness to hold between two graphs using the chromatic number of the one in terms of the other, and plays an important role in the sequel.

Remark 8. Let F and G be graphs. If $F \asymp_n G$ for some integer $n \geq 2$, then $\chi(F : -G) \leq 2$ and $\chi(G : -F) \leq 2$.

Proof. Suppose that $\chi(G : -F) \geq 3$. Then it follows from Corollary 3.2.9 of [1] that the G-free chromatic number of any graph H and the F-free chromatic number of H cannot both be equal to n, contradicting $F \asymp_n G$. Since this assumption is symmetric with respect to the positions of F and G (due to the fact that the relation $\asymp_n$ is reflexive), the other inequality immediately follows. $\square$

Now, for *any* pair F and G of graphs, it holds that $\chi(F : -G) \leq 2$ or $\chi(G : -F) \leq 2$. The added requirement that F and G are related now forces *both* these chromatic numbers down to at most two. Intuitively speaking, this means that related graphs lie closely together in the sense that the chromatic number of one in terms of the other *and* vice versa cannot be too large. These matters are elaborated upon later.

The following result is derived from the above remark by taking G equal to K_2. In this case, relatedness of a graph F with K_2 is shown to have direct implications for the structure of F.

Corollary 1. *Let F be a graph. If $F \asymp_n K_2$ for some integer $n \geq 2$, then F is bipartite.*

13 A Chromatic Metric on Graphs   341

Proof. Applying Corollary 8 with $G = K_2$ yields $\chi(F) \leq 2$, which is equivalent to F being bipartite. Note that the other statement, $\chi(K_2 : -F) \leq 2$, is trivial in this situation, since for all graphs F other than K_2 itself, the F-free chromatic number of K_2 is 1 in any case. $\hfill\square$

It is later shown that the complete bipartite graphs are all related to K_2. Although this could not be proved for general bipartite graphs, no example of a bipartite graph alien to K_2 has yet been found, and we strongly suspect that relatedness with K_2 characterizes the property of being bipartite. Implications of this conjecture are examined in Sect. 13.4.

The following corollary to Remark 4 is a useful "pinching" result: If a graph is wedged (in terms of the induced subgraph relation) between two others, and the outer graphs are related, then all three are mutually related.

Remark 9. Let F and H be graphs with $F \vartriangleleft H$ and let, for some $n \geq 2$, $F \asymp_n H$. Let G be any graph with $F \vartriangleleft G \vartriangleleft H$. Then $F \asymp_n G \asymp_n H$. Also, if $F \asymp H$ and G is a graph with $F \vartriangleleft G \vartriangleleft H$, then $F \asymp G \asymp H$.

Let H be a nontrivial graph related to K_2. Then the remark applies, with $F = K_2$. The statement then becomes the following. Every induced subgraph G of a graph H, which is related to K_2, is related to H (and, of course, also to K_2). In this case, we know that H must be bipartite and infer that any induced subgraph of H – which must again be bipartite – must be related to K_2.

Under certain circumstances, relatedness between graphs can survive disjoint unions of the graphs in the following sense. If a graph F is related to some graphs G_i, then F is also related to the disjoint union of the G_i. Theorem 1 gives the conditions for this to happen.

Theorem 1. *Let F and G be connected graphs, with $F \asymp_n G$ for some integer $n \geq 2$. Let G_i be a graph with $G_i \vartriangleleft G$, for each $i = 1, 2, \ldots, k$, and let $\mathcal{G} = (\bigcup_{i=1}^{k} G_i) \bigcup G$. Then $F \asymp_n \mathcal{G}$.*

Proof. Let H_n be a graph establishing the n-chromatic relatedness between F and G. Let $H = (k(n-1) + 1)H_n$. Then it follows from Remark 3.4.2 of [1] that $\chi(H : -F) = n$, since F is connected. The same is now shown for the $\mathcal{G}$-free chromatic number of H:

Suppose that $\chi(H : -\mathcal{G}) \leq n - 1$ and let a $\mathcal{G}$-free coloring $\{V_1, V_2, \ldots, V_{n-1}\}$ in $n - 1$ colors be given for H. Let $\{V_1^l, V_2^l, \ldots, V_{n-1}^l\}$ be the induced $n - 1$-coloring on each H_n, for $l = 1, 2, \ldots, k(n-1) + 1$. Since $\chi(H_n : -\mathcal{G}) = n$, this is not a $\mathcal{G}$-free coloring, hence a copy of G occurs in at least one color class. This happens for each copy H_n in H. Should more than k copies of G occur in any given color class V_j of H, a monochromatic copy of $\mathcal{G}$ would be induced, since $\mathcal{G}$ contains all the graphs G_i as induced subgraphs. There are $n - 1$ color classes, hence at most $k(n-1)$ copies of G can occur in any color class without inducing a copy of $\mathcal{G}$. However, there are $k(n-1) + 1$ copies of H_n in $\mathcal{G}$, making the occurrence of at least one monochromatic copy of $\mathcal{G}$ inevitable.

This proves that $\chi(H : -\mathcal{G}) \geq n$. To prove equality, a $\mathcal{G}$-free coloring for H in n colors is given: color each H_n G-freely in n colors. An n-coloring of H is obtained by taking the union of the jth color classes, for each $j = 1, 2, \ldots, n$. This coloring is $\mathcal{G}$-free, for suppose a copy of $\mathcal{G}$ occurs in some color class. Since $G \triangleleft \mathcal{G}$, a copy of G also occurs in this color class. However, this is impossible. Due to the connectedness of G, if it occurs somewhere, it has to be completely contained in one of the copies of H_n, whereas the fact that the coloring for each H_n is G-free prevents the monochromatic occurrence of G in H_n. Hence

$$\chi(H : -\mathcal{G}) = \chi(H : -F) = n,$$

and thus F and $\mathcal{G}$ are n-chromatically related. $\qquad\square$

Corollary 2. *Let F and G be connected graphs and let $F \asymp G$. Let G_i be a graph with $G_i \triangleleft G$, for each $i = 1, 2, \ldots, k$, and let $\mathcal{G} = (\bigcup_{i=1}^{k} G_i) \bigcup G$. Then $F \asymp \mathcal{G}$.*

Corollary 3. *Let F be a connected graph and let $k \geq 2$ be some integer. Then $F \asymp kF$.*

Corollary 4. *Let F and G be connected graphs and let $k \geq 2$ be some integer. If $F \asymp G$, then $F \asymp kG$.*

In the first part of the proof of this theorem it was not necessary to know that the graph G forms part of the disjoint union. However, to find a coloring for H in n colors, it was necessary for G to be an induced subgraph of $\mathcal{G}$ and for this, G has to be in the union.

A result similar to the previous theorem holds for the join of graphs. This is given next.

Corollary 5. *Let F and G be graphs not having some property of graphs of the form $M + N$. Let $F \asymp_n G$ for some integer $n \geq 2$. Let $G_i \triangleleft G$, for each $i = 1, 2, \ldots, k$, and let $\mathcal{G} = (\sum_{i=1}^{k} G_i) + G$. Then $F \asymp_n \mathcal{G}$.*

Proof. Consider the graphs $\overline{F}$ and $\overline{G}$. They are connected; they are n-related, and the graphs $\overline{G_i}$ are induced subgraphs of $\overline{G}$ for all i. Theorem 1 can thus be applied to deduce that $\overline{F}$ is n-chromatically related to $(\bigcup_{i=1}^{k} \overline{G_i}) \bigcup \overline{G}$. Taking complements again, we find that $F \asymp_n \mathcal{G}$. $\qquad\square$

The reader is referred to the examples concluding Sect. 3.4 of [1] for examples of the property that occurs in this corollary. One obvious property is that of disconnectedness, which emphasizes the complementarity between the theorem and the corollary.

Example 4. The requirement in Theorem 1 that the graphs F and G be connected, is necessary. An example showing this is furnished by the following facts about null graphs to be proved later: N_n is related to N_{2n-2}, but N_n is chromatically alien to $2N_{2n-2}$. Another example is K_n, being related to K_{2n-2} but not to $K_{2n-2} + K_{2n-2}$. This shows the necessity in the corollary for requiring the graphs not to have some property of graphs of the form $M + N$.

13 A Chromatic Metric on Graphs

In the following results, conditions are given for n-chromatic relatedness between graphs to imply m-chromatic relatedness for integers m and n with $m \leq n$ or $m \geq n$. The first is deduced from the general result, Remark 5, holding for classes of graphs.

Remark 10. Let F and G be graphs. If $F \vartriangleleft G$, then $F \asymp_n G$ implies that $F \asymp_m G$ for every $m \leq n$.

Proof. Since $F \vartriangleleft G$ if and only if $-F \subseteq -G$, the remark can be applied to prove the result. □

Recall that the concepts n-relatedness and relatedness were different for classes of graphs. There exist classes of graphs that are n-related, but not m-related for some integer $m \neq n$. However, an analogous example could not be found for graphs, so that it is possible that, at least for graphs, these concepts coincide. In this event, the conclusion of the above result would of course follow without any condition having to be met. A full discussion of this possibility is deferred to Sect. 13.4.

A result that deduces m-relatedness in the other direction can be proved using other conditions on the graphs. Recall that a graph with the girth-endvertex property has girth at least five and no two endvertices have a common neighbor.

Theorem 2. *Let F and G be connected graphs with the girth-endvertex property and let $F \ntriangleleft G$ and $G \ntriangleleft F$. Let $n \geq 1$ be some integer. Then $F \asymp_n G$ implies that $F \asymp_m G$ for every $m \geq n$.*

Proof. By assumption, a $-F$ n-chromatic and $-G$ n-chromatic graph G_n exists, establishing the n-relatedness between F and G. Let

$$G_n^F = F[K_1, G_n, \ldots, G_n].$$

Then $\chi(G_n^F : -F) = n + 1$ by Lemma 3.5.6 of [1]. Furthermore, we also have $\chi(G_n^F : -G) = n$. A $-G$ n-coloring of G_n^F is obtained by coloring each copy of G_n in G_n^F G-freely in n colors and using any of these colors to color the single vertex K_1. By the same lemma, this coloring is indeed G-free, since $G \ntriangleleft F$.

Analogously, a graph G_n^G can be constructed with the properties that $\chi(G_n^G : -F) = n$ and $\chi(G_n^G : -G) = n+1$. But then the graph $G_{n+1} = G_n^F \cup G_n^G$ is a graph for which the following holds. $\chi(G_{n+1} : -F) = \chi(G_{n+1} : -G) = n+1$ by Remark 3.4.2 of [1]. This means that F and G are $n + 1$-related. Thus, for any integer $m \geq n$, $m - n$ applications of this procedure will show that the graphs F and G are m-related. □

This result has the following important implication.

Corollary 6. *For any two connected graphs F and G with the girth-endvertex property and satisfying $F \ntriangleleft G$ and $G \ntriangleleft F$, F is chromatically related to G.*

Proof. By induction: K_1 is a graph establishing 1-chromatic relatedness between F and G. By the theorem, if F and G are n-chromatically related for some integer $n \geq 1$, they are also $n + 1$-chromatically related. □

The following relatedness can, for example, be deduced from this result. All cycles on at least five vertices, all paths on at least four vertices, and all trees satisfying the condition that no two endvertices have a common neighbour, are pairwise related to each other.

13.4 Towards a Characterization of Relatedness

In Remark 8, a necessary condition for two graphs F and G to be n-related for some $n \geq 2$ was deduced: If $F \times_n G$ for some integer $n \geq 2$, then $\chi(F : -G) \leq 2$ and $\chi(G : -F) \leq 2$. Let us consider in detail what it means for two graphs F and G to satisfy $\chi(F : -G) \leq 2$ and $\chi(G : -F) \leq 2$. Only the following four options are open:

- $\chi(F : -G) = 1$ and $\chi(G : -F) = 1$, which means that $G \ntriangleleft F$ and $F \ntriangleleft G$. Although this may surely hold for structurally quite different graphs, additional conditions on F and G may ensure that F and G are related.
- $\chi(F : -G) = 1$ and $\chi(G : -F) = 2$. Then $G \ntriangleleft F$ and $F \triangleleft G$, but F does not occur in G as an induced subgraph to any great extent, since a partition of $V(G)$ in two color classes is sufficient to avoid monochromatic copies of F.
- $\chi(F : -G) = 2$ and $\chi(G : -F) = 1$. This is the same as the previous case, with the roles of F and G interchanged: Now $G \triangleleft F$ and $F \ntriangleleft G$.
- $\chi(F : -G) = 2$ and $\chi(G : -F) = 2$. In this case, both $G \triangleleft F$ and $F \triangleleft G$, which means that $F \cong G$.

Only from the last case does the converse of the implication in the above mentioned corollary follow easily – two isomorphic graphs are of course always related. Corollary 6 provides a partial converse in the first of the above cases. If $F \ntriangleleft G$ and $G \ntriangleleft F$, and, additionally, F and G both satisfy the girth-endvertex property, then they are related.

The question whether the condition on the graphs F and G that $\chi(F : -G) \leq 2$ and $\chi(G : -F) \leq 2$ is also a sufficient condition for relatedness to hold between F and G, has a bearing on two other open questions:

Firstly, as a special case of the result quoted above, relatedness of a graph with K_2 implies that it is bipartite. No examples have been found to prove that the converse does not hold, and it is conjectured that being bipartite is indeed equivalent to relatedness with K_2.

Secondly, requiring that two graphs F and G be related is a seemingly much more severe condition than requiring only n-relatedness to hold between them for some $n \geq 2$; however, no pair of alien graphs has as yet been found for which n-relatedness holds for some $n \geq 2$. Again, one is led to conjecture that these two concepts may be equivalent.

The following interrelationships can be established between these three statements. For ease of reference, we quote them as

Conjecture 1. If F and G are graphs and $n \geq 2$ is an integer such that $F \asymp_n G$, then $F \asymp G$.

Conjecture 2. If F and G are graphs with $\chi(F : -G) \leq 2$ and $\chi(G : -F) \leq 2$, then $F \asymp G$.

Conjecture 3. If G is a bipartite graph, then $G \asymp K_2$.

Theorem 3. *Conjecture 1 is equivalent to Conjecture 2 for connected graphs.*

Proof. Let $\chi(F : -G) \leq 2$ and $\chi(G : -F) \leq 2$. Then the graph $F \cup G$ establishes 2-relatedness between F and G. Since F is connected,

$$\chi(F \cup G : -F) = \max\{\chi(F : -F), \chi(G : -F)\} = \max\{2, 2\} = 2$$

and similarly, $\chi(F \cup G : -G) = 2$. Now, Conjecture 1 can be invoked to deduce that $F \asymp G$.

For the converse, let $F \asymp_n G$ for some integer $n \geq 2$. Then, by Corollary 8, $\chi(F : -G) \leq 2$ and $\chi(G : -F) \leq 2$ and Conjecture 2 implies that $F \asymp G$. $\quad\square$

Theorem 4. *Conjecture 2 implies Conjecture 3.*

Proof. Let G be bipartite. Then $\chi(G) = \chi(G : -K_2) \leq 2$. Also, $\chi(K_2 : -G) = 1$ for any bipartite graph G except K_2; in that case, the value is 2. Thus Conjecture 2 implies that $G \asymp K_2$. $\quad\square$

13.5 Some Specific Relatednesses

Relatedness of the complete graphs will be basic to the main results of this chapter. These graphs exhibit a clustering behavior, the graphs in each cluster being mutually related and the size of the cluster depending on the order of the smallest graph contained in it.

The following inequality, valid for all integers $n \geq 2$,

$$\chi(G : -K_{2n-1}) \leq \left\lceil \frac{\chi(G : -K_n)}{2} \right\rceil$$

shows that two complete graphs are alien if the difference in their orders is large enough. From this result, nothing can be inferred about the relatedness or non-relatedness of the graphs in between. If it were known that two complete graphs are related, then Remark 9 could be used to show that all complete graphs with orders lying between these two are also related to them and each other. It is now shown, using a result of Broere and Frick (Corollary 5 of [6]) that all complete graphs whose orders remain inside these limits are indeed chromatically related.

346   G. Benadé

Theorem 5 ([6]). *Let n, k, and r be positive integers with $n \geq r \geq 2$ and $k \geq 2$. Then there exists a graph G such that*

$$\chi(G : -K_r) = \chi(G : -K_{r+1}) = \cdots = \chi(G : -K_n) = k$$

if and only if $n \leq 2r - 2$.

Corollary 7. *Let n, k and r be positive integers with $n \geq r \geq 2$ and $k \geq 2$. Then the graphs K_r, K_{r+1}, $\ldots$, K_n are all mutually k-chromatically related if and only if $n \leq 2r - 2$.*

Corollary 8. *Let n and r be positive integers with $n \geq r \geq 2$. Then the graphs K_r, K_{r+1}, $\ldots$, K_n are all mutually chromatically related if and only if $n \leq 2r - 2$.*

This shows that the above inequality is the best possible. A useful reformulation of this result is the following characterization of relatedness between complete graphs.

Theorem 6. *Let m and n be positive integers. Then $K_m \asymp K_n$ if and only if there exists an integer $r \geq 2$ such that $r \leq m \leq 2r - 2$ and $r \leq n \leq 2r - 2$.*

Taking complements, the corresponding result for the null graphs is immediately apparent.

Corollary 9. *Let m and n be positive integers. Then $N_m \asymp N_n$ if and only if there exists an integer $r \geq 2$ such that $r \leq m \leq 2r - 2$ and $r \leq n \leq 2r - 2$.*

If n is taken to be 2 in these results, we get $K_2 \asymp K_2$ and the corresponding triviality for N_2. For $n = 3$, the statement is that $K_3 \asymp K_4$, and of course K_2 and K_3 have to be alien, since their orders are too far apart to satisfy the condition of the theorem. Thus the first two clusters consist of K_2 only, and K_3 and K_4.

The next type of graph for which relatedness is considered, is the class of complete m-partite graphs for some integer $m \geq 2$. Such a graph is proved to be related to the complete graph with order m.

Theorem 7. *Let $n_1, n_2, \ldots, n_m$, with $m \geq 2$, be positive integers. Then $K_{n_1, n_2, \ldots, n_m} \asymp K_m$.*

Proof. For each integer $l \geq 2$, define the graph $G_l = K_{p(n)}$, the complete p-partite graph with n vertices in each partition class, where we define

$$p = (m - 1)(l - 1) + 1$$
$$n \geq (l - 1)a$$
$$a = \max\{n_1, n_2, \ldots, n_m\}.$$

Then $\chi(G_l : -K_m) \leq l$: A $-K_m$ l-coloring for G_l is obtained by gathering the p partition classes of G_l in groups of $m - 1$ and taking these groups as color classes.

13 A Chromatic Metric on Graphs   347

In this manner, $l(m-1)$ partition classes can be colored with l colors without inducing a monocolored K_m. Since $l(m-1) = lm - l \geq lm - m - l + 2 = (m-1)(l-1) + 1 = p$ (recall that $m \geq 2$), this is a coloring of all the vertices of G_l. Also, $\chi(G_l : -K_{n_1,n_2,\dots,n_m}) \geq l$.

Suppose that a $K_{n_1,n_2,\dots,n_m}$-free coloring in $l-1$ colors is given for G_l. Denote the p partition classes of G_l by $W_1, W_2, \dots, W_p$, and the $l-1$ color classes in this coloring by $V_1, V_2, \dots, V_{l-1}$. Then the coloring induces a partition of each partition class W_i in $l-1$ parts. Of these parts, at least one will contain at least a vertices, since $|W_i| = n \geq (l-1)a$ for every i. Thus, the first $(m-1)(l-1)$ partition classes W_i each contains at least a vertices all having the same color. However, no more than $m-1$ such sets of a vertices can occur in a given color class, else a monochromatic copy of $K_{a,a,\dots,a}$ will be induced, and hence also a monochromatic copy of $K_{n_1,n_2,\dots,n_m}$. (This follows from the definition of a.)

Now there are exactly $(m-1)(l-1)$ such sets of a vertices each and $l-1$ colors; therefore each color class has to contain *exactly* $m-1$ of these sets. The partition class W_p, however, also contains a monochromatic set of a vertices, occurring say in the color class V_i. In this color class now a monochromatic copy of $K_{a,a,\dots,a}$ is induced, which contradicts the assumption that the partition $V_1, V_2, \dots, V_{l-1}$ is a $K_{n_1,n_2,\dots,n_m}$-free coloring of G_l.

This proves that $\chi(G_l : -K_{n_1,n_2,\dots,n_m}) \geq l$. Since $K_m \vartriangleleft K_{n_1,n_2,\dots,n_m}$, we have

$$l \geq \chi(G_l : -K_m) \geq \chi(G_l : -K_{n_1,n_2,\dots,n_m}) \geq l,$$

and equality holds throughout. Hence, $K_m \asymp_l K_{n_1,n_2,\dots,n_m}$ holds for all integers $l \geq 2$ and this establishes relatedness between K_m and $K_{n_1,n_2,\dots,n_m}$. $\square$

By taking complements in this theorem, a null graph with order m is found to be related to any disjoint union of m complete graphs, for any integer $m \geq 2$.

Corollary 10. *Let* $n_1, n_2, \dots, n_m$, *with* $m \geq 2$, *be positive integers. Then* $K_{n_1} \cup K_{n_2} \cup \dots \cup K_{n_m} \asymp N_m$.

By taking complements throughout, the results proved for complete graphs were translated to the corresponding results about null graphs. The following rather surprising theorem connects these two classes of graphs by establishing relatedness between any complete graph and the null graph with the same order.

Theorem 8. *For all integers* $m, n \geq 2$, $N_m \asymp K_n$.

Let $l \geq 2$ be any integer. The graph $G_{m,n}^l$ establishing the desired l-chromatic relatedness is defined as follows. $G_{m,n}^l$ consists of two copies of $K_{(l-1)(n-1)}$ and two copies of $N_{(l-1)(m-1)}$, together with all edges between vertices of the two null graphs and all edges between the vertices of the pair of graphs $K_{(l-1)(n-1)}$ and $N_{(l-1)(m-1)}$, for both pairs. Now $\omega(G_{m,n}^l) = (l-1)(n-1) + 1$. Thus, $\chi(G_{m,n}^l : -K_n) \geq \lceil \frac{(l-1)(n-1)+1}{n-1} \rceil = l$. For the reverse inequality, we observe that a K_n-free coloring of $G_{m,n}^l$ in l colors exists. Color the copies of $K_{(l-1)(n-1)}$ K_n-freely in $l-1$ colors and the two null graphs in the lth color. This proves that $\chi(G_{m,n}^l : -K_n) = l$.

Now consider the graph $\overline{G_{m,n}^l}$. This is the graph $G_{n,m}^l$, obtained by interchanging the numbers m and n in the above definition of $G_{m,n}^l$. By the same arguments as above, $\chi(G_{n,m}^l : -K_m) = l$. By taking complements in this equation, one obtains that $\chi(\overline{G_{n,m}^l} : -N_m) = \chi(G_{m,n}^l : -N_m) = l$. Thus $G_{m,n}^l$ is a graph for which the K_n-free and the N_m-free chromatic numbers are both l, proving that K_n and N_m are related.

To conclude the section, we prove the main result relating any connected non-complete graph to some complete graph.

Theorem 9. $G \asymp K_n$ for every connected noncomplete graph G and every integer $n > \omega(G)$.

Proof. For any $m \geq 2$, we know from Folkman's theorem that there is a graph G_m with $\chi(G_m : -G) = m$ and $\omega(G_m) = \omega(G)$. Let $H_m = G_m \cup K_r$, where $r = (m-1)(n-1) + 1$. Since $\chi(K_r : -G) = 1$ and G is connected, it follows from Remark 3.4.2 of [1] that

$$\chi(H_m : -G) = \max\{\chi(G_m : -G), \chi(K_r : -G)\} = m$$

Also,

$$\begin{aligned}
\chi(H_m : -K_n) &= \max\{\chi(G_m : -K_n), \chi(K_r : -K_n)\} \\
&= \max\{1, \chi(K_r : -K_n)\} \\
&= \max\left\{1, \left\lceil \frac{r}{n-1} \right\rceil\right\} \\
&= \max\{1, m\} \\
&= m.
\end{aligned}$$

Here, the second equality follows from $n > \omega(G) = \omega(G_m)$, the third from Remark 3.3.2 of [1], and the fourth from the choice of r. Thus we have m-chromatic relatedness between G and K_n for all $m \geq 2$, proving chromatic relatedness. □

Corollary 11. $G \asymp N_n$ for every nontrivial graph G for which the complement $\overline{G}$ is connected and for every integer $n > \beta(G)$.

Proof. Let G be nontrivial and let $\overline{G}$ be connected, with $n > \beta(G)$. Then $\overline{G}$ is noncomplete and $\omega(\overline{G}) = \beta(G) < n$. The theorem then yields $\overline{G} \asymp K_n$ and thus $G \asymp N_n$. □

In this theorem, relatedness of a graph with K_n is deduced whenever the clique number of the graph is less than n. This condition is not necessary, however, since there do exist (noncomplete) graphs that are related to complete graphs with order both equal to and less than their clique numbers.

Example 5. In the next section it is proved that $P_3 \asymp K_2$, whereas $\omega(P_3) = 2$.

Example 6. An example of a noncomplete graph G with $G \asymp K_n$ and $n < \omega(G)$, is $G = K_n \cup K_{n+j}$ for any $1 < j \leq n - 2$: By Theorem 6, $K_n \asymp K_{n+j}$, and since $K_n \triangleleft K_{n+j}$, it follows by Corollary 2 that $K_n \asymp K_n \cup K_{n+j}$. However, the clique number of K_{n+j} is strictly greater than n.

The following result gives a partial converse of Theorem 9. A bound on the clique number is derived if the graph is known to be related to a complete graph.

Theorem 10. *Let G be a graph with $G \asymp K_n$ for some integer $n \geq 2$. Then $\omega(G) \leq 2n - 2$.*

Proof. Let $\chi(H_l : -G) = \chi(H_l : -K_n) = l$ for some graph H_l, for each integer $l \geq 2$. Suppose that $\omega(G) > 2n - 2$. Then, as $K_{\omega(G)} \triangleleft G$, it follows that $\chi(H_l : -G) \leq \chi(H_l : -K_{\omega(G)})$. Furthermore, since $\omega(G) > 2n - 2 \geq n$, we also deduce that $K_n \triangleleft K_{\omega(G)}$. Thus $\chi(H_l : -K_{\omega(G)}) \leq \chi(H_l : -K_n)$. From this it follows for each l that $\chi(H_l : -K_n) = \chi(H_l : -K_{\omega(G)})$ for each l, and hence $K_n \asymp K_{\omega(G)}$. Now, applying Theorem 6, the existence of an integer $r \geq 2$ is inferred for which $r \leq n, \omega(G) \leq 2r - 2$. This is a contradiction. Thus $\omega(G) \leq 2n - 2$. $\qquad\square$

In the special case of $n = 2$, if $G \asymp K_2$, one gets $\omega(G) \leq 2$ on applying the theorem. This means that such graphs G are triangle-free, corroborating the known fact that relatedness with K_2 implies that the graph is bipartite.

One may ask whether a small difference between the clique numbers of two graphs is necessary or sufficient for them to be related. That this is not true is seen by referring to Theorem 8. There, for all integers $m, n \geq 2$, K_n was found to be related to N_m. By making n arbitrarily large, the clique number of K_n can be made arbitrarily large, whereas the clique number of N_m is fixed at 1.

13.6 Distantly Related Graphs

The relations considered up till now are not transitive. In the following definition the transitive closure of the chromatic relatedness–relation is taken to obtain a relation which *is* transitive and therefore an equivalence relation on the class of graphs.

Definition 3. Let $k \geq 0$ be an integer. Two classes P and Q are called k-**distantly related**, denoted by $P \sim_k Q$, if there exist classes $R_0, R_1, \ldots, R_k$ such that

$$P = R_0 \asymp R_1 \asymp \ldots \asymp R_k = Q.$$

P and Q are called **distantly related** if $P \sim_k Q$ for some k, denoted by $P \sim Q$.

It is not immediately evident that such a chain of relatednesses linking any two classes exists. However, if attention is restricted to classes of the form $-F$ for some graph F, then it is proved that, for an arbitrary pair of classes, even a finite number of classes can be found that constitutes a link in this manner. Although the following general remarks may also be formulated for classes, henceforth the theory is developed for this special case only. Applying the above definition to classes of the form $-F$, the term distant relatedness is defined for graphs.

Definition 4. Let $k \geq 0$ be an integer. Two graphs F and G are called k-**distantly related**, denoted by $F \sim_k G$, if there exist graphs $R_0, R_1, \ldots, R_k$ such that

$$P \cong R_0 \asymp R_1 \asymp \ldots \asymp R_k \cong Q.$$

F and G are called **distantly related**, denoted by $F \sim G$, if $F \sim_k G$ for some k.

If F and G are 0-related, they are isomorphic; if they are 1-related, they are related; being 2-related and not 1-related implies that they are alien.

As remarked above, the main goal of this section is to prove that any two graphs on at least two vertices are distantly related. Keeping in mind that the relation $\sim$ is an equivalence relation, another way of formulating this is to say that such graphs all lie in the same equivalence class of $\sim$; this means that there is only one equivalence class, being the class $\mathbf{G} \setminus \{K_1, \emptyset\}$. (Recall that none of the definitions of relatedness makes sense for the graphs K_1 and $\emptyset$.) To prove this assertion, distant relatedness between any two complete graphs is first established. Then, using Theorem 9, distant relatedness is established between connected graphs and complete graphs. By taking complements, the corresponding result holds for disconnected graphs and null graphs. Finally, the link given by Theorem 8 between the complete graphs and the null graphs completes the chain of relatednesses.

Remark 11. If two graphs F and G are k-distantly related for some integer k, then they are l-distantly related for all integers $l \geq k$, since the chain of relatednesses between F and G may be lengthened to an arbitrary length by inserting isomorphic copies of some graph in the chain.

Remark 12. If F and G are graphs and $F \asymp G$, then $F \sim_1 G$ and Remark 11 implies that $F \sim_k G$ for every $k \geq 1$.

We now start on the program outlined above by proving that distant relatedness holds between any two complete graphs.

Lemma 1. *Let m and n be integers with $m, n \geq 3$. Then $K_m \sim K_n$.*

Proof. By Theorem 6, $K_r \sim K_s$ for any two integers r and s with $r \geq 2$ and $2 \leq s \leq 2r - 2$. Hence $K_n \asymp K_{n+1}$ for every $n \geq 3$ and a (finite) chain of relatednesses can thus be found between any two complete graphs with order at least three. □

This does not include the case where one of the complete graphs is K_2, as K_2 is in a "cluster" of its own: Theorem 6 links it to no other complete graph by relatedness; in fact, this theorem asserts that it is alien to all the larger complete graphs. However,

13 A Chromatic Metric on Graphs

K_2 is 2-distantly related to K_3 and thus, by the lemma, distantly related K_n for $n \geq 3$. From the next two results, a chain establishing this 2-distant relatedness is seen to be $K_2 \asymp P_3 \asymp K_3$.

Lemma 2. $K_2 \asymp P_3$.

Proof. Let $m \geq 1$ be an integer. The graph establishing m-chromatic relatedness is a complete m-partite graph $K_{n_1, n_2, \dots, n_m}$. By Remark 3.3.1 of [1], the K_2-free chromatic number of this graph is given by

$$\chi(K_{n_1, n_2, \dots, n_m} : -K_2) = \left\lceil \frac{m}{2-1} \right\rceil = m$$

and by Theorem 3.3.10, the P_3-chromatic number of the same graph is

$$\chi(K_{n_1, n_2, \dots, n_m} : -P_3) = \min_{1 \leq j \leq m} \{m, m - j + n_j\}.$$

Hence, it is sufficient to choose the numbers $n_1, n_2, \dots n_m$ in such a way that $m - j + n_j = m$. This is achieved by letting $n_j = j$ for $j = 1, 2, \dots, m$. Then, for all $m \geq 1$, $\chi(K_{n_1, n_2, \dots, n_m} : -P_3) = \chi(K_{n_1, n_2, \dots, n_m} : -K_2) = m$. $\square$

Lemma 3. $K_3 \asymp P_3$.

Proof. Let p be an even, positive integer. Again,

$$\chi(K_{n_1, n_2, \dots, n_p} : -K_3) = \left\lceil \frac{p}{3-1} \right\rceil = \frac{p}{2}$$

and

$$\chi(K_{n_1, n_2, \dots, n_p} : -P_3) = \min_{1 \leq j \leq p} \{p, p - j + n_j\}.$$

These two chromatic numbers will be equal if $p - j + n_j \geq \frac{p}{2}$, with equality for some j. Let

$$n_j = j \quad \text{if } 1 \leq j \leq \frac{p}{2}$$

$$n_j = j - \frac{p}{2} \quad \text{if } \frac{p}{2} < j \leq p$$

Then $p - j + n_j = \frac{p}{2}$ for all $j > \frac{p}{2}$, thus we obtain the value $\frac{p}{2}$ for the P_3-free chromatic number of $K_{n_1, n_2, \dots, n_p}$ for every even $p \geq 4$. By choosing $m = \frac{p}{2}$ for all even $p \geq 4$, we get

$$\chi(K_{n_1, n_2, \dots, n_p} : -P_3) = \chi(K_{n_1, n_2, \dots, n_p} : -K_3) = m$$

for all $m \geq 2$. Therefore the graphs P_3 and K_3 are chromatically related. $\square$

The preceding lemmas are now summarized as Theorem 11. By taking complements, the corresponding result about the null graphs is also obtained.

352 G. Benadé

Theorem 11. $K_m \sim K_n$ *for all* $m, n \geq 2$.

Corollary 12. $N_m \sim N_n$ *for all* $m, n \geq 2$.

Considering Theorem 9 and its corollary together with the above results, thus far any two connected graphs have been proved distantly related through the complete graphs and any two disconnected graphs are distantly related using the null graphs. The link between the complete graphs and the null graphs, establishing distant relationship between any two graphs, is given by Theorem 8. This proves the final result of this section.

Theorem 12. *Let* F *and* G *be any two graphs with order at least two. Then* F *and* G *are distantly related.*

13.7   The Chromatic Metric

By Theorem 12, a finite chain of relatednesses exists between any two graphs. It now makes sense to try to determine the length of a shortest such chain, which could be called a chromatic distance between the two graphs. This motivates the following definition.

Definition 5. Let F and G be graphs of order at least two. The **chromatic distance** between F and G, denoted by $d_c(F, G)$, is the smallest number k for which F and G are k-distantly related.

By Theorem 12, this is a well-defined function on the class of all pairs of graphs of order two or more, taking integer values k for $k \geq 0$. If, for two graphs F and G, the chromatic distance between them is 0, they are isomorphic; if it is 1, F and G are related; and for values of $d_c(F, G)$ larger than 1 F and G are alien.

Since applying this distance function to any two graphs presupposes that they can be used to define chromatic numbers, the convention agreed upon earlier still holds and, in particular, **any graph occurring as an argument of** d_c **will be assumed to have at least two vertices**.

Remark 13. The chromatic distance is a metric on the class of all graphs of order at least two.

Proof. Let F, G and H be graphs. Then $d_c(F, G) = 0$ if and only if F and G are isomorphic; $d_c(F, G) = d_c(G, F)$ (the function d_c is symmetric); and the triangle inequality holds: $d_c(G, H) \leq d_c(G, F) + d_c(F, H)$.   □

On the strength of this remark, the chromatic distance function is also called the **chromatic metric**.

Remark 14. Let F and G be graphs, with $d_c(F, G) = n$ for some integer n. Then $d_c(\overline{F}, \overline{G}) = n$.

13 A Chromatic Metric on Graphs   353

Proof. If $d_c(F, G) = n$, then there exist graphs $R_0, R_1, \ldots, R_n$ satisfying $F \cong R_0 \asymp R_1 \asymp \cdots \asymp R_n \cong G$.

By taking complements in this chain of relatednesses, one obtains $\overline{F} \cong \overline{R_0} \asymp \overline{R_1} \asymp \cdots \asymp \overline{R_n} \cong \overline{G}$. This proves that $d_c(\overline{F}, \overline{G}) \leq n$. Since n is the smallest integer for which such a chain of relatednesses of length n exists, if $S_0, S_1, \ldots, S_m$ is a sequence of graphs with $m < n$ and $F \cong S_0$, $S_m \cong G$, then there must an index $i \in \{1, 2, \ldots m\}$ for which $S_{i-1} \not\asymp S_i$. Suppose now that $d_c(\overline{F}, \overline{G}) < n$. By taking complements in a chain of relatednesses of length less than n linking $\overline{F}$ and $\overline{G}$, a sequence $S_0, S_1, \ldots, S_m$ as above is obtained, linking F and G. Thus, for some index i, non-relatedness must hold between S_{i-1} and S_i, and thus a non-relatedness is also found in the chain of relatednesses linking $\overline{F}$ and $\overline{G}$. This is a contradiction, proving that n is the length of a shortest such chain between $\overline{F}$ and $\overline{G}$. $\square$

We now combine our previous results by proving that $d_c(F, G) \leq 3$ for all graphs F and G of order at least two.

Theorem 13. *Let F and G be graphs and let $m, n \geq 2$ be integers. Then the following hold.*

(1)
$$\left.\begin{array}{c} d_c(K_m, K_n) \\ d_c(N_m, N_n) \end{array}\right\} = \begin{cases} 0 \ \text{if } n = m \\ 1 \ \text{if } n \neq m \text{ and there is an } r \text{ such} \\ \quad\quad \text{that } r \leq m, n \leq 2r - 2. \\ 2 \ \text{if } m > 2n - 2 \text{ or } n > 2m - 2 \end{cases}$$

(2) $d_c(K_m, N_n) = 1$.

(3) If F is connected and noncomplete, then
$$d_c(F, K_n) \begin{cases} = 1 \ \ \text{if } \omega(F) < n \\ \leq 2 \ \ \text{if } n \leq \omega(F) \leq 2n - 2. \\ \leq 3 \ \ \text{if } 2n - 2 < \omega(F) \end{cases}$$

(4) If $\overline{F}$ is connected and F is nontrivial, then
$$d_c(F, N_n) \begin{cases} = 1 \ \ \text{if } \beta(F) < n \\ \leq 2 \ \ \text{if } n \leq \beta(F) \leq 2n - 2. \\ \leq 3 \ \ \text{if } 2n - 2 < \beta(F) \end{cases}$$

(5) If both F and G are connected and noncomplete, or if they are both nontrivial and have connected complements, then $d_c(F, G) = 2$.

(6) If F is noncomplete and G is nontrivial and $\overline{F}$ and $\overline{G}$ are connected, then $d_c(F, G) \leq 3$.

Proof. (1) These equalities follow directly from Theorem 6 and its corollary.

(2) By Theorem 8, K_m and N_n are related for all integers $m, n \geq 2$.

(3) Let F be connected and noncomplete. Then $F \asymp K_n$ for any integer n for which $\omega(F) < n$. In the second case, the graph $K_{\omega(G)+1}$ may act as a link between K_n and F, because $K_n \asymp K_{\omega(F)+1}$ and $F \asymp K_{\omega(F)+1}$. Lastly, any connected and noncomplete graph with clique number less than some integer n

is related to K_n and also to $K_{\omega(F)+1}$, which is again related to F. In this case, also, $d_c(K_n, F) \geq 2$, since $K_n \asymp F$ implies that $\omega(F) \leq 2n - 2$ (Theorem 10).

(4) By taking complements in the previous statement, the corresponding result for nontrivial graphs with connected complements is obtained.

(5) If F and G are connected and noncomplete, then each is related to some complete graph with order more than its clique number. Hence a complete graph on any number of vertices more than the maximum of the clique numbers of F and G is chromatically related to both graphs, and the distance between them is two if they are not related. Taking complements yields the corresponding statement for nontrivial graphs with connected complements.

(6) For the last case, consider a connected noncomplete graph F and a nontrivial graph G of which the complement is connected. Then it is possible to find an integer n such that $F \asymp K_n$ and $G \asymp N_n$: simply choose n to be more than the maximum of the two numbers $\omega(F)$ and $\beta(G)$. Since $n \geq 2$, it follows from Theorem 8 that $K_n \asymp N_n$, thus the chain $F \asymp K_n \asymp N_n \asymp G$ of relatednesses exists between F and G. This proves the upper bound of three for the distance between F and G. $\qquad\square$

Recall that, for all graphs F and G, at least one of the two numbers $\chi(F : -G)$ and $\chi(G : -F)$ will be at most 2. It was conjectured that if *both* are at most 2, then F and G are related; that is, $d_c(F, G) = 1$. It might thus seem possible that a sufficiently large value for one of these two numbers might result in the distance between F and G also increasing. This theorem now shows that the distance can increase to at most 3, and indeed there are graphs for which the chromatic number of one in terms of the other can be made arbitrarily large, whilst the chromatic distance between them is 2:

For a suitable choice of positive integers m and n, the chromatic number $\chi(K_m : -K_n) = \lceil \frac{m}{n-1} \rceil$ can be made arbitrarily large, whereas $d_c(K_m, K_n) \leq 2$.

Furthermore, although this metric certainly does assume the value 2, so far it is not known whether there exist graphs F and G for which $d_c(F, G) = 3$. By this theorem, such graphs, if they do exist, are in one of the following three categories:

- F is connected and noncomplete and $G = K_n$ for some integer n with $\omega(F) > 2n - 2$.
- $\overline{F}$ is connected and F is nontrivial and $G = N_n$ for some integer n with $\beta(F) > 2n - 2$.
- F and $\overline{G}$ are connected.

However, we suspect that such a pair of graphs does not exist and therefore surmise that the following is true.

Conjecture 4. For all graphs F and G of order at least two, $d_c(F, G) \leq 2$.

Translating this statement to relatednesses, we have that, for any pair of graphs F and G, there exists a graph H such that $F \asymp H \asymp G$. This implies the following:

For any two graphs F and G there exists a graph H such that all four of the chromatic numbers $\chi(F : -H)$, $\chi(H : -F)$, $\chi(H : -G)$, and $\chi(G : -H)$ are at most 2.

13 A Chromatic Metric on Graphs

The converse is not known to hold – that it does, would follow from Conjecture 2 – so this may be a weaker condition on a pair of graphs F and G than requiring that $d_c(F, G) \leq 2$. However, proving the conjecture via this weaker version and Conjecture 2 might turn out to be a more tractable option. Furthermore, as in a direct proof of the conjecture, here also only graphs from the above three categories need to be considered. It now transpires that this weaker condition is indeed true for all graphs F and G, thus making the conjecture a direct corollary to Conjecture 2.

Lemma 4. *For any two graphs F and G there exists a graph H such that $\chi(F : -H) \leq 2$, $\chi(H : -F) \leq 2$, $\chi(H : -G) \leq 2$, and $\chi(G : -H) \leq 2$.*

Proof. For all graphs F and G not in one of the three categories mentioned above, it follows from Theorem 13 that $d_c(F, G) \leq 2$ and hence that there exists a graph H such that $F \asymp H \asymp G$. This implies that these four chromatic numbers are at most two.

Now consider graphs from these three categories (where the third category is dealt with in three cases):

- Let F be a noncomplete and connected graph and let $G = K_n$ for some integer $n \geq 2$ with $2n - 2 < \omega(F)$. Let $H = N_k$, where $k = v(F)$. Then $H \not\preceq F$, $F \not\preceq H$, $G \not\preceq H$, and $H \not\preceq G$ and therefore all four of the chromatic numbers $\chi(F : -H)$, $\chi(H : -F)$, $\chi(H : -G)$, and $\chi(G : -H)$ are 1.
- Let F be nontrivial and $\overline{F}$ connected, and let $G = N_n$ for some integer $n \geq 2$ with $2n - 2 < \beta(F)$. Taking complements in the previous situation, it is seen that $H = K_k$, for $k = v(F)$, is a graph with the desired properties.
- Let F be noncomplete and let F and $\overline{G}$ be connected. Let $H = K_k$ with $k > v(F), v(G)$. Then again, $H \not\preceq F$, $F \not\preceq H$, $H \not\preceq G$, and $G \not\preceq H$.
- Let F be complete, G nontrivial, and $\overline{G}$ connected. Suppose that $F = K_n$ for some integer $n \geq 2$. If $H = N_k$, where $k > v(F), v(G)$, the relevant chromatic numbers are again all equal to 1.
- Let F be complete and G trivial. This is item (5) of the previous theorem; accordingly, $d_c(F, G) = 1$ and the requirement of the lemma is satisfied.

Thus, in all the cases there exists a graph satisfying the conditions of the lemma, and the proof is complete. □

Theorem 14. *Conjecture 2 implies Conjecture 4.*

13.8 Conclusion

The preceding sections culminate in the definition of the chromatic metric, defined on all graphs as the length of a shortest chain of relatednesses between two graphs. It is a remarkable fact that the maximum distance between any two graphs is at most 3. Our results are tied together by the following chain of conjectures. We suspect that two graphs are related if and only if the chromatic number of each in terms of the other is at most two. This would imply two other conjectures: that the bipartite graphs are characterized by relatedness to K_2, and that the maximum distance between any two graphs, as measured with the metric defined above, is actually 2.

These conjectures still await resolution. It would furthermore be worthwhile to investigate a refining of these concepts, leading to metrics that allow a wider range of values. Deducing information about the structure of a graph depending on its distance to known graphs, as exemplified above, would provide a useful tool in the structural analysis of graphs.

References

1. Benadé JG (1990) Some aspects of generalised graph colourings. Ph.D thesis, Department of Mathematics, Rand Afrikaans University
2. Benadé G, Broere I (1988) Generalized colourings: existence of uniquely colourable graphs. Verslagreeks van die Departement Wiskunde, RAU, no. 6/88
3. Benadé G, Broere I (1990) A construction of uniquely C_4-free colourable graphs. Quaest Math 13:259–264
4. Benadé G, Broere I (1992) Chromatic relatedness of graphs. Ars Combinatoria 34:326–330
5. Bollobás B, Thomason AG (1977) Uniquely partitionable graphs. J Lond Math Soc 16: 403–410
6. Broere I, Frick M (1988) A characterization of the sequence of generalized chromatic numbers of a graph. In: Proceedings of the 6th international conference on the theory and applications of graphs, Kalamazoo
7. Broere I, Frick M (1990) On the order of uniquely colourable graphs. Discrete Math 82(3): 225–232
8. Brown JI (1987) A theory of generalized graph colouring. Ph.D. thesis, Department of Mathematics, University of Toronto
9. Brown JI, Corneil DG (1987) On generalized graph colourings. J Graph Theory 11:87–99
10. Burger M (1984) The cochromatic number of a graph. Ph.D. thesis, Department of Mathematics, Rand Afrikaans University
11. Chartrand G, Geller DP (1968) On uniquely colorable planar graphs. J Combin Theory 6: 265–271
12. Chartrand G, Lesniak L (1986) Graphs and digraphs, 2nd edn. Wadsworth, Belmont
13. Folkman J (1970) Graphs with monochromatic complete subgraphs in every edge colouring. SIAM J Appl Math 18:19–24
14. Frick M (1987) Generalised colourings of graphs. Ph.D thesis, Department of Mathematics, Rand Afrikaans University
15. Greenwell D, Lovász L (1974) Applications of product colouring. Acta Math Acad Sci H 25:335–340
16. Harary F (1985) Conditional colorablility of graphs. In: Harary F, Maybee J (eds) Graphs and applications. Proc 1st Col Symp Graph theory. Wiley, New York, pp 127–136
17. Lesniak-Foster L, Straight HJ (1977) The cochromatic number of a graph. Ars Combinatoria 3:39–45
18. Mynhardt CM, Broere I (1985) Generalized colorings of graphs. In: Alavi Y et al (eds) Graph theory and its applications to algorithms and computer science. Wiley, New York, pp 583–594

Chapter 14
Some Applications of Eigenvalues of Graphs

Sebastian M. Cioabă

Abstract The main goal of spectral graph theory is to relate important structural properties of a graph to its eigenvalues. In this chapter, we survey some old and new applications of spectral methods in graph partitioning, ranking, epidemic spreading in networks and clustering.

Keywords Eigenvalues · Graph · Partition · Laplacian

MSC2000: Primary 15A18; Secondary 68R10, 05C99

14.1 Introduction

The study of eigenvalues of graphs is an important part of combinatorics. Historically, the first relation between the spectrum and the structure of a graph was discovered in 1876 by Kirchhoff when he proved his famous matrix-tree theorem. The key principle dominating spectral graph theory is to relate important invariants of a graph to its spectrum. Often, such invariants such as chromatic number or independence number, for example, are difficult to compute so comparing them with expressions involving eigenvalues is very useful. In this chapter, we present some connections between the spectrum of a graph and its structure and some applications of these connections in fields such as graph partitioning, ranking, epidemic spreading in networks, and clustering. For other applications of eigenvalues of graphs we recommend the surveys [44] (expander graphs), [51] (pseudorandom graphs), or [61, 62] (spectral characterization of graphs).

To an undirected graph G of order n, one can associate the following matrices:

- The adjacency matrix $A = A(G)$.

S.M. Cioabă (✉)
Department of Mathematical Sciences, University of Delaware, 501 Ewing Hall,
Newark, DE 19716-2553, USA
e-mail: cioaba@math.udel.edu

M. Dehmer (ed.), *Structural Analysis of Complex Networks*,
DOI 10.1007/978-0-8176-4789-6_14, © Springer Science+Business Media, LLC 2011

This is an n-by-n matrix whose rows and columns are indexed after the vertices of G. For each $u, v \in V(G)$, $A(u, v)$ equals the number of edges between u and v.

- The Laplacian matrix $L = L(G)$.
  It is also known as the combinatorial Laplacian of G and it equals $D - A$, where D is the diagonal matrix containing the degrees of the vertices of G and A is the adjacency matrix of G.
- The normalized Laplacian matrix $\mathcal{L} = \mathcal{L}(G)$.
  This equals $D^{-\frac{1}{2}} L D^{-\frac{1}{2}} = I_n - D^{-\frac{1}{2}} A D^{-\frac{1}{2}}$.

Given a real and symmetric matrix M of order n, we denote its eigenvalues by $\lambda_1(M) \geq \lambda_2(M) \geq \cdots \geq \lambda_n(M)$.

If G is an undirected graph, then all the previous matrices are symmetric and consequently, their eigenvalues are real numbers.

We use the following notation throughout this paper. The eigenvalues of the adjacency matrix $A(G)$ are indexed in nonincreasing order:

$$\lambda_1(G) \geq \lambda_2(G) \geq \cdots \geq \lambda_n(G) \tag{14.1}$$

The eigenvalues of the combinatorial Laplacian matrix $L(G)$ are listed in nondecreasing order:

$$\mu_1(G) \leq \mu_2(G) \leq \cdots \leq \mu_n(G) \tag{14.2}$$

The eigenvalues of the normalized Laplacian $\mathcal{L}(G)$ are listed in nondecreasing order:

$$\theta_1(G) \leq \theta_2(G) \leq \cdots \leq \theta_n(G) \tag{14.3}$$

If G is d-regular, the previous three matrices are related as follows:

$$\mathcal{L}(G) = \frac{1}{d} L(G) = I_n - \frac{1}{d} A(G). \tag{14.4}$$

This implies that the eigenvalues of these matrices satisfy the following equation:

$$\theta_i(G) = \frac{\mu_i(G)}{d} = 1 - \frac{\lambda_{n+1-i}(G)}{d}. \tag{14.5}$$

We list below some basic properties of eigenvalues of graphs. For more details on eigenvalues of graphs see the monographs of Cvetković et al. [24, 25] (for eigenvalues of the adjacency matrix), the survey of Mohar [55] (for eigenvalues of the Laplacian), the monograph of Godsil and Royle [38] (for eigenvalues of the adjacency matrix and of the Laplacian), or the book of Chung [18] (for eigenvalues of the normalized Laplacian). The close relation between eigenvalues and the edge distribution of a graph is outlined in the following section.

For any real and symmetric matrix M of order n, its eigenvalues are real and they can be described as follows.

14 Some Applications of Eigenvalues of Graphs

Theorem 1 (Courant–Fisher). *Let M be a real and symmetric matrix of order n. Then*

$$\lambda_1(M) = \max_{x \in \mathbb{R}^n, x \neq 0} \frac{x^t M x}{x^t x}$$

For any $j \in \{2, \ldots, n\}$,

$$\lambda_j(M) = \min_{u_1,\ldots,u_{j-1} \in \mathbb{R}^n} \max_{\substack{x \in \mathbb{R}^n, x \neq 0 \\ x \perp u_1,\ldots,u_{j-1}}} \frac{x^t M x}{x^t x}$$

$$= \max_{v_1,\ldots,v_{n-j} \in \mathbb{R}^n} \min_{\substack{x \in \mathbb{R}^n, x \neq 0 \\ x \perp v_1,\ldots,v_{n-j}}} \frac{x^t M x}{x^t x}.$$

14.2 Eigenvalues of the Adjacency Matrix

The eigenvalues of the adjacency matrix were studied in 1957 in a paper [23] by Collatz and Sinogowitz. In [23], the authors determined the eigenvalues of the following graphs:

- The complete graph K_n: spectrum $n - 1$ and -1 with multiplicity $n - 1$.
- The path P_n: spectrum

$$2 \cos \frac{\pi j}{n + 1}, j \in \{1, \ldots, n\}. \tag{14.6}$$

- The cycle C_n: spectrum

$$2 \cos \frac{2\pi j}{n}, j \in \{1, \ldots, n\}. \tag{14.7}$$

Collatz and Sinogowitz also showed that the largest eigenvalue of the adjacency matrix of a graph G with n vertices satisfies the following inequalities:

$$2 \cos \frac{\pi}{n + 1} \leq \lambda_1(G) \leq n - 1. \tag{14.8}$$

Equality holds in the first inequality if and only if $G = P_n$ and equality holds in the second inequality if and only if $G = K_n$.

A walk of length r in G is a sequence of vertices $u_0, u_1, \ldots, u_r$ such that u_i is adjacent to u_{i+1} for each $0 \leq i \leq r - 1$. The previous walk is closed if $u_0 = u_r$.

The following lemma can be easily proved by induction.

Lemma 1. *The (u, v)-th entry of A^r equals the number of walks of length r which start at u and end at v.*

Let $W_r(G)$ denote the number of closed walks of length r in G. An easy consequence of the previous result is the following lemma which is the basis of many important results involving eigenvalues of the adjacency matrix. This is often used when studying the eigenvalues of random graphs.

Lemma 2. *For any integer $r \geq 1$,*

$$W_r(G) = \mathrm{tr}A^r = \sum_{i=1}^{n} \lambda_i^r. \tag{14.9}$$

A simple connection between the structure of a graph and its eigenvalues is given by the following result.

Lemma 3. *A graph G is bipartite if and only if the spectrum of its adjacency matrix is symmetric with respect to 0.*

Proof. The proof follows using the previous result. $\qquad\square$

For regular graphs, we have more information regarding the extreme eigenvalues.

Lemma 4. *Let G be a connected d-regular graph on n vertices. Then*

(i) $d = \lambda_1 > \lambda_2 \geq \cdots \geq \lambda_n \geq -d$.
(ii) G is bipartite if and only if $\lambda_n = -d$.

As mentioned earlier, there is a close connection between the eigenvalues of a graph and its structure. The following result, also known as the expander mixing lemma (cf. [44]), exemplifies this connection.

Theorem 2. *Let G be a connected d-regular graph and $\lambda = \max(|\lambda_2|, |\lambda_n|)$. If $S, T \subset V(G)$, then*

$$\left| |E(S,T)| - \frac{d}{n}|S||T| \right| \leq \lambda \frac{\sqrt{|S|(n-|S|)|T|(n-|T|)}}{n}. \tag{14.10}$$

This result implies that if λ is small compared to d, then the edge distribution of G is close to the edge distribution of the random graph with the same edge density as G. The graphs with small λ are called expanders and are very important in many areas of mathematics and computer science (see the excellent survey of Hoory et al. [44] on expander graphs and their applications).

14.3 Eigenvalues of the Laplacian

The first application of the Laplacian of a graph is the matrix-tree theorem or Kirchhoff's theorem [49] (see [9], Chap. II for more details). If L is the Laplacian matrix of a graph G and $i \neq j$ are two vertices of G, then let $L_{(ij)}$ be the matrix obtained from L by deleting row i and column j.

14 Some Applications of Eigenvalues of Graphs 361

Theorem 3 (Matrix-Tree Theorem). *If* $i \neq j$ *are two vertices of a connected graph* G, *then the number of spanning trees of* G *equals the absolute value of* $\det(L_{(ij)})$. *Also, the number of spanning trees of* G *equals* $\frac{\mu_2 \ldots \mu_n}{n}$.

We list now some simple properties of the eigenvalues of the Laplacian of a graph.

Lemma 5. *Let* G *be a graph. Then*

(i) *The Laplacian matrix of* G *is a positive semidefinite matrix.*
(ii) *The smallest eigenvalue* $\mu_1(G)$ *of the Laplacian of* G *equals* 0 *and its multiplicity equals the number of components of* G.
(iii) *The graph* G *is connected if and only if* $\mu_2(G) > 0$.

Proof. Orient the edges of G arbitrarily. Consider a signed incidence matrix N of G with respect to the orientation of the edges. The rows of N are indexed by the vertices of G, the columns of N are indexed by the edges of G, and the entries of N are defined as follows:

$$N(i,e) = \begin{cases} +1, \text{if } i \text{ is the head of } e \\ -1, \text{if } i \text{ is the tail of } e \\ 0, \text{otherwise.} \end{cases}$$

By a simple calculation, it follows that $L(G) = NN^t$. This implies that the Laplacian of G is a positive semidefinite matrix. Thus, all its eigenvalues are nonnegative. Also, for any vector $x \in \mathbb{R}^n$,

$$x^t L x = x^t N N^t x = (N^t x)^t (N^t x) = \sum_{ij \in E(G)} (x_i - x_j)^2. \tag{14.11}$$

Let $H_1, \ldots, H_k$ denote the components of G. For each $i \in \{1, \ldots, k\}$, consider the vector v_i that is the characteristic vector of the component H_i. It is easy to see that v_i is an eigenvector of L corresponding to the eigenvalue 0. Also, $v_1, \ldots, v_k$ are linearly independent which implies that the multiplicity of 0 is at least k.

Let y be an arbitrary eigenvector corresponding to the eigenvalue 0. It follows that $y^t L y = 0$. Using (14.11), it follows that $\sum_{ij \in E(G)} (y_i - y_j)^2 = 0$ which means that the entries of y are constant on each component of G. This implies y is a linear combination of $v_1, \ldots, v_k$ which shows that the multiplicity of 0 as an eigenvalue of $L(G)$ equals the number of components of G. The last part of the theorem follows easily. $\square$

A very important property of the eigenvalues of the Laplacian is that they control the edge distribution in the graph. Chung [19] proved the following result.

Theorem 4. *Let* G *be a connected graph on* n *vertices with average degree* d. *Then for any subsets* $S, T \subset V(G)$,

$$\left| |E(S,T)| - \frac{d}{n} |S||T| \right| \leq \frac{\max_{i \neq 0} |d - \mu_i|}{n} \sqrt{|S|(n - |S|)|T|(n - |T|)}. \tag{14.12}$$

14.4 Eigenvalues of the Normalized Laplacian

The normalized Laplacian was introduced by Chung [18]. We list now some simple properties of the eigenvalues of the normalized Laplacian of a graph. These properties are very similar to those of the Laplacian of G. The eigenvalues of the normalized Laplacian seem to relate better to parameters related to random walks on graphs (cf. [19]).

Lemma 6. *Let G be a graph. Then*

   (i) *The normalized Laplacian matrix of G is a positive semidefinite matrix.*
   (ii) *The smallest eigenvalue $\theta_1(G)$ of the Laplacian of G equals 0 and its multiplicity equals the number of components of G.*
  (iii) *The graph G is connected if and only if $\theta_2(G) > 0$.*

In [19], Chung proved a matrix-tree theorem for the normalized Laplacian.

Theorem 5. *If G is a connected graph, the number of its spanning trees equals*

$$\frac{\prod_{i \in V(G)} d_i}{\sum_{i \in V(G)} d_i} \prod_{i \neq 1} \theta_i$$

The eigenvalues of the normalized Laplacian also influence the edge distribution in the graph. If $S \subset V(G)$, define vol(S) as $\sum_{i \in S} d_i$. Chung [19] proved the following result.

Theorem 6. *If S and T are subsets of vertices in a connected graph G, then*

$$\left| |E(S,T)| - \frac{\mathrm{vol(S)vol(T)}}{\mathrm{vol(G)}} \right| \leq \bar{\theta} \frac{\sqrt{\mathrm{vol(S)vol(\overline{S})vol(T)vol(\overline{T})}}}{\mathrm{vol(G)}}$$

where $\bar{\theta} = \max_{i \neq 0} |1 - \theta_i|$.

14.5 Graph Partitioning Using Eigenvalues and Eigenvectors

There are many examples of graph-theoretic questions which can be formulated as the problem of partitioning the vertices of a graph $G = (V, E)$ into a fixed number $k \geq 2$ of disjoint nonempty subsets $V_1, \ldots, V_k$ such that some objective function $f(V_1, \ldots, V_k)$ is maximized or minimized.

The famous MAX-CUT problem is concerned with determining a partition of G into two parts V_1 and V_2 such that the number of edges between these parts $e(V_1, V_2)$ is maximum.

14 Some Applications of Eigenvalues of Graphs

Finding the edge-connectivity of G is equivalent to finding a partition of G into two parts V_1 and V_2 such that $e(V_1, V_2)$ is minimum. Finding the vertex connectivity of G means determining a partition of G into three parts V_1, V_2, V_3 such that $e(V_1, V_3) = 0$ and $|V_2|$ is minimum.

Given a subset of vertices S, let $\Phi(S) = \frac{e(S,S^c)}{\min(|S|,|S^c|)}$ and $\Psi(S) = \frac{|N(S)\setminus S|}{|S|}$. Determining the edge-expansion constant of a graph means finding the minimum of $\Phi(S)$ taken over all subsets S of $V(G)$. The vertex-expansion of G is the minimum of $\Psi(G)$ over all subsets of S with $|S| \leq \frac{|V(G)|}{2}$.

The idea of using eigenvalues to study graph-partitioning problems originated with Donath and Hoffman [29]. Their work was based on previous results of Hoffman and Wielandt [43]. A partition of the vertex set of a graph G into k nonempty subsets $V_1, \ldots, V_k$ is called a k-partition of G.

Theorem 7 ([29]). *Let $V_1, \ldots, V_k$ be a k-partition of a G such that $|V_i| = n_i$ for each $i \in \{1, \ldots, k\}$ and $n_1 \geq n_2 \geq \cdots \geq n_k \geq 1$. Then*

$$\sum_{1 \leq i < j \leq k} e(V_i, V_j) \geq \frac{1}{2} \sum_{l=2}^{k} n_l \mu_l(G). \tag{14.13}$$

Proof. The proof uses the Hoffman–Wielandt inequality [43] which states that is A and B are two real and symmetric matrices of the same order m, then

$$\mathrm{tr}(AB^t) \leq \sum_{i=1}^{n} \lambda_i(A)\lambda_i(B) \tag{14.14}$$

Taking A to be the Laplacian of G and B the direct sum of all one matrices of order $n_1, n_2, \ldots, n_k$ yields the required result. $\square$

Actually Donath and Hoffman proved some stronger results in [29]. They showed that the previous result is true when one replaces the Laplacian by any matrix of the form $F - A$ where F is a diagonal matrix whose entries sum up to twice the number of edges of G.

If the sizes n_i are not equal, then Donath and Hoffman prove the following improvement of the previous theorem.

Theorem 8. *Let $G = (V, E)$ be a graph and $V = V_1 \cup V_2 \cup \cdots \cup V_k$ be a partition of G into k parts such that $|V_i| = n_i$ and $n_1 \geq n_2 \geq \cdots \geq n_k$. Let $y_2 \geq \cdots \geq y_k$ be the roots of*

$$\left(\sum_i n_i\right) x^{k-1} - 2\sum_{i<j} n_i n_j x^{k-2} + 3\sum_{i<j<l} n_i n_j n_l x^{k-3} - \cdots = 0$$

Then

$$\sum_{1 \leq i < j \leq k} e(V_i, V_j) \geq \sum_{i=2}^{k} y_i \mu_i(G)$$

If k divides n, a k-partition $V_1, \ldots, V_k$ of G is called balanced if $|V_i| = \frac{n}{k}$ for any $i \in \{1, \ldots, k\}$. The k-section width $sw_k(G)$ of a graph G is defined as

$$sw_k(G) = \min \sum_{1 \leq i < j \leq k} e(V_i, V_j),$$

where the minimum is taken over all balanced k-partitions of G. The 2-section width is also called the bisection width. The calculation of the bisection width of a graph G is NP-hard [34], even when it is restricted to the class of d-regular graphs [16].

The result of Donath and Hoffman implies that for any graph G on n vertices,

$$sw_k(G) \geq \frac{n \sum_{l=2}^{k} \mu_l(G)}{2k} \qquad (14.15)$$

In particular, for $k = 2$, the bisection width satisfies the inequality

$$sw_2(G) \geq \frac{n \mu_2(G)}{4} \qquad (14.16)$$

Motivated by questions in parallel computation, Elsässer et al. [31] studied balanced k-partitions of graphs. The authors gave a new proof of inequality (14.15) and characterized the equality case. Note that Theorem 1 of [31] is a particular case of Corollary 4.3.18 from [45].

Theorem 9. *Let G be a connected graph on n vertices and let $V_1, \ldots, V_k$ be a balanced k-partition of G of size $sw_k(G)$. Let $x_1, \ldots, x_k$ be eigenvectors corresponding to the first k smallest eigenvalues of $L(G)$. If*

$$sw_k(G) = \frac{n \sum_{l=2}^{k} \mu_l(G)}{2k}$$

then

(i) For any $i \in \{1, \ldots, k\}$, if $s, t \in V_i$, then $x_j(s) = x_j(t)$ for any $j \in \{1, \ldots, k\}$.
(ii) For any $i \neq j \in \{1, \ldots, k\}$ and any two vertices $s, t \in V_i$, the number of neighbours of s in V_j equals the number of neighbours of t in V_j.

The authors of [31] also provide examples of simple graphs for which the bound from Theorem 7 is far from optimal. We describe some of their examples below.

Given two graphs G and H, the Cartesian product $G \square H$ has vertex set $V(G) \times V(H)$ and its edges are defined as follows:

$$(a_1, b_1) \sim (a_2, b_2)$$

if and only if $a_1 \sim a_2$ and $b_1 = b_2$ or $a_1 = a_2$ and $b_1 \sim b_2$. The Laplacian matrix of $G \square H$ equals $L(G) \otimes I_{|V(H)|} + I_{|V(G)|} \otimes L(H)$. Thus, the eigenvalues of the Laplacian of $G \square H$ are of the form $\mu_i(G) + \mu_j(H)$ for $i \in \{1, \ldots, |V(G)|\}$ and $j \in \{1, \ldots, |V(H)|\}$ (see also [55]).

14 Some Applications of Eigenvalues of Graphs

The $r \times r$ torus graph is the Cartesian product $C_r \square C_r$ of two cycles of length r. This graph is a 4-regular graph and thus, the eigenvalues of its Laplacian are related to the eigenvalues of its adjacency matrix as pointed out by (14.5). Using (14.5) and (14.7), a $\sqrt{n} \times \sqrt{n}$-torus has $\mu_2 = \mu_3 = \mu_4 = 2 - 2 \cos\left(\frac{2\pi}{\sqrt{n}}\right)$. Thus, the right side of Theorem 7 yields a lower bound of $\frac{3\pi^2}{2}$ for sw_4. However, the 4-section width of the previous graph is $4\sqrt{n}$.

When $k = 2$, Bezrukov et al. [8] characterized the graphs for which equality is attained in (14.16).

Theorem 10. *Let $G = (V, E)$ be a connected graph on an even number of vertices. The following statements are equivalent:*

(i) $sw_2(G) = \frac{n\mu_2(G)}{4}$.
(ii) There is an eigenvector corresponding to $\mu_2(G)$ which has only -1 and $+1$ entries.
(iii) In any optimal bisection, $V(G) = V_0 \cup V_1$, any vertex is incident to exactly $\frac{\mu_2(G)}{2}$ edges.

There are many graphs whose bisection width equals $\frac{n\mu_2}{4}$. Such examples are the complete graphs K_n on an even number of vertices (they have $sw_2 = \frac{n^2}{4}$ and $\mu_2 = 4$), the Petersen graph (it has $n = 10, sw_2 = 5$, and $\mu_2 = 2$), and the d-dimensional hypercube Q_d (it has $n = 2^d, sw_2 = 2^{d-1}$, and $\mu_2 = 2$).

However, there are many graphs for which inequality (14.16) is weak. Guattery and Miller [39] constructed some examples of such graphs. We describe one of their examples below. For $k \geq 1$, the graph G_k has vertex set $\{1, \dots, 4k\}$ and consists of two disjoint paths on $2k$ vertices (with vertex set $\{1, \dots, 2k\}$ and $\{2k + 1, \dots, 4k\}$, respectively). Also, for any $1 \leq j \leq k$, the vertex $k + j$ is adjacent to the vertex $3k + j$. The graph G_k is planar and looks like a ladder with $2k$ steps from which the bottom k steps have been removed.

The graph G_k has $4k$ vertices. It has a bisection width 2 since removing the edges $(k, k + 1)$ and $(3k, 3k + 1)$ yields two disjoint components of order $2k$, namely the ones induced by $\{1, \dots, k, 2k + 1, \dots, 3k\}$ and $\{k + 1, \dots, 2k, 3k + 1, \dots, 4k\}$.

Guattery and Miller [39] showed that $\mu_2(G) \leq 4 \sin^2\left(\frac{\pi}{2k}\right)$ and that the spectral partition produces a bisection width of size k (the bisection width given by the spectral method is $\{1, \dots, 2k\}$ and $\{2k + 1, \dots, 4k\}$).

In [8], the authors also improve the Donath–Hoffman lower bound on the bisection width for a graph with specific level structure. We briefly describe their results below. Let $V = V_0 \cup V_1$ be a bisection of a graph G with cut size σ. Let V_0^1 denote the subset of vertices in V_0 that are incident to a cut edge. For $i \geq 1$, let V_0^i denote the set of vertices in V_0 at distance $i - 1$ from a vertex in V_0^1. One can define the sets V_1^i similarly. Also, let $E_\epsilon^i = E(V_\epsilon^i, V_\epsilon^{i+1})$ for $i \geq 1$ and $\epsilon \in \{0, 1\}$. If $g : \mathbb{N} \to \mathbb{N}$, then $LS(g, \sigma)$ denotes the class of graphs which have a bisection of cut size σ and a level structure as above such that $|E_\epsilon^i| \leq \sigma g(i)$ for $\epsilon \in \{0, 1\}$ and all $i \geq 1$.

Theorem 11. *If $G \in LS(g, \sigma)$, then there exists a function $\gamma : \mathbb{R}^+ \to \mathbb{R}$ with $\gamma(x) \to 0$ as $x \to \infty$ such that*

- *If $A := 1 + 2 \sum_{i=2}^{\infty} \frac{1}{g(i-1)} < \infty$, then $\sigma \geq \frac{n\mu_2}{4} \cdot A \left(1 + \gamma \left(\frac{n}{\sigma}\right)\right)$.*
- *If $g(i) = i + 1$, then $\sigma \geq \frac{n\mu_2}{4} \cdot \text{LambertW} \left(\frac{4}{\mu_2}\right) \left(1 + \gamma \left(\frac{n}{\sigma}\right)\right)$, where LambertW$(x)$ is the inverse function of xe^x.*
- *$g(i) = (i + 1)^\alpha$ and $0 \leq \alpha < 1$, then $\sigma \geq \frac{n\mu_2}{4}^{\frac{1+\alpha}{2}} \cdot f(\alpha) \left(1 + \gamma \left(\frac{n}{\sigma}\right)\right)$, where $f(\alpha) = \dfrac{1+\alpha}{2((1-\alpha)(3-\alpha))^{\frac{1+\alpha}{2}}}$.*

In [8], the authors show that there are graphs for which the bounds from the previous theorem are tight up to a constant factor. If G is a graph of maximum degree d, then $G \in LS(g, sw_2(G))$, where $g(i) = (d - 1)^i$. This is because $\max(|V_0^i|, |V_1^i|) \leq sw_2(G)(d - 1)^{i-1}$. In this case, the previous theorem implies that $sw_2(G) \geq \frac{n\mu_2}{4} \cdot \frac{d}{d-2}(1 - o(1))$ as $\frac{n}{sw_2(G)} \to \infty$.

Recall that a connected d-regular graph is called Ramanujan if $|\lambda_i| \leq 2\sqrt{d - 1}$ for each $\lambda_i \neq \pm d$. In [8], the authors use the previous theorem to improve the bisection width of Ramanujan graphs. The Donath–Hoffman bound implies a lower bound of $0.0042n$ for the bisection width of a 3-regular Ramanujan graph and of $0.133n$ for the bisection width of a 4-regular Ramanujan graph. They improve the previous bound to $0.082n$ and $0.176n$, respectively. In the opposite direction, Monien and Preis [56] gave upper bounds on the bisection width of $\left(\frac{1}{6} + \epsilon\right)n$ for 3-regular graphs and of $(0.4 + \epsilon)n$ for 4-regular graphs, for any $\epsilon > 0$, when n is larger than some function of the chosen ϵ. For more recent results regarding the bisection width of random regular graphs, see [28].

Using some eigenvalue interlacing results of Haemers [40], Bollobás and Nikiforov [10] proved the following result which also implies inequality 14.15.

Theorem 12. *Let $V_1, \dots, V_k$ be a k-partition of a graph G on n vertices. Then*

$$\sum_{l=0}^{k-1} \mu_{n-l}(G) \geq \sum_{1 \leq i < j \leq k} e(V_i, V_j) \left(\frac{1}{|V_i|} + \frac{1}{|V_j|}\right) \geq \sum_{l=2}^{k} \mu_l(G). \qquad (14.17)$$

Fiedler [32, 33] used the eigenvalues of the Laplacian in connection with the connectivity of a graph. From the Courant–Fisher theorem we know that

$$\mu_2(G) = \min_{\substack{x \in \mathbb{R}^n, x \neq 0 \\ x \perp 1}} \frac{\sum_{ij \in E(G)} (x_i - x_j)^2}{\sum_{i \in V(G)} x_i^2} \qquad (14.18)$$

In [32], Fiedler called $\mu_2(G)$ the algebraic connectivity of the graph G because of its connections with the usual vertex- and edge-connectivity of G. Recall that the vertex connectivity $k(G)$ of a connected graph G is the minimum number of vertices whose deletion disconnects G. By convention $k(K_n) = n - 1$. The edge

14 Some Applications of Eigenvalues of Graphs

connectivity of a connected graph G is the minimum number of edges whose removal disconnects G. The following result is well known (see [67, 68] for more details).

Lemma 7 ([68]). *If G is a connected graph with minimum degree $\delta(G)$, then*

$$1 \leq k(G) \leq k'(G) \leq \delta(G)$$

Fiedler [32] proved the following important theorem.

Theorem 13 ([32]). *Let G be a connected graph. Then*

$$\mu_2(G) \leq k(G) \leq k'(G).$$

Proof. The proof follows after showing that deleting any vertex from a connected graph decreases the algebraic connectivity by at most 1. More precisely, if i is vertex of a connected graph G on n vertices and $H = G \setminus \{i\}$, then

$$\mu_2(G) \leq \mu_2(H) + 1$$

This can be proved using the Courant–Fisher theorem. □

Kirkland et al. [50] characterized the connected graphs for which $\mu_2(G) = k(G)$. If G_1 and G_2 are distinct graphs, then the join, $G_1 \vee G_2$ of G_1 and G_2 is the graph obtained from the union of G_1 and G_2 by adding all edges between $V(G_1)$ and $V(G_2)$.

Theorem 14 ([50]). *Let G be a connected graph on n vertices. Then $\mu_2(G) = k(G)$ if and only if G can be written as a join $G_1 \vee G_2$ where G_1 is a disconnected graph on $n - k(G)$ vertices and G_2 is a graph on $k(G)$ vertices with $\mu_2(G_2) \geq 2k(G) - n$.*

For regular graphs, Fiedler's results were improved by Krivelevich and Sudakov [51].

Theorem 15 ([51]). *Let G be a connected, d-regular graph. If $\mu_2(G) = d - \lambda_2(G) \geq 2$, then $k'(G) = d$. Also,*

$$k(G) \geq d - \frac{36\lambda^2(G)}{d}$$

Here, $\lambda(G) = \max |\lambda_i(G)|$ where the maximum is taken over all eigenvalues $\lambda_i(G) \neq \pm d$.

The first inequality of the previous theorem was recently improved by the author (see [22] for more details). In [51], Krivelevich and Sudakov show that the error term in the second inequality is tight up to a constant factor.

A graph G of order n is called strongly regular with parameters (n, d, a, b) if it is d-regular, any two adjacent vertices have exactly a common neighbours, and any two nonadjacent vertices have exactly b common neighbours. From the definition, it follows that if A is the adjacency matrix of an (n, d, a, b) strongly regular graph, then

$$A^2 = dI + aA + b(J - I - A)$$

where J is the all one matrix. This implies that the eigenvalues of an (n, d, a, b) strongly regular graph are

$$d, \frac{a - b \pm \sqrt{(a - b)^2 + 4(d - b)}}{2}$$

Strongly regular graphs are well studied and have many connections to finite geometry and algebra.

Brouwer and Mesner [15] showed that the vertex-connectivity of a strongly regular graph equals its degree. Brouwer and Haemers [12] showed that the edge-connectivity of a distance-regular graph equals its degree.

These results were recently improved by Brouwer and Koolen [14] who showed that the vertex-connectivity of a distance-regular graph equals its degree d and that the only disconnecting subsets of size d are the vertex neighborhoods.

In [32], Fiedler obtained other inequalities relating the connectivity of a graph to the eigenvalue of the Laplacian.

Theorem 16 ([32]). *Let G be a connected graph with n vertices and maximum degree $\Delta(G)$. Let $\omega = \frac{\pi}{n}$. Then*

$$\mu_2(G) \geq 2k'(G)(1 - \cos \omega)$$

and

$$\mu_2(G) \geq 2k'(G)(\cos \omega - \cos 2\omega) - 2\Delta(G) \cos \omega (1 - \cos \omega)$$

In another seminal paper [33], Fiedler studied the eigenvectors of the Laplacian of a graph.

Theorem 17 ([33]). *Let G be a connected graph, and let u_2 be an eigenvector corresponding to the eigenvalue $\mu_2(G)$. For any $\beta \in \mathbb{R}$, let $V_+(\beta) = \{i \in V(G) : u_2(i) \geq \beta\}$ and $V_-(\beta) = \{j \in V(G) : u_2(j) \leq \beta\}$. Then for any $\beta \geq 0$, the subgraph induced by $V_+(-\beta)$ is connected and the subgraph induced by $V_-(\beta)$ is connected as well.*

The entries of an eigenvector u_2 corresponding to $\mu_2(G)$ can be used to construct graph-partitioning algorithms. The basic idea of spectral partitioning is to find a splitting value β and partition the graph into $V_-(\beta) = \{i : u_2(i) \leq \beta\}$ and $V(G) \setminus V_-(\beta) = \{j : u_2(j) > \beta\}$. Choosing the value of β depends on the specific application. Some popular choices are the following.

- **Bisection width:** β is the median of $u_2(1), \dots, u_2(n)$.
- **Edge expansion:** β is the value that minimizes $\Phi(S)$, where $S = V_-(\beta)$.
- **Vertex expansion:** β is the value that minimizes $\Psi(S)$, where $S = V_-(\beta)$.

14 Some Applications of Eigenvalues of Graphs

Since the mid-1980s, many researchers (see [1,6,54,60] for example) have studied the connections between the expansion of a graph and its Laplacian eigenvalues. The following result is due to Mohar [54].

Theorem 18 ([54]). *If G is a connected graph, then*

$$\frac{\mu_2(G)}{2} \leq \Phi(G) \leq \sqrt{(2\Delta(G) - \mu_2(G))\mu_2(G)}. \tag{14.19}$$

In [18], Chung proved a similar result involving the eigenvalues of the normalized Laplacian and the expansion properties of a graph. Recall that if S is a subset of vertices of a graph G, vol(S) is defined to be $\text{vol(S)} = \sum_{i \in S} d_i$. The Cheeger ratio of S is defined to be $h_S = \frac{|E(S,\overline{S})|}{\min(\text{volS},\text{vol}\overline{S})}$. This is very similar to the definition of $\Phi(S)$. Also, note that $h_S = h_{\overline{S}}$. The Cheeger constant of G is

$$h_G = \min_{S \subset V(G)} h_S. \tag{14.20}$$

Recall that θ_i is the i-th smallest eigenvalue of the normalized Laplacian $I - D^{-\frac{1}{2}} A D^{-\frac{1}{2}}$. Chung [18] proved the following result (see also [20] for a simpler proof).

Theorem 19. *If G is a connected graph, then*

$$2h_G \geq \theta_2 \geq \frac{s_G^2}{2} \geq \frac{h_G^2}{2}. \tag{14.21}$$

Here s_G is the minimum Cheeger ratio of subsets S_k, consisting of vertices with the largest k values in the eigenvector associated with θ_2, for all $1 \leq k \leq n - 1$.

In a recent paper [21], Chung studied local cuts and local graph partitioning algorithms based on eigenvalues of the normalized Laplacian. More precisely, given a connected graph $G = (V, E)$ and a subset of vertices S, the local Cheeger constant c_S of S is defined as

$$c_S = \min_{T \subset S} \frac{|E(T, \overline{T})|}{\text{vol(T)}}.$$

The closure S^* of S is formed by the vertices of S and the vertices adjacent to a vertex of S. A function $f : S^* \to \mathbb{R}$ satisfies the Dirichlet boundary condition if $f(i) = 0$ for all $i \in S^* \setminus S$. The Dirichlet eigenvalue θ_S of S is defined as

$$\theta_S = \min \frac{\sum_{ij \in E(G)} (f(i) - f(j))^2}{\sum_{i \in S} f^2(i)d_i},$$

where the minimum is taken over all nonzero functions $f : S^* \to \mathbb{R}$ satisfying the Dirichlet boundary condition.

Chung [21] proves the following local Cheeger inequality.

Theorem 20. *Let G be a connected graph and S be a subset of vertices of G such that $G[S]$ is connected. Then*

$$c_S \geq \theta_S \geq \frac{c_S^2}{2}.$$

The proof of this theorem yields a simple algorithm (which is based on the eigenvector corresponding to θ_S) for finding a local cut. The previous theorem will guarantee that this cut is within a quadratic of the optimum.

Spielman and Teng [58, 59] have proved that the spectral partitioning method works well for planar graphs.

Theorem 21 ([58]). *Let G be a connected planar graph on n vertices and maximum degree $\Delta(G)$. Then*

$$\mu_2(G) \leq \frac{8\Delta(G)}{n}.$$

Proof. From the Courant–Fisher theorem, we know that

$$\mu_2(G) = \min_{x \perp 1} \frac{\sum_{ij \in E(G)} (x_i - x_j)^2}{\sum_{i \in V(G)} x_j^2}.$$

It follows that

$$\mu_2(G) = \min \frac{\sum_{ij \in E(G)} ||v_i - v_j||^2}{\sum_{i \in V(G)} ||v_i||^2}, \tag{14.22}$$

where the minimum is taken over all vectors $v_1, \ldots, v_n \in \mathbb{R}^n$ such that $\sum_{i \in V(G)} v_i = \vec{0}$.

Spielman and Teng now use the following *kissing disk* theorem of Koebe, Andreev, and Thurston (see [58, 59] for more details).

Theorem 22 (Koebe–Andreev–Thurston). *If G is a planar graph with vertex set $\{1, \ldots, n\}$, then there exist a set of disks $\{D_1, \ldots, D_n\}$ in the plane such that (i, j) is an edge of G if and only if D_i touches D_j.*

A cap is the intersection of a half-space with a sphere and its boundary is a circle. Using a stereographic projection to map the kissing disk embedding of the graph G to a kissing cap embedding of G, Spielman and Teng show that we can represent the planar graph G by kissing caps on the unit sphere such that the centroid of the centers of the caps is the center of the sphere.

Let v_i be the center of the cap corresponding to vertex i. One can assume that $\sum_{i \in V(G)} v_i = \vec{0}$. Denote by r_i the radius of the cap corresponding to vertex i. If cap i touches cap j, then $||v_i - v_j|| < r_i + r_j$ by the triangle inequality. This implies that

$$\sum_{ij \in E(G)} ||v_i - v_j||^2 < \sum_{ij \in E(G)} (r_i + r_j)^2 \leq \sum_{ij \in E(G)} 2(r_i^2 + r_j^2)$$

$$\leq 2\Delta(G) \sum_{i \in V(G)} r_i^2.$$

14 Some Applications of Eigenvalues of Graphs 371

Since the caps do not overlap, it follows that the area of the unit sphere is larger than the sum of the areas of the caps which implies that

$$4\pi \geq \sum_{i \in V(G)} \pi r_i^2.$$

Using inequality (14.22), the desired result follows. □

The following theorem of Mihail [53] (see [54] for related results and see [18–20] for similar results for the normalized Laplacian) shows that one can use eigenvectors corresponding to μ_2 (or approximations of such eigenvectors) to find subsets S with small $\Phi(S)$.

Theorem 23. *Let G be a connected graph with maximum degree $\Delta(G)$, and let*

$$\Phi_G = \min_{S \subset V(G)} \Phi(S).$$

Then for any vector $v \in \mathbb{R}$ with $\mathbf{1} \perp v$, we have that

$$\Phi_G^2 \leq 2\Delta(G) \frac{v^t L(G)v}{v^t v}.$$

Also, there exists $\beta \in \mathbb{R}$ such that $\Phi(S) \leq \sqrt{2\Delta(G) \frac{v^t L(G)v}{v^t v}}$, where $S = V_-(\beta)$.

Combining the previous two results, one can deduce that the edge-expansion constant of a planar graph G on n vertices is at most $\frac{4\Delta(G)}{\sqrt{n}}$. Also, there is a polynomial time algorithm for finding a subset S such that $\Phi(S) \leq \frac{4\Delta(G)}{\sqrt{n}}$. By a classical result of Lipton and Tarjan [52], it follows that the previous result is tight up to a constant factor.

The genus of a graph G is the smallest g such that G can be embedded in a surface of genus g without any edge crossings. Planar graphs are graphs of genus 0.

Kelner [47] extended the previous results of Spielman and Teng to graphs of genus g.

Theorem 24. *Let G be a graph with n vertices, genus g, and bounded degree. Then*

$$\mu_2(G) \leq O\left(\frac{g}{n}\right) \tag{14.23}$$

where the constant in the O-notation depends on $\Delta(G)$.

Combining this result with Mihail's theorem yields a polynomial time algorithm for finding a subset S of a graph G of order n, genus g, and bounded maximum degree such that $\Phi(S) \leq O\left(\sqrt{gn}\right)$. Using a method described in the appendix of [59], one can use this algorithm to find a bisection of size $O(\sqrt{gn})$.

These results are tight up to a constant factor as shown by the examples found by Gilbert et al. [35]. The authors described a class of bounded degree graphs with no bisection of size less than $O(\sqrt{gn})$.

Clustering is the partitioning of data into groups of similar items. A clustering algorithm performs well if items that are similar are assigned to the same cluster and items that are not similar are assigned in different clusters. This situation can be modeled by a weighted graph in which the weight of an edge w_{ij} measures the similarity between the vertices i and j.

Kannan et al. [46] suggested the following measure of the quality of a clustering of a graph. Given a connected weighted graph $G = (V, E)$, a partition $V = V_1 \cup \cdots \cup V_k$ is called an (a, ϵ)-clustering if

- $\Phi_{G[V_i]} \geq a$ (the subgraph induced by each cluster has weighted edge expansion at least a).
- $\sum_{1 \leq i < j \leq k} |E(V_i, V_j)| \leq \epsilon |E(G)|$ (the weight of the intercluster edges is at most a times the weight of the edges in G).

The following optimization problem is studied in [46]: given a, find an (a, ϵ)-clustering that minimizes ϵ. In [46], the authors use a recursive algorithm based on the spectral partitioning method described above to find an approximate algorithm for the previous problem. They show that if G has an (a, ϵ)-clustering, then using the spectral partitioning method, one can find an $\left(\frac{a^2}{72 \log^2(n/\epsilon)}, 20\sqrt{\epsilon} \log(n/\epsilon) \right)$-clustering.

The idea of using eigenvectors for ranking goes back to Kendall [48] and Wei [66]. Brin and Page [11] introduced the notion of PageRank in their seminal paper on Web search. The Web pages are classified according to their importance scores given by PageRank which are computed from the graph structure of the Web. The PageRank importance of a Web page is determined by the PageRank importance of the Web pages linking to it.

The Web is regarded as a graph with nodes being Web pages and edges being hyperlinks. The basic idea of PageRank is that links from important vertices should weigh more than links from less important vertices.

Consider a connected graph G with adjacency matrix A and let D denote the diagonal matrix containing the degrees of the vertices of G. Define $W = D^{-1}A$. Thus,

$$W(i, j) = \begin{cases} \frac{1}{d_i}, & \text{if } i \sim j \\ 0, & \text{otherwise.} \end{cases}$$

The matrix W can be regarded as the transition probability matrix of a random walk on the vertices of G. The stationary distribution of this random walk is the row vector $\pi = \left(\frac{d_1}{\text{vol}(G)}, \ldots, \frac{d_n}{\text{vol}(G)} \right)$.

The PageRank vector $\text{pr}(a, s)$ of a graph G is the unique solution of the equation

$$\text{pr}(a, s) = as + (1 - a)\text{pr}(a, s)W,$$

14 Some Applications of Eigenvalues of Graphs

where $a \in (0, 1]$ is a jumping constant and s is a starting vector. The PageRank vector associated with search ranking has $s = \frac{1}{n}\mathbf{1}$. PageRank vectors whose starting vectors are concentrated on a small number of vertices are called personalized PageRank vectors and were introduced by Haveliwala [41]. The PageRank vector can be used to design graph partitioning algorithms (see Andersen et al. [7] and Chung [20]).

An important problem in communication networks is the following: given a connected graph G and a set of pairs of vertices (s_i, t_i), $1 \leq i \leq r$, find r edge-disjoint paths $Q_1, \ldots, Q_r$, where Q_i connects s_i to t_i. In [4], Alon and Capalbo used the connections between the edge distribution of a graph and its eigenvalues to prove the following result.

Theorem 25. *Let G be a connected d-regular graph and let $\lambda = \max(|\lambda_2|, |\lambda_n|)$. Assume that $d > 8\lambda$ and let $c > 0$ be a constant and $r := c\frac{nd \log(d/4\lambda)}{\log n}$. Given r pairs of vertices (s_i, t_i) such that no vertex of G appears more than $\frac{d}{3}$ times as s_i or t_i, there exists a polynomial time algorithm that finds r edge-disjoint paths Q_i such that Q_i joins s_i to t_i.*

The questions of finding paths of logarithmic length between each pair remains open.

14.6 Epidemic Spreading in Networks

It is well known that graphs can be used as abstract models of various networks that appear in computer science, biology, and sociology among others. The problem of virus propagation has been studied in these areas and various models have been proposed.

The susceptible–infective–susceptible (SIS) model assumes that each node of a network (graph) can be in one of two states: healthy but susceptible (S) to infection, or infected (I). An infected node can spread infection along the network to susceptible nodes. An infected node can be cured locally and it becomes susceptible again. A directed edge from node i to node j means that i can infect j. A rate of infection β is associated with each edge and a virus curing rate, δ, is associated with each infected node.

The epidemic threshold of a graph G is the value τ such that if $\frac{\beta}{\delta} < \tau$, then the viral outbreak dies out over time and if $\frac{\beta}{\delta} < \tau$, then the infection survives.

Recently, Wang et al. [65] found connections between the eigenvalues of a graph and the epidemic threshold in the SIS model.

Consider a connected network (graph) $G = (V, E)$. The model considered in [65] assumes discrete time. During each time interval, an infected node i tries to infect its neighbours with probability β. At the same time, the node i can be cured with probability δ.

Recall that $\lambda_1(G)$ denotes the largest eigenvalue of the adjacency matrix of G. The main result of [65] is the following theorem whose proof we sketch below.

Theorem 26. *The epidemic threshold of a graph G equals $\frac{1}{\lambda_1(G)}$.*

Proof. Let $p_{i,t}$ denote the probability that i is infected at time t and $q_{i,t}$ denote the probability that i will not be infected by its neighbours at time t.

A node i is healthy at time t if

- i was healthy at time $t-1$ and did not receive infections from its neighbours at t.
- i was infected before t, cured at t, and did not receive infections from its neighbours at t.
- i was infected before t, received and ignored infections from its neighbours at time t, and was cured at time t.

Assume that the probability that a curing event at node i takes place after infection from neighbours is 50%. This means that

$$1 - p_{i,t} = (1 - p_{i,t-1})q_{i,t} + \delta p_{i,t-1}q_{i,t} + \frac{1}{2}\delta p_{i,t-1}(1 - q_{i,t}).$$

Let P_t denote the column vector $(p_{1,t}, \ldots, p_{n,t})$. From the previous equation, one can obtain that

$$P_t = ((1 - \delta)I_n + \beta A(G))\, P_{t-1}. \tag{14.24}$$

For the infection to die off, the vector P_t should tend to zero as t gets large. This will happen when for each i, the i-th eigenvalue of $((1 - \delta)I_n + \beta A(G))^t$ tends to 0 as t gets large. It follows that $1 - \delta + \beta\lambda_1(G) < 1$ which means that $\tau = \frac{1}{\lambda_1(G)}$. $\square$

Using the previous argument, it is shown in [65] that when $\frac{\beta}{\delta}$ is below the epidemic threshold, the number of infected nodes decays exponentially over time.

The result from [65] motivated further research. In [64], the authors studied the problem of minimizing the spectral radius of a connected graph of order n and diameter D. In [64], the authors solved this problem when $D \in \{1, 2, n - 3, n - 2, n - 1\}$, but many questions remain open (see [63, 64] for more details).

Another model for epidemic spreading in networks is the susceptible-infective-removed (SIR) model. Consider again a graph G with n vertices. Each vertex can be in one of three possible states, susceptible (S), infective (I), or removed (R). Again, we assume discrete time. We assume that the initial set of infective vertices at time 0 is nonempty, and the rest of the vertices are susceptible at time 0.

Let $X_i(t)$ denote the indicator that the vertex i is infected at time t and $Y_i(t)$ the indicator that i is removed at time t. Each vertex that is infected tries to infect each of its neighbours; each infection attempt is successful with probability β independent of other infection attempts. Each infected node is removed at the end of the time slot. It follows that the probability that a vertex i becomes infected at the end of time t is $1 - \prod_{j \sim i}(1 - \beta X_i(t))$. The evolution stops when there are no more infective vertices in the graph. One would like to know how many vertices are removed at this time.

This model was studied by Draief et al. [30] who proved the following theorem.

14 Some Applications of Eigenvalues of Graphs 375

Theorem 27. *Assume that $\beta\lambda_1(G) < 1$. Then, the total number of vertices removed $|Y(\infty)|$ satisfies the inequality*

$$E[|Y(\infty)|] \leq \frac{\sqrt{n|X(0)|}}{1 - \beta\lambda_1(G)}, \qquad (14.25)$$

where $X(0)$ is the number of initial infective vertices.

14.7 Eigenvalues and Other Graph Invariants

Finding the chromatic number of a graph is also a graph partitioning problem. Among the first results connecting the eigenvalues of a graph to its chromatic and independence number were the following theorems due to Wilf [69], Delsarte [27], and Hoffman [42]. These are classical results with many applications in discrete mathematics and also more recent applications in quantum computing (see [26,36]). For extensions of these results and other applications, see Haemers [40], Nikiforov [57], or Godsil and Newman (see [37] and [36]).

Theorem 28 ([69]). *If G is a connected graph with chromatic number $\chi(G)$, then*

$$\chi(G) \leq 1 + \lambda_1(G).$$

Theorem 29 ([27,42]). *If G is a connected graph of order n, then*

$$\chi(G) \geq 1 + \frac{\lambda_1}{-\lambda_n}.$$

If G is d-regular and $\alpha(G)$ denotes the independence number of G, then

$$\alpha(G) \leq \frac{-n\lambda_n}{d - \lambda_n}.$$

Godsil and Newman [37] have obtained similar results for graphs containing loops and used these results to find bounds for the independence number of the Erdös–Rényi graphs.

We note here the results of Alon et al. [5] which provide inequalities in the opposite direction.

Theorem 30. *Let G be a connected d-regular graph and let $\lambda = \max(|\lambda_2|, |\lambda_n|)$. Then for any subset S of vertices of G, the subgraph $G[S]$ induced by S contains an independent set of size*

$$\alpha(G[S]) \geq \frac{n}{2(d - \lambda)} \ln\left(\frac{|S|(d - \lambda)}{n(\lambda + 1)} + 1\right).$$

Also, the chromatic number of G satisfies the inequality

$$\chi(G) \leq \frac{6(d - \lambda)}{\ln\left(\frac{d-\lambda}{\lambda+1} + 1\right)}.$$

For d-regular graphs with $\lambda = O(\sqrt{d})$, the previous result implies $\chi(G) = O(d/\ln d)$. As described in [5, 51], there are many graphs with this property.

As mentioned earlier, the MAX-CUT problem is an example of a graph partitioning problem. Alon [2] used the following result to find tight bounds for the maximum cut of several families of graphs such as triangle-free graphs. Given a graph G, let $f(G)$ denote the maximum number of edges in a bipartite subgraph of G.

Theorem 31. *If G is a d-regular graph of order n, then*

$$f(G) \leq \frac{n(d - \lambda_n)}{4}.$$

Alon [2] showed that if G is a triangle-free graph with e edges, then it contains a bipartite subgraph with at least $\frac{e}{2} + ce^{\frac{4}{5}}$ edges and this result is tight up to the constant c. This result was extended by Alon et al. in [3] who showed that graphs with girth at least $r \geq 4$ contain a bipartite subgraph with at least $\frac{e}{2} + c'e^{\frac{r}{r+1}}$ edges and this result is tight up to a constant factor for $r = 4, 5$.

Butler and Chung [17] extended previous results of Krivelevich and Sudakov [51] and found an eigenvalue condition that implies the existence of a Hamiltonian cycle in a graph.

Theorem 32. *Let G be a connected graph with average degree d. If there exists a positive constant C such that*

$$|d - \mu_i| \geq C \frac{(\log \log n)^2}{\log n (\log \log \log n)} d$$

for $i > 1$ and n is sufficiently large, then G contains a Hamiltonian cycle.

Brouwer and Haemers [13] conjectured that any strongly regular graph (except the Petersen graph) is Hamiltonian. They have verified this conjecture for graphs with at most 99 vertices.

14.8 Conclusions

As our knowledge and technology advance, the complexity of the social, communication, and biological networks surrounding us is increasing rapidly. Many important combinatorial parameters of large networks are often hard to calculate or

14 Some Applications of Eigenvalues of Graphs

approximate. Eigenvalues provide an effective and efficient tool for studying properties of large graphs which arise in practice. In this chapter, we presented some applications of eigenvalues of graphs. Spectral graph theory is a very dynamic area that will continue to grow. We believe that more applications and tighter connections between graph eigenvalues and other graph invariants will be found in the future.

Acknowledgments This work is supported by a start-up grant from the Department of Mathematical Sciences at the University of Delaware. The author is grateful to the referees for their comments.

References

1. Alon N (1986) Eigenvalues and expanders. Combinatorica 6:83–96
2. Alon N (1996) Bipartite subgraphs. Combinatorica 16:301–311
3. Alon N, Bollobás B, Krivelevich M, Sudakov B (2003) Maximum cuts and judicious partitions of graphs without short cycles. J Combin Theory B 88:329–346
4. Alon N, Capalbo M (2007) Finding disjoint paths in expanders deterministically and online. In: FOCS, pp 518–524
5. Alon N, Krivelevich M, Sudakov B (1999) List coloring of random and pseudo-random graphs. Combinatorica 19:453–472
6. Alon N, Milman V (1985) λ_1, Isoperimetric inequalities for graphs and superconcentrators. J Combin Theory B 38:73–88
7. Andersen R, Chung F, Lang K (2006) Local graph partitioning using pagerank vectors. In: Proceedings of the 47th annual IEEE symposium on foundation of computer science, pp 475–486
8. Bezrukov S, Elsässer R, Monien B, Preis R, Tillich J-P (2004) New spectral lower bounds on the bisection width of graphs. Theor Comput Sci 320:155–174
9. Bollobás B (1998) Modern graph theory. Graduate texts in mathematics, vol 184. Springer, New York
10. Bollobás B, Nikiforov V (2004) Graphs and Hermitian matrices: eigenvalue interlacing. Discrete Math 289:119–127
11. Brin S, Page L (1998) The anatomy of a large-scale hypertextual Web search engine. Comput Netw ISDN Syst 30:107–117
12. Brouwer A, Haemers W (2005) Eigenvalues and perfect matchings. Lin Algebra Appl 395:155–162
13. Brouwer A, Haemers W (2010) Spectra of graphs. Monograph in preparation +200pp, available at http://homepages.cwi.nl/~aeb/math/ipm.pdf
14. Brouwer A, Koolen J (2009) The vertex connectivity of a distance regular graph. Eur J Combinator 30:668–673
15. Brouwer A, Mesner DM (1985) The connectivity of strongly regular graphs. Eur J Combinator 6:215–216
16. Bui TN, Chaudhuri S, Leighton FT, Sipser M (1987) Graph bisection algorithms with good average case behaviour. Combinatorica 2:171–191
17. Butler S, Chung FRK (2010) Small spectral gap in the combinatorial Laplacian implies Hamiltonian, Annals of Combinatorics 13(4):403–412
18. Chung FRK (1997) Spectral graph theory. American Mathematical Society, Providence, RI
19. Chung FRK (2004) Discrete isoperimetric inequalities. Surveys in differential geometry IX. International Press, Somerville, MA, pp 53–82
20. Chung FRK (2007) Four proofs of the Cheeger inequality and graph paritition algorithms. In: Proceedings of ICCM, vol II

21. Chung FRK (2007) Random walks and local cuts in graphs. Lin Algebra Appl 423:22–32
22. Cioabă SM (2010) Eigenvalues and edge-connectivity of regular graphs. Lin Algebra Appl 432:458–470
23. Collatz L, Sinogowitz U (1957) Spektren endlicher Grafen. Abh Math Sem Univ Hamburg 21:63–77
24. Cvetković D, Doob M, Sachs H (1980) Spectra of graphs – theory and application. Academic, New York, 3rd edn, Johann Ambrosius Barth (1995)
25. Cvetković D, Rowlinson P, Simic S (1997) Eigenspaces of graphs. Cambridge University Press, Cambridge
26. de Klerk E, Pasechnik DV (2007) A note on the stability of an orthogonality graph. Eur J Combinator 28:1971–1979
27. Delsarte P (1973) An algebraic approach to the association schemes of coding theory. Philips Res Rep Suppl 10:1–97
28. Diáz J, Do N, Serna MJ, Wormald NC (2003) Bounds on the max and min bisection of random cubic and random 4-regular graphs. Theor Comput Sci 307:531–548
29. Donath WE, Hoffman AJ (1973) Lower bounds for the partitioning of the graphs. IBM J Res Dev 17:420–425
30. Draief M, Ganesh A, Massoulié L (2008) Threshold for virus spread on networks. Ann Appl Probab 18(2):359–378
31. Elsässer R, Lücking T, Monien B (2003) On spectral bounds for the k-partitioning of graphs. Theor Comput Syst 36:461–478
32. Fiedler M (1973) Algebraic connectivity of graphs. Czech Math J 23:298–305
33. Fiedler M (1975) A property of eigenvectors of non-negative symmetric matrices and its applications to graph theory. Czech Math J 85:619–633
34. Garey MR, Johnson DS, Stockmeyer L (1976) Some simplified NP-complete graph problems. Theor Comput Sci 1:237–267
35. Gilbert J, Hutchinson J, Tarjan R (1984) A separation theorem for graphs of bounded genus. J Algorithm 5:391–407
36. Godsil CD, Newman MW (2008) Coloring an orthogonality graph. SIAM J Discrete Math 22:683–692
37. Godsil CD, Newman MW (2008) Eigenvalue bounds for independent sets. J Combin Theory B 98:721–734
38. Godsil CD, Royle G (2001) Algebraic graph theory. Springer, Berlin
39. Guattery S, Miller GL (1998) On the quality of spectral separators. SIAM J Matrix Anal Appl 19:701–719
40. Haemers W (1995) Interlacing eigenvalues and graphs. Lin Algebra Appl 226/228:593–616
41. Haveliwala TH (2003) Topic-sensitive PageRank: a context-sensitive ranking algorithm for Web search. IEEE Trans Knowl Data Eng 15:784–796
42. Hoffman AJ (1970) On eigenvalues and colorings of graphs. In: Graph theory and its applications. (Proceedings of an advanced seminar, mathematical research center, University of Wisconsin, Madison, WI, 1969). Academic, New York, pp 79–91
43. Hoffman AJ, Wielandt H (1953) The variation of the spectrum of a normal matrix. Duke Math J 29:37–39
44. Hoory S, Linial N, Wigderson A (2006) Expander graphs and their applications. Bull AMS 43:439–561
45. Horn RA, Johnson CR (1985) Matrix analysis. Cambridge University Press, Cambridge
46. Kannan R, Vempala S, Vetta A (2004) On clusterigs: good, bad and spectral. J ACM 51:497–515
47. Kelner JA (2004) Spectral partitioning, eigenvalue bounds, and circle packings for graphs of bounded genus. In: Proceedings of the symposium on the theory of Computing (STOC)
48. Kendall MG (1955) Further contributions to the theory of paired comparisons. Biometrics 11:43–62
49. Kirchhoff G (1847) Über die Auflösung der Gleichungen auf welche man bei der Untersuchung der Linearen Vertheilung galvanischer Ströme geführt wird. Ann Phys Chem 72:497–508

14 Some Applications of Eigenvalues of Graphs 379

50. Kirkland SJ, Molitierno JJ, Neumann M, Shader BL (2002) On graphs with equal algebraic and vertex connectivity. Lin Algebra Appl 341:45–56
51. Krivelevich M, Sudakov B (2006) Pseudo-random graphs. In: More sets, graphs and numbers. Bolyai Society Mathematical Studies, vol 15, pp 199–262
52. Lipton RJ, Tarjan RE (1979) A separator theorem for planar graphs. SIAM J Appl Math 36:177–189
53. Mihail M (1989) Conductance and convergence of Markov chains – a combinatorial treatment of expandes. In: Proceedings of the 30th annual symposium on foundations of computer science. IEEE Computer Society, Washington, DC, pp 526–531
54. Mohar B (1989) Isoperimetric numbers of graphs. J Combin Theory B 47:274–291
55. Mohar B (1997) Some applications of Laplace eigenvalues of graphs. In: Hahn G, Sabidussi G (eds) Graph symmetry. Kluwer, Dordrecht, pp 225–275
56. Monien B, Preis R (2001) Upper bounds on the bisection width of 3 and 4-regular graphs. In: Pultr A, Sgall J, Kolman P (eds) Mathematical foundations of computer science, vol 2136. Lecture Notes in Computer Science. Springer, pp 524–536
57. Nikiforov V (2001) Some inequalities for the largest eigenvalue of a graph. Combinator Probab Comput 11:179–189
58. Spielman D, Teng SH (1996) Spectral partitioning works: planar graphs and finite element meshes. In: Proceedings of the 37th annual symposium on foundations of computer science, pp 96–106
59. Spielman D, Teng SH (2007) Spectral partitioning works. Lin Algebra Appl 421:284–305
60. Tanner RM (1984) Explicit construction of concentrators from generalized N-gons. SIAM J Algebra Discr 5:287–294
61. van Dam E, Haemers W (2003) Which graphs are determined by their spectrum? Lin Algebra Appl 373:241–272
62. van Dam E, Haemers W (2009) Developments on spectral characterization of graphs. Discrete Math 309:576–586
63. van Dam E, Jamakovic A, Kooij RE, Van Mieghem P (2006) Robustness of networks against viruses: the role of spectral radius. In: Proceedings of the 13th annual symposium of the IEEE/CVT Benelux, Liége, Belgium, pp 35–38
64. van Dam E, Kooij RE (2007) The minimal spectral radius of graphs with give diameter. Lin Algebra Appl 423:408–419
65. Wang Y, Chakrabarti D, Wang C, Faloutsos C (2003) Epidemic spreading in real networks: an eigenvalue viewpoint. In: Proceedings of the symposium on reliable distributed computing, Florence, Italy, pp 25–34
66. Wei TH (1952) The algebraic foundations of ranking theory. Cambridge University Press, London
67. West DB (2001) Introduction to graph theory, 2nd edn. Prentice Hall, Englewood Cliffs, NJ
68. Whitney H (1932) Congruent graphs and the connectivity of graphs. Am J Math 54:150–168
69. Wilf H (1967) The eigenvalues of a graph and its chromatic number. J Lond Math Soc 42:330–332

Chapter 15
Minimum Spanning Markovian Trees: Introducing Context-Sensitivity into the Generation of Spanning Trees

Alexander Mehler

Abstract This chapter introduces a novel class of graphs: *Minimum Spanning Markovian Trees* (MSMTs). The idea behind MSMTs is to provide spanning trees that minimize the costs of edge traversals in a Markovian manner, that is, in terms of the path starting with the root of the tree and ending at the vertex under consideration. In a second part, the chapter generalizes this class of spanning trees in order to allow for damped Markovian effects in the course of spanning. These two effects, (1) the sensitivity to the contexts generated by consecutive edges and (2) the decreasing impact of more antecedent (or "weakly remembered") vertices, are well known in cognitive modeling [6, 10, 21, 23]. In this sense, the chapter can also be read as an effort to introduce a graph model to support the simulation of cognitive systems. Note that MSMTs are not to be confused with branching Markov chains or Markov trees [20] as we focus on generating spanning trees from given weighted undirected networks.

Keywords Markovian trees · Minimum spanning trees · Cohesion trees · Linguistic networks · Semiotic networks

MSC2000: Primary 05C05, 68R10; Secondary 05C12, 05C38, 05C75

15.1 Introduction

According to [24] a *Network Optimization Problem* (NOP) is generally described as follows. Given a weighted graph $G = (V, E, \mu)$ whose edges are weighted by an edge weighting function $\mu : E \to \mathbb{R}$, the task is to find a subgraph of G that satisfies a set of well-defined properties by optimizing (i.e., minimizing or maximizing)

A. Mehler (✉)
Goethe-University Frankfurt am Main, Senckenberganlage 31,
60325 Frankfurt am Main, Germany
e-mail: Mehler@em.uni-frankfurt.de

M. Dehmer (ed.), *Structural Analysis of Complex Networks*,
DOI 10.1007/978-0-8176-4789-6_15, © Springer Science+Business Media, LLC 2011

a certain function of μ. A very prominent example of a NOP is given by *Minimum (weight) Spanning Trees* (MST) that have been defined in order to derive tree-like structures from graphs subject to edge weights as representations of costs of edge traversal [1, 24]. Mehler [18] shows how MSTs and their relatives in the form of *Shortest Path Trees* (SPT) can be used to span the kernels of so-called generalized trees [2,3]. The idea behind this approach is to provide a model of graphs that range from tree- to graph-like structures so that they share the efficiency of trees (in terms of information processing) with the expressiveness of graphs. A basic outcome of Mehler [18] is that the choice of the kernel of a generalized tree determines the semantics of its edge types in a nontrivial manner so that we get a variety of generalized trees suitable for different tasks. This raises the question of spanning trees as alternative kernels of generalized trees beyond the well-known concepts of an MST and an SPT.

The present chapter focuses on this question. Its starting point is Mehler [14] who extends the notion of an MST by means of so-called *Dependency Trees* (DT) and *Cohesion Trees* (CT). These classes of spanning trees have been further developed by Mehler [16] by means of *bipartite* cohesion trees as a graph model of browsing in the semantic web [5]. The basic principle behind these tree-like structures is to introduce *context-sensitivity* into the generation of spanning trees. From the point of view of a vertex to be processed, this context-sensitivity is induced by its candidate predecessor (as in the case of DTs) or by the path to be continued by that vertex (as in the case of CTs). Take, for example, the cognitive process of spreading activation in association networks in the course of discourse comprehension [11,23]: depending on the starting point (i.e., as determined by the discourse already processed) the course of spreading activation, its intensity, and its direction differ so that different subgraphs of the association network are activated if this starting point is varied. It is this sort of context-sensitivity that is in the focus of dependency trees and especially of cohesion trees.

Dependency trees were originally introduced in computational linguistics to structure (dis-)similarity relations of signs in a tree-like manner [7,12,15,22]. Generally speaking, a dependency tree is generated as follows: for a distinguished vertex r of a graph $G = (V, E, \mu)$, vertices are inserted into the dependency tree T spanned over G starting with r in ascending order of their geodesic distance from r where the predecessor of any vertex $v \in V$ to be inserted into T is chosen to be the vertex $w \in V$ that in terms of μ is closest to v among all vertices already inserted into T. Cohesion trees extend this principle of context-sensitive spanning by additionally evaluating the candidate paths to be continued by v starting with r. In this sense, cohesion trees are more context-sensitive than dependency trees which, in turn, are more context-sensitive than MSTs. Mehler [14] shows that DTs and CTs are actually different from MSTs, while Mehler [15] relates these two notions to the area of context-sensitive clustering. By distinguishing the layer of dominating topoi on the one hand and the layer of dominated textual instances of these topoi on the other, Mehler [16] gains an additional layer of control of spanning tree-like structures and, thus, of context-sensitivity in the generation of such trees. In Markovian terms: while cohesion trees relate to Markovian trees in a manner to be specified in this chapter, their bipartite extension relates to the notion of a *hidden* minimum spanning Markovian tree – we leave the latter extension to future work and therefore concentrate on MSMTs.

In this chapter we generalize the class of dependency trees and of cohesion trees by means of minimum spanning Markovian trees. In this way, we no longer demand that the underlying graph be a completely connected weighted graph (as has been done by Mehler [14, 15]). We loosen this precondition and demand instead that the underlying graph be connected. Further, we show that Markovian trees of the sort introduced here are based on three types of parameters: (1) the operative notion of path sensitivity, (2) their measure of vertex relatedness, and (3) the choice of their root.

The basic idea of MSMTs runs as follows. If in the course of spanning an MSMT $T = (V, E, r)$ we have to decide where to insert a newly encountered vertex v, then we select that vertex w as the end of the unique path P starting with the root r that minimizes the costs of edge traversals among all candidate paths to which w might be attached according to the topology of the graph to be spanned. Starting with its end vertex v, suffixes of such paths P are evaluated whose length corresponds to the Markov order of the MSMT to be generated. Thus, if the order of an MSMT is larger than one it is always more sensitive to paths than any corresponding MST or DT. This approach departs from dependency trees [22] and similarity trees [12] in two respects:

1. We start with more general connected graphs.
2. We view paths as context-building units by recursively accounting for the predecessor's vertices in a Markovian manner to be specified in this chapter.

Although our basic motivation behind introducing MSMTs as a novel class of graphs is to find alternative kernels of generalized trees, we disregard the impact of our findings on the typing of the edges of such trees – for a first account of such a semantics that analyzes the impact of MSTs and SPTs on the edge types of generalized trees see Mehler [18]. In contrast to this, we concentrate on the class of MSMTs itself and leave a proof–theoretical analysis of its impact on generalized trees to future work.

The chapter is organized as follows. Section 15.2.1 generalizes the notion of a dependency tree in terms of so-called minimum spanning Markovian trees. Further, Sect. 15.2.2 generalizes the notion of a cohesion tree by means of damping Markovian transitions in MSMTs along the paths starting with the corresponding root. Finally, Sect. 15.3 concludes, ending with a discussion of the prospects for future work.

15.2 Context-Sensitive Spanning Trees

In order to pave the way for our graph model we start with graph-theoretical preliminaries used throughout this chapter.[1]

---

[1] See Mehler [18].

Definition 1 (Preliminaries I). Let $G = (V, E, \mathcal{L}_V, \mathcal{L}_E, \mu)$ be a connected weighted undirected graph whose vertices are uniquely labeled by the function $\mathcal{L}_V : V \to L_V$ for the set of vertex labels L_V and whose edges are uniquely labeled by $\mathcal{L}_E : E \to L_E$ for the set of edge labels L_E. Throughout this chapter we assume that $L_V \subset \mathbb{N}_0$ and $L_E \subset \mathbb{N}_0$; that is, vertices and edges are labeled by ordinal numbers and this numbering is consecutive. By $\leq_V \subset V^2$ ($\leq_E \subset E^2$) we refer to the natural order of $L_V \subset \mathbb{N}_0$ ($L_E \subset \mathbb{N}_0$) such that for all $a, b \in V, a \neq b$ $(e, f \in E, e \neq f)$: $a <_V b$ $(e <_E f)$ iff $\mathcal{L}_V(a) < \mathcal{L}_V(b)$ $(\mathcal{L}_E(e) < \mathcal{L}_E(f))$. This allows us to define the order relation

$$\leq_a = \leq_V \cup \leq_E$$

on the set of vertices *and* edges. Without loss of generality we assume that $\mu : E \to \mathbb{R}^+ \setminus \{0\}$ is an edge weighting function that represents costs of traversing edges in E. Think of μ, e.g., as a function of the loss of coherence induced by following hyperlinks. Now let $\mathbb{P}(G)$ be the set of all simple paths in G and

$$P = (v_{i_0}, e_{j_1}, v_{i_1}, \ldots, v_{i_{m-1}}, e_{j_m}, v_{i_m}) \in \mathbb{P}(G)$$

a simple path such that $\forall 1 \leq k \leq m : e_{j_k} = \{v_{i_{k-1}}, v_{i_k}\} \in E$. Then,

$$V(P) = \{v_{i_0}, v_{i_1}, \ldots, v_{i_{m-1}}, v_{i_m}\} \subseteq V$$

is the set of all vertices,

$$E(P) = \{e_{j_1}, \ldots, e_{j_m}\} \subseteq E$$

the set of all edges, and

$$VE(P) = V(P) \cup E(P)$$

the set of all constituents of P. Further, $V_{\text{end}}(P) = \{v_{i_0}, v_{i_m}\}$ is the set of both end vertices and $V_{\text{in}}(P) = V(P) \setminus V_{\text{end}}(P)$ the set of all inner vertices of P. Next, by

$$[P_{v_{i_0} v_{i_m}}]^{-p} = (v_{i_0}, e_{j_1}, v_{i_1}, \ldots, v_{i_{m-p-1}}, e_{j_{m-p}}, v_{i_{m-p}}) \in \mathbb{P}(G),$$

$p \leq m$, we denote the *prefix* of P that consists of $m - p + 1$ vertices. Note that if G is a tree then for each $v, w \in V$ the simple path ending at v and w is unique. Such paths are denoted as P_{vw} indexed by their end vertices v and w. Now we can define the order relation

$$\leq_a \subseteq \mathbb{P}(G)^2$$

on the set of paths $\mathbb{P}(G)$ of G such that for $P = (v_{i_1}, e_{i_2}, \ldots, e_{i_{m_i-1}}, v_{i_{m_i}}) \in \mathbb{P}(G)$ and $P' = (v_{j_1}, e_{j_2}, \ldots, e_{j_{m_j-1}}, v_{j_{m_j}}) \in \mathbb{P}(G)$, $P \neq P'$: $P \leq_a P'$ iff $\exists r < \min(m_i, m_j) \forall k \in \{1, \ldots, r\} : VE(P) \ni x_{i_k} = x_{j_k} \in VE(P') \wedge VE(P) \ni x_{i_{r+1}} <_a$

$x_{j_{r+1}} \in VE(P')$. Further, by $\mathbb{P}_G(v, w)$ we denote the set of all simple paths in G ending at v and w. Finally, for $v_{i_{m_i}} = v_{j_1}$ we define the concatenation

$$P \circ P' = (v_{i_1}, e_{i_2}, \ldots, e_{i_{m_i}-1}, v_{i_{m_i}}, e_{j_2}, \ldots, e_{j_{m_j}-1}, v_{j_{m_j}})$$

of P and P'.

We need to define order relations on the sets of vertices, edges, and paths in order to account for the redundancy that is omnipresent in semiotic systems. In graph-theoretical terms, this redundancy is manifested by multiple edges so that the shortest path between two vertices, for example, is not necessarily unique in a graph corresponding to a semiotic system. Think of a graph with exactly two vertices and $n \gg 1$ multiple edges ending at these vertices each of the same weight. In this case we have n different MSTs so that we have to decide which one to use if uniqueness is required. This is exactly the task of the order relations specified by Definition 1. Next, we introduce further preliminaries that repeat some other well-known concepts of graph theory:

Definition 2 (Preliminaries II). Let $G = (V, E, \mathcal{L}_V, \mathcal{L}_E, \mu)$ be a weighted connected graph according to Definition 1. Then, we extend μ as a function of $\mathbb{P}(G)$, that is,

$$\mu \colon \mathbb{P}(G) \to (0, \infty)$$

such that for each $P = (v_{i_0}, e_{j_1}, v_{i_1}, \ldots, v_{i_{m-1}}, e_{j_m}, v_{i_m}) \in \mathbb{P}(G)$ we set

$$\mu(P) = \sum_{k=1}^{m} \mu(e_{j_k})$$

Based on μ we define the *geodesic path* $GP_\mu(v, w)$ *between v and w in G* as

$$GP_\mu(v, w) = \inf_{\leq_a} \left\{ \underset{P \in \mathbb{P}_G(v,w)}{\arg\min} \ \mu(P) \right\}$$

Next, the *geodesic distance* $\hat{\mu} \colon V \times V \to [0, \infty)$ between $v, w \in V$ is defined as

$$\hat{\mu}(v, w) = \begin{cases} 0 & v = w \\ \mu(GP_\mu(v, w)) & v \neq w \end{cases}$$

Finally, for any weighted graph $G = (V, E, \mu)$ we define

$$\mu(G) = \sum_{e \in E} \mu(e)$$

Remark 1. In order to prevent negative cycles [24, p. 85] we henceforth assume that μ is a function from E to $\mathbb{R}^+ \setminus \{0\}$. Further, throughout this chapter we only deal

386   A. Mehler

with finite graphs. Finally, in order to keep the formalization simple we concentrate on undirected graphs and therefore disregard their directed orientations.

Remark 2. Throughout this chapter we always assume the existence of the order relation $\leq_a$ as introduced by Definition 1 without explicitly noting this in the subsequent definitions of graphs. The reason for this omission is to keep the formalism simple.

We start with developing our formal apparatus for the class of connected labeled weighted undirected graphs.

Definition 3 (Spanning Pattern). Let $G = (V, E, \mathcal{L}_V, \mathcal{L}_E, \mu)$ be an undirected graph according to Definition 1. Let further $x \in V$ be a distinguished vertex of G and

$$\Omega_x = \{\leq_x\} \cup \{\leq_{\mathbb{P}_{xy}} \mid y \in V \setminus \{x\}\}$$

be a set – called a *spanning pattern* on G – such that for each $y \in V \setminus \{x\}$: $\leq_{\mathbb{P}_{xy}}$ is a *well-ordering* on the set of neighbors $N_G(y)$ of y in G that is conditioned (according to Line 14 of Algorithm 15.1) by the subset $\mathbb{P}_{xy} \subseteq \mathbb{P}(G)$ of paths ending at x and y, respectively. Further, $\leq_x$ is a *well-ordering* on V with infimum

$$x = \inf_{\leq_x} V$$

such that for each $v, w \in V \setminus \{x\}$:

$$v \leq_x w \Leftrightarrow$$

$$\hat{\mu}(x, v) < \hat{\mu}(x, w) \ \vee \ (\hat{\mu}(x, v) = \hat{\mu}(x, w) \wedge \mathcal{L}_V(v) < \mathcal{L}_V(w)) \ \vee \ v = w$$

We write $v <_x w$ to denote that $v \neq w \wedge v \leq_x w$. Analogously, we write $v <_{\mathbb{P}_{xy}} w$ to denote that $v \neq w \wedge v \leq_{\mathbb{P}_{xy}} w$.

Definition 4 (Spanning Pattern-Based Subgraph). Let Ω_x be a spanning pattern for a distinguished vertex x of a graph $G = (V, E, \mathcal{L}_V, \mathcal{L}_E, \mu)$ that is defined according to Definition 3. Then, the graph

$$\text{span}(G, x, \Omega_x) = (V', E', \mathcal{L}'_V, \mathcal{L}'_E, x, \mu')$$

computed by Algorithm 15.1 is called a *spanning pattern-based subgraph* of G subject to Ω_x.

Theorem 1. *For any V_i, $1 < i < |V|$, and any vertex $w \in V$ selected according to Lines 13 and 14 of Algorithm 15.1 it always holds that $N_G(w) \cap V_i \neq \emptyset$.*

Proof. Obviously, $\forall v \in V_i : v \leq_x w$. Further, according to Lines 12–24 of Algorithm 15.1 there is no $y \in V \setminus V_i$, $y \neq w$, such that $y \leq_x w$. Thus, there exists a path $P_{xu} \in \mathbb{P}(G)$ such that $V(P_{xu}) \subseteq V_i$, $\{u, w\} \in E$ so that $N_G(w) \cap V_i \neq \emptyset$. Otherwise $w = \inf_{\leq_x} V \setminus V_i$ would not hold according to the definition of the geodesic distance $\hat{\mu}$. $\square$

15 Minimum Spanning Markovian Trees

Require: A graph $G = (V, E, \mathcal{L}_V, \mathcal{L}_E, \mu)$ and a spanning pattern Ω_x for a distinguished vertex $x \in V$ according to Definition 3.
Ensure: $\text{span}(G, x, \Omega_x) = (V', E', \mathcal{L}'_V, \mathcal{L}'_E, x, \mu')$ according to Definition 4.

1: **procedure** PATTERNBASEDSUBGRAPH(G, x, Ω_x)
2:    **for** $y = \inf_{\leq_x} V \setminus \{x\}$ **do**
3:       $e_{j_1} \leftarrow \{x, y\} \in E$
4:       $V_2 \leftarrow \{x, y\}$
5:       $E_2 \leftarrow \{e_{j_1}\}$
6:       $\mathcal{L}_{V_2} \leftarrow \mathcal{L}|_{V_2}$
7:       $\mathcal{L}_{E_2} \leftarrow \mathcal{L}|_{E_2}$
8:       $\mu_2 \leftarrow \mu|_{E_2}$
9:       $G_2 \leftarrow (V_2, E_2, \mathcal{L}_{V_2}, \mathcal{L}_{E_2}, \mu_2)$
10:   **end for**
11:   **for** $1 < i < |V|$ **do**
12:      **if** $w = \inf_{\leq_x} V \setminus V_i$ **then**
13:         $\mathbb{P}_{xw} \leftarrow \{P_{xw} \in \mathbb{P}(G) \mid V_{\text{in}}(P_{xw}) \setminus \{w\} \subseteq V_i \wedge V_{\text{end}}([P_{xw}]^{-1}) \cap$
             $N_G(w) \neq \emptyset\}$ $\triangleright$ *the path context of w in G constrained by V_i*
14:         **if** $v = \inf_{\leq_{\mathbb{P}_{xw}}} N_G(w) \cap V_i$ **then**
15:            $e_{j_i} \leftarrow \{v, w\}$
16:            $V_{i+1} \leftarrow V_i \cup \{w\}$
17:            $E_{i+1} \leftarrow E_i \cup \{e_{j_i}\}$
18:            $\mathcal{L}_{V_{i+1}} \leftarrow \mathcal{L}|_{V_{i+1}}$
19:            $\mathcal{L}_{E_{i+1}} \leftarrow \mathcal{L}|_{E_{i+1}}$
20:            $\mu_{i+1} \leftarrow \mu|_{E_{i+1}}$
21:            $G_{i+1} \leftarrow (V_{i+1}, E_{i+1}, \mathcal{L}_{V_{i+1}}, \mathcal{L}_{E_{i+1}}, \mu_{i+1})$
22:         **end if**
23:      **end if**
24:   **end for**
25:   **for** $G_n = (V_n, E_n, \mathcal{L}_{V_n}, \mathcal{L}_{E_n}, \mu_n), n = |V|$ **do**
26:      $V' \leftarrow V_n = V \wedge E' \leftarrow E_n \wedge \mathcal{L}'_V \leftarrow \mathcal{L}_{V_n} \wedge \mathcal{L}'_E \leftarrow \mathcal{L}_{E_n} \wedge \mu' \leftarrow \mu_n$
27:   **end for**
28:   **return** $\text{span}(G, x, \Omega_x)$
29: **end procedure**

Algorithm 15.1: Computing spanning pattern-based subgraphs

Theorem 2. *For any graph $G = (V, E, \mathcal{L}_V, \mathcal{L}_E, \mu)$ and any $\Omega_x = \{\leq_x\} \cup \{\leq_{\mathbb{P}_{xy}} \mid y \in V \setminus \{x\}\}$, both defined according to Definition 3 and any $1 < i < |V|$, the graph $G_{i+1} = (V_{i+1}, E_{i+1}, \mathcal{L}_{V_{i+1}}, \mathcal{L}_{E_{i+1}}, \mu_{i+1})$ generated according to Algorithm 15.1, Line 21 is a tree.*

Proof. In order to prove this theorem we need to show that $G_{i+1} = (V_{i+1}, E_{i+1}, \mathcal{L}_{V_{i+1}}, \mathcal{L}_{E_{i+1}}, \mu_{i+1})$ is connected and has $|E_{i+1}| = i$ edges where $|V_{i+1}| = i + 1$ [4,9]. These two properties follow directly from the greedy nature of Algorithm 15.1. Firstly, connectedness follows from the fact that Lines 12 and 14 select – by analogy to Prim's algorithm – exactly one vertex of $V \setminus V_i$ and exactly one vertex of V_i in order to connect them. Thus, each iteration of the **for** loop (Line 11) generates exactly one edge to connect one more vertex where initially (see Lines 2–10 of Algorithm 15.1) G_2 is of order 2 with a single edge. Thus, $|E_{i+1}| = i$. $\square$

388   A. Mehler

Remark 3. Because of its prominent role in determining the spanning pattern Ω_x as the starting point for spanning the tree $\text{span}(G, x, \Omega_x) = (V', E', \mathcal{L}'_V, \mathcal{L}'_E, x, \mu')$, the vertex x is used as the distinguished root vertex of this tree.

Remark 4. Theorem 2 tells us something about the type of output generated by Algorithm 15.1 in spite of the fact that the relations $\leq_{\mathbb{P}_{xw}}$ are not yet fully defined beyond the claim that they are well-orderings on the set of vertices. Thus, in conjunction with Theorem 2, Definition 4 provides a scheme based on the spanning pattern Ω_x as the scheme-building variable that allows us to derive different types of trees whose gestalt can only be fully determined if the latter variable is instantiated. From that point of view, the present chapter introduces a certain instantiation of a spanning pattern that defines MSMTs.

The following corollary follows directly from Theorem 2.

Corollary 1. $\text{span}(G, x, \Omega_x) = G_n = (V_n, E_n, \mathcal{L}_{V_n}, \mathcal{L}_{E_n}, \mu_n)$, $n = |V|$, *is a tree.*

More important is the following corollary. It includes a statement about the path context of a vertex to be inserted at some iteration of the **for** loop (see Line 11) of Algorithm 15.1. Obviously, this corollary follows from the fact that all graphs G_{i+1} generated by this algorithm are trees.

Corollary 2. *For each neighbor* $v \in N_G(w)$ *of vertex w identified according to Lines 13 and 14 of Algorithm 15.1 the path context* $\mathbb{P}_{xw}$ *of w in G constrained by* V_i *contains exactly one path* $P_{xw} \in \mathbb{P}_{xw}$ *such that* $V_{\text{end}}([P_{xw}]^{-1}) = \{x, v\}$.

Remark 5. We can think of Algorithm 15.1 as a model of a construction–integration process based on priming relations in an association network in the sense of Kintsch [10].

1. *Construction:* The order relation $\leq_x$ maps the construction phase. It defines the initial selection of all vertices $w \in V$ to be inserted into the tree to be integrated. The smaller the geodesic distance of w to the root-forming vertex x of $\text{span}(G, x, \Omega_x)$, the earlier this vertex is processed. *To recapitulate this in semiotic terms:* $\leq_x$ models priming relations induced when using x as a prime. Starting with x, this priming does not evolve chaotically but according to an order determined by the geodesic distances of the *primed* vertices to x in the underlying association network: units that are closer to x in this network are primed earlier and are, therefore, processed faster within the subsequent integration phase. $\leq_x$ maps this order that determines the "time" at which vertices are inserted into the tree to be spanned.
2. *Integration:* Order relations of the type $\leq_{\mathbb{P}_{xw}}$ provide orderings of the neighborhoods $N_G(w)$ of vertices w to be integrated into the output graph so that a tree-like structure emerges. *To recapitulate this in semiotic terms:* $\leq_{\mathbb{P}_{xw}}$ models effects of *context* priming as induced by the path to which w is attached. That is, w is attached to the end vertex of a path to which it fits best in terms of the operative association relations of the underlying association network. Context

effects of this sort are not mapped by $\leq_x$ but by $\leq_{\mathbb{P}_{xw}}$. To which degree this context effect takes place is specified by the Markovian order of the MSMT to be introduced as a special sort of spanning pattern-based subgraph.

In a nutshell: the spanning pattern Ω_x models a sort of context-sensitivity as exemplified by priming relations and addresses certain aspects of structure formation based thereon.

One objection to this semiotic interpretation might be that tree-like structures do not adequately model the kind of structure formation based on priming relations and that general graphs provide more adequate models. Actually, we respond to this objection by pointing to the aim of introducing tree-like kernels of generalized trees that because of their expressiveness can do exactly this: mapping graph-like structures.

Remark 6. $\leq_x$ is defined by means of the geodesic distance $\hat{\mu}$ since we do not deal with completely connected graphs (as this has been done by Mehler [14]), but more generally with connected graphs. Therefore, we need to leave the grounds of Euclidean spaces and related geometric conceptions of semantic spaces in order to enter the area of weighted, possibly sparse but connected graphs.

15.2.1 Minimum Spanning Markovian Trees

A class of spanning trees that can be derived by instantiating the spanning pattern of Definition 3 is given by *Minimum Spanning Markovian Trees* (MSMT). The idea behind this notion is to generalize the concept of *(minimum spanning) Dependency Trees* (DT) [14]. This concept has been defined in order to include context-sensitivity into the generation of spanning trees that goes beyond minimizing the weights of edges selected to generate *Minimum Spanning Trees* (MST) [24]. MSMTs are context-sensitive as they condition this selection to the paths generated by the greedy Algorithm 15.1. MSMTs are formally introduced by the Definitions 5–7.

Definition 5 (Degree of Markovian Connectivity). Let $G = (V, E, \mathcal{L}_V, \mathcal{L}_E, \mu)$ be a graph, $x \in V$ a distinguished vertex, and $\leq_x$ defined according to Definition 3. Let further $m \in \mathbb{N} \setminus \{0\}$. Then, for each $y \in V \setminus \{x\}$ and

$$P_{xy} = P_{v_{i_0} v_{i_{t+1}}} = (v_{i_0}, e_{j_1}, v_{i_1}, \ldots, v_{i_t}, e_{j_{t+1}}, v_{i_{t+1}}) \in \mathbb{P}(G),$$

$v_{i_0} = x$, $v_{i_{t+1}} = y$, $e_{j_k} = \{v_{i_{k-1}}, v_{i_k}\} \in E$, $k \in \{1, \ldots, t+1\}$, the *degree of end vertex connectivity* $c_m(P_{xy})$ of y to P_{xy} *of order* m is defined as:

$$c_m(P_{xy}) = \sum_{k=\max(1,(t+1)-m+1)}^{t+1} \mu(e_{j_k})$$

Based on this notion we define the *degree of Markovian connectivity of order m* of y as an end vertex of $P_{xy} = P_{v_{i_0}v_{i_{t+1}}}$ as:

$$C\left(X_{t+1} = v_{i_{t+1}} \mid X_t = v_{i_t}, X_{t-1} = v_{i_{t-1}}, \ldots, X_0 = v_{i_0}\right)$$
$$= C\left(X_{t+1} = v_{i_{t+1}} \mid X_t = v_{i_t}, \ldots, X_{t-m+1} = v_{i_{t-m+1}}\right)$$
$$= C\left(X_{t+1} = y \mid X_t = v_{i_t}, \ldots, X_{t-m+1} = v_{i_{t-m+1}}\right)$$
$$\leftarrow c_m(P_{xy})$$

Remark 7. As the Markovian connectivity of a vertex y can be evaluated with respect to paths P_{xy} of varying length the lower bound of the sum in the definition of $c_m(P_{xy})$ is specified by means of max. This provides the flexibility to deal with paths of variable length. Note further that in order to evaluate the Markovian connectivity of order m of an end vertex of a path we have to evaluate m edges.

Obviously, the Markovian connectivity of order $t + 1$ of y to $P_{xy} = P_{v_{i_0}v_{i_{t+1}}}$ equals $\mu(P_{v_{i_0}v_{i_{t+1}}})$. Further, for any $m > t+1$ it holds that $c_m(P_{xy}) = c_{t+1}(P_{xy})$. Note that the notion of Markovian connectivity is reminiscent of the notion of a Markov process of order m. However, Definition 5 does not deal with probabilities but, more generally, with weighted edges of a graph defined according to Definition 3. These weights do not necessarily denote probabilities. They may also represent membership degrees of a fuzzy relation represented as a graph. This generalization helps bridging the notion of a Markov process on the one hand and that of a spanning tree on the other as becomes clear in the following pages.

In order to instantiate the schema of spanning trees introduced by Definition 4 we need to find appropriate instances of the well-orderings $\leq_{P_{xy}}$ mentioned in Definition 3 (of spanning patterns) and also used in Algorithm 15.1. *Appropriate* means that the instantiation of these order relations has to be related to the notion of Markovian connectivity as defined above. This is done by means of the notion of a Markovian neighborhood ordering.

Definition 6 (Markovian Neighborhood Ordering). Let $G = (V, E, \mathcal{L}_V, \mathcal{L}_E, \mu)$, be a graph according to Definition 3. Let further $x, y \in V$ and $\mathbb{P}_{xy}$ be a subset of paths ending at x and y where for each $v \in N_G(y)$ there is at most one path $P(v) \in \mathbb{P}_{xy}$ such that v is an end vertex of $[P(v)]^{-1}$. Further, let $N_G'(y) \subseteq N_G(y)$ be the subset of all neighbors of y for which the latter condition holds. For any $w \in N_G'(y)$ we write

$$[P(w)]^{-1} = (v_{i_{w_0}}, e_{w_{j_1}}, v_{i_{w_1}}, \ldots, v_{i_{w_{t-1}}}, e_{j_{w_t}}, v_{i_{w_t}}), v_{i_{w_0}} = x, v_{i_{w_t}} = w$$

in order to denote this unique path. Now, let $m \in \mathbb{N} \setminus \{0\}$. Then, for each $y \in V \setminus \{x\}$ and each $v, w \in N_G(y)$ we define the relation

$$\leq_{\mathbb{P}_{xy}}^{[m]} \subseteq (N_G(y))^2$$

15 Minimum Spanning Markovian Trees

such that

$$
v \leq_{\mathbb{P}_{xy}}^{[m]} w \Leftrightarrow
\begin{cases}
v, w \in N'_G(y) \wedge \\
C(X_{t+1} = y \mid X_t = v_{i_{v_t}}, \ldots, X_{t-m+1} = v_{i_{v_{t-m+1}}}) < \\
C(X_{t+1} = y \mid X_t = v_{i_{w_t}}, \ldots, X_{t-m+1} = v_{i_{w_{t-m+1}}}) \\
\vee\, v, w \in N'_G(y) \wedge \\
C(X_{t+1} = y \mid X_t = v_{i_{v_t}}, \ldots, X_{t-m+1} = v_{i_{v_{t-m+1}}}) = \\
C(X_{t+1} = y \mid X_t = v_{i_{w_t}}, \ldots, X_{t-m+1} = v_{i_{w_{t-m+1}}}) \wedge \\
\mathcal{L}_V(v) <_V \mathcal{L}_V(w) \\
\vee\, v \in N'_G(y) \wedge w \notin N'_G(y) \\
\vee\, v, w \notin N'_G(y) \wedge \mathcal{L}_V(v) <_V \mathcal{L}_V(w) \\
\vee\, v = w
\end{cases}
$$

$\leq_{\mathbb{P}_{xy}}^{[m]}$ is called *Markovian neighborhood ordering of order m of $N_G(y)$ constrained by $\mathbb{P}_{xy}$.*

Theorem 3. *For any $x, y \in V$ and $m \in \mathbb{N} \setminus \{0\}$, $\leq_{\mathbb{P}_{xy}}^{[m]}$ is a well-ordering on $N_G(y)$.*

Proof. We notice that for any $v, w \in N_G(y)$ the conditions enumerated by the disjunction in Definition 6 are exhaustive and mutually exclusive so that $\leq_{\mathbb{P}_{xy}}^{[m]}$ is a total relation on $N_G(y)$. Further, we easily verify that $\leq_{\mathbb{P}_{xy}}^{[m]}$ is reflexive – if $v = w$, antisymmetric – if, e.g., $v, w \in N'_G(y)$, then always either $C(X_{t+1} = y \mid X_t = v_{i_{v_t}}, \ldots, X_{t-m+1} = v_{i_{v_{t-m+1}}}) < C(X_{t+1} = y \mid X_t = v_{i_{w_t}}, \ldots, X_{t-m+1} = v_{i_{w_{t-m+1}}})$ or $\mathcal{L}_V(v) <_V \mathcal{L}_V(w)$ – and transitive (since $=$, $<$, and $<_V$ are transitive). $\qquad\square$

Note that the set $\mathbb{P}_{xy}$ in Definition 6 is defined by analogy to its counterpart with the same name in Definition 3. This correspondence is the starting point of bridging the notion of a Markovian neighborhood in a graph with that of a spanning tree.

Definition 7 (Minimum Spanning Markovian Tree). Let $G = (V, E, \mathcal{L}_V, \mathcal{L}_E, \mu)$ be a graph according to Definition 3 and $\Omega_x^{[m]} = \{\leq_x\} \cup \{\leq_{\mathbb{P}_{xy}}^{[m]} \mid y \in V \setminus \{x\}\}$ be a spanning pattern based on the Markovian neighborhood orderings $\leq_{\mathbb{P}_{xy}}^{[m]}$ of order $m \in \mathbb{N} \setminus \{0\}$ for a distinguished vertex x. Then,

$$
G\left(\Omega_x^{[m]}\right) = \mathrm{span}\left(G, x, \Omega_x^{[m]}\right)
$$

is called a *Minimum Spanning Markovian Tree (MSMT) of order m spanned over G by means of $\Omega_x^{[m]}$ starting with x.*

Remark 8. Note that the expression $G(\Omega_x^{[m]}) = \mathrm{span}(G, x, \Omega_x^{[m]})$ means that $G(\Omega_x^{[m]})$ is generated by Algorithm 15.1 where the input parameter Ω_x is instantiated by $\Omega_x^{[m]}$, that is, $\Omega_x \leftarrow \Omega_x^{[m]}$.

Corollary 3. *Because of Theorem 2 it readily follows that under the conditions of Definition 7, $G(\Omega_x^{[m]})$ is a (spanning) tree (of G).*

Now, we prepare a theorem that shows in which sense the attribute *minimum* is appropriately added to the name of an MSMT.

Definition 8 (Costs of Trailing in a Weighted Rooted Tree). Let $T = (V, E', \mathcal{L}_V', \mathcal{L}_E', x, \mu')$ be a weighted rooted tree and $m \in \mathbb{N} \setminus \{0\}$. The *costs* $c_x^{[m]}(T)$ *of trailing in* T *induced by* x are defined as

$$c_x^{[m]}(T) = \sum_{v \in V} c_m(P_{xv}(T))$$

where $P_{xv}(T)$ is the unique simple path in T ending at root x and vertex v.

Remark 9. In the case where T in Definition 8 is an MSMT of a corresponding graph, the notion of trailing costs includes two central ingredients:

- Firstly, the vertex x from whose perspective the tree is trailed and which determines the order of vertices being processed according to their geodesic distance $\hat{\mu}(v, x)$ to x,
- Secondly, the degree m of Markovian connectivity that affects $c_x^{[m]}(T)$ by specifying the length of the suffix of P_{xv} that is taken into consideration.

In this way, Definition 8 reflects both perspectives: (1) the one induced by the root of the MSMT to be spanned and (2) the one induced by the path to which vertices are attached when instantiating the input parameter Ω_x of Algorithm 15.1 by $\Omega_x^{[m]}$. Obviously, this concept of costs departs from its classical counterpart in terms of minimum spanning trees as it introduces the sort of context sensitivity induced by both perspectives.

The way in which MSMTs actually minimize costs is now specified by proving the following theorem.

Theorem 4. *Let* $G = (V, E, \mathcal{L}_V, \mathcal{L}_E, \mu)$ *be a graph according to Definition 3, $x \in V$ a distinguished vertex of G, and $m \in \mathbb{N} \setminus \{0\}$. Let T be any spanning tree of G such that*

$$\forall w \in V \exists! P_{xw} \in \mathbb{P}(T) \forall v \in V_{\text{in}}(P_{xw}): v \leq_x w$$

In this case the following inequality holds:

$$c_x^{[m]}(T) \geq c_x^{[m]}(G(\Omega_x^{[m]}))$$

Proof. Let $G = (V, E, \mathcal{L}_V, \mathcal{L}_E, \mu)$ be a graph according to Definition 3, $x \in V$ a distinguished vertex of G, $m \in \mathbb{N} \setminus \{0\}$, and $G(\Omega_x^{[m]}) = (V, E', \mathcal{L}_V', \mathcal{L}_E', x, \mu')$ be the MSMT of order m spanned over G by means of $\Omega_x^{[m]}$ starting with x. Let further

$T = (V, E'', \mathcal{L}''_V, \mathcal{L}''_E, x, \mu'')$ be a spanning tree of G rooted in the same vertex x and assume that

$$c_x^{[m]}(T) < c_x^{[m]}\left(G\left(\Omega_x^{[m]}\right)\right)$$

Without loss of generality we consider the sum

$$\sum_{v_{i_1},\ldots,v_{i_{|V|}}} c_m(P_{xv_{i_k}}(T))$$

such that $x = v_{i_1} \leq_x \cdots \leq_x v_{i_{|V|}}$ (as we deal with sums we can always arrange vertices in this order). Suppose now that j, $1 < j \leq |V|$, is the smallest index such that

$$c_m(P_{xv_{i_j}}(T)) < c_m\left(P_{xv_{i_j}}\left(G\left(\Omega_x^{[m]}\right)\right)\right)$$

Then, according to the precondition of Theorem 4 we state, firstly, that for each $u \in V_{in}(P_{xv_{i_j}}(T)) : u \leq_x v_{i_j}$. Secondly, we observe that there have to be two vertices $v \neq w$ such that $v \leq_x v_{i_j}$, $w \leq_x v_{i_j}$, $\{v, v_{i_j}\} \in E''$, $\{w, v_{i_j}\} \in E'$, $\{v, v_{i_j}\} \in E$, and $\{w, v_{i_j}\} \in E$. That is, we are in a situation that looks as follows where dashed lines denote paths:[2]

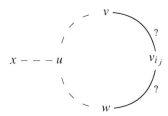

This situation is obviously in contradiction to the principle of computing $G(\Omega_x^{[m]})$ since in this case $v <_{\mathbb{P}_{xv_{i_j}}}^{[m]} w$ (note that $v \neq w$). □

Corollary 4. *Among all candidate spanning trees of a graph G defined according to Definition 3 rooted in the vertex x, the MSMT rooted in that vertex is of minimal costs of trailing starting with x.*

This is a simple consequence of the proof of Theorem 4.

Remark 10. Note that we did not define the trailing cost $c_x^{[m]}(T)$ of a spanning tree $T = (V, E', \mathcal{L}'_V, \mathcal{L}'_E, x, \mu')$ of $G = (V, E, \mathcal{L}_V, \mathcal{L}_E, \mu)$ as

$$c_x^{[m]}(T) = \sum_{v \in V} \mu'(x, v) + c_m(P_{xv}(T))$$

---

[2] Note that $P_{xv_{i_j}}(T)$ and $P_{xv_{i_j}}(G(\Omega_x^{[m]}))$ may have a common subpath ending at x and u.

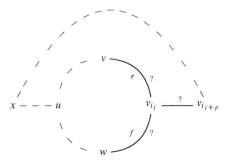

Fig. 15.1 A scenario in which vertex v_{i_j} has three candidates to which it might be attached: vertex v, w, and $v_{i_{j+p}}$

That is we assume that minimum spanning Markovian trees do not minimize this sum. *Why?* Look at Fig. 15.1 which extends the diagram of the preceding proof. Suppose that although
$$w <^{[m]}_{\mathbb{P}_{xv_{i_j}}} v$$
we get the following inequality of the corresponding geodesic distances
$$\hat{\mu}'(x,v) < \hat{\mu}'(x,w)$$
a situation that might easily occur. In this case we can generate a spanning tree by selecting the edge $e = \{v, v_{i_j}\}$ while skipping the edge $f = \{w, v_{i_j}\}$ under the condition that
$$\hat{\mu}'(x,w) - \hat{\mu}'(x,v) \gg c_m(P_{x..vv_{i_j}}) - c_m(P_{x..wv_{i_j}})$$
where $P_{x..wv_{i_j}} = GP_\mu(x,w) \circ (w, f, v_{i_j})$ and $P_{x..vv_{i_j}} = GP_\mu(x,v) \circ (v, e, v_{i_j})$. For $m \gg 1$ such a situation may easily occur. In this case, the corresponding MSMT would select the edge e and skip the edge f – in contrast to what is expected if the sum $\sum_v \hat{\mu}'(x,v)$ has to be minimized too. That is, MSMTs are Markovian in a real sense as they *look back* to a degree specified by their order without simultaneously optimizing a *look ahead* as implied by minimizing the sum $\sum_{v \in V} \hat{\mu}'(x,v)$.

Remark 11. Note that in Theorem 4 we demand that $v \leq_x w$ for all $v \in V_{\text{in}}(P_{xw})$, $P_{xw} \in \mathbb{P}(T)$, $w \in V$. The reason to concentrate on the inner vertices of this sort is to prevent us from considering paths that lead away from the root vertex x. Look again at Fig. 15.1 and suppose that there is a vertex $v_{i_{j+p}} \neq v_{i_j}$ such that $v_{i_j} \leq_x v_{i_{j+p}}$. When considering candidate vertices to which v_{i_j} should be attached, the vertex $v_{i_{j+p}}$ is ruled out as a candidate even if
$$v_{i_{j+p}} <^{[m]}_{\mathbb{P}_{xv_{i_j}}} v \wedge v_{i_{j+p}} <^{[m]}_{\mathbb{P}_{xv_{i_j}}} w$$

15 Minimum Spanning Markovian Trees

Although attaching v_{i_j} to $v_{i_{j+p}}$ may induce lower trailing costs, we face the risk of generating a disconnected graph when selecting $\{v_{i_j}, v_{i_{j+p}}\}$ instead of $\{w, v_{i_j}\}$ – in contrast to the aim of finding a spanning tree.

These remarks may help to explain what exactly MSMTs minimize. The answer is: the degree of Markovian connectivity of paths subject to the processing order of the vertices as a function of their geodesic distance to the root of the MSMT. As mentioned above this notion is reminiscent of the construction–integration theory of discourse comprehension [10].

Corollary 5. *Let* $G = (V, E, \mathcal{L}_V, \mathcal{L}_E, \mu)$, $|E| = \frac{|V|(|V|-1)}{2}$, *be a completely connected weighted graph. Let further*

$$\Omega_x^{[1]} = \{\leq_x\} \cup \left\{ \leq_{\mathbb{P}_{xy}}^{[1]} \mid y \in V \setminus \{x\} \right\}$$

a spanning pattern according to Definition 3 based on the Markovian neighborhood orderings $\leq_{\mathbb{P}_{xy}}^{[1]}$ *of order 1. Then, the MSMT* $G(\Omega_x^{[1]})$ *of order 1 spanned over* G *by means of* $\Omega_x^{[1]}$ *starting with* x *is a dependency tree spanned over* G *according to Mehler [14].*

This corollary simply follows from the fact that dependency trees as formalized by Mehler [14] are just MSMTs of minimal order 1 based on completely connected graphs. Thus, MSMTs are generalizations of the class of simpler dependency trees and similarity trees in these two respects. MSMTs are more general than dependency trees as they cover Markovian neighborhoods of vertices beyond their immediate neighborhood. This introduces a sort of Markovian memory of edge traversals or a corresponding sort of transitivity whose scope is determined by the order of the MSMT. To put it more strikingly: *Where you are going* (in the sense of which vertex is used to continue a given path) *depends on where you are from* (in the sense of which vertices precede that vertex in the path). Markovian spanning trees model exactly this kind of dependency of a vertex on its antecedent path where the higher the Markovian order of the MSMT, the longer the antecedent path on which this vertex is dependent. Finally, we present an estimation of the time complexity of generating instances of this class of spanning trees.

Theorem 5. *The time complexity to compute an MSMT of a given order m spanned over a sparse graph in which the average number of neighbors of vertices grows in a logarithmic manner is on the order of* $\mathcal{O}(|V| \log |V|)$.

Proof. We refer to Algorithm 15.1 and suppose that we deal with sparse graphs as given by complex semiotic networks [17]. Thus, we assume that $|E| \ll |V|^2$. Then, we have to firstly compute the order relation $\leq_x$. This can be done by a breadth-first search that is on the order of $\mathcal{O}(|V| + |E|)$. Having specified the order of vertices to be processed we have to repeat the **for** loop (see Line 11) $|V| - 2$ times. In this case we assume that every vertex in $y \in V$ is assigned a vector $\mathbf{v}$ with m dimensions so that whenever we build an edge $\{v, w\}, v \in V_i, w = \inf_{\leq_x} V \setminus V_i$ (Line 15) we set

$$\mathbf{w}[1] = \mu(\{v, w\})$$
$$\mathbf{w}[2] = \mathbf{v}[1] + \mu(\{v, w\})$$
$$\cdots$$
$$\mathbf{w}[m] = \mathbf{v}[m-1] + \mu(\{v, w\})$$

That is, by means of $\mathbf{w}[m-1]$ we can check the connectivity of a vertex y to be attached to w by simply computing $\mu\{w, y\} + \mathbf{w}[m-1]$. In this way, we put in memory the degrees of Markovian connectivity per inserted vertex so that we do not need to recompute them by going back over the path ending at w.[3] As the latter operation neither depends on $|V|$ nor on $|E|$ its time complexity is constant (it does not grow with $|V|$ or $|E|$). Using this format of representing degrees of Markovian connectivity we have to check in the worst case the $[m-1]$ entry of $\frac{|V|}{2}$ many vertices so that we can estimate the time complexity of computing an MSMT of order m by

$$\mathcal{O}\left(|V| + |E| + (|V| - 2) \cdot \frac{|V|}{2}\right) = \mathcal{O}(|V|^2)$$

This complexity can be reduced by arranging the values $\mathbf{v}[m-1]$ in a way that enables efficient searches in the order of a complexity smaller than $\frac{|V|}{2}$. Another way of seeing how this complexity is reduced is by assuming that we deal with sparse graphs in which the average number of neighbors of vertices grows with $|V|$ in a logarithmic manner. In this case the complexity of computing an MSMT of order m is reduced to

$$\mathcal{O}(|V| + |E| + (|V| - 2) \log |V|) = \mathcal{O}(|V| \log |V|)$$

$\square$

15.2.2 Damped Markovian Trees

Mehler [14] has introduced an extension of dependency trees that maps weakly transitive distance effects in spanning trees of the sort of Markovian trees. Here, the attribute *weakly transitive* denotes the effect of a damped impact of preceding vertices as a function of their geodesic distance to the end vertex of the path to be processed. Trees spanned by means of this principle have been called *Cohesion Trees* (CT) in order to recall the linguistic notion of *cohesive ties* [8, 13] that in the present case are manifested by interlinked vertices of a graph. The idea was to relax

---

[3] Note that we use a C++-like notation in order to denote accesses to vectors.

15 Minimum Spanning Markovian Trees

the Markovian dependence on mediately linked vertices as a function of their distance in the corresponding path. According to the notion of cohesion, the cohesive force of cohesive ties is indeed a decreasing function of their distance in discourse – it is this effect that we want to model by means of damped minimum spanning Markovian trees. From a semiotic point of view this approach is self-explanatory as it is known that processes of spreading activation decay with the distance they cover in the corresponding network. In this section we generalize the notion of a minimum spanning Markovian tree by means of the notion of *decaying* degrees of Markovian connectivity. This is done as follows.

Definition 9 (Damped Degree of Markovian Connectivity). Let $G = (V, E, \mathcal{L}_V, \mathcal{L}_E, \mu)$ be a graph, $x \in V$ a distinguished vertex, and $\leq_x$ be defined according to Definition 3. Further, let $m \in \mathbb{N} \setminus \{0\}$ and $d : \{1, \dots, m\} \to [0, 1]$ be a function. Then, for each $y \in V \setminus \{x\}$ and

$$P_{xy} = P_{v_{i_0} v_{i_{t+1}}} = (v_{i_0}, e_{j_1}, v_{i_1}, \dots, v_{i_t}, e_{j_{t+1}}, v_{i_{t+1}}) \in \mathbb{P}(G),$$

$v_{i_0} = x$, $v_{i_{t+1}} = y$, $e_{j_k} = \{v_{i_{k-1}}, v_{i_k}\} \in E$, $k \in \{1, \dots, t+1\}$, the *damped degree of end vertex connectivity* $\hat{c}_m(P_{xy})$ of y to P_{xy} *of order* m is defined as:

$$\hat{c}_m(P_{xy}, y) = \sum_{k=\max(1,(t+1)-m+1)}^{t+1} d(t + 1 - k + 1) \cdot \mu(e_{j_k})$$

We call d a *damping function*. Based on this notion we define the *damped degree of Markovian connectivity of order* m of y as an end vertex of $P_{xy} = P_{v_{i_0} v_{i_{t+1}}}$ as:

$$
\begin{aligned}
\hat{C}\left(X_{t+1} = v_{i_{t+1}} \mid X_t = v_{i_t}, X_{t-1} = v_{i_{t-1}}, \dots, X_0 = v_{i_0}\right) \\
= \hat{C}\left(X_{t+1} = v_{i_{t+1}} \mid X_t = v_{i_t}, \dots, X_{t-m+1} = v_{i_{t-m+1}}\right) \\
= \hat{C}\left(X_{t+1} = y \mid X_t = v_{i_t}, \dots, X_{t-m+1} = v_{i_{t-m+1}}\right) \\
\leftarrow \hat{c}_m(P_{xy})
\end{aligned}
$$

Figure 15.2 exemplifies candidate functions of d. They range from constant functions (Case A) to functions decaying in a logarithmic manner (Case F). Note that Fig. 15.2 does not show real cases, but simplifies the presentation in order to hint at prototypical cases:

1. Case A shows a constant function d of m that because of mapping each degree $n \leq m$ onto 1 leads to degrees of Markovian connectivity according to Definition 5.
2. Case B exemplifies a function that apart from the first vertex (as the immediate predecessor of the end vertex of the path under consideration) damps all other vertices down to zero. This kind of damping leads to degrees of Markovian connectivity in the line of MSMTs of order 1.

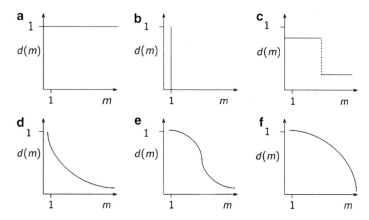

Fig. 15.2 Six prototypical instances of the damping function d (see Definition 9)

3. Case C varies the latter case by considering damping ratios smaller than 1. Obviously, it damps vertices in the manner of a step function.
4. Case D shows a function that demonstrates, so to speak, an exponential growth of the degree by which more distant units are damped. In other words: Case D valuates the importance of more distant units in a way that decays exponentially. Thus, the connectivity of a distant vertex (i.e., if $k \gg 1, k \in \{1, \ldots, m\}$) to its predecessor and successor nodes in a path has a much smaller effect than the one of a closer vertex (i.e., if $k \ll m$) even in the case of identical connectivity values. In this case, larger distances to closer vertices count more than those to more distant vertices. Functions of this sort model an effect of a declining impact as a function of the distance of vertices in a path.
5. Case E demonstrates a reversed S-shaped function for which less distant vertices are damped to a much lower degree than more distant ones. That is, in this case we observe a transition from a concave to a convex part of the damping function.
6. Finally, Case F demonstrates a concave function that slowly decays the damping effect for growing distances.

Which function d should be preferred depends on the application area. In the present case we might argue that impact decays by analogy to similarity by an exponential function of distance (see [6]) so that Case D would be preferred. In the area of Web mining this is confirmed by the so-called *link-content conjecture* of Menczer [19] who states that the content of a page is similar to the one of the pages that link to it. Menczer [19] presents data in support of this conjecture that point at an exponential decay of this similarity induced by following hyperlinks between pages. So when modeling trails in the Web by means of MSMTs one should prefer this sort of exponential damping in order to account for the decay of similarity of mediately linked documents. From this semiotic perspective the values $d(k), k \in \{1, \ldots, m\}$, may also be viewed as degrees of *salience* (see Table 15.1) that decays as a function of the distance to the topical vertex. However, other applications may decide differently. In any case, Definition 9 is general enough to map a wide range of instantiations of the damping function d.

15 Minimum Spanning Markovian Trees

Table 15.1 Three classes of spanning trees in relation to MSTs from the point of view of context-sensitivity

Level	Graph model	Support of context
0	MST	$\emptyset$
1	MSMT-1	Root, predecessor
2	MSMT-m, $m > 1$	Root, path of length m
3	DMSMT	Root, path of length m, degree of salience

The next step of defining damped minimum spanning Markovian trees is to utilize damped degrees of Markovian connectivity in the definition of Markovian neighborhood orderings.

Definition 10 (Damped Markovian Neighborhood Ordering). Let $G = (V, E, \mathcal{L}_V, \mathcal{L}_E, \mu)$, $x, y \in V$, $\mathbb{P}_{xy}$, $N'_G(y)$, $[P(v)]^{-1}$ and m all defined as in Definition 6. Then, for each $y \in V \setminus \{x\}$ and $v, w \in N_G(y)$ we define

$$\leq_{\mathbb{P}_{xy}}^{[m],d} \subseteq (N_G(y))^2$$

such that

$$v \leq_{\mathbb{P}_{xy}}^{[m],d} w \Leftrightarrow \begin{cases} v, w \in N'_G(y) \wedge \hat{c}_m(P_{x..vy}) < \hat{c}_m(P_{x..wy}) \\ \vee\, v, w \in N'_G(y) \wedge \hat{c}_m(P_{x..vy}) = \hat{c}_m(P_{x..wy}) \wedge \mathcal{L}_V(v) <_V \mathcal{L}_V(w) \\ \vee\, v \in N'_G(y) \wedge w \notin N'_G(y) \\ \vee\, v, w \notin N'_G(y) \wedge \mathcal{L}_V(v) <_V \mathcal{L}_V(w) \\ \vee\, v = w \end{cases}$$

where $P_{x..vy} = GP_\mu(x, v) \circ (v, \{v, y\}, y)$ and $P_{x..wy} = GP_\mu(x, w) \circ (w, \{w, y\}, y)$. $\leq_{\mathbb{P}_{xy}}^{[m],d}$ is called the *damped Markovian neighborhood ordering of order m of $N_G(y)$ constrained by $\mathbb{P}_{xy}$*.

According to Definitions 3 and 4 we need to show that the relations $\leq_{\mathbb{P}_{xy}}^{[m],d}$ are well-orderings.

Corollary 6. *For any $x, y \in V$, $m \in \mathbb{N} \setminus \{0\}$, and any damping function $d : \{1, \ldots, m\} \to [0, 1]$, $\leq_{\mathbb{P}_{xy}}^{[m],d}$ is a well-ordering on $N_G(y)$.*

Proof. We can prove the latter corollary simply by analogy to Theorem 3. We notice that for any $v, w \in N_G(y)$ the conditions enumerated by the disjunction in Definition 10 are exhaustive and mutually exclusive so that $\leq_{\mathbb{P}_{xy}}^{[m],d}$ is a total relation on $N_G(y)$. Further, we verify that $\leq_{\mathbb{P}_{xy}}^{[m],d}$ is reflexive, antisymmetric (if, e.g., $v, w \in N'_G(y)$, then always either $\hat{c}_m(P_{x..vy}) < \hat{c}_m(P_{x..wy})$ or $\mathcal{L}_V(v) <_V \mathcal{L}_V(w)$) and transitive (since $=$, $<$, and $<_V$ are transitive). $\square$

The final step is to define damped minimum spanning Markovian trees by utilizing $\leq_{\mathbb{P}_{xy}}^{[m],d}$ as the constitutive neighborhood-related well-ordering.

Definition 11 (Damped Minimum Spanning Markovian Tree). Let $G = (V, E, \mathcal{L}_V, \mathcal{L}_E, \mu)$ be a graph according to Definition 3. Let further $d : \{1, \ldots, m\} \to [0, 1]$ be a damping function and

$$\hat{\Omega}_x^{[m],d} = \{\leq_x\} \cup \left\{ \leq_{\mathbb{P}_{xy}}^{[m],d} \mid y \in V \setminus \{x\} \right\}$$

for a distinguished vertex $x \in V$, some $m \in \mathbb{N} \setminus \{0\}$, and a damping function d. Then,

$$G(\hat{\Omega}_x^{[m],d}) = \text{span}\left(G, x, \Omega_x^{[m],d} \right)$$

is called a *damped minimum spanning Markovian tree (DMSMT) of order m spanned over G by means of $\hat{\Omega}_x^{[m],d}$ starting from x and damped by d*.

Corollary 7. *Because of Theorem 2 it readily follows that under the conditions of Definition 11, $G(\Omega_x^{[m]})$ is a (spanning) tree (of G).*

We do not consider damped MSMTs in terms of a proof–theoretical analysis further but hint at the fact that they may help to extend the kind of Markovian spanning trees as introduced by MSMTs.

15.3 Conclusion

In this chapter we have introduced the notion of a minimum spanning Markovian tree together with its extension in the form of damped minimum spanning Markovian trees. Table 15.1 summarizes these concepts and relates them to classical minimum spanning trees. It shows that by changing over to MSMTs we gain context-sensitivity of graph modeling by taking paths as context-building units into account. Further, the extensions of MSMTs in the form of DMSMTs provide a more realistic model of salience or memory than ordinary m-order MSMTs. We have argued that the motivation for introducing these kinds of structures comes from linguistics where we have to model context effects of trailing in semiotic networks (e.g., association networks and semantic memories or networks of Web documents). A related example is browsing in a semantic web that – because of its bipartite structure – requires an extension of MSMTs in the form of *hidden* minimum spanning Markovian trees. This extension will be the task of future work.

Acknowledgments Financial support of the German Federal Ministry of Education (BMBF) through the research project *Linguistic Networks* and of the German Research Foundation (DFG) through the Excellence Cluster 277 *Cognitive Interaction Technology* (via the Project *Knowledge Enhanced Embodied Cognitive Interaction Technologies* (KnowCIT)), the SFB 673 *Alignment in Communication* (via the Project X1 *Multimodal Alignment Corpora: Statistical Modeling and Information Management*), the Research Group 437 *Text Technological Information Modeling* (via the Project A4 *Induction of Document Grammars for Webgenre Representation*), and the LIS-Project *Content-Based P2P-Agents for Thematic Structuring and Search Optimization in Digital Libraries* at Bielefeld University is gratefully acknowledged.

References

1. Caldarelli G, Vespignani A (eds) (2007) Large scale structure and dynamics of complex networks. World Scientific, Hackensack, NJ
2. Dehmer M, Mehler A (2007) A new method of measuring the similarity for a special class of directed graphs. Tatra Mt Math Publ 36:39–59
3. Dehmer M, Mehler A, Emmert-Streib F (2007) Graph-theoretical characterizations of generalized trees. In: Proceedings of the 2007 international conference on machine learning: models, technologies, and applications (MLMTA'07), Las Vegas, pp 25–28
4. Diestel R (2005) Graph theory. Springer, Heidelberg
5. Fensel D, Hendler J, Lieberman H, Wahlster W (2003) Spinning the semantic Web. Bringing the World Wide Web to its full potential. MIT Press, Cambridge, MA
6. Gärdenfors P (2000) Conceptual spaces. MIT Press, Cambridge, MA
7. Gritzmann P (2007) On the mathematics of semantic spaces. In: Mehler A, Köhler R (eds) Aspects of automatic text analysis, vol 209. Studies in fuzziness and soft computing. Springer, Berlin, pp 95–115
8. Halliday MAK, Hasan R (1976) Cohesion in english. Longman, London
9. Jungnickel D (2008) Graphs, networks and algorithms. Springer, Berlin
10. Kintsch W (1998) Comprehension: a paradigm for cognition. Cambridge University Press, Cambridge
11. Kintsch W (2001) Predication. Cognitive Sci 25:173–202
12. Lin D (1998) Automatic retrieval and clustering of similar words. In: Proceedings of the COLING-ACL '98, pp 768–774
13. Martin JR (1992) English text: system and structure. John Benjamins, Philadelphia, PA
14. Mehler A (2002) Hierarchical analysis of text similarity data. Künstliche Intelligenz (KI) 2: 12–16
15. Mehler A (2002) Hierarchical orderings of textual units. In: Proceedings of the 19th international conference on computational linguistics (COLING '02), Taipei, San Francisco, pp 646–652
16. Mehler A (2005) Lexical chaining as a source of text chaining. In: Patrick J, Matthiessen C (eds) Proceedings of the 1st computational systemic functional grammar conference, University of Sydney, Australia, pp 12–21
17. Mehler A (2008) Structural similarities of complex networks: a computational model by example of wiki graphs. Appl Artif Intell 22(7&8):619–683
18. Mehler A (2009) Generalized shortest paths trees: a novel graph class applied to semiotic networks. In: Dehmer M, Emmert-Streib F (eds) Analysis of complex networks: from biology to linguistics. Wiley-VCH, Weinheim, pp 175–220
19. Menczer F (2004) Lexical and semantic clustering by web links. J Am Soc Inf Sci Technol 55(14):1261–1269
20. Menshikov MV, Volkov SE (1997) Branching Markov chains: qualitative characteristics. Markov Proc Relat Fields 3:1–18
21. Murphy GL (2002) The big book of concepts. MIT Press, Cambridge
22. Rieger BB (1978) Feasible fuzzy semantics. In: 7th International conference on computational linguistics (COLING-78), pp 41–43
23. Sharkey AJC, Sharkey NE (1992) Weak contextual constraints in text and word priming. J Mem Lang 31(4):543–572
24. Tarjan RE (1983) Data structures and network algorithms. Society for Industrial and Applied Mathematics, Philadelphia

Chapter 16
Link-Based Network Mining

**Jerry Scripps, Ronald Nussbaum, Pang-Ning Tan,
and Abdol-Hossein Esfahanian**

Abstract Network mining is a growing area of research within the data mining community that uses metrics and algorithms from graph theory. In this chapter we present an overview of the different techniques in network mining and suggest future research possibilities in the direction of graph theory.

Keywords Network mining · Link mining · Data mining

MSC2000: Primary 91D30; Secondary 94C15

16.1 Introduction

Since the early 1990s when it began to coalesce as a discipline, data mining has grown in scope and depth. Although its foundations are in computer science, statistics and machine learning, data mining has forged its way into such diverse areas as medicine, biology, chemistry, sociology, and other humanities as well as business and engineering. In this chapter we discuss how concepts in graph theory have been absorbed into data mining, allowing it to expand into important new directions.

Data mining is the search for hidden knowledge within large data sets. The data consist of a collection of objects. In a small data set, it is often possible to extract meaningful patterns and models by applying traditional statistics and data analysis methods. However, with large data sets, it is necessary to employ more scalable and sophisticated techniques from data mining to deal with the high volume, high dimensionality, noisy, and potentially distributed nature of the data. Some of the techniques, such as classification, are predictive, while other techniques, such as

---

J. Scripps (✉)
School of Computing and Information Systems, 1 Campus Drive, Grand Valley State University, Allendale, MI 49401, USA
e-mail: scrippsj@gvsu.edu

M. Dehmer (ed.), *Structural Analysis of Complex Networks*,
DOI 10.1007/978-0-8176-4789-6_16, © Springer Science+Business Media, LLC 2011

clustering and association analysis, are descriptive. These techniques rely heavily on the attributes of objects. It is the attributes that distinguish one object from another and provide a basis for making predictions or forming descriptions.

The objects introduced above are considered to be independent of each other. For classification, the class of one object does not depend on the data from any other object. This independence assumption allows data mining techniques to make use of statistical models that depend upon the samples being independent and identically distributed. In most domains though, the objects are related. Taking into account the relationships between the objects has led to the emergence of an area called link-based network mining or *link mining* [26]. Link mining techniques have found applications in diverse types of network data, including social networks, protein interaction networks, food web, telecommunication, and transportation networks.

Link mining research has focused on various issues, such as understanding how networks are formed, how their structural properties can be characterized, what are the underlying hidden patterns, and how to make sound inferences based on models derived from the data. Recent advances in this area have led to the development of novel techniques for link prediction, node classification, node ranking, influence maximization, etc. A key challenge for these techniques is to make use of both the node attributes and links in a way that will produce better results than using either the attributes or link information alone.

The remainder of this chapter is organized as follows. In Sect. 16.2, we discuss the problem of characterizing the properties of a network. Section 16.3 reviews the link mining techniques that have been proposed in the literature. Finally, in Sect. 16.4, we present some of the open problems and ongoing research in this area.

16.2 Networks and Their Properties

A *network* is a collection of *nodes* (people, publications, companies, etc.). Each node can be assigned attribute values. In addition to having attributes, the nodes can have relationships with other nodes. The relationships are represented as links, normally a binary value to indicate the presence or absence of a relationship. In some cases, the links are weighted according to the strength of their connections. Some links are also directed to denote asymmetric relationships between objects (e.g., influence of an individual over another or hyperlink from one Web page to another). Many of the techniques described in this chapter assume that the network is static. In a static network, a link exists between a pair of nodes if they have interacted at any point during the period of data collection. This assumption helps to make the link mining tasks computationally more tractable. Increasingly, the focus in link mining is on networks that are dynamic where nodes and links can be added or removed over time from the network.

16 Link-Based Network Mining

Table 16.1 Summary of network metrics

Type	Metric	Description	References
Node	Degree	Number of links	[10, 69]
	Closeness	Mean shortest path to all other nodes	[10, 69]
	Betweenness	Number of shortest paths that include node	[10, 69]
	Authority	Nodes are assigned values from the dominant eigenvector of adjacency matrix	[41, 53]
Node-pair	Distance	Shortest path between two nodes	[14, 62]
	Min-cut	Minimum number of edges, when removed separate a pair of nodes	[14]
	Common neighbors	The number of common neighbors between two nodes	[51]
	Jaccard's coefficient	Number of common neighbors divided by the total neighbors of the two nodes	[64]
	Adamic/Adar	Number of common neighbors divided by the total neighbors of the two nodes	[1]
	Katz	The weighted sum of all paths between two nodes	[39]
Network	Density	Ratio of links to total number of node pairs	[14]
	Clustering coefficient	Similar to density (see reference)	[70]
	Average path length	Mean shortest path length over all node pairs	[14]

16.2.1 Metrics

Metrics from graph theory and social network analysis are useful for many of the techniques that are described shortly. Table 16.1 lists examples of metrics that are often used. Some of the metrics are applicable to single nodes, while others are used to characterize node-pairs or the entire network.

16.2.2 Network Characterization

One of the active research areas in link mining is to understand characteristics of real-world networks and to identify the generative mechanism that leads to the formation of such networks. For example, many social and physical networks have been observed to exhibit a power-law degree distribution. Such scale-free networks, as they are known, can be explained using a generative mechanism known as preferential attachment [6]. Another example is the small-world network, which has properties such as high clustering coefficient and low average path length.

Figure 16.1 provides a visual depiction of the different network types. The properties of these network types are summarized in Table 16.2 in terms of well-known metrics such as degree distribution, clustering coefficient, and path length.

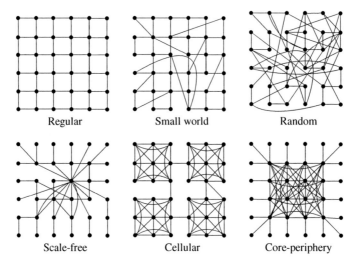

Fig. 16.1 Network types

Table 16.2 Summary of network types

Network type	Properties	References
Regular	Uniform degree distribution	
Random	Poisson degree distribution	[21]
Small world	High clustering coefficient and low average path length	[70]
Scale-free	Power-law degree distribution	[6]
Cellular	Tightly connected cells with sparse interconnections	[2, 24]
Core-periphery	Single, tightly connected core of nodes with periphery points connected only to the core	[2, 9]
Forest fire	Composite approach to model networks with shrinking diameter and densification	[45]

In addition to the metric-based approaches, Sole and Velverde [63] and Dehmer and Emmert-Streib [16] proposed characterizing networks using information-theoretic and statistical approaches. Although the metrics shown in Table 16.2 are useful to compare different network types, it is also helpful to examine some of the invariant properties of a network. As an example, the following theorem describes the nature of density within a network.

Theorem 1. *For any graph $G = (V, E)$, and for any fixed integer k, $1 < k \leq |V|$, the average density over all k-induced subgraphs of G is the same as the density of G.*

Proof. Let $G = (V, E)$ be a graph. We begin by considering the set of all k-induced subgraphs of G, for an arbitrary value of k. An induced subgraph of G is one containing a subset of the nodes in G, along with every link in G that has both endpoints in the subset. A k-induced subgraph is one with k nodes. There are $\binom{|V|}{k}$ of these, and each link in G will appear in exactly $\binom{|V|-2}{k-2}$ of them. The probability that a given edge appears in a particular k-induced subgraph equals

$$\frac{\binom{|V|-2}{k-2}}{\binom{|V|}{k}} = \frac{\frac{(|V|-2)!}{(k-2)!(|V|-k)!}}{\frac{|V|!}{k!(|V|-k)!}} = \frac{k!(|V|-2)!}{|V|!(k-2)!} = \frac{k(k-1)}{|V|(|V|-1)}. \tag{16.1}$$

By definition, the maximum possible number of links in a graph with k nodes is $\frac{k(k-1)}{2}$. So the average density of all k-induced subgraphs of G is

$$\frac{|E|\frac{k(k-1)}{|V|(|V|-1)}}{\frac{k(k-1)}{2}} = \frac{2|E|}{|V|(|V|-1)}, \tag{16.2}$$

thus proving the theorem. $\qquad\square$

In addition to the existing network generation models, there are growing interests in understanding the dynamics of networks as nodes and edges are being added or removed. For example, Hanneke and Xing [32] presented an extension of the exponential random graph model to represent network evolution over time. Leskovec et al. [45] developed a new model to account for certain properties that were revealed in their study of real evolving networks. Their model, which is based on a "forest fire" spreading process, exhibits characteristics such as shrinking network diameter, increasing densification, as well as power-law degree distribution.

16.3 Techniques of Link Mining

Link mining has the same goal as data mining – finding hidden knowledge – but in a different setting. Some of the techniques have morphed from data mining to link mining, such as clustering to community finding and classification to link-based classification. Additionally, new techniques have emerged such as influence maximization and link prediction. Here we describe these techniques in detail.

16.3.1 Community Finding

The technique of *community finding*, also called group detection [26], positional analysis [20, 69], or blockmodelling [69], is the process of placing nodes into groups in such a way that the nodes within a group are "similar" to each other and "dissimilar" to nodes in other groups. This is equivalent to clustering [64] and graph partitioning [37]. From the graph theory perspective, the problem of community finding is to remove links from the graph so that the remaining graph has the "desired" components.

Community finding is ill-posed; there is no agreed-upon metric for evaluation. Metrics that are commonly used can be separated into supervised, where the original community assignments are known, and unsupervised, where they are not known. Supervised metrics, such as purity, entropy, and normalized mutual information, measure – in different ways – how well the "found" communities reflect the original communities. Unsupervised metrics generally measure the cohesion (the similarity of the nodes within communities) and/or the separation (the distance between nodes from different communities). A recently proposed unsupervised graph-based metric from Newman and Girvan [50], called modularity, is based on the fraction of links within a community to those between communities. An additional challenge to community finding is scalability. Networks such as the World Wide Web or online social networks can have millions, even billions, of nodes. Using an agglomerative hierarchical algorithm with a complexity of $O(n^3)$ – where n is the number of nodes – can be infeasible.

The most common approach to community finding is to segment the entire network into disjoint groups, where each node is assigned to exactly one community. Traditional clustering algorithms such as k-means, DBScan, Chameleon, etc. [64] can be applied to generate such communities. Graph partitioning algorithms are also applicable. For example, spectral clustering [61] divides a network into balanced components based on the eigenvalues of its Laplacian matrix. This approach is equivalent to finding a partitioning that minimizes the normalized cut criterion [18]. Karypis and Kumar [38] developed a multilevel graph partitioning approach that can accommodate different heuristic functions for coarsening, partitioning, and refining the clusters. Although these algorithms were not specifically designed for networks, their application is straightforward. An approach that was specifically designed for networks, from Girvan and Newman [28], uses the edge betweenness metric to remove edges iteratively. It is intuitively appealing since high betweenness edges would appear to be bottlenecks between communities; however, it is slow [55, 68].

A variant of finding disjoint communities is to discover a hierarchy of communities. This approach allows the communities to be nested and organized as a tree structure called a dendrogram. Agglomerative hierarchical clustering methods such as single-link and complete-link can be used to find hierarchies in networks. More recently, Clauset et al. [15] proposed a method of extracting hierarchies based on maximum likelihood methods and Markov chain Monte Carlo sampling.

Algorithms that find disjoint communities are popular but do not allow for situations when nodes can belong to more than one community. This is often the case in social networks where, for example, an individual can join two or more communities. An extreme case is to use fuzzy clustering, where every node belongs to every community with an associated weight. Another overlapping clustering method, developed by Banerjee et al. [5], is based on a mixture model of exponential family distributions. One of the challenges of overlapping communities is excess overlap. A paper by Scripps et al. [57] attempts to minimize the number of nodes with overlapping communities by isolating the bridge-nodes – nodes that are linked to more than one community – using a graph min-cut algorithm.

16 Link-Based Network Mining

Some community finding methods do not try to completely cluster the entire network. Instead they form communities from a given root set of nodes. This approach is helpful when only a portion of the network is known or the network is large and a complete grouping is unnecessary. As examples, consider Web page search where a set of related pages can be useful or in a large bibliographic database, where one only wants to know the community of researchers to which an author belongs. A Markov chain approach by Gibson et al. [27], specific to the World Wide Web, starts with a core set of pages, adds a fixed number of nearby pages, then forms the communities from the authoritative pages in the expanded set. Min-cut has also been adapted [23] to find a community by using the targeted node as the source and adding a virtual sink connected to all nodes in the graph.

More recently, progress in community finding has focused along several directions. Semi-supervised learning methods have become popular in the clustering literature, where side information is available in the form of constraints on pairs of nodes that should or should not be grouped together. The side information can be obtained from the similarity between node attributes, partial knowledge of the class labels, etc. The side information may improve community finding in many ways: to aid in the cluster initialization, to guide the clustering process toward finding better partitions, and to learn the appropriate distance metric consistent with the domain expectations [7, 13, 25]. Another trend is finding communities in dynamic networks, where the nodes, links, and attributes change over time. Backstrom et al. [4] studied how the structural features of communities affect how nodes join and leave communities. A paper by Tantipathananandh et al. [65] proposed a new framework for tracking community changes in dynamic networks by modeling it as a graph coloring problem. Communities are identified by approximately solving a combinatorial optimization problem using dynamic programming.

16.3.2 Node Ranking

Ranking is the process of creating a total ordering of the nodes in a network. The rank of a node reflects the measurement of some particular structural property of the network, with respect to the node, which conveys a semantic meaning such as importance, popularity, authority, etc. As an end in itself, rankings can also be used to look for well-connected or *central* nodes in a network.

In link mining, ranking is done using *centrality* measures [69]. The first, *degree* centrality, is simply the degree of the node; in directional graphs it can be indegree or outdegree. In a social network, the degree quantifies a node's popularity. In a bibliographic dataset the indegree is a measure of a paper's authority, while the outdegree measures the number of papers that it cites.

Closeness centrality is the average shortest distance between a node and all other (reachable) nodes in the network:

$$closeness(v) = \frac{1}{|V| - 1} \sum_{u \in V \setminus v} d(v, u)$$

where V is the set of nodes and $d(v, u)$ is the distance from v to u. Lower values of closeness indicate a more centrally located node. In a social network, a node with a low closeness rank indicates a person who generally has short communication paths to others, for instance, a CEO in a corporation. In a terrorist network it could help to identify a cell leader.

Another centrality measure is *betweenness*, which is the number of shortest paths between all pairs of nodes that go through it:

$$betweenness(v) = \sum_{s \in V, t \in V, s \neq t \neq v} \frac{g_{st}(v)}{g_{st}}$$

where g_{st} is the number of shortest paths from s to t and $g_{st}(v)$ is the number of shortest paths from s to t that go through v. Betweenness can be defined in the same way for edges. In a social network a node with a high betweenness score is considered important because it is likely that it will be encountered as members navigate the network. In a road map network, where nodes represent intersections or small communities, nodes with high betweenness are likely to be points of high congestion.

A popular ranking method for large directed networks like the World Wide Web is the *eigenvector* method [41, 53]. In this method a node's rank is the sum of the ranks of its incoming neighbors. Given a network with n nodes and an adjacency matrix A where A_{ij} is 1 if there is a directed edge from i to j and zero otherwise, for the node v_i, the rank r_i is defined as:

$$r_i = \frac{1}{\lambda} \sum_{j=1}^{n} A_{ij} r_j$$

where λ is a constant. Written in matrix form it becomes the eigenvector equation $\lambda r = Ar$. The dominant eigenvector of A provides an effective measure of authority rank. Google's PageRank is an example of this ranking method. Nodes with a high rank are said to be authoritative as many other nodes refer to them. This method of ranking effectively stifles the problem of manipulation. In the World Wide Web, for which this was proposed, unscrupulous Web hosts would create fake Web pages linked to their main page to raise its rank. However, since these bogus Web pages themselves have a low rank it does not increase the main page's rank very much. Unfortunately, this formulation has problems with graph cycles. The rank for nodes in a cycle will grow unabated. Page and Brin [53] solved this problem by adding a decay factor E yielding the equation $\lambda r = Ar + E$. Other approaches, similar to PageRank, include the algorithms HITS [41] and SALSA [43].

In addition to centrality measures, nodes can be ranked using other graph metrics such as eccentricity, but the practical significance of such metrics is less clear than, say, degree. Recently, there has been considerable interest in assigning rank values to nodes based on their community belongingness. Guimera et al. [31] introduced a metric called participation coefficient, which measures to what degree

16 Link-Based Network Mining 411

a node participates in other communities. Their approach requires the communities in a network to be identified first using an algorithm that optimizes a modularity function of the network partition. As a consequence, the ranks of the nodes are sensitive to the choice of community finding algorithm. Scripps et al. [59] introduced an alternative metric called *rawComm* for assigning ranks and roles to nodes without applying a community finding algorithm. The metric rawComm estimates the number of communities to which a node v_i belongs based on its local neighborhood structure:

$$\text{rawComm}(v_i) = \sum_{v_j \in N(v_i)} \frac{1}{1 + n_1 p + n_2(1 - q)}$$

where $N(v_i)$ is the set of nodes that are connected to v_i, n_1 is the number of common neighbors of v_i and v_j, and n_2 is $|N(v_i)| - n_1$. The values p and q represent the probabilities that two linked nodes are in the same community and two non-linked nodes are in different communities, respectively. Nodes with higher rawComm values are connected to more communities, making them good *ambassador* nodes.

Node ranking in dynamic networks remains an important but largely unexplored area of research. Desikan et al. [17] has recently developed an incremental approach to adjusting PageRank scores in evolving graphs without recomputing the ranks. Another promising direction would be to detect interesting trends in a dynamic network, for example, finding nodes with rapidly increasing authority scores. Finally, the problem could also be extended to ranking communities or groups of nodes.

16.3.3 Influence Maximization

Closely related to ranking is the technique of *influence maximization* (also known as *diffusion of innovation*), which is important in the areas of epidemic spread and viral marketing. The goal is to find influential nodes – nodes that will spread their influence quickly through the network. Influence is assumed to spread using a particular model of *diffusion*. In these models, nodes become activated (contracted a virus or bought a product) and can, in turn, activate their neighbors.

Diffusion models include the families of threshold and cascade models. In the threshold models [30] a node becomes activated when a certain percentage of its neighbors become activated. Newly activated nodes under the cascade models [29] have a one-time chance to activate neighbors with a given probability. Most of the models are probabilistic in nature. Without probability (e.g., if nodes are activated with certainty) every graph component with an activated node would end up with all nodes activated. Using appropriate probabilities ensures that activated nodes will only activate some of their neighbors and that the spread will stop before the entire network is activated.

The problem then is to choose nodes that will maximize a particular utility function. The most apparent utility function is the spread of activation to as many nodes

Fig. 16.2 Choosing a node to maximize influence

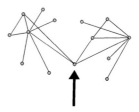

as possible. For example, in viral marketing, a company may want to offer a small number of free or discounted products to influential people in the hopes that they will inspire others to purchase the product.

One might first consider activating only the highest degree nodes to obtain the optimal solution. However, one can quickly imagine that if the high degree nodes are all neighbors, the spread of influence will be less than if lower degree nodes, more spread out, were chosen. For example, in Fig. 16.2, to maximize the number of nodes activated, the selected node is likely the best choice even though it is not the highest degree. Another challenge is that the link information may not be reliable; for example, in an online network, links between users are easy to add but do not always reflect genuine friendship. Furthermore, given the size of many real-world networks, simulating the activation process repeatedly to find the optimal solution is computationally expensive.

Kempe et al. [40] showed that the problem is NP-complete under the specific diffusion models of independent cascade and linear thresholds. They then propose a greedy strategy based on submodular functions [48], which guarantees a solution that is provably within 63% of optimal for these same models. In their experiments, the greedy strategy always performs better than the alternative strategies of selecting the nodes with the highest degree or lowest closeness scores.

An alternative to the problem was suggested by Scripps et al. [58] where the network is assumed to have latent communities. The problem then becomes choosing the nodes that will maximize the number of communities with activated nodes. For example, a viral marketer may be interested in offering free samples to a small group of individuals who will promote the product to as many demographic groups (sports fans, poetry lovers, etc.) as possible. The authors propose using the *rawComm* metric described in Sect. 16.3.2. Their experiments show that *rawComm* does better than greedy, degree, or closeness at maximizing the spread to latent communities.

Domingos and Richardson [19] proposed a cost/benefit approach to the influence maximization problem. They assume there is a cost for activating nodes and a revenue associated with activated nodes. The problem then becomes choosing a subset of nodes to activate that will maximize the expected lift in profit (i.e., revenue minus cost). A solution to the influence maximization problem in the face of competition was proposed by Bharathi et al. [8]. For example, multiple companies with similar products may attempt to influence the buying decisions of a targeted group of customers. Extending influence maximization to dynamic networks, where

16 Link-Based Network Mining

nodes may join or leave the network, is another open research problem. Variations of the problem in dynamic networks include finding the nodes that are most influential for new nodes or identifying the nodes whose influence spread is increasing or decreasing.

16.3.4 Link Prediction

The *link prediction* problem can be stated as follows. Given a network, can we infer the node pairs that are likely to be linked together? Link mining techniques are applicable to static networks (to infer missing links in an incomplete network) or dynamic networks (to predict new interactions that will occur in the near future). Examples of link prediction problems include detecting covert ties between criminal suspects or identifying future collaboration between researchers.

Link prediction is a challenging problem due to the sparsity of many networks. Predicting which nonlinked node pairs will become linked has so far yielded very low accuracies [46]. Rattigan and Jensen [56] have shown that this is due to the skewed class distribution; as networks grow and evolve, the number of nonlinked pairs increases quadratically while the number of linked pairs often grows only linearly. Research in social sciences has suggested the tendency of individuals to establish friendship ties with others who have similar interests (attributes) [36]. In addition, individuals may also become friends because they share common friends (link structure) or belong to similar groups (communities). Integrating these different sources of information to improve link prediction is a challenge.

Liben-Nowell and Kleinberg [46] compared a large number of graph metrics as link predictors. They tested the metrics on bibliographic data sets using only the link structure and ignoring the node attributes. This work has been expanded to include both link and attribute data [33, 46] by using binary classifiers. Another approach is to use probabilistic generative models, where the goal is to learn the joint probability density of the nodes, links, subgroups, etc., and to predict the missing links by applying Bayes's theorem [49, 67]. Because of the sparsity of networks, Rattigan and Jensen [56] proposed a variation to the problem known as anomalous link discovery, where the goal is to find links that are anomalous. Recent works have also considered the changes in the network over time. Potgieter et al. [54] combined the metrics from the Liben-Nowell study with temporal metrics such as return, moving average, and recency. In another work by O'Madadhain et al. [52], a time-evolving probabilistic classifier is constructed from training data sampled over many time periods. Hanneke and Xing [32] developed an extension of the exponential random graph model to account for the evolution of networks over time. A recent study by Backstrom et al. [4] on the evolution of communities in large social networks suggested that community structures and link formation are closely related. Making use of latent community structures for link prediction is another possible direction for future research.

16.3.5 Link-Based Classification

The classification problem is to predict the class of an object (which is simply an attribute of interest). A training set of objects whose class is known is given. A classifier is trained on the training set and used to predict the class of objects whose class is unknown. In a social network, for example, we may know the marital status of many of the members but not all. A classifier may help to identify combinations of other attributes (such as age group and income) that can be used to infer the marital status of those for whom it is unknown.

Traditional classifiers make a simplifying assumption that the objects are independent of each other. Researchers have recently begun to take advantage of the clearly defined relationships (links) within networks to improve classification. In the social network example above, the marital status of a person can potentially be inferred from the marital status of his or her friends. The challenge for link-based classification is integrating the attribute and link data. Using the attributes of neighbors has been shown to actually be detrimental in some cases [12]; however, using the class of the neighbor has been shown to be helpful [12, 47]. A related challenge is to recognize and utilize the structures inherent in the network. The study by Yang et al. [71] identified the existence of certain regularities in networks. For example, some networks exhibit encyclopedia regularity where nodes of one class link to nodes of the same class.

Some researchers have concentrated on utilizing a local approach to node classification. For example, Chakrabarti et al. [12] have developed a technique for Web page categorization that exploits link information in a small neighborhood around the Web pages. They showed that, by using both the attributes of a node and the class of its neighbors, the error rate of an attribute-based classifier can be reduced up to 70%. In another work by Lu and Getoor [47], two classifiers were trained, one on the attribute data and the other using neighborhood class statistics of neighbors. They showed that the combined classifiers result in improved predictions. The problem of propagating the class of known nodes through a network to the nodes with unknown classes is analogous to the label propagation problem in graph-based semi-supervised learning literature [67]. Unlike link-based classification, the graph used for label propagation is constructed based on the attribute similarity between objects.

Probabilistic models have also been used for link-based classification. Taskar et al. [66] proposed a probabilistic relational model using conditional Markov networks. They showed that the collective classification of multiple related entities can be inferred from the learned model. Similarly, Neville and Jensen [49] proposed a generative model that simultaneously learns the latent communities and the conditional probabilities associated with them.

Link-based classification can be extended to make use of the temporal information of an evolving network. For example, the class distribution of the nodes may change over time, and thus, can be exploited to improve the prediction. In some applications, a node can be assigned to multiple classes (e.g., a Web page having multiple tags or a gene with more than one functional class). Therefore, another new direction is to learn all the classes associated with a given node, a problem that is known as multilabel learning. In another direction, although some of the above approaches have shown improvements by using the class information of neighbors, the information will be less helpful for some nodes than for others. Scripps et al. [58] have shown that the role that a node plays in the network can guide the classifier to use the neighborhood information when it is likely to help. For example, neighborhood information is less predictive for nodes linked to many communities.

16.4 Conclusion

In this chapter we have reviewed characteristics of networks and described some of the techniques that have been proposed in the link mining area. These techniques have been applied to a wide variety of networks such as the World Wide Web [41, 53], terrorist networks [60], viral marketing [19], organizational structure [42], and even to some domains that do not immediately appear to be structured like a network, such as macroeconomics [35].

Although recent progress in link mining research has been significant, there are many new directions for growth. Consider the temporal aspect. Techniques such as link prediction can assume that the network is changing over time; however, most of the algorithms are still designed to work with static networks. To account for changes in the network over time, concepts such as *selection* (people preferring to make friends with others having similar attributes) and *influence* (changing attributes to align more with friends) [11] are beginning to become integrated into new models.

Another new direction is inspired by the recent explosion in online social networks and their available APIs. New Web and desktop applications could utilize a user's social network information (with the permission of the user). As an example, a retailer with access to a person's social network could make more meaningful product recommendations. Third-party access to network data reveals a number of new problems such as handling partial network data and blending networks. One negative aspect of online social networks is the concern for privacy. Although many sites provide security settings for users, many complacently use the defaults, which can allow strangers to view their personal data. As shown by Backstrom et al. [3], even in anonymized networks, it is still possible to identify users. Techniques to safeguard against abuse in social networks would be well received.

Considering the massive size of many real-world networks, sampling from large graphs is another promising research direction [44]. Sampling allows inferences to be made from a representative subgraph, thus enabling current algorithms to scale up to massive-sized networks. However, in many domains, the sample may be biased as a result of the data collection process (e.g., when crawling the pages of an online social networking Web site). Issues such as boundary effects of the subgraph and biases in parameter estimation are some of the challenges that must be addressed [22].

Finally, improvements in network visualization [34] would be another interesting growth direction. A display of a small graph of a hundred nodes can be very helpful. However, consider the users in Facebook, who often have over 200 friends. Providing an informative, visual representation of such large and complex networks is an important new challenge.

As online social networks continue to proliferate and become integrated into more traditional applications the opportunity for new techniques will also continue to grow accordingly. Graph theory will undoubtedly play a role in this growth.

References

1. Adamic L, Adar E (2003) Friends and neighbors on the web. Soc Networks 25:211–230
2. Airoldi EM, Carley KM (2005) Sampling algorithms for pure network topologies. SIGKDD Explorations 7:13–22
3. Backstrom L, Dwork C, Kleinberg J (2007) Wherefore art thou r3579x? Anonymized social networks, hidden patterns, and structural steganography. In: Proceedings of the 16th international World Wide Web conference
4. Backstrom L, Huttenlocher D, Kleinberg J, Lan X (2006) Group formation in large social networks: membership, growth, and evolution. In: Proceedings of the 12th ACM SIGKDD international conference on knowledge discovery and data mining
5. Banerjee A, Krumpelman C, Ghosh J, Basu S, Mooney R (2005) Model based overlapping clustering. In: Proceedings of the 11th ACM SIGKDD international conference on knowledge discovery and data mining
6. Barabási A-L, Bonabeau E (2003) Scale-free networks. Sci Am 288:50–59
7. Basu S, Bilenko M, Mooney R (2004) A probabilistic framework for semi-supervised clustering. In: Proceedings of the 10th ACM SIGKDD international conference on knowledge discovery and data mining, Seattle, WA
8. Bharathi S, Kempe D, Salek M (2007) Competitive influence maximization in social networks. In: Deng X, Graham FC (eds) Proceedings of WINE 2007. Springer, Heidelberg
9. Borgatti SP, Everett MG (1999) Models of core/periphery structures. Soc Networks 21: 375–395
10. Brandes U, Erlebach T (2005) Network analysis. Lecture Notes in Computer Science. Springer, Berlin
11. Burk W, Steglich CEG, Snijders TAB (2007) Beyond dyadic interdependence: actor-oriented models for co-evolving social networks and individual behaviors. Int J Behav Dev 31:397
12. Chakrabarti S, Dom B, Indyk P (1998) Enhanced hypertext categorization using hyperlinks. In: Proceedings of the SIGMOD international conference on management of data. ACM, New York, pp 307–318
13. Chang H, Yeung D-Y (2008) Robust path-based spectral clustering. Pattern Recogn 41: 191–203
14. Chartrand G, Oellermann O (1992) Applied and algorithmic graph theory. McGraw-Hill, New York
15. Clauset A, Moore C, Newman MEJ (2006) Structural inference of hierarchies in networks. In: Statistical network analysis: models, issues, and new directions, vol 4503, pp 1–13
16. Dehmer M, Emmert-Streib F (2008) Structural information content of networks: graph entropy based on local vertex functionals. Comput Biol Chem 32:131–138
17. Desikan P, Pathak N, Srivastava J, Kumar V (2005) Incremental page rank computation on evolving graphs. In: Proceedings of the 14th international World Wide Web conference (Special interest tracks and posters)

18. Dhillon IS, Guan Y, Kulis B (2004) Kernel k-means: spectral clustering and normalized cuts. In: Proceedings of the 10th ACM SIGKDD international conference on knowledge discovery and data mining
19. Domingos P, Richardson M (2001) Mining the network value of customers. In: Proceedings of the 7th ACM SIGKDD international conference on knowledge discovery and data mining. ACM, New York, pp 57–66
20. Doreian P, Batagelj V, Ferligoj A (2005) Positional analysis of sociometric data. In: Carrington P, Scott J, Wasserman S (eds) Models and methods in social network analysis, Cambridge, New York
21. Erdős P, Rényi A (1960) On the evolution of random graphs, vol 5. Publications of the institute of Mathematics, Hungarian Academy of Science, pp 17–61
22. Fienberg S (2006) Panel discussion from statistical network analysis: models, issues, and new directions. In: Proceedings of the ICML 2006 workshop on statistical network analysis, Pittsburgh, PA, USA
23. Flake G, Tsioutsiouliklis K, Tarjan R (2002) Graph clustering techniques based on minimum cut trees. Technical Report, NEC, Princeton, NJ
24. Frantz T, Carley KM (2005) A formal characterization of cellular networks. Technical Report CMU-ISRI-05-109, School of Computer Science, Carnegie Mellon University
25. Gao J, Tan PN, Cheng H (2006) Semi-supervised clustering with partial background information. In: Proceedings of SDM'06: SIAM international conference on data mining
26. Getoor L, Diehl CP (2005) Link mining: a survey. SIGKDD Explorations 7:3–12
27. Gibson D, Kleinberg J, Raghavan P (1998) Inferring web communities from link topology. In: Proceedings of the 9th ACM conference on hypertext and hypermedia
28. Girvan M, Newman M (2002) Community structure in social and biological networks. Proc Natl Acad Sci 99:7821–7826
29. Goldenberg J, Libai B, Muller E (2001) Using complex systems analysis to advance marketing theory development: modeling heterogeneity effects on new product growth through stochastic cellular automata. Academy of Marketing Science Review
30. Granovetter M (1978) Threshold models of collective behavior. Am J Sociol 83:1420–1443
31. Guimerà R, Sales-Pardo M, Amaral L (2007) Classes of complex networks defined by role-to-role connectivity profiles. Nat Phys 3:63–69
32. Hanneke S, Xing E (2006) Discrete temporal models of social networks. In: Proceedings of the 23rd international conference on machine learning workshop on statistical network analysis
33. Al Hasan M, Chaoji V, Salem S, Zaki M (2006) Link prediction using supervised learning. In: Proceedings of SDM'06: SIAM data mining conference workshop on link analysis, counter-terrorism and Security
34. Heer J, Boyd D (2005) Vizster: visualizing online social networks. In: Proceedings of IEEE symposium on information visualization. IEEE Press, Minneapolis, MN
35. Jackson M (2008) Social networks in economics. In: Benhabib J, Bisin A, Jackson MO (eds) Handbook of social economics. Elsevier, Amsterdam
36. Kandel D (1978) Homophily, selection, and socialization in adolescent friendships. Am J Sociol 84:427–436
37. Karypis G, Kumar V (1995) Analysis of multilevel graph partitioning. Supercomputing
38. Karypis G, Kumar V (1999) A fast and high quality multilevel scheme for partitioning irregular graphs. SIAM J Sci Comput 20:359–392
39. Katz L (1953) A new status index derived from sociometric analysis. Psychometrika 18: 39–43
40. Kempe D, Kleinberg J, Tardos E (2003) Maximizing the spread of influence through a social network. In: Proceedings of the 9th ACM SIGKDD international conference on knowledge discovery and data mining, pp 137–146
41. Kleinberg J (1999) Sources in a hyperlinked environment. J ACM 46:604–632
42. Krackhardt D, Hanson JR (1993) Informal networks: the company behind the chart. Harvard Bus Rev 71:104–111

43. Lempel R, Moran S (2001) Salsa: the stochastic approach for link-structure analysis. ACM Trans Inf Syst 19:131–160
44. Leskovec J, Faloutsos C (2006) Sampling from large graphs. In: SIGKDD
45. Leskovec J, Kleinberg J, Faloutsos C (2005) Graphs over time: densification laws, shrinking diameters and possible explanations. In: Proceedings of the 11th ACM SIGKDD international conference on knowledge discovery in data mining
46. Liben-Nowell D, Kleinberg J (2003) The link prediction problem for social networks. In: Proceedings of the 12th international conference on information and knowledge management, New Orleans, LA
47. Lu Q, Getoor L (2003) Link-based classification. In: Proceedings of the 20th international conference on machine learning, ICML
48. Nemhauser GL, Wolsey LA, Fisher ML (1978) An analysis of approximations for maximizing submodular set functions. Math Program 14:265–294
49. Neville J, Jensen D (2005) Leveraging relational autocorrelation with latent group models. In: Proceedings of the 5th IEEE international conference on data mining
50. Newman M, Girvan M (2004) Finding and evaluating community structure in networks. Phys Rev E 69:026113
51. Newman MEJ (2001) Clustering and preferential attachment in growing networks. Phys Rev E 64:025102
52. O'Madadhain J, Hutchins J, Smyth P (2005) Prediction and ranking algorithms for event-based network data. SIGKDD Explorations 7:23–30
53. Page L, Brin S, Motwani R, Winograd T (1998) Pagerank citation ranking: bringing order to the web. Technical report, Stanford University
54. Potgieter A, April K, Cooke R, Osunmakinde IO (2006) Temporality in link prediction: understanding social complexity. J Trans Eng Manag
55. Radicchi F, Castellano C, Cecconi F, Loreto V, Parisi D (2004) Defining and identifying communities in networks. Proc Natl Acad Sci USA 101:2658–2663
56. Rattigan M, Jensen D (2005) The case for anomalous link discovery. SIGKDD Explorations 7:41–47
57. Scripps J, Tan PN (2006) Clustering in the presence of bridge-nodes. In: Proceedings of SDM'06: SIAM international conference on data mining, Bethesda, MD
58. Scripps J, Tan PN, Esfahanian A-H (2007) Exploration of link structure and community-based node roles in network. Technical report, Michigan State University
59. Scripps J, Tan PN, Esfahanian A-H (2007) Exploration of link structure and community-based node roles in network analysis. In: Proceedings of the 7th IEEE international conference on data mining
60. Senator T (2002) Darpa: evidence extraction and link discovery program. DARPATech
61. Shi J, Malik J (2000) Normalized cuts and image segmentation. IEEE Trans Pattern Anal 22(8):888–905
62. Skorobogatov VA, Dobrynin AA (1988) Metrical analysis of graphs. MATCH 23:105–155
63. Solé RV, Valverde S (2004) Information theory of complex networks: on evolution and architectural constraints. In: Lecture notes in physics, vol 650, pp 189–207
64. Tan P, Steinbach M, Kumar V (2005) Introduction to data mining. Addison Wesley, Boston, MA
65. Tantipathananandh C, Berger-Wolf TY, Kempe D (2007) A framework for community identification in dynamic social networks. In: Proceedings of the 13th ACM SIGKDD international conference on knowledge discovery and data mining, pp 717–726
66. Taskar B, Abbeel P, Koller D (2002) Discriminative probabilistic models for relational data. In: Proceedings of the 18th conference on uncertainty in artificial intelligence (UAI02)
67. Taskar B, Wong MF, Abbeel P, Koller D (2003) Link prediction in relational data. In: Neural information processing systems conference (NIPS03)
68. Tyler JR, Wilkinson DM, Huberman BA (2003) Email as spectroscopy: automated discovery of community structure within organizations. In: Proceedings of the 5th international conference on communities and technologies

69. Wasserman S, Faust K (1994) Social network analysis: methods and applications. Cambridge University Press, Cambridge
70. Watts DJ, Strogatz SH (1998) Collective dynamics of 'small-world' networks. Nature 393:440–442
71. Yang Y, Slattery S, Ghani R (2002) A study of approaches to hypertext categorization. J Intell Inf Syst 18:219–241

Chapter 17
Graph Representations and Algorithms in Computational Biology of RNA Secondary Structure

Stefan Washietl and Tanja Gesell

Abstract The analysis of RNA structures is an important problem in computational biology. In this chapter we review various algorithms to predict and compare RNA secondary structures. These algorithms are based on graph theory and use representations of RNA secondary structure as outerplanar graphs and trees.

Keywords Bioinformatics · RNA folding · Outerplanar graphs · Tree editing

MSC2000: Primary 94C15; Secondary 92C40, 92D20, 05C30

17.1 Introduction

RNA secondary structure is an important level in the structural hierarchy of RNA molecules. The basic principle of RNA secondary structure – discrete building blocks connected by molecular interactions – naturally leads to graph-theoretical approaches for their formal analysis. Therefore, graph and tree algorithms have a long tradition in RNA structural biology and some of them are among the pioneering work in the field of computational biology [19].

However, only in more recent years has it become evident that the biological importance of RNA has been vastly underestimated for a long time. The first description of "ribozymes" showed that RNA can catalyze biochemical processes, an activity which was only known for protein enzymes before [5]. The discovery of micro RNAs led to a paradigm shift in our understanding of gene regulation [1]. New high-throughput experimental techniques as well as computational predictions suggest that there are tens of thousands of so-far-unrecognized RNAs in mammalian

---

T. Gesell (✉)
Center for Integrative Bioinformatics Vienna, Max F. Perutz Laboratories,
Dr. Bohr-Gasse 9, 1030 Vienna, Austria
and
University of Vienna, Medical University of Vienna and University of Veterinary Medicine,
Vienna, Austria
e-mail: tanja.gesell@univie.ac.at

M. Dehmer (ed.), *Structural Analysis of Complex Networks*,
DOI 10.1007/978-0-8176-4789-6_17, © Springer Science+Business Media, LLC 2011

cells [4, 27, 28]. Although it is still unclear whether it is justified to proclaim a modern "RNA world", there is no doubt that RNAs have to be considered as important key players in the cell and that the structural biology of RNAs will be of particular importance in the next years.

The goal of the chapter is to give an overview of some of the most fundamental principles and algorithms of RNA secondary structures with a particular focus on their representation as graphs. There are different ways to encode RNAs as graphs which allow for attractive algorithms for their analysis.

We start with some background information on biochemical properties of RNAs (Sect. 17.2). Next, we show how these basic biochemical properties as well as the commonly used classification of structural elements in RNAs can be formalized by a definition of RNA secondary structures as outerplanar graphs (Sects. 17.3 and 17.4). The RNA folding problem, i.e., the prediction of the secondary structure for a given RNA sequence, is the main topic of this chapter. In Sect. 17.5, we introduce a simple recursive algorithm to the combinatorial problem of counting secondary structures, which is the basis for the different RNA folding algorithms reviewed in Sects. 17.6–17.8. In the last two sections, we address the problem of comparing RNA structures. We show that RNA secondary structures can be encoded as trees and present tree editing as a special type of graph-matching problem to calculate distances between RNA structures. The chapter finishes with a summary and conclusion.

17.2 Structural Properties of RNA Molecules

RNA is a polymer made of individual units called *nucleotides* [17]. Nucleotides consist of a ribose group, a phosphate group, and one of four different bases adenine (A), cytosine (C), guanine (G), and uracil (U). The succession of the four different bases of the nucleotides defines the *primary structure* or *sequence* of the molecule (Fig. 17.1). Adjacent nucleotides in the primary sequence are connected by covalent bonds, i.e., strong chemical bonds that do not open under normal conditions. These bonds build the "backbone" of the molecule. RNA is generally single stranded but complementary regions in the molecule can fold back onto themselves and form double helices similar to the well-known DNA helix. In RNA, we usually find the so-called Watson–Crick pairs CG and AU as well as GU "wobble pairs". The intramolecular base-pairing results in a pattern of double helical stretches interspersed with unpaired regions which is called the *secondary structure*. Unlike the covalent bonds of the backbone this base-pairing is realized by weaker hydrogen bonds that can be opened and closed under physiological conditions. The arrangement of secondary structure elements in space finally forms the three-dimensional *tertiary structure*.

RNA folding is a hierarchical process. The secondary structure usually forms before and independently of the tertiary structure and contributes most of the stabilizing energy. The formation of the tertiary structure usually does not induce changes in the secondary structure.

17 Graph Representations and Algorithms in Computational Biology

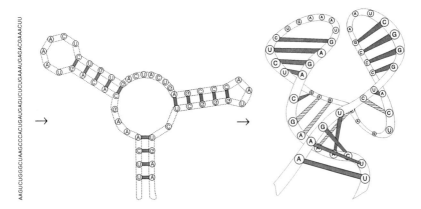

Fig. 17.1 Principles of RNA structure. The primary structure (*left*) is defined by the succession of the four different nucleotide types A,C,G,U. Pairing patterns among AU, GC, and GU form the secondary structure (*middle*). The secondary structure elements interact with each other in a complex three-dimensional pattern, the tertiary structure (*right*). Note that in the tertiary structure nonstandard base-pairs can occur (*dashed*) that are usually not considered in algorithms analyzing the secondary structure. The example shows a so-called hammerhead ribozyme, a short self-cleaving RNA

The function of the molecule is ultimately dependent on the tertiary structure. However, a secondary structure can serve as a coarse-grained approximation and is an extremely useful level on which to understand RNA function.

17.3 Secondary Structure as Outerplanar Graph

An early graph-based definition of RNA secondary structure is due to Waterman [29].

Definition 1. A secondary structure is a vertex labeled graph on n vertices with an adjacency matrix $A = (a_{ij})$ fulfilling:

(a) $a_{i,i+1} = 1$ for $1 \leq i \leq n-1$.
(b) For each i, $1 \leq i \leq n$ there is at most one $a_{ij} = 1$ where $j \neq i \pm 1$.
(c) If $a_{ij} = a_{kl} = 1$ and $i < k < j$ then $i < l < j$.

The first part defines the continuous "backbone" of the primary structure of the molecule. The second part defines the secondary structure interactions and allows each nucleotide (vertex) to be paired with at most one other nucleotide not immediately adjacent in the backbone. An edge of this type (i, j), $j \neq i \pm 1$, is called a *base-pair*. A vertex i connected only to $i \pm 1$ is called *unpaired*. The third part of the definition excludes interactions that are (somewhat arbitrarily) classified as tertiary structure interactions. In particular, this rule excludes structures known as "pseudoknots".

The molecule geometry in RNA does not allow sharp bends with unpaired regions shorter than three. In practical applications, one usually adds additional biological reality to this definition by requiring $a_{ij} = 0$ if $1 < j - i \leq 3$.

Secondary structure graphs following this definition are *outerplanar*; i.e., they have an embedding in the plane such that all vertices lie on the boundary of its exterior region (Fig. 17.2). The edges representing the base-pairs lie inside and do not cross.

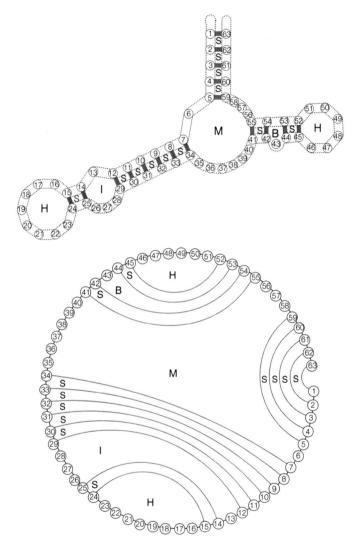

Fig. 17.2 Graph representation and structural elements of RNA secondary structures. H: hairpin I: interior loop, S: stacked pair, B: bulge loop, M: multi loop. *Top:* Conventional drawing of the structures as used by biochemist and molecular biologist. *Bottom:* Circle representation emphasizing the graph-like nature of the secondary structure. The circle represents the backbone of the RNA. Each nucleotide is connected to its immediate neighbours within the backbone. In addition each nucleotide can form one (and only one) base-pair to another nucleotide (*arcs*) or stay unpaired. The definition of RNA secondary structures excludes pseudoknotted structures; i.e., the arcs are not allowed to cross. The faces of the graph correspond to the different substructure elements

17 Graph Representations and Algorithms in Computational Biology   425

Similar types of graphs have been studied in a nonbiological context. For example, they are closely related to Touchard's *linked diagrams* [26] and a generalization of Yan's "bamboo-shoot graphs" [32].

17.4 Classification of Structural Elements

To describe and understand complex RNA secondary structures, biologists distinguish between different structural elements. Also for formal treatment it is helpful to identify and classify the basic building blocks in a secondary structure. For Zuker's structure prediction algorithm ([38], Sect. 17.7) the so-called *k-loop decomposition* [37] is used.

Definition 2. A base k is called immediately interior of the base-pair (i, j) if $i < k < j$ and there is no other base-pair (p, q) such that $i < p < k < q < j$.

Definition 3. The base-pair (i, j) and all bases immediately interior to (i, j) is called a loop closed by (i, j). The number of base-pairs contained in the loop (including the closing base-pair) is called the degree of the loop.

Loops correspond to the faces of the outerplanar secondary structure graph (Fig. 17.2). Commonly used structural elements of RNA secondary structures can be defined using this formalism.

Definition 4. Classification of structural elements:

- A loop of degree 1 is called a ***hairpin***.
- A loop of degree 2 is called an ***interior loop***. Let (i, j) be the closing base-pair and (p, q) the base-pair immediately interior. There are two special cases of interior loops:

  - ***stacked pair*** if $p - i = 1$ and $j - q = 1$.
  - ***bulge*** if $p - i > 1$ or $j - q > 1$ but not both.

- A loop of degree ≥ 3 is called a ***multiloop***.

17.5 Counting Secondary Structures

The combinatorial problem of counting the number of secondary structures that can be formed by a sequence of a given length is of particular interest. Its recursive solution was first realized by Waterman [29, 30] 30 years ago and it is the basis for many of the folding algorithms we describe later in this chapter.

Let x be a sequence of n nucleotides $x_i \in \{A,C,G,U\}$, $1 \leq i \leq n$. If we assume a specific sequence not all positions can pair, but only those following the base-pairing rules for RNA structures (Sect. 17.2). We use the base-pairing matrix Π

with the entries $\Pi_{ij} = 1$ if sequence positions i and j can form a base-pair; i.e., if (i, j) is in the set of allowed base-pairs $B = \{GC, CG, AU, UA, GU, UG\}$, and $\Pi_{ij} = 0$ otherwise. Further, let x_{ij} be the subsequence from i to j, and N_{ij} the number of secondary structures that can be formed by x_{ij}.

To calculate N_{ij}, we assume that we already know $N_{i+1,j}$, i.e., the number of structures of a subsequence shorter by one base. A newly added base can either be unpaired or form a base-pair with some other base k. In the first case, the unpaired base is followed by any possible structure in subsequence $x_{i+1,j}$. In the latter case, the new base-pair divides the sequence in two subsequences $x_{i+1,k-1}$ and $x_{k+1,j}$. Since base-pairs do not cross (Definition 1, part (c)), both subsequences can be treated independently and their numbers can be simply multiplied. These considerations lead to the following recursion:

$$N_{ij} = N_{i+1,j} + \sum_{\substack{i+1 \le k \le j \\ \Pi_{ik}=1}} N_{i+1,k-1} N_{k+1,j} \qquad (17.1)$$

with $N_{ii} = 1$.

RNA secondary structure graphs lead to many other interesting combinatorial questions (e.g., [10] and references therein) which are, however, not of immediate relevance for most practical applications in bioinformatics.

17.6 Structure Prediction Using Simple Base-Pairing Rules

The "RNA folding" problem, i.e., the prediction of the secondary structure for a given primary sequence, is without doubt the most relevant problem for practical applications. Experimental determination of structures can be laborious and is not feasible in large scale. Computational predictions are therefore widely used in the everyday analysis of RNAs.

Thermodynamic methods for RNA folding are the most established and most frequently used methods today. Put in simple terms, the goal is to find the structure with the most favourable folding energy. Usually, the free energy of folding ΔG relative to the unfolded sequence is considered. Paired regions add stabilizing (by convention negative) energy contributions to ΔG while unpaired regions add destabilizing (positive) energy terms.

The first attempts to calculate optimal secondary structures for simplified energy models are due to Nussinov and co-workers [18, 19]. In the simplest case, one can assign each type of base-pair a negative and fixed energy contribution. Then the problem reduces to finding the structure with the maximum number of base-pairs. In a more sophisticated (but still largely unrealistic scenario), one assigns each type of base-pair (i, j) a specific energy contribution β_{ij}. The overall energy of a fold is the sum of all base-pair energies. In this model, we can find the minimal energy F_{ij} of a sequence x_{ij} using a very similar strategy as used for enumerating all structures in (17.1). Adding one base at a time, either the new base is unpaired or it

```
function Backtrack(i, j)
begin
    if i > j then return
    if K_ij = 0 then Backtrack(i + 1, j) else
        output: (i, K_i,j)
        Backtrack(i + 1, K_ij − 1)
        Backtrack(K_ij + 1, j)
    end
end
```

Algorithm 17.1: Recursive backtracking procedure to retrieve the list of base-pairs in the optimal structure

forms a pair with some base k. The overall minimum is the minimum of these two cases. To obtain the minimum of the latter case in which i forms a base-pair, all possible base-pairs (i, k) are evaluated. Each base-pair (i, k) separates the subsequence in two intervals and due to the independence for the minimum free energy we obtain the following recursion:

$$F_{ij} = \min \left\{ F_{i+1,j}, \min_{\substack{i+1 \le k \le j \\ \Pi_{ik}=1}} \left\{ F_{i+1,k-1} + F_{k+1,j} + \beta_{ik} \right\} \right\} \tag{17.2}$$

This is an example of a dynamic programming algorithm frequently encountered in bioinformatics. A matrix containing the optimal solution for all possible subsequences is filled and the entry $F_{1,n}$ finally contains the optimal solution for the whole sequence of length n. The algorithmic complexity of this procedure is $\mathcal{O}(n^2)$ and $\mathcal{O}(n^3)$ in memory and CPU, respectively.

However, evaluating (17.2) gives only the minimum free energy and not the structure itself. A so-called *backtracking* or *backtracing* procedure is used to get the list of base-pairs corresponding to the optimal energy. A helper matrix K is filled during the recursion. We set $K_{ij} = k$, where k is the base which gives the optimal secondary structure when paired with i for a subsequence from i to j. If i is unpaired in the optimal structure we set $K_{ij} = 0$. We can then retrieve the list of base-pairs of the optimal structure using a simple recursive procedure as shown in Algorithm 17.1. We start with input $(i, j) = (1, n)$; i.e., we consider $K_{1,n}$ which holds the pairing partner k of position 1 in the optimal structure of the whole sequence of length n. $K_{1,n} = k$ divides the sequence in two independent subsequences which are evaluated by recursively calling the same function again.

17.7 Structure Prediction Using the Loop-Based Energy Model

Although this procedure clearly gives the optimal structure in an algorithmic sense, structure predictions obtained this way are generally not very accurate. The energy model based on simple base-pairing rules only poorly reflects the biophysical properties of real RNA molecules. Most of the stabilizing energy in RNA secondary structures comes from stacking interactions of neighbouring base-pairs. A realistic energy model thus needs to consider the *loops* in a structure (Sect. 17.4). The so-called loop-based energy model or nearest-neighbour model assigns each loop l in a structure S a free energy ΔG. The total free energy of the structure is the sum of all loops:

$$\Delta G(S) = \sum_{l \in S} \Delta G(l) \tag{17.3}$$

The energy rules used in current state-of-the art prediction programs are quite complex [15, 31] and it would be out of scope to present them in detail here. Generally, the energy depends on the type of the loop, the size of the loop, the closing base-pairs, and the bases immediately interior of the closing base-pair. The energy values have been determined empirically using melting experiments that measure the energy which is required to open specific structural elements. Only stacks and some other small loops are tabulated exhaustively. The energy rules for other types of loops usually contain extrapolations and other approximations.

In principle, one can find the minimum free energy model using a similar strategy as shown before. However, it is not sufficient to distinguish only two cases in the recursion. Instead, all possible loop types have to be considered in a systematic decomposition procedure. Recursions for this problem have first been proposed by Zuker and Stiegler [38]. Here we show a version following reference [11] that decomposes structures in such a way that each substructure is considered exactly once.

Figure 17.3 shows a graphical outline of the decomposition steps. The procedure requires four matrices. F_{ij} contains the free energy of the overall optimal structure of the subsequence x_{ij}. The newly added base can be unpaired or it can form a pair. For the latter case, we introduce the helper matrix C_{ij}, that contains the free energy of the optimal substructure of x_{ij} under the constraint that i and j are paired. This structure closed by a base-pair can either be a hairpin, an interior loop, or a multiloop. The hairpin case is trivial because no further decomposition is necessary. The interior loop case is also simple because it reduces again to the same decomposition step. The multiloop step is more complicated. The energy of a multiloop depends on the number of components, i.e., substructures that emanate from the loop. To implicitly keep track of this number there is need for an additional two helper matrices. M_{ij} holds the free energy of the optimal structure of x_{ij} under the constraint that x_{ij} is part of a multiloop with at *least one* component. M_{ij}^1 holds the free energy of the optimal structure of x_{ij} under the constraint that x_{ij} is part of a multi-loop and has *exactly one* component closed by pair (i, k) with $i \leq k < j$. The idea is to decompose a multiloop in two arbitrary parts of which the first is a multiloop with at least one component and the second a multiloop with exactly one component and starting with a base-pair. These two parts corresponding to M and M^1 can further be decomposed into substructures that we already know, i.e., unpaired intervals,

17 Graph Representations and Algorithms in Computational Biology

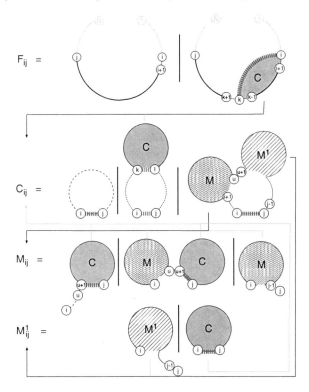

Fig. 17.3 Illustration of the recursive structure decomposition steps in Zuker's folding algorithm. The property of a sequence with chain length N is built up recursively from the properties of smaller segments under the assumption that the contributions are additive. The procedure requires four matrices: F_{ij}, C_{ij}, M_{ij}, and M^1_{ij} (cf. (17.4)). *Bold dashed lines* indicate base-pairs, *dotted lines* indicate unpaired substructures, and *solid black lines* indicate arbitrary structures. Please refer, to the text for a detailed description of the procedure

substructures closed by a base-pair, or (shorter) multiloops. We can summarize the recursion as follows:

$$\begin{aligned}
F_{ij} &= \min\left\{ F_{i+1,j},\ \min_{i<k\le j}\left(C_{ik} + F_{k+1,j}\right)\right\} \\
C_{ij} &= \min\left\{ \mathcal{H}(i,j),\ \min_{i<k<l<j}\left(C_{kl} + \mathcal{I}(i,j;k,l)\right),\right. \\
&\qquad \left. \min_{i<u<j}\left(M_{i+1,u} + M^1_{u+1,j-1} + a\right)\right\} \\
M_{ij} &= \min\left\{ \min_{i<u<j}\left((u-i+1)c + C_{u+1,j} + b\right),\right. \\
&\qquad \left. \min_{i<u<j}\left(M_{i,u} + C_{u+1,j} + b\right),\ M_{i,j-1} + c\right\} \\
M^1_{ij} &= \min\left\{ M^1_{i,j-1} + c,\ C_{ij} + b\right\},
\end{aligned}$$

(17.4)

$\mathcal{H}(i, j)$ is the energy for a hairpin closed by base-pair (i, j) and $\mathcal{I}(i, j; k, l)$ the energy for an interior loop closed by the two base-pairs (i, j) and (k, l). Multiloop energies are approximated by a simple linear relationship: $E_{\mathrm{ML}} = a + b \cdot \mathrm{degree} + c \cdot \mathrm{size}$. Multiloops are generally considered destabilizing. The constant a is used to penalize opening a multiloop in the first place. The constants b and c penalize the number of components and the size of unpaired intervals, respectively.

Using these recursions, the minimum free energy and – using an appropriate backtracking procedure – the optimal structure under the full loop-based energy model can be found. This approach is currently the most widely used method to predict RNA secondary structures. The most popular implementations are `mfold` [36] and `RNAfold` from the Vienna RNA package [9].

17.8 Prediction of Base-Pairing Probabilities

At room temperature, the energy contributions from the base-pairing in a molecule is on the same order of magnitude as the thermal energy. As a consequence, base-pairs can open and close and an RNA molecule does not only fold into a single structure but forms an *ensemble* of different structures. Following basic principles of thermodynamics, the probability of a given structure S is proportional to its Boltzmann factor:

$$\mathrm{Prob}(S) = \frac{\exp(-\Delta G(S)/RT)}{Z} \tag{17.5}$$

where T is the absolute temperature and R the universal gas constant. The normalization factor Z is a particularly important quantity. It is the Boltzmann weighted sum over all possible structures and is called the *partition function*:

$$Z = \sum_{S} \exp(-\Delta G(S)/RT) \tag{17.6}$$

As shown by McCaskill [16], the partition function can be calculated using similar recursions and dynamic programming algorithms as used for calculating the minimum free energy. For the simple base-pair energy model, the recursion to calculate the partition function can be formulated as follows:

$$Z_{ij} = Z_{i+1,j} + \sum_{\substack{i+1 \leq k \leq j \\ \pi_{ik} = 1}} Z_{i+1,k-1} Z_{k+1,j} \exp(-\beta_{ik}/RT) \tag{17.7}$$

Please note the analogy to (17.2). We can simply replace the minimum by the sum, the sums with multiplications, and the energy contribution by its Boltzmann factor. The value of the partition function by itself is usually not of immediate interest. In practice, the most interesting information is the probability of a

17 Graph Representations and Algorithms in Computational Biology

specific base-pair within the equilibrium ensemble, or more precisely the probability $p_{ij} = \sum_{(i,j)\in\mathcal{S}} \text{Prob}(\mathcal{S})$ of observing a structure $\mathcal{S}$ that contains the base-pair (i, j). To calculate p_{ij} we need to know the partition function over all structures forming (i, j) and the total partition function Z:

$$p_{ij} = \widehat{Z}_{ij} Z_{i+1,j-1} \exp(-\beta_{ij}/RT)/Z \tag{17.8}$$

The helper quantity $\widehat{Z}_{ij}$ is the partition function over all structures *outside* the interval x_{ij}. Using similar considerations as for the "forward" recursion one arrives at

$$\widehat{Z}_{ij} = \widehat{Z}_{i,j+1} + \sum_{\substack{1 \le k < i \\ \Pi_{k,j+1}=1}} \widehat{Z}_{k,j+1} \exp(-\beta_{k,j+1}/RT) Z_{k+1,i-1}$$

$$+ \sum_{\substack{j+2 \le k \le n \\ \Pi_{k,j+1}=1}} \widehat{Z}_{i,k} \exp(-\beta_{k,j+1}/RT) Z_{j+2,k-1} \tag{17.9}$$

A common way to summarize the structural properties of an RNA molecule in the thermodynamic ensemble is to calculate the probability matrix of all possible base-pairs. This matrix can be conveniently visualized as "dot-plot". Figure 17.4 shows an example of a pairing matrix for a short hammerhead RNA calculated with the program RNAfold of the Vienna RNA package [9] that implements the partition function calculations described here for the full loop-based energy model.

17.9 Tree Representations of Secondary Structures

Trees are graphs with special structural properties which have been commonly used to represent RNA secondary structures [14]. An undirected graph is called a tree if the graph in question is cycle free and connected. Further, in an undirected rooted tree T, there is always a designated vertex r called the root of T for which every edge is directed away from r. Each vertex in T is uniquely accessible from r. In an ordered tree an ordering is specified for the children of each node. RNA secondary structures can be encoded as rooted, ordered, labeled trees (Fig. 17.5).

In the full tree representation [8] each internal node represents a base-pair, while leaves represent unpaired bases. The root vertex does not correspond to a physical part of the RNA. Shapiro et al. used a more abstract encoding [21, 22] in which internal nodes correspond to the different loop types (stack, interior loop, bulge, multiloop, hairpin). Depending on the type of representation the labels have different meaning. In the following, we assume that the labels are from some finite alphabet Σ.

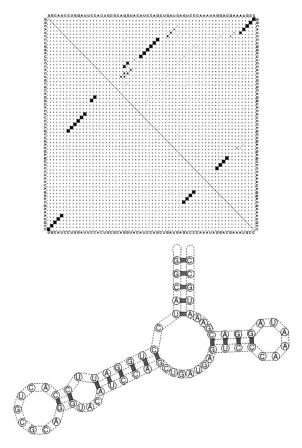

Fig. 17.4 Base-pairing probability matrix. The area of the dots in the *upper right triangle* of the matrix is proportional to the probability that a specific base-pair forms in the thermodynamic equilibrium. The *lower left triangle* shows the pairing pattern in the optimal structure of minimum free energy. Again, a hammerhead RNA is shown as an example (*conventional drawing below*). The structure was calculated using the program RNAfold

17.10 Comparing Secondary Structures Using Tree Editing

Comparison of secondary structures is a common problem in RNA bioinformatics and used, for example, to classify RNAs into families or to detect evolutionarily conserved RNA secondary structures [33].

From a graph-theoretical point of view we are facing a graph similarity problem. Generally, approaches to determine the structural similarity of graphs can be divided into two major categories: Exact graph matching and inexact graph matching. To match two relational structures exactly means that one has to determine isomorphic or subgraph isomorphic relations. Historically, Zelinka [34] was the first to determine the distances between isomorphism classes. Then, Kaden [12] and Sobik

17 Graph Representations and Algorithms in Computational Biology

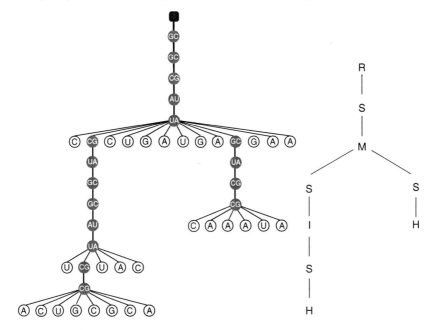

Fig. 17.5 Tree representations of RNA secondary structures. *Left:* The "full tree" representation [8]. *Right:* Shapiro-style tree [21, 22]. R,S,M,I,H denote root, stem, multi-loop, interior loop, and hairpin loop nodes, respectively

[23,24] extended the resulting metric to labelled graphs as well as graphs of different orders. A well-known graph metric from the inexact graph matching paradigm was developed by Bunke [3]. For coping with structural errors, he transformed a graph G_1 into G_2 by calculating the edit costs of certain graph edit operations. Finally, the so-called *graph edit distance* is defined by the minimal edit costs to transform G_1 to G_2.

Here we describe *tree editing* [20, 25] as a way to compare two secondary structures represented as trees. We consider three basic edit operations on a labeled tree T. The *relabel* operation changes the label of a node v. The *delete* operation removes a node v with parent v', making the children of v the children of v'. The *insert* operation is the complement of the delete operation, inserting a new node v as child of v' making the children of v' the children of v. We follow the presentation in [2], and write $(l_1 \to l_2)$ for an edit operation and use λ as a special blank symbol. If $l_1 \neq \lambda$ and $l_2 \neq \lambda$ it is a relabeling operation. If $l_2 = \lambda$ it is a deletion and if $l_1 = \lambda$ it is an insertion.

Each edit operation is assigned a *cost* γ which is a metric on the alphabet $\Sigma \cup \{\lambda\}$. A sequence of edit operations that transforms one tree T_1 into another tree T_2 is called an *edit script*. The cost of an edit script is the sum of the costs of its edit operations. The edit distance $\delta(T_1, T_2)$ is the cost of the minimum cost edit script.

A *mapping* is another way of representing the editing operations between two trees. The mapping between T_1 and T_2 is a binary relation $M \in V(T_1) \times V(T_2)$ between the vertex sets of the two trees such that for any pair $(v_1, w_1), (v_2, w_2) \in M$ holds:

(a) $v_1 = v_2$ iff $w_1 = w_2$ (one-to-one condition).
(b) v_1 is an ancestor of v_2 iff w_1 is an ancestor of w_2 (ancestor condition).
(c) v_1 is left to v_2 iff w_1 is left to w_2 (sibling condition).

By this definition each $v \in T_1$ has either a unique partner $w \in T_2$ with $(v, w) \in M$ or no partner at all. Let N_1 and N_2 be the set of nodes in T_1 and T_2, respectively, that have no partner in the other tree. The cost of M is then given by:

$$\gamma(M) = \sum_{(v,m) \in M} \gamma(v \to w) + \sum_{v \in N_1} \gamma(v \to \lambda) + \sum_{w \in N_2} \gamma(\lambda \to w) \qquad (17.10)$$

Mappings are relations and can be composed. Let T_1, T_2, T_3 be three trees and M_1 and M_2 the mapping of T_1 to T_2 and T_2 to T_3, respectively. Then

$$M_1 \circ M_2 = \{(v, w) \mid \exists u \in V(T_2) \text{ that } (v, u) \in M_1 \text{ and } (u, w) \in M_2\} \qquad (17.11)$$

is a mapping of T_1 to T_3. The function γ is a metric and one can show that the minimum cost mapping is equivalent to the minimum cost edit script, i.e., the edit distance δ.

To calculate δ, we need to calculate the minimum cost mapping. As it turns out, this problem can be solved using again simple recursions and a dynamic programming algorithm. Formally, we need to extend the definitions above to *forests* as roots can be deleted turning the ordered tree into an ordered forest. Likewise two trees in a forest can be merged by inserting a new root. Let F be a forest and v be a node in F. In the following we write $F - v$ for the forest that is obtained be deleting v, $F - T(v)$ the forest that is obtained by deleting v and its descendants, and $F(v)$ for the set of trees that have the children of v as their roots.

Let F_1 and F_2 be two forests and v and w the root of the rightmost tree in F_1 and F_2, respectively. M is the optimal mapping between F_1 and F_2. There are three possibilities: (1) v has no partner in M, then the optimal mapping is the mapping between $F_1 - v$ and F_2 and a deletion of v. (2) w has no partner in M then the optimal mapping is the mapping between F_1 and $F_2 - w$ with w inserted. (3) Both v and w have partners. Using the definition of mappings and considering the fact that both v and w are by construction the rightmost roots it is easy to show that this implies $(v, w) \in M$. The edit distance can be computed using the following recursions corresponding to the three different cases.

$$\delta(F_1, F_2) = \min \begin{cases} \delta(F_1 - v, F_2) + \gamma(v \to \lambda) \\ \delta(F_1, F_2 - w) + \gamma(\lambda \to w) \\ \delta(F_1(v), F_2(w)) + \delta(F_1 - T_1(v), F_2 - T_2(w)) + \gamma(v \to w) \end{cases}$$

17 Graph Representations and Algorithms in Computational Biology

The recursion is initialized with the edit distance between two empty forests $\delta(\theta, \theta) = 0$. The cases of one tree being empty is handled as follows

$$\delta(F_1, \theta) = \delta(F_1 - v, \theta) + \gamma(v \to \lambda)$$

$$\delta(\theta, F_2) = \delta(\theta, F_2 - w,) + \gamma(\lambda \to w)$$

These recursions can be used as the basis for a dynamic programming algorithm to compute the optimal mapping. It can be shown that the time complexity is bounded by $\mathcal{O}(|F_1|^2|F_2|^2)$. Various improvements can speed up the algorithm [13, 35]. Implementations for the algorithms are provided by the program RNAdistance of the Vienna RNA package [9].

17.11 Summary and Conclusion

Structural analysis of RNAs is a common problem in computational biology. RNA secondary structure can be regarded as a coarse-grained approximation of its three-dimensional structure that captures important biological features of the molecule.

Secondary structures can be encoded as labeled outerplanar graphs. In its simplest form, Nussinov's algorithm finds the graph with the maximum number of edges for a given labelling. This algorithm is a simple way of addressing the RNA folding problem. In practice, the more sophisticated Zuker algorithm is used that does not consider base-pairs independently but the structural elements formed by neighbouring base-pairs. In other words, it considers the faces of the graph rather than the mere edges. Empirical folding energy contributions are assigned to the faces and a recursive algorithm calculates the structure with the optimal folding energy. McCaskill's algorithm is a closely related variant that calculates the partition function of all structures in a thermodynamic ensemble. This quantity can be used to calculate equilibrium base-pair probabilities.

In order to compare two secondary structures it turned out to be useful to encode the structures as trees. This allows us to use different tree distance measures like tree editing for their comparison.

The topic presented in this chapter is a relatively rare example from computational biology where concepts from discrete mathematics can be seamlessly combined with biophysical knowledge to successfully address complex problems of highly practical relevance like the prediction of molecule structures. There are alternative ways to address the problems presented in this chapter, most notably approaches based on stochastic context-free grammars [6, 7]. Such probabilistic machine-learning-based approaches are becoming increasingly popular. However, although these methods follow a completely different paradigm for RNA modelling, they all build upon similar graph representations and – at their core – surprisingly similar algorithms are at work.

Acknowledgements Funding from the Austrian GEN-AU projects "noncoding RNA" and "Bioinformatics Integration Network" as well as financial support to the CIBIV institute from the Wiener Wissenschafts-, Forschungs- and Technologiefonds (WWTF) is gratefully acknowledged.

References

1. Bartel DP (2004) MicroRNAs: genomics, biogenesis, mechanism, and function. Cell 116(2): 281–297
2. Bille P (2005) A survey on tree edit distance and related problems. Theor Comput Sci 337 (1–3):217–239
3. Bunke H (1983) What is the distance between graphs? Bull EATCS 20:35–39
4. Carninci P et al (2005) The transcriptional landscape of the mammalian genome. Science 309(5740):1559–1563
5. Doudna JA, Cech TR (2002) The chemical repertoire of natural ribozymes. Nature 418(6894):222–228
6. Dowell RD, Eddy SR (2004) Evaluation of several lightweight stochastic context-free grammars for RNA secondary structure prediction. BMC Bioinformatics 5:71
7. Eddy SR, Durbin R (1994) RNA sequence analysis using covariance models. Nucleic Acids Res 22(11):2079–2088
8. Fontana W, Konings DA, Stadler PF, Schuster P (1993) Statistics of RNA secondary structures. Biopolymers 33:1389–1404
9. Hofacker IL, Fontana W, Stadler PF, Bonhoeffer LS, Tacker M, Schuster P (1994) Fast folding and comparison of RNA secondary structures. Monatsh Chem 125:167–188
10. Hofacker IL, Schuster P, Stadler PF (1998) Combinatorics of RNA secondary structures. Discrete Appl Math 89:177–207
11. Hofacker IL, Stadler PF (2007) RNA secondary structures. In: Lengauer T (ed) Bioinformatics: from genomes to therapies, vol 1. Wiley-VCH, Weinheim, Germany, pp 439–489
12. Kaden F (1982) Graphmetriken und Distanzgraphen. ZKI-Informationen Akad Wiss DDR 2(82):1–63
13. Klein P (1998) Computing the edit distance between unrooted ordered trees. In: Bilardi G, Italiano GF, Pietracaprina A, Pucci G (eds) Algorithms – ESA '98, Proceedings of 6th annual European symposium, Venice, Italy, 24–26 August 1998. Lecture Notes in Computer Science, vol 1461. Springer, Heidelberg, pp 91–102
14. Le SY, Nussinov R, Maizel JV (1989) Tree graphs of RNA secondary structures and their comparisons. Comput Biomed Res 22(5):461–473
15. Mathews DH, Sabina J, Zuker M, Turner H (1999) Expanded sequence dependence of thermodynamic parameters provides robust prediction of RNA secondary structure. J Mol Biol 288:911–940
16. McCaskill JS (1990) The equilibrium partition function and base pair binding probabilities for RNA secondary structure. Biopolymers 29:1105–1119
17. Nelson DL, Cox MM (2004) Lehninger principles of biochemistry, 4th edn. W.H. Freeman, New York
18. Nussinov R, Jacobson AB (1980) Fast algorithm for predicting the secondary structure of single-stranded RNA. Proc Natl Acad Sci USA 77(11):6309–6313
19. Nussinov R, Piecznik G, Griggs JR, Kleitman DJ (1978) Algorithms for loop matching. SIAM J Appl Math 35(1):68–82
20. Selkow SM (1977) The tree-to-tree editing problem. Inf Process Lett 6(6):184–186
21. Shapiro BA (1988) An algorithm for comparing multiple RNA secondary stuctures. CABIOS 4:387–393
22. Shapiro BA, Zhang K (1990) Comparing multiple RNA secondary structures using tree comparisons. CABIOS 6:309–318

17 Graph Representations and Algorithms in Computational Biology

23. Sobik F (1982) Graphmetriken und Klassifikation strukturierter Objekte. ZKI-Informationen Akad Wiss DDR 2(82):63–122
24. Sobik F (1986) Modellierung von Vergleichsprozessen auf der Grundlage von Ähnlichkeitsmaßen für Graphen. ZKI-Informationen Akad Wiss DDR 4:104–144
25. Tai K (1979) The tree-to-tree correction problem. J ACM 26:422–433
26. Touchard J (1952) Sur un problème de configurations et sur les fractions continues. Can J Math 4(6894):2–25
27. Washietl S et al (2007) Structured RNAs in the ENCODE selected regions of the human genome. Genome Res 17(6):852–864
28. Washietl S, Hofacker IL, Lukasser M, Hüttenhofer A, Stadler PF (2005) Mapping of conserved RNA secondary structures predicts thousands of functional non-coding RNAs in the human genome. Nat Biotechnol 23:1383–1390
29. Waterman MS (1978) Secondary structure of single-stranded nucleic acids. Studies on foundations and combinatorics. Advances in Mathematics Supplementary Studies. Academic Press, New York, 1:167–212
30. Waterman MS, Smith TF (1978) RNA secondary structure: a complete mathematical analysis. Math Biosci 42:257–266
31. Xia T, SantaLucia J, Burkard ME, Kierzek R, Schroeder SJ, Jiao X, Cox C, Turner DH (1998) Thermodynamic parameters for an expanded nearest-neighbor model for formation of RNA duplexes with watson-crick base pairs. Biochemistry 37(42):14719–14735
32. Yan L (1995) A family of special outerplanar graphs with only one triangle satisfying the cycle basis interpolation property. Discrete Math 143(1–3):293–297
33. Yao Z, Weinberg Z, Ruzzo WL (2006) CMfinder – a covariance model based RNA motif finding algorithm. Bioinformatics 22(4):445–452
34. Zelinka B (1975) On a certain distance between isomorphism classes of graphs. Časopis pro pěst Mathematiky 100:371–373
35. Zhang K, Shasha D (1989) Simple fast algorithms for the editing distance between trees and related problems. SIAM J Comput 18:1245–1262
36. Zuker M (2003) Mfold web server for nucleic acid folding and hybridization prediction. Nucleic Acids Res 31(13):3406–15
37. Zuker M, Sankoff D (1984) RNA secondary structures and their prediction. Bull Math Biol 46:591–621
38. Zuker M, Stiegler P (1981) Optimal computer folding of larger RNA sequences using thermodynamics and auxiliary information. Nucleic Acids Res 9:133–148

Chapter 18
Inference of Protein Function from the Structure of Interaction Networks

Oliver Mason, Mark Verwoerd, and Peter Clifford

Abstract We consider the problem of using graph-theoretical techniques to predict the function of unannotated proteins in an organism's proteome. Specifically, we present an overview of the major methods for predicting protein function based on interaction network structure and describe an abstract framework within which these methods can be treated in a unified fashion. We also present a comparison of the proposed methods and highlight some open theoretical and practical questions in the area.

Keywords Protein function prediction · Graph algorithms · Graph multicuts · Markov random fields

MSC2000: Primary 46N60; Secondary 05C85, 05C90

18.1 Introduction: Functional Classification and Protein Interaction Networks

The past decade has witnessed enormous advances in experimental methodologies within the life sciences [4, 37, 39]. These advances have significant implications for the future practice of biology and related disciplines. The systems being studied and the questions being addressed in contemporary biology are more intricate and complex than ever before. With modern experimental techniques, it is possible to investigate living systems in a manner and at a level of detail that would have been unthinkable 50 years ago. In turn, this has generated a need for novel ways of thinking about the systems being considered and of assimilating the data being made available.

Arguably the most noticeable aspect of recent developments in biology is that the volume and nature of the data being generated are unprecedented. This latter

---

O. Mason (✉)
Hamilton Institute, NUI Maynooth, Maynooth, Ireland
e-mail: oliver.mason@nuim.ie

M. Dehmer (ed.), *Structural Analysis of Complex Networks*, 439
DOI 10.1007/978-0-8176-4789-6_18, © Springer Science+Business Media, LLC 2011

fact has led in a natural way to the recognition within the life sciences of a pressing need for a closer and mutually beneficial interaction between computational and mathematical scientists on the one hand and biologists on the other. It is important to appreciate that these developments pose challenges and opportunities for both experimental and theoretical scientists. Although modern biology now needs the computational methods and theoretical frameworks provided by the mathematical sciences, it is equally true that mathematicians and computational scientists need to recognise that biological systems are substantially different from the physical and engineering systems that have historically been the main focus of applied mathematics. Old methods and tools cannot simply be dusted off and applied to this new project [22, 36]; novel approaches and techniques are required, specifically tailored to and inspired by biological problems.

Our purpose in the current chapter is to provide an introduction to one such problem that has been the focus of much attention in the field of computational biology in the recent past. Specifically, we are concerned with the problem of assigning functions to newly discovered genes and proteins [18, 35]. Although the number of completely sequenced genomes has grown steadily over the past decade, many fundamental questions still remain. One of the most significant of these relates to determining the biological function of the gene products or proteins identified in this process. At the time of writing, even for simple organisms such as *S. cerevisiae* approximately 20% of the organism's proteins have no assigned function [35]; such proteins are said to be unannotated.

In the following section, we expand on what is to be understood by the notion of protein function. For now, we simply note that, in view of the significant numbers of unannotated proteins even for simple organisms, a major challenge for computational biology is to devise methods to reliably assign functions to these proteins based on plausible biological hypotheses and the known functions of annotated proteins. Unsurprisingly, there has been a considerable amount of interest in this problem since the publication of sequenced genomes, and a variety of approaches have been proposed. Early approaches made use of tools such as BLAST or PSI-BLAST [1, 2] to predict function based on sequence similarity to proteins of known function. Other methods based on phylogenetic profiles and gene co-expression patterns have also been proposed [28, 43]. In keeping with the overall theme of this book, we focus here on graph-theoretical methods that seek to exploit recently compiled networks of protein–protein interactions (PPI) to assign protein function. We describe some of the main such methods and the principles on which they are based in Sects. 18.3 and 18.4 below. In the interests of brevity, we shall largely focus on direct methods and not describe techniques based on clustering methods such as those presented in [8, 29, 30] (although we shall briefly describe one such method in Sect. 18.5).

The structure of the chapter is as follows. In the next section, we describe the biological background to the problem of protein function assignment, and then present a mathematical formulation of the problem in Sect. 18.3. In Sect. 18.4 we describe several studies that have provided evidence of a connection between network topology and protein function. Section 18.5 is concerned with discussing some of the

18 Inference of Protein Function from the Structure of Interaction Networks   441

main graph-based algorithms for protein function assignment and highlighting some difficulties and open questions to which they give rise. In Sect. 18.6, we discuss a number of numerical studies comparing the various methods described throughout the chapter. Finally, in Sect. 18.7 we present some concluding remarks about the computational and graph-theoretical challenges arising from the problem of protein function assignment.

18.2 Functional Annotation and Protein Interaction Networks

To facilitate the application of computational methods in biology, there is a clear need for consistent terminology and notation suitable for automated methods. The problems that arise when different databases employ inconsistent notations, unsuitable for computation, are clear. It is patently obvious that effective algorithms for assigning functions to proteins cannot be developed if there is ambiguity in the meaning of the term "function" itself. With the completion of whole-genome sequencing projects, attention naturally focussed on providing a consistent, unambiguous terminology for classifying the biological function of all proteins within an organism. This in itself is not a straightforward task and the various issues and difficulties involved in developing an unambiguous functional terminology are discussed in the paper [18]. At the time of writing, a variety of comprehensive annotation schemes have been developed [3, 10, 21, 32]. To illustrate the basic ideas behind these, we describe two of the most widely adopted schemes: namely the Gene Ontology (GO) scheme [3] and the Functional Catalogue (FunCat) [32].

18.2.1 The Gene Ontology Consortium and MIPS
           Functional Catalogue

The GO classification scheme consists of three basic categories, which describe different aspects of a gene product's functionality within the cell. The first of these is *molecular function* and is defined in [3] as being the biochemical activity of the protein or gene product. Examples of general molecular function include "enzyme" or "transporter" for example. Essentially, molecular function concerns the specific action of the given protein without reference to the higher level biological processes in which it plays a role. These are the concern of the second category within the GO scheme: *biological process*. The terms in this category define higher level activities usually comprised of several molecular functions, and examples include "cell growth" and "signal transduction". Finally, the third category within the scheme is "cellular component", which describes where in the cell a particular gene product is active. Every term in the GO categories has a unique numerical identifier or key of the form "GO:xxxxxxx". The GO project itself is the result of a collaboration between numerous databases around the world and has been running since 1998. The project is ongoing and is added to and updated on a regular basis. For more information, the interested reader should consult either the original reference [3] or the excellent documentation on the GO website http://www.geneontology.org/.

The second annotation scheme – FunCat – was initially developed as part of the *S. cerevisiae* genome project at MIPS but has since been broadened to include annotations for several other eukaryotic and prokaryotic organisms [32]. FunCat is structured differently to the GO scheme and is essentially organised as a number of tree-like hierarchies. At the coarsest level, there are 28 different basic functional categories including "metabolism", "storage", "protein synthesis", and "cell type differentiation". Each of these corresponds to a tree consisting of successively more specific functions. Each function within the annotation scheme is assigned a unique numerical identifier that indicates which broad category it belongs to and its place in the hierarchy. For instance, the broad category metabolism is assigned the identifier 01; at the next level down the hierarchy amino acid metabolism is assigned the identifier 01.01; at one level lower again, we find "metabolism of the glutamate group" with the identifier 01.01.03; the more specific "metabolism of glutamine" is assigned 01.01.03.01 while at the most specific level biosynthesis of glutamine and degradation of glutamine are assigned 01.01.03.01.01 and 01.01.03.01.02, respectively.

Constructing comprehensive and convenient annotation schemes is only one step, albeit a vital one, in the annotation of gene products; the proteins of an organism must next be assigned to some functional category within the scheme. Essentially, once the genome of an organism has been sequenced each protein identified must have some collection of terms in either the GO scheme or the FunCat scheme assigned to it. As mentioned in the introduction, even for simple organisms this second task is far from complete; however, for many organisms a significant fraction of their proteome is annotated. Our interest is in combining this information with knowledge of protein interactions to make sensible predictions for the functions of unannotated proteins.

18.2.2 Protein–Protein Interaction Networks

A significant outcome of the experimental advances made in recent years has been the construction of large-scale networks detailing the physical and biochemical interactions among the genes, proteins, and metabolites within the cell [25]. In particular, the development of high-throughput techniques for detecting PPI such as yeast-2-hybrid [17] or tandem-affinity purification with mass spectrometry (TAP-MS) [14] has led to the generation of detailed maps of PPI for a growing number of simple organisms. Essentially a PPI network is an undirected graph with nodes representing the proteins in an organism and edges representing physical interactions between proteins. At the time of writing, the organisms for which PPI networks have

18 Inference of Protein Function from the Structure of Interaction Networks   443

been constructed include *S. cerevisiae*, *D. melanogaster*, *C. elegans*, and *H. pylori* [15, 24, 26, 31, 40] and data on these networks can be obtained from a variety of regularly updated databases such as DIP [41] or MIPS [26].

The construction of PPI networks has generated considerable interest in elucidating the structural properties of these networks, identifying common features, and relating network structure to biological phenomenology. In particular, properties such as degree distributions, clustering, and modularity have attracted attention [5]. On the biological implications of network structure, the possible connection between the structural role of a protein within the network and its biological importance, as measured by its essentiality to the organism's survival or its impact on growth rate, has been thoroughly investigated [19, 42]. The work discussed in the remainder of this chapter is along similar lines and concerns relating the structure of PPI networks to biological function.

More specifically, we have already mentioned that, even for simple organisms, significant numbers of proteins remain unannotated and that assigning functions to these proteins is a central problem in modern biology. Recently, several researchers have attempted to exploit graph-based algorithms to predict functions for unannotated proteins based on their position within the PPI network and the functions of annotated proteins in the network. The experimental evidence supporting the assumption that interaction network structure contains functionally relevant information is presented in Sect. 18.4. In the following section, we present some mathematical background as well as a more formal statement of the problem considered.

18.3   Mathematical Formulation and Preliminaries

18.3.1   Notation and Mathematical Background

Throughout, $\mathbb{R}$ denotes the field of real numbers and $\mathbb{R}_+$ denotes the set of nonnegative real numbers. For finite sets S, T, $S \times T$ denotes the usual Cartesian product of S and T, while $|S|$ denotes the cardinality of S and 2^S represents the power set of S consisting of all subsets of S. For a function $f : S \to T$ and a subset $R \subset S$, $f|_R$ denotes the restriction of f to R. Also, for sets S and T, $S \backslash T = \{x \in S : x \notin T\}$ denotes the set difference of S and T, while $S \triangle T$ denotes the symmetric difference of S and T given by $(S \cup T) \backslash (S \cap T)$.

The PPI networks we consider in this chapter are naturally modelled as finite undirected graphs. For background on basic concepts and notation, consult [12]. Unless stated otherwise, we use the notation $G = (V, E)$ for a finite undirected graph where V is the vertex set and E denotes the set of edges. Also, the notation *uv* denotes the edge between vertices u and v in V if such an edge exists. In this case, we also say that u and v are *neighbours* or *level-1 neighbours*.

444   O. Mason et al.

For an undirected graph, $G = (V, E)$ and a vertex, $u \in V$, the set of all neighbours of u is denoted $N(u)$ and given by

$$N(u) = \{v \in V : uv \in E\}.$$

We denote paths by the list of vertices occurring along the path. For instance, $u = v_1, \ldots, v_m = v$ denotes a path of length $m - 1$ between u and v, where $v_i v_{i+1} \in E$ for $i = 1, \ldots, m - 1$. As usual, we say that a graph is connected if there is a path between any pair (u, v) of vertices in V. When dealing with real PPI network data, it is typical to work with the largest connected component of the graph. In fact, most data sets consist of a single very large component with a number of other components of significantly smaller size. Thus, in the interests of simplifying terminology, we shall always assume that the graphs we deal with are connected.

The distance $d(u, v)$ between any two nodes in V is defined to be the length of the shortest path between u and v (by default $d(u, u) = 0$). Given a node u and a positive integer k, the k-neighbourhood of u is the set of all nodes $v \neq u$ in V with $d(u, v) \leq k$, and is denoted by $N^k(u)$. We refer to nodes v with $d(u, v) = k$ as level-k neighbours of u.

18.3.2 Protein Function Prediction: Formal Statement

The problem of protein function prediction (PFP) is a rich source of challenging graph-theoretical questions, many of which are open, with interesting connections to other branches of applied graph theory. However, although various graph-based algorithms have been proposed, as yet no standard formal notation for the problem has been adopted. In the interests of clarity and to highlight how the various approaches described later fit within the one framework, we now state in a formal way the problem of PFP.

Let the following four items be given:

- An undirected connected graph $G = (V, E)$ representing the PPI network
- A finite set $F = \{f_1, \ldots, f_k\}$ representing the distinct terms in some annotation scheme such as FunCat or GO
- A partition of the vertex set $V = V_a \cup V_u$ in which V_a contains the annotated proteins and V_u contains the unannotated proteins in the network
- A mapping $\psi : V_a \rightarrow 2^F \backslash \{\emptyset\}$ defined on V_a taking nonempty values in the power set of F representing the functions assigned to the annotated proteins in the network.[1]

---

[1] See Sect. 18.6 for a remark concerning the completeness of the annotation given by ψ.

The essence of the problem of PFP is to extend ψ to the whole of V, formally, to define $\hat{\psi} : V \to 2^F \backslash \{\emptyset\}$ such that the restriction $\hat{\psi}|_{V_a} = \psi$. We use the notation $\hat{\Psi}$ to denote the set of all such extensions.

Throughout the remainder of this chapter, the notation $G = (V, E)$, V_a, V_u, F, ψ, $\hat{\psi}$, $\hat{\Psi}$ is used exclusively in this way.

Clearly the set $\hat{\Psi}$ is going to be too large for practical purposes and more criteria on $\hat{\psi}$ need to be specified in order to generate biologically plausible extensions. These additional criteria should be based on biologically plausible assumptions for which there is experimental evidence. In the next section, we describe the fundamental principles that underpin all of the graph-based approaches to PFP proposed in the recent past.

For now, we continue to develop the abstract framework of PFP. Most of the algorithms that we consider later follow a similar basic scheme. For each unannotated protein $u \in V_u$, a score function $\rho_u : F \to \mathbb{R}_+$ is defined with $\rho_u(f)$ indicating the likelihood that u actually has the function f. The functions f in F are then ranked in descending order of $\rho_u(f)$ and the extension $\hat{\psi}$ can be defined in one of two ways:

$$\hat{\psi}(u) = \{f \in F : \rho_u(f) \geq \tau\} \qquad (18.1)$$

for some specified threshold τ, or else by choosing the functions corresponding to the highest m values of $\rho_u(f)$, so that $|\hat{\psi}(u)| = m$ and

$$\min\{\rho_u(f) : f \in \hat{\psi}(u)\} \geq \max\{\rho_u(f) : f \notin \hat{\psi}(u)\}. \qquad (18.2)$$

What distinguishes the various algorithms is the principles used to generate the score function ρ and corresponding ranking scheme on F. In Sect. 18.5 we discuss several different such ranking schemes and their corresponding algorithms for functional annotation. However, in the next section we discuss the evidence for a connection between network topology and protein function and identify the key assumptions underlying all graph-based PFP algorithms.

18.4 Topology and Protein Function

The fundamental idea behind the use of PPI to predict function is quite simple; proteins that interact with each other are likely to share functionality. In essence this is analogous to the concept that individuals within an organization who meet regularly are more likely to be involved in performing similar tasks. This core idea is in itself biologically plausible and its validity has been investigated by a number of researchers. Specifically, attention has focussed on the tendency of a protein u within a PPI network $G = (V, E)$ to share functions with the proteins or nodes belonging to the k-neighbourhoods $N^k(u)$ of u for different (positive integer) values of the parameter k.

An early, and influential, study of this kind appeared in the paper [34]. Specifically, the authors of this paper used publicly available data on PPI in yeast and the annotation scheme of the Yeast Proteome Database [10] to investigate the tendency of a protein to share functions with its immediate neighbours. The YPD classification scheme used was based on cellular role and is similar to the "biological process" categorization in the GO scheme described above. In [34], the authors first focussed on the annotated proteins within the network; in our terminology, on the induced subgraph on the set of vertices V_a. For each $u \in V_a$, the functions of the neighbouring vertices in $N(u) \cap V_a$ were ranked in descending order of frequency and the three most commonly occurring functions were postulated as potential functions for u. If any of these predicted functions corresponded with a known function of u, the prediction was deemed to be correct. Promisingly, this simple scheme led to "correct" predictions for 72% of the vertices in V_a. Moreover, to test the degree to which protein function was dependent on the actual topology of the network, the authors "shuffled" or scrambled links between the nodes in the network in a random manner. They report that on average for the randomized networks, the rate of correct predictions was as low as 12%.

In addition to simply checking if an annotated protein in V_a shares function with some of its immediate neighbours, some slightly more sophisticated measures of functional similarity can be investigated. For instance, the *functional similarity* [9] of u, v in V_a has been defined as

$$FS(u, v) = \frac{|\psi(u) \cap \psi(v)|}{|\psi(u) \cup \psi(v)|}. \tag{18.3}$$

Using functional annotation data obtained from the FunCat database in January 2008, we have investigated the average of $FS(.,.)$ over pairs of interacting and non-interacting proteins in a network of protein interactions in yeast obtained from the Database of Interacting Proteins [41] in January 2008. The average taken over non-interacting pairs was 0.015042, while the average for interacting pairs of proteins was an order of magnitude higher at 0.15155.

The last observation and the results reported in [34] lend experimental support to the hypothesis that interacting proteins tend to share functions and to the general idea of exploiting interaction data to predict protein function. Further evidence of this nature was subsequently presented in [16] and then more recently in [9]. This latter paper also investigated more closely the relationship between a protein's functions and those of its level-2 neighbours in $N^2(u) \backslash N(u)$. Using the FunCat scheme of annotation, and interaction data obtained from the Munich Information Center for Protein Sequences, the authors of [9] found that proteins are significantly more likely to share functions exclusively with level-2 neighbours than they are to share function exclusively with their level-1 neighbours. The findings reported in [9] add to earlier evidence described in [7, 33] that the functions of level-2 neighbours $N^2(u) \backslash N(u)$ of a protein are also potentially useful indicators of the functions of u. It should also be noted that the authors of [9] introduce a novel topological

18 Inference of Protein Function from the Structure of Interaction Networks 447

measure for the degree of functional similarity between two proteins u and v in V. We describe this and an associated algorithm for function prediction in more detail in the following section.

The discussion in the previous two paragraphs should have made fairly clear the general principle on which graph-based protein function prediction is based; namely:

> Given an unannotated u in V_u, the functions assigned in $\hat{\psi}(u)$ should be "similar" to the assigned functions of the level-1 and level-2 neighbours of u.

The above principle is still stated in a relatively loose fashion; it is the particular interpretation of the word "similar" that determines the unique approach of each specific algorithm for PFP. In the next section, we describe in detail several of the more significant of these to have emerged recently.

18.5 Graph-Based Algorithms for PFP

18.5.1 Majority Rule

In light of the observations made above, the simplest approach to predicting functions for an unannotated protein u in an interaction network $G = (V, E)$ would be to rank functions based on the frequency with which they occur among the annotated neighbours of u. This is the approach taken in the so-called majority rule described in [34].

The score function for the majority rule is defined as follows. For $u \in V_u$ and $f \in F$, $\rho_u(f)$ is defined to be the total number of occurrences of f among the annotated neighbours of u. Formally,

$$\rho_u(f) = |\{v \in V_a \cap N(u) : \psi(v) = f\}|.$$

In keeping with the general framework outlined in Sect. 18.3, the functions with the highest values of $\rho_u(f)$ are assigned to u.

The two principal advantages of the majority rule are its simplicity and the fact that it directly relates to the core biological hypothesis that interacting proteins should have similar functionality. However, it also has several obvious drawbacks. As it only considers annotated neighbours of unannotated proteins, it is unable to make any prediction for proteins with no annotated neighbours. A further significant disadvantage also arises from the fact that it only takes into account interactions between vertices in V_u and V_a. This can indirectly lead to the very principle on which it is based being violated. To see this consider the fragment shown in Fig. 18.1.

Here the solid nodes are annotated and belong to V_a while the two unshaded nodes are unannotated and belong to V_u. Suppose the annotated neighbour of u is annotated with a function f_i, while the two annotated neighbours of v and w are annotated with a function $f_j \neq f_i$. Then, following the majority rule, we would assign the function f_i to u and f_j to v and w. But then u would have been assigned

Fig. 18.1 Difficulties with the majority rule

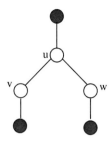

the function f_i while the majority of its neighbours would be annotated with the different function f_j. Although this example is overly simplistic, it illustrates this fundamental problem with the majority rule. Another problem with this approach is that it fails to take into account all of the information known about the annotated part of the network. In particular, the frequency with which the various functions actually occur for the annotated nodes is ignored. As this frequency can vary considerably from function to function, this information should play some role in determining the most likely functions for the unannotated nodes in V_u.

18.5.2 Chi-Squared Scheme

An attempt to address the final issue mentioned in the previous paragraph was made in [16]. The algorithm described in this paper allows nodes in a k-neighbourhood of an unannotated protein to be taken into account and ranks functions based on a chi-squared score rather than a simple count. However, as noted in [9], the reliability of predictions tends to decline dramatically when nodes at a distance greater than 2 are used to predict function. For this reason, and in the interests of simplicity, we describe the method of [16] for the case of immediate neighbours only. In any case, the extension to the more general case is obvious.

For each function $f \in F$, let π_f denote the fraction of annotated nodes in the network that are annotated with f:

$$\pi_f = \frac{|\{w \in V_a : \psi(w) = f\}|}{|\{V_a\}|}.$$

Also, for any $u \in V_u$, let $n_u = |V_a \cap N(u)|$, $n_u^f = |\{v \in V_a \cap N(u) : \psi(v) = f\}|$. Then the score function for the chi-squared scheme is given by

$$\rho_u(f) = \frac{(n_u^f - \pi_f n_u)^2}{\pi_f n_u}.$$

As with the majority rule, the functions f assigned to u are those corresponding to the largest values of $\rho_u(f)$.

18 Inference of Protein Function from the Structure of Interaction Networks   449

Before proceeding to more complicated approaches to PFP, at this point it is appropriate to make some observations on the majority and chi-squared schemes.

1. Although the chi-squared scheme allows annotated nodes other than immediate neighbours to be used, it still suffers from the limitation of only considering annotated nodes in constructing the extension $\hat{\psi}$ of ψ. This can lead to similar problems to that illustrated in Fig. 18.1. Also, although the chi-squared scheme may appear to address some of the limitations of the basic majority scheme, the justification for the use of the chi-squared statistic in this context is unclear.

2. It is worth keeping in mind that all of the methods described here are essentially trying to accomplish the same task: namely ranking the potential functions in F for an unannotated node in V_u. A number of theoretical questions naturally arise as to how the different ranking schemes relate to each other. For example, when will the methods give rise to the same predictions for an unannotated protein? It is trivial to see that if for $u \in V_u$, the ratio $\frac{n_u^f}{\pi_f n_u}$ is constant for all $f \in F$, then both the majority rule and the chi-squared scheme will give rise to the same ranking on F. In fact, if the sum

$$\frac{n_u^f}{\pi_f n_u} + \frac{\pi_f n_u}{n_u^f}$$

is constant, the same is true.

3. Another important question relates to how sensitive the various ranking schemes are to inaccurate network data. This is particularly vital in view of the known issues with high-throughput techniques such as yeast-2-hybrid which have been used to generate the interaction networks. Attempts to evaluate the robustness of ranking schemes to data inaccuracies have been made in [7, 9, 34] but there are no substantial theoretical studies on this topic at the time of writing.

18.5.3 Maximally Consistent Assignments and Functional Flow

We now discuss three basic extensions of the majority rule assignment, which we shall refer to respectively as (V) the Vazquez method, (K) the Karaoz method, and (N) the Nabieva method. Methods (V) and (K) pose the protein assignment problem as an optimization problem, the objective of which is to minimize the number of "inconsistent" edges. The methods differ mainly in the way they define inconsistency. In the Vazquez method, the notion of an inconsistent edge is defined with respect to an *assignment* ψ (given an assignment ψ, an edge $e = (u, v)$ is said to be inconsistent if $\psi(u) \cap \psi(v) = \emptyset$), whereas in the Karaoz method, it is defined with respect to a function f (given a function $f \in F$, an edge (u, v) is said to be consistent if either $f \in \psi(u) \cap \psi(v)$ or $f \notin \psi(u) \cup \psi(v)$ and inconsistent otherwise). This observation suggests that, even though these methods are sometimes considered variations of the same idea [27], they may in fact not be as closely related as they would seem

450 O. Mason et al.

at first sight. The Nabieva method is somewhat different from the other two, in the sense that it is not an optimization-based method. Rather, it employs the notion of functional flow to describe how nodes that are sufficiently close to one another can inherit (a fraction of) each other's functional annotation.

The Vazquez Method

The Vazquez method was introduced in [38]. The basic idea is as follows. Let ψ : $V_a \mapsto 2^F \setminus \emptyset$ be given and let $\hat{\Psi}$ denote the set of all extensions of ψ to the whole of V. Consider the functional $E : \hat{\Psi} \mapsto \mathbb{R}_+$,

$$E(\hat{\psi}) := \sum_{v \in V_u} \left| \{ u \in N(v) : \hat{\psi}(u) \cap \hat{\psi}(v) \neq \emptyset \} \right|.$$

We say that an assignment $\hat{\psi}'$ is maximally consistent if $E(\hat{\psi}) \leq E(\hat{\psi}')$ for all $\hat{\psi} \in \hat{\Psi}$. In general, finding maximally consistent assignments is computationally intractable. Regardless, let us suppose we can compute the set of all maximally consistent assignments and let this set be given as $\{\hat{\psi}_1, \hat{\psi}_2, \ldots, \hat{\psi}_m\}$. Then we can define a scoring function $\rho : F \mapsto \mathbb{R}_+$, as follows:

$$\rho_u(f) := \frac{1}{m} \left| \{ i : \hat{\psi}_i(u) = f \} \right|.$$

To avoid the computational problems associated with determining maximally consistent assignments, the authors of [38] employ a stochastic optimization technique called simulated annealing to compute approximations to the maximally consistent assignments. Using this technique they generate 100 assignments $\{\tilde{\psi}_1, \ldots, \tilde{\psi}_{100}\}$ at random (note that the method does not guarantee that $\tilde{\psi}_i \neq \tilde{\psi}_j$ for all (i, j) so the solution set may contain multiple copies of the same solution) and then define a scoring function $\tilde{\rho}$ as follows:

$$\tilde{\rho}_u(f) := \frac{1}{100} \left| \{ i : \tilde{\psi}_i(u) = f \} \right|.$$

To assess the performance of the method, the authors follow a common approach. They apply the algorithm to a test set $V_{\text{test}} \subset V_a$ comprising only annotated nodes, which, for the purpose of assessment, are treated as if unannotated. For a given node $v \in V_{\text{test}}$, a prediction is considered successful if $\tilde{\psi}(v) \in \psi(v)$. The rate of success is defined as the ratio of the number of successful predictions to the total number of predictions, $|V_{\text{test}}|$ (this is analogous to the specificity measure (18.7) defined below). The authors show that, roughly speaking, the likelihood of the method making a successful prediction increases with the degree of a node. The results indicate that the Vazquez method outperforms the Majority Rule in terms of overall success rate, although the percentage difference between the two is on average no more than 10%, and 20% at best.

18 Inference of Protein Function from the Structure of Interaction Networks

The Karaoz Method

The Karaoz method was introduced in [20]. The basic idea is as follows. Let $\psi :$ $V_a \mapsto 2^F \setminus \{\emptyset\}$ be given. We define $\sigma : V_a \times F \mapsto \{0, 1\}^{|V_a|}$, as follows:

$$\sigma(u, f) := \begin{cases} 1 & \text{if } f \in \psi(u); \\ 0 & \text{otherwise.} \end{cases}$$

Let $\hat{\sigma} : V \mapsto \{0, 1\}^{|V|}$ be an extension of σ to the whole of V, and denote by $\hat{\Sigma}$ the set of all such extensions. We define the cost functional $E : \hat{\Sigma} \times F \mapsto \mathbb{R}_+$

$$E(\hat{\sigma}, f) := \sum_{v \in V} \left| \{u \in N(v) : \hat{\sigma}(u, f) = \hat{\sigma}(v, f)\} \right|$$

We say that an assignment $\hat{\sigma}' \in \hat{\Sigma}$ is maximally consistent with respect to a function $f \in F$ if $E(\hat{\sigma}, f) \leq E(\hat{\sigma}', f)$ for all $\hat{\sigma} \in \hat{\Sigma}$. In theory, the Karaoz method would assign a function f to a protein $v \in V_u$ if there exists a maximally consistent assignment $\hat{\sigma}$ such that $\hat{\sigma}(v, f) = 1$. As in the case of the Vazquez method, however, finding such maximally consistent assignments is computationally challenging. In the paper the authors use a heuristic approach to generate approximate solutions, which form the input to a leave-one-out cross-validation procedure. For results, we refer to the original paper. We would like to offer two comments. The first comment concerns the optimization criterion. In the paper, the authors point out that maximizing E corresponds to maximizing the (weighted) sum of consistent edges, where an edge is called consistent (with respect to a function f), if either both the endpoints have the function f or neither of the endpoints have the function f. The first possibility (both the endpoints have the function f) would seem more natural than the second (neither of the endpoints have the function f), and one would expect this requirement to carry more weight than the other. However, as it is, they are given equal importance. The second comment concerns the heuristic optimization procedure employed by the authors, details of which can be found in the paper. The said procedure involves an update rule, which is applied serially to each node in the network. It is not clear whether, and to what extent, the order in which the update rule is applied affects the overall outcome of the procedure. More importantly, it is not clear how close the solutions obtained by this procedure are to the maximally consistent solutions (the bound given in the paper is rather loose). Lastly, even if some or all solutions obtained by this method are maximally consistent, there is no guarantee that it will find all possible maximally consistent solutions. In other words, there is no way of knowing that the sample is representative. This is important when assigning multiple functions to proteins as it affects the scoring function.

The Nabieva Method

Unlike the previous two methods, the functional flow algorithm [27], which we describe next, does not involve the optimization of a cost functional. In this method, a protein is modelled as a "source" of functions. Associated with each protein there is a number of reservoirs, one for each function, each of which holds a certain amount of function. The amount of function in the reservoirs can change as a result of functional flow between neighbouring proteins. The direction and the magnitude of this flow are determined by the functional gradient and the link capacity, respectively. More formally, let $R_v^f(k)$ denote the amount of function f at protein v at time step k. For all edges $uv \in E$ and all functions $f \in F$, we assume that the flow $g_{uv}^f(k)$ of function f at time k along the edge uv is given by

$$g_{uv}^f(k) := \begin{cases} 0 & \text{if } R_u^f(k) < R_v^f(k) \\ \min\left\{w_{uv}, \frac{w_{uv}}{\sum_{y \in N(u)} w_{uy}}\right\} & \text{otherwise} \end{cases}$$

where w_{uv} denotes the capacity or weight of the edge uv. A simple flow balance analysis shows that $R_u^f(k)$ satisfies the recurrence equation

$$R_v^f(k) = R_v^f(k-1) + \sum_{u \in N(v)} \left(g_{uv}^f(k) - g_{vu}^f(k)\right).$$

It is assumed that the amount of function $R_v^f(0)$ stored in the reservoir for function f at protein v at time $k = 0$ is "∞" if $v \in V_u$ and $f \in \psi(v)$ and 0 otherwise. The functional score for function f at node v is defined as the total inflow of f during a fixed number of iteration steps d, where d is taken to be half of the diameter of the graph:

$$\rho_v(f) = \sum_{k=1}^{d} \sum_{u \in N(v)} g_{uv}^f(k).$$

Functional flow is certainly a novel approach to the problem of protein annotation and the results presented in [27] are promising. In particular, these indicate that it outperforms the majority and the chi-squared methods, among others. On the other hand, there are some fundamental theoretical questions concerning the algorithm. The biological foundation for the concept of "flow" and for considering annotated proteins as infinite sources needs to be clarified. From a more numerical point of view, the algorithm is iterative in nature and in [27] is terminated after a finite number of steps.[2] However, no unambiguous criterion for determining when to terminate the algorithm is provided; a loose justification based on network diameter

---

[2] Functional flow clearly cannot converge in its current form.

18 Inference of Protein Function from the Structure of Interaction Networks 453

is suggested. Before functional flow can be reliably employed, the impact of the choice of stopping time on the predictions made needs to be more fully understood. Moreover, a clear way of determining in advance when to stop the algorithm is obviously necessary for its wider use.

18.5.4 *Markov Random Fields and Level-2 Neighbours*

Markov Random Field Approaches

Another promising approach to the problem of PFP has been proposed in [11, 23]. We focus on the details of the scheme described in [11]. Here, the authors adopt the technique of Markov random fields, which was originally developed in the area of image reconstruction. This is a probabilistic approach to the problem and the output of the algorithm is a distribution function for each individual function $f \in F$, which specifies the probability $Pr(f \in \hat{\psi}(u))$ that a node $u \in V_u$ has the function f. Specifically, in the scheme of [11], the nodes of V are labelled as $u_1, \ldots, u_p, u_{p+1}, \ldots, u_{p+q}$ where $u_i \in V_u$ for $i = 1, \ldots, p$, $u_i \in V_a$ for $i = p + 1, \ldots, p + q$ and $p + q = n$. Then for a given function $f \in F$, the random variable $X = (X_1, \ldots, X_n)$ on V is defined as

$$X_i = \begin{cases} 1 & \text{if node } i \text{ is annotated with } f \\ 0 & \text{if node } i \text{ is not annotated with } f. \end{cases} \qquad (18.4)$$

Then the initial data specified by the annotation $\psi : V_a \to 2^F \setminus \{\emptyset\}$ is combined with the network topology to construct a prior distribution $Pr(X|\theta)$ where θ represents the parameters of the model. Using this prior distribution, an iterative Bayesian approach is used to estimate the posterior probabilities

$$Pr(X_1, \ldots, X_p | X_{p+1} = x_{p+1}, \ldots, X_{p+q} = x_{p+q}). \qquad (18.5)$$

After this is completed for each $f \in F$ and each node $u \in V_u$, we will have computed a probability $Pr(u, f)$ indicating the likelihood that the node u has the function f. In keeping with the general framework outlined in Sect. 18.3, this corresponds to the score function $\rho_u(f)$ and can naturally be used to define a ranking on the functions in F and to generate predictions for the possible functions of an unannotated protein $u \in V_u$.

The authors of [11] tested the performance of their algorithm on PPI network data for *S. cerevisiae* obtained from the MIPS database and used the yeast functional annotation in the Yeast Proteome Database (YPD). We discuss the details of their results and their validation method in the following section.

Algorithms Based on Level-2 Neighbours

The author of [18] suggested that the number of common interacting partners two proteins possess be used as a measure of functional similarity. Building on this idea, the same research group developed the PRODISTIN algorithm for PFP and clustering of interaction networks in [7]. The core idea behind this algorithm is to use the so-called Czekanowski–Dice (CD) distance on a PPI network $G = (V, E)$ as a measure of functional similarity. Formally, for $u \in V$, define the extended neighbourhood of u as $N_e(u) = N(u) \cup \{u\}$. Then for any pair of nodes $u, v \in V$, the CD distance is given by

$$CD(u, v) = \frac{|N_e(u) \triangle N_e(v)|}{|N_e(u) \cup N_e(v)| + |N_e(u) \cap N_e(v)|}. \tag{18.6}$$

The CD distance provides a quantitative measure of how significant an overlap there is between the interacting partners of u and those of v. If $N_e(u) = N_e(v)$, then $CD(u, v) = 0$, while on the other extreme if $N_e(u)$ and $N_e(v)$ are disjoint, then $CD(u, v) = 1$.

In [7], the matrix of CD distances was used to cluster the proteins in the network and proteins belonging to the same cluster are assumed to have similar functions. This allows the functions of annotated proteins within a cluster to be used to predict the functions of unannotated nodes within the same cluster. Using the YPD classification scheme and a validation scheme based on artificially denoting annotated proteins as unannotated, the PRODISTIN algorithm was found to significantly outperform the majority rule.

A more complicated extension of the CD distance, *Functional Similarity Weight* (FS-Weight) was introduced in [9]. Using this concept, the authors of [9] define a likelihood score $\rho_u(f)$ that a given protein u has a function $f \in F$, which fits more naturally into the framework described here. As with the MRF approach, functions for which $\rho_u(f)$ lies above a prespecified threshold are assigned to unannotated proteins u. The authors of [9] compared the performance of their approach to that of the chi-squared, majority, PRODISTIN, and functional flow schemes using the FunCat classification. They found that the FS-Weight approach outperforms all of the other methods. Note, however, that MRF-based approaches and the other optimisation schemes were not included in this comparison. An alternative approach based on level-2 neighbours has also been proposed in [33].

18.6 Validation and Comparison of Methods

Given the variety of different approaches to network-based PFP available, it is important to have a clear understanding of the relative strengths and drawbacks of the various methods proposed. We have already spent some time discussing theoretical issues that arise in connection with the methods described here; however,

18 Inference of Protein Function from the Structure of Interaction Networks 455

the question of how well these methods actually perform in practice is clearly of paramount importance. In this section, we review and summarise the results of several comparative studies published in the literature and illustrate the key ideas behind the metrics used in these comparisons.

The first question to be addressed in comparing the performance of different methods is which metric or metrics to use to perform the comparison. In [11], the "leave one out" validation method and the measures *sensitivity* and *specificity* were introduced and used to compare the performance of their MRF method to the majority and chi-squared schemes. This is one of the earliest systematic comparative studies of its kind and the core methodology of this paper has subsequently been used by other authors [9].

The idea behind "leave one out" is simply to treat each annotated protein as unannotated in turn, run the algorithm and compare the predicted functions to the known functions of the protein. Thus, the validation process only considers the annotated part of the network (the induced subgraph on V_a) and treats one annotated protein u as unannotated. The algorithm is then run on V_a and functional predictions are made for u using either a prediction threshold τ as in (18.1) or by choosing the highest m scoring functions as in (18.2). In this way, predictions are generated for all annotated proteins in the network, which can be compared to their known functions.

A number of standard metrics have been introduced to assess the performance of classifiers and it is natural to ask if these can be adapted to evaluate PFP algorithms. The notions of true-positive rate (tp-rate) and false positive rate (fp-rate) and Receiver Operating Characteristic (ROC) graphs are basic in evaluating classifier performance [13]. These concepts are usually defined for classification problems with two classes – positive and negative. However, if we consider the set of all ordered pairs (v, f) where $v \in V_a$ and $f \in F$, and define a *positive* as a pair for which $f \in \psi(v)$ and a negative as a pair for which $f \notin \psi(v)$, it is possible to introduce definitions of fp-rate and tp-rate for PFP algorithms. As a simple illustration, we include a ROC graph based on these definitions comparing the majority and chi-squared methods in Fig. 18.2. This plot was generated using functional data from the FunCat annotation scheme and using PPI data obtained from the Database of Interacting Proteins [41] in January 2008. For low values of fp-rate, the majority rule outperforms the chi-squared rule, but the results are far from conclusive. Note that the values of fp-rate are extremely low because, in the definition of positives and negatives, the number of negatives will be considerably larger than the number of positives [13]. An alternative definition of true and false positives was given in [27] (which we discuss below) but in general it is not straightforward to apply ROC techniques to PFP algorithms as the number of classes is far greater than two and each protein can be assigned to more than one class.

We next describe some specific metrics introduced for assessing PFP algorithms and discuss the results reported in the literature. In [11] the concepts *specificity*:

$$ SP = \frac{\sum_{u \in V_a} |\hat{\psi}(u) \cap \psi(u)|}{\sum_{u \in V_a} |\hat{\psi}(u)|} \tag{18.7} $$

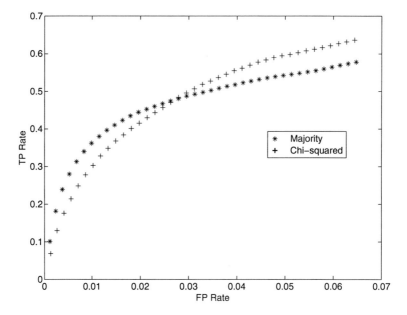

Fig. 18.2 TP-rate fP-rate plot for majority and chi-squared rules

and sensitivity:

$$SN = \frac{\sum_{u \in V_a} |\hat{\psi}(u) \cap \psi(u)|}{\sum_{u \in V_a} |\psi(u)|}, \qquad (18.8)$$

were introduced. Specificity measures the ratio of the total number of correct predictions to the total number of predictions while sensitivity measures the ratio of the total number of correct predictions to the total number of known functions in the initial assignment.

Notions of sensitivity and specificity are standard in the evaluation of classifier performance. To avoid confusion, we briefly point out how the above definitions relate to those in the ROC literature. Sensitivity and specificity are usually defined for classification problems with two classes, positive and negative. The sensitivity of a method is defined to be the ratio of the total number of correctly classified positives to the total number of positives. As above, if we consider the set of all ordered pairs (v, f) where $v \in V_a$ and $f \in F$, and define a positive as a pair for which $f \in \psi(v)$, it is clear that the definition of sensitivity given above corresponds to the usual one in this case. Specificity is usually defined as the ratio of the total number of correctly classified negatives to the total number of negatives. This is slightly different from the definition given above. To make it possible to discuss results from the published literature, we have chosen to work with the definitions that have been adopted in studying PFP.

18 Inference of Protein Function from the Structure of Interaction Networks

When the threshold τ is low or when a large value of m is chosen, the scheme will generate a large number of predictions, and on the average a large number of correct predictions, leading to a higher value of SN but as the denominator of SP will also be large in this case, the specificity is low. This essentially means that many spurious predictions are generated at low thresholds, which is intuitively obvious.

The idea behind the use of these measures in comparing protein prediction algorithms is the following. If two methods are being compared and for any fixed value of specificity method 1 has a higher value of sensitivity than method 2, then method 1 is performing better than method 2. In practice, a plot of sensitivity against specificity is generated over a range of threshold values for the various methods being compared and if the plot of one method lies above that of another, the former method is deemed to be more accurate. To illustrate the idea we include in Fig. 18.3 a simple plot of sensitivity versus specificity for the majority and chi-squared schemes. This plot was generated using functional data from the FunCat annotation scheme and using PPI data obtained from the Database of Interacting Proteins [41] in January 2008. Although the results plotted here cannot be said to be conclusive, they do indicate that the majority rule outperforms the chi-squared rule over a range of specificity values between 0.08 and 0.35. The chi-squared rule gives higher sensitivity values when specificity is lower than 0.08 but at this stage the specificity is arguably too low for any practical purposes anyway.

These results essentially agree with the comparison presented in [11] as to the relative performance of the chi-squared and majority schemes; however, in [11], the YPD classification scheme was used and the PPI data were obtained from the MIPS

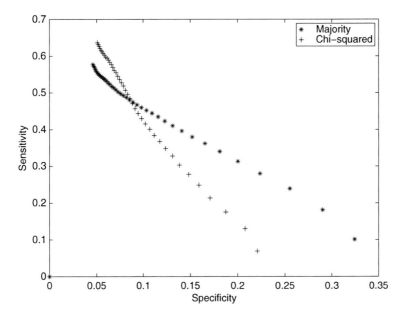

Fig. 18.3 Sensitivity-specificity plot for majority and chi-squared rules

database. The primary interest of that study was to compare the MRF scheme to the majority and chi-squared schemes. For each of the three basic categories in the YPD classification scheme the sensitivity-specificity curve for the MRF algorithm was found to lie above those of both the majority and the chi-squared schemes, indicating that the MRF is the most accurate of these three PFP algorithms. It is worth noting that in the results of [11] the majority rule appears to significantly outperform the chi-squared scheme.

A more comprehensive comparative study was performed in [9]. Here using PPI data from the GRID database [6], the majority, chi-squared, PRODISTIN, functional flow and Functional Similarity Weight (FSW) methods were compared using the MIPS FunCat classification scheme. Moreover, in a separate study in the same paper, all of the aforementioned algorithms and the MRF scheme of Deng et al. [11] were compared using the GO annotations. As in the study in [11], sensitivity-specificity plots were used as a comparative tool; however, in the FunCat scheme, only *informative functional categories* were used. A functional category $f \in F$ is informative if $|\{v \in V_a : f \in \psi(v)\}| \geq 30$ (at least 30 proteins are annotated with f). Based on the FunCat scheme, the results indicate that the PRODISTIN and FSW schemes significantly outperform the others, with FSW a clear winner. Of the other methods included in the study, the majority rule performs slightly better than the rest but the difference is considerably less than that between majority and FSW. There is no significant difference in performance between any of the other methods.

The MRF method was included in the comparison in [9] based on the GO scheme and, with the exception of the FSW approach performed best in each of the three basic categories of biochemical function, cellular role, and subcellular location. No clear difference could be observed between the performances of any of the other methods in any of the three schemes. As pointed out in the recent survey paper [35], given that the FSW method was proposed by the authors of [9], their study can only be viewed as an independent comparison of the methods other than FSW. As such, it provides evidence that the MRF scheme outperforms other competing schemes, with little significant difference between the results for majority, chi-squared, PRODIS-TIN, and functional flow.

It should be noted here that another comparative study was presented in [27] based on a different metric to the specificity-sensitivity plot described above. In this paper, functional flow was compared to Genmulticut – a variation of the optimisation approach described in [38] – the majority rule, and the chi-squared scheme. The PPI data used were obtained from the GRID database [6], while the FunCat annotations were used; however, only categories at the second level in the hierarchy were used in contrast to the study in [9], which employed the finest classification in the FunCat database. In [27], to evaluate the performance of the algorithms, the annotated proteins in V_a were divided into two sets, V_a^1, V_a^2, and the functional annotations of the proteins in V_a^1 were used to predict functions for V_a^2 using the top-ranked functions as in (18.2). A ROC curve of true positives against false positives was then generated as the parameter m was varied. A prediction for a protein was deemed true if more than half of the predicted function are true and false otherwise. It is not easy to rigorously justify this definition of true and false positive

18 Inference of Protein Function from the Structure of Interaction Networks 459

and there are clear problems with it. For instance, a protein which has only been annotated with three functions can never generate a true positive if we choose to predict seven functions per protein, even if the three true functions are the top three ranked functions by the method. This is clearly unsatisfactory, and adapting the standard ROC approach along the lines suggested above appears to be a preferable way to proceed. On a straight comparison, functional flow and majority were seen to perform better than the chi-squared and the Genmulticut methods, with functional flow performing marginally better than the majority method.

18.7 Discussion and Conclusions

The aim of this chapter has been to introduce the biological problem of PFP, provide a starting point into the literature on this topic, and highlight potential areas of research for applied graph theorists to which this problem gives rise. For this reason, we have provided the biological background to the problem, describing the difficulties inherent in giving a consistent, unambiguous meaning to the term "function" and outlining how some of the main databases of functional annotation deal with these issues. We have also provided a formal abstract framework for discussing network-based function prediction and have formulated the major algorithms in the literature within this framework using common notation throughout. In this final section, we briefly discuss some general issues with the problem of PFP itself and with the various algorithms described here.

First of all, it has been noted that there is a definite need for a systematic and uniform evaluation of the performance of the different techniques [9]. Currently, most comparisons in the published literature, such as those discussed above in Sect. 18.6, appear in papers by authors describing and, understandably, supporting their own method and a comprehensive, independent comparison would be very welcome. Also, the PPI data and the functional classification scheme used for evaluation purposes vary from one paper to another making it yet more difficult to make an objective judgement on the relative performance of the algorithms. To further complicate the situation, a set of universally accepted measures that can be used to rate performance is also lacking at the moment, with different measures again being used in different papers. The sensitivity and specificity measures discussed above seem to have attained some level of acceptance but it is far from universal at the time of writing. Furthermore, the measures that have been proposed should be subjected to a more thorough theoretical analysis so as to understand precisely what they have to say about algorithm performance.

One fundamental issue that applies to all of the algorithms discussed here has been previously highlighted in [11]. All current approaches, and the framework outlined in Sect. 18.3 assume that the annotations of the annotated proteins in V_a are complete. This assumption is clearly questionable as many proteins in V_a may possess functions other than those in the list specified by the current annotation ψ.

Although this is a limitation of the methods, it is difficult to see how it can be circumvented, as we can only ever work with the data available at the time.

From a theoretical point of view, two promising areas for research suggest themselves. First, the robustness of function prediction algorithms to data inaccuracies should be studied. Results along the lines of those obtained for some centrality measures used to predict essentiality would be very interesting and useful. Second, as we have previously mentioned in Sect. 18.3, theoretically investigating how the predictions of the various schemes relate to one another is also an important and challenging question.

Acknowledgements This work was partially supported by Science Foundation Ireland (SFI) grant 03/RP1/I382 and the Irish Higher Education Authority (HEA) PRTLI Network Mathematics grant. Neither Science Foundation Ireland nor the Higher Education Authority is responsible for any use of data appearing in this publication.

References

1. Altschul SF et al (1990) Basic local alignment search tool. J Mol Biol 215:403–410
2. Altschul SF et al (1997) Gap-blast and psi-blast: a new generation of protein database search programs. Nucleic Acids Res 25(17):3389–3402
3. Ashburner M et al (2000) Gene ontology: tool for the unification of biology. Nat Genet 25: 25–29
4. Baldi P, Hatfield GW (2002) DNA microarrays and gene expression. Cambridge University Press, Cambridge
5. Barabasi L, Oltvai Z (2004) Network biology: understanding the cell's functional organization. Nat Rev Genet 5:101–113
6. Breitkreutz BJ et al (2003) The GRID: the general repository for interaction datasets. Genome Biol 4:R23
7. Brun C et al (2003) Functional classification of proteins for the prediction of cellular function from a protein–protein interaction network. Genome Biol 5:R6
8. Bu D et al (2003) Topological structure analysis of the protein–protein interaction network in budding yeast. Nucleic Acids Res 31(9):2443–2450
9. Chua H, Sung W, Wong L (2006) Exploiting indirect neighbours and topological weight to predict protein function from protein–protein interactions. Bioinformatics 22(13):1623–1630
10. Costanzo M et al (2001) YPD, PombePD and WormPD: model organism volumes of the bioknowledge library, an integrated resource for protein information. Nucleic Acids Res 29(1):75–79
11. Deng M et al (2003) Prediction of protein function using protein–protein interaction data. J Comput Biol 10(6):947–960
12. Diestel R (2000) Graph theory. Springer, Berlin
13. Fawcett T (2005) An introduction to ROC analysis. Pattern Recogn Lett 27(8):861–874
14. Gavin A et al (2002) Functional organization of the yeast proteome by systematic analysis of protein complexes. Nature 415:141–147
15. Giot L et al (2003) A protein interaction map of *Drosophila melanogaster*. Science 302: 1727–1736
16. Hishigaki H et al (2001) Assessment of prediction accuracy of protein function from protein–protein interaction data. Yeast 18:523–531
17. Ito T et al (2001) A comprehensive two-hybrid analysis to explore the yeast protein interactome. Proc Natl Acad Sci USA 98(8):4569–4574

18 Inference of Protein Function from the Structure of Interaction Networks 461

18. Jacq B (2001) Protein function from the perspective of molecular interactions and genetic networks. Brief Bioinform 2(1):38–50
19. Jeong H, Mason S, Barabasi A, Oltvai Z (2001) Lethality and centrality in protein networks. Nature 411:41–42
20. Karaoz U et al (2004) Whole-genome annotation by using evidence integration in functional-linkage networks. Proc Natl Acad Sci USA 101:2888–2893
21. Karp P et al (2002) The ecoCyc database. Nucleic Acids Res 30(1):56–58
22. Kitano H (2002) Systems biology: a brief overview. Science 295:1662–1664
23. Letovsky S, Kasif S (2003) Predicting protein function from protein/protein interaction data: a probabilistic approach. Bioinformatics 19:i197–i204
24. Li S et al (2004) A map of the interactome network of the metazoan *C. elegans*. Science 303:540–543
25. Mason O, Verwoerd M (2007) Graph theory and networks in biology. IET Syst Biol 1(2): 89–119
26. Mewes H et al (2002) MIPS: a database for genomes and protein sequences. Nucleic Acids Res 30(1):31–34
27. Nabieva E et al (2005) Whole-proteome prediction of protein function via graph-theoretic analysis of interaction maps. Bioinformatics 21:i302–i310
28. Pellegrini M et al (1999) Assigning protein function by comparative genome analysis: protein phylogenetic profiles. Proc Natl Acad Sci USA 96(8):4285–4288
29. Pereira-Leal J, Enright A, Ouzounis C (2004) Detection of functional modules from protein interaction networks. Protein Struct Funct Bioinform 54:49–57
30. Przulj N, Wigle D, Jurisica I (2004) Functional topology in a network of protein interactions. Bioinformatics 20(3):340–348
31. Rain J et al (2001) The protein–protein interaction map of *Heliobacter pylori*. Nature 409: 211–215
32. Ruepp A et al (2004) The FunCat, a functional annotation scheme for systematic classification of proteins from whole genomes. Nucleic Acids Res 32(18):5539–5545
33. Samanta M, Liang S (2003) Predicting protein functions from redundancies in large-scale protein interaction networks. Proc Natl Acad Sci USA 100(22):12579–12583
34. Schwikowski B, Uetz P, Fields S (2000) A network of protein–protein interactions in yeast. Nat Biotechnol 18:1257–1261
35. Sharan R, Ulitsky I, Shamir R (2007) Network-based prediction of protein function. Mol Syst Biol, 3:88
36. Sontag E (2004) Some new directions in control theory inspired by systems biology. IET Syst Biol 1:9–18
37. Twyman RM (2004) Principles of proteomics. Garland Science/BIOS Scientific Publishers (Advanced Text Series), Taylor and Francis, London
38. Vazquez A et al (2003) Global protein function prediction from protein–protein interaction networks. Nat Biotechnol 21(6):697–700
39. Venter C et al (2001) The sequence of the human genome. Science 291:1304–1351
40. Von Mering C et al (2002) Comparative assessment of large-scale data sets of protein–protein interactions. Nature 417:399–403
41. Xenarios I et al (2000) DIP: the database of interacting proteins. Nucleic Acids Res 28(1): 289–291
42. Yu H et al (2004) Genomic analysis of essentiality within protein networks. Trends Genet 20(6):227–231
43. Zhou X, Kao M, Wong W (2002) Transitive functional annotation by shortest-path analysis of gene expression data. Proc Natl Acad Sci USA 99(20):1278312788

Chapter 19
Applications of Perfect Matchings in Chemistry

Damir Vukičević

Abstract Perfect matchings or one factors in mathematics correspond to Kekulé structures in chemistry. In this chapter, we present methods for determination of the existence and enumeration of perfect matchings. The Pfaffian method of enumeration of perfect matchings in planar graphs is presented. The importance of the enumeration of perfect matchings (Kekulé structures) is illustrated with several different chemical applications. A method for coding Kekulé structures which enables efficient storing in the computer is presented. Also, the recently introduced notion of algebraic Kekulé structures is explained and its role in the classification of Kekulé structures according to their significance is discussed. The concept of the resonance graph is presented and its role in the study of fullerene molecules is commented.

Keywords Perfect matching · Kekulé structure · Pfaffian · Enumeration · Resonance graph · Anti-Kekulé number

MSC2000: Primary 05C70; Secondary 05C90, 05C85

19.1 Introduction

Let $G = (V, E)$ be a graph. Matching (or independent set of edges) $M \subseteq E(G)$ in G is a set of edges such that no two edges are adjacent (i.e., incident to the same vertex). Vertex $v \in V(G)$ is covered by M if it is incident to some edge in M. Matching M is a perfect matching if it covers all vertices of G.

An alternative way to define the perfect matching in G is to say that it is a 1-regular subgraph of G. Note that these two definitions are not equivalent, because

---

D. Vukičević (✉)
Faculty of Mathematics and Natural Sciences, University of Split, Nikole Tesle 12,
HR-21000 Split, Croatia
e-mail: vukicevi@pmfst.hr

M. Dehmer (ed.), *Structural Analysis of Complex Networks*,
DOI 10.1007/978-0-8176-4789-6_19, © Springer Science+Business Media, LLC 2011

the first one defines matching as a set of edges and the latter one as a graph. However, these two notions are often incorrectly identified.

Kekulé structure is the term standardly used for perfect matchings in the chemical literature. Throughout this chapter the terms perfect matchings and Kekulé structures are used: perfect matchings in more mathematical contexts and Kekulé structures in more chemical contexts.

In this chapter, several important mathematical methods that give relevant chemical information are illustrated:

1. An algorithm that establishes the existence of the perfect matching (or Kekulé structure in graph) is presented. The existence of Kekulé structure(s) of some chemical compound gives important information about its stability. It is well known that benzenoids that have Kekulé structures (so-called aromatic benzenoids) can be stable, while those that do not have Kekulé structures are not stable.
2. Algorithms for the enumeration of Kekulé structures are presented. It is described how the number of perfect matchings of graphs (and some of their subgraphs) can be used in chemistry. Three applications are described: estimation of resonance energy, estimation of π-electron energy, and estimation of bond lengths.
3. Algebraic Kekulé structures are presented and chemical information incorporated in them is analyzed. Also, the information content of algebraic Kekulé and classic Kekulé structures is compared.
4. Possibilities of coding and storing information about Kekulé structures are presented.
5. Several classification methods of Kekulé structures are analyzed. It is shown that classifications obtained by these methods correlate well with the chemical significance of Kekulé structures.
6. A resonance graph is presented and its application in nanotechnology is commented.

The main goal of this chapter is to present a multidisciplinary study of Kekulé structures. A dominant role is played by three sciences: mathematics, chemistry, and computer science. Models presented here give an insight into the nature of some chemical compounds. The most interesting feature of these models is the possibility to predict the properties of chemical compounds that have never existed nor been synthesized. Hence, this theory gives us a glimpse beyond our real world, a glimpse into the world of almost infinite possibilities.

Here, we present only a very limited scope of mathematical analyses and possible chemical applications of this research. Much other information about chemical structures can be extracted from the enumeration and classification of Kekulé structures. Therefore, Kekulé structures attracted much attention in both the mathematical (see [1] and references within) and chemical literature (see [2] and references within).

Remark 1. Note that loops are not included in any matching, hence they are of no interest for matching theory. Therefore throughout this chapter by the word graph, we imply loopless graph.

19 Applications of Perfect Matchings in Chemistry

19.2 Existence and Enumeration of Perfect Matchings

Let us start by saying a few words about enumeration and existence of perfect matchings. As described in the following sections, existence and number of perfect matchings of a molecular graph (and some of its subgraphs) are used to give very good estimates of much chemically relevant data. Existence of perfect matchings can be determined in polynomial time [3] using the Hungarian method. In order to present this method, we need the concept of the enlarged path. Let M be any matching in graph G. An M-alternated path is a path whose edges alternate in sets M and $E \backslash M$. An M-alternated path is an M-augmented path if its first and last vertex are not covered by M.

Let us describe the Hungarian method algorithm [3,4] for bipartite graphs:

Input: bipartite graph with bipartition (X, Y), $|X| = |Y|$.
Output: perfect matching or a set $S \subseteq X$ such that $|N(S)| < |S|$.

Step 1: Let M be any matching, e.g. $M = \varnothing$.
Step 2: If all vertices in X are covered by M stop (M is perfect matching). Otherwise, let $u \in X$ be a vertex not covered by M. Denote $S = \{u\}$, $T = \varnothing$.
Step 3: If $N(S) = T$, stop ($|N(S)| < |S|$, hence G has no perfect matching). Otherwise, let $y \in N(S) \backslash T$.
Step 4: If y is covered by M, let $yz \in M$. Replace $S \leftrightarrow S \cup \{z\}$ and $T \leftrightarrow T \cup \{y\}$ and return to step 3. If y is not covered by M, let P be an M-augmented (u, y)-path. We replace M by $M' = M \Delta E(P)$ and return to step 2.

This algorithm can be easily modified [4] such that it solves the problem for all graphs. Unfortunately, enumeration of perfect matchings is an NP-hard problem. Simple (but nonpolynomial) recursive algorithms that enumerate all perfect matchings in graphs are based on the following theorems.

Theorem 1. *Let G be a graph and $uv \in E(G)$. Then, the number $\mu(G)$ of matchings of G is*

$$\mu(G) = \mu(G - u - v) + \mu(G - uv).$$

Proof. The first summand is the number of perfect matchings M such that $uv \in M$ and the second summand is the number of perfect matchings M such that $uv \notin M$. $\square$

Theorem 2. *Let G be a graph and $u \in V(G)$. Then,*

$$\mu(G) = \sum_{v \in V(G): uv \in E(G)} \mu(G - u - v).$$

Proof. Trivial. $\square$

These recursive algorithms are very efficient for small graphs, but inefficient for large graphs. Unfortunately, it can be shown that in general there is no efficient method to enumerate perfect matchings. Let B be any 0/1 matrix of type $n \times n$. Let $G(B)$ be a graph with bipartition $(\{u_1, \ldots, u_n\}, \{v_1, \ldots, v_n\})$ such that b_{ij} is the number of edges connecting vertices u_i and v_j. It can be easily seen that the number of perfect matchings in B is equal to the permanent of matrix B. Hence, the problem of calculation of permanents of matrices of type $n \times n$ can be reduced to the problem of finding perfect matchings in a graph with n vertices. From paper [5], it follows that calculation of the permanent is an NP-hard problem. Therefore, it is of interest to provide an algorithm that efficiently enumerates perfect matchings in some special (interesting) classes of graphs. Such an algorithm is constructed by Kasteleyin [6] for simple planar graphs. Here we present this algorithm following the exposition style of [1, 6, 7].

Let G be a graph. For the sake of simplicity, we assume that $V(G) = \{0, \ldots, n-1\}$. Note that:

Lemma 1. *If M and M' are two perfect matchings in G, then $M \cup M'$ is a collection of single edges and even (i.e., even length) cycles.*

Let C be an even cycle in G and $\overrightarrow{G}$ an orientation of G. Cycle C is oddly oriented if there is an odd number of co-oriented edges when the cycle is traversed in either direction. An orientation $\overrightarrow{G}$ of G is Pfaffian if every cycle in $M' \cup M$ is oddly oriented for every pair of perfect matchings M and M'.

The skew adjacency matrix $A_s\left(\overrightarrow{G}\right)$ is defined by

$$a_{ij} = \begin{cases} 1, & (i, j) \in E\left(\overrightarrow{G}\right); \\ -1, & (j, i) \in E\left(\overrightarrow{G}\right); \\ 0, & \text{otherwise.} \end{cases}$$

Denote by $\overleftrightarrow{G}$ the directed graph obtained from G by replacing each edge ij by a pair of antiparallel edges (i, j) and (j, i). An even cycle cover of $\overleftrightarrow{G}$ is a collection SC of even directed cycles C in $\overleftrightarrow{G}$ such that every vertex of G is contained in exactly one cycle in SC.

Let us prove:

Lemma 2. *There is a bijection between (ordered) pairs of perfect matchings in G and even cycle covers in $\overleftrightarrow{G}$.*

Proof. Let (M, M') be a pair of perfect matchings in G. For each edge in $M \cap M'$ take both directed edges in $\overleftrightarrow{G}$. Now, orient each cycle in $M \cup M'$ (with length ≥ 4). in such a way that an edge in M incident to the vertex of the lowest number in C is oriented away from that vertex. The resulting collection of directed cycles is an even cycle cover in G. It can be easily checked that this assignment is bijective. $\square$

19 Applications of Perfect Matchings in Chemistry

Let us prove the Kasteleyn theorem.

Theorem 3. *For any Pfaffian orientation $\overrightarrow{G}$ of G, the number of perfect matchings of G is equal to $\sqrt{\det A_s\left(\overrightarrow{G}\right)}$.*

Proof. From the last lemma, it follows that it is sufficient to show that $\det A_s\left(\overrightarrow{G}\right)$ is equal to the even cycle covers in $\overleftrightarrow{G}$. Note that

$$\det A_s\left(\overrightarrow{G}\right) = \sum_{\pi \in S_n} sgn\left(\pi\right)\prod_{i=0}^{n-1} a_{i,\pi(i)}, \qquad (*)$$

where S_n is the group of all permutations of $\{0,\ldots,n-1\}$ .Consider permutation π and its decomposition to the cycle $\pi = \gamma_1\cdots\gamma_k$ where γ_j acts on some set $V_j \subseteq V$. The corresponding product $\prod_{i\in V_j} a_{i,\pi(i)}$ is nonzero if and only if the edges $\{(i,\pi(i)) : i \in V_j\}$ form a directed cycle.

Let us prove that the sum remains unchanged if we restrict it to permutation with only even length cycles. Suppose that $\pi = \gamma_1\cdots\gamma_k$ is a permutation with γ_j corresponding to the odd cycle. Let $\pi' = \gamma_1\cdots\gamma_{j-1}\left(\gamma_j\right)^{-1}\gamma_{j+1}\cdots\gamma_n$. Then $\prod_{i=0}^{n-1} a_{i,\pi(i)} = -\prod_{i=0}^{n-1} a_{i,\pi'(i)}$ and $sgn(\pi) = sgn\left(\pi'\right)$. Hence, permutations π and π' cancel out. Thus, we may pair up permutations with odd cycles so that they cancel each other.

Now, consider permutation $\pi = \gamma_1\ldots\gamma_k$ with only even cycles. We have $sgn(\pi) = sgn\left(\gamma_1\right)\ldots sgn\left(\gamma_n\right)$. Hence,

$$sgn\left(\pi\right)\prod_{i=0}^{n-1} a_{i,\pi(i)} = \prod_{j=1}^{k}\left[\left(\prod_{i\in V_j} a_{i,\pi(i)}\right)\cdot sgn\left(\gamma_j\right)\right].$$

Since $\prod_{i\in V_j} a_{i,\pi(i)} = -1$ and $sgn(\gamma_j) = -1$, it follows that π contributes 1 to $(*)$. $\square$

Using Euler's formula, it can be proved that:

Lemma 3. *Let $\overrightarrow{G}$ be a connected planar digraph embedded in the plane. Suppose every face except the (outer) infinite face, has an odd number of edges that are oriented clockwise. Then, in any simple cycle C, the number of edges oriented clockwise is of opposite parity to the number of vertices of $\overrightarrow{G}$ inside C. In particular $\overrightarrow{G}$ is Pfaffian.*

Proof. Let C be any cycle in G. Denote by v, k, c, f, and e numbers of vertices inside C, vertices on C, edges on C oriented clockwise, faces inside C, and edges inside C, respectively. Denote by c_i the number of clockwise-oriented edges on boundaries of faces in the interior of C for $i = 0, \ldots, f - 1$. According to Euler's formula

$$\underbrace{(v + k)}_{\text{\# vertices}} + \underbrace{(f + 1)}_{\text{\# faces}} - \underbrace{(e + k)}_{\text{\# edges}} = 2$$

which implies

$$e = v + f - 1.$$

Assumption $c_i \equiv 1 \pmod 2$ implies

$$f \equiv \sum_{i=0}^{f-1} c_i \equiv c + e \equiv c + (v + f - 1) \pmod 2$$

Hence, $c + v$ is odd. $\qquad\square$

It remains to prove that:

Theorem 4. *Every planar graph has a Pfaffian orientation.*

Proof. Let us assume that G is connected since otherwise we may treat each component separately. We prove the claim by induction on the number of edges. If G is a tree, then any orientation is Pfaffian. Now, let G be a graph such that $|E| \geq |V|$. There is an edge e such that its deletion merges two faces in one. By induction hypothesis graph $G \setminus e$ has a Pfaffian orientation. It is sufficient to orient e in such way that one of these two faces has an odd number of edges oriented clockwise (the same will automatically hold for the other one, too). $\qquad\square$

Also, chemists and mathematicians have developed many enumeration methods for enumerating special classes of graphs and for some classes of graphs explicit formulas have been obtained. Benzenoids (or benzenoid graphs) are connected graphs inserted in the plane in such way that all their faces except infinite one are regular congruent hexagons and two faces either have empty intersection or their intersection is a joint edge incident to two joint vertices. We say that two hexagons are adjacent if they share an edge. The degree of a hexagon is the number of its adjacent hexagons. A chain benzenoid is a benzenoid in which every hexagon has degree at most two.

Here, we present (without proof) an algorithm for enumeration of Kekulé structures of chain benzenoids [8, 9]:

1. Start from an arbitrary side of the chain. Write an external numeral one.
2. Enter unity in all hexagons until the first kink.
3. Right after each kink the entry to be written into the next hexagon is the sum of the numerals for the preceding linear segment.
4. The same numeral is the entered into all hexagons until the next kink, and so on.

Fig. 19.1 Benzenoid structure with 79 Kekulé structures

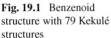
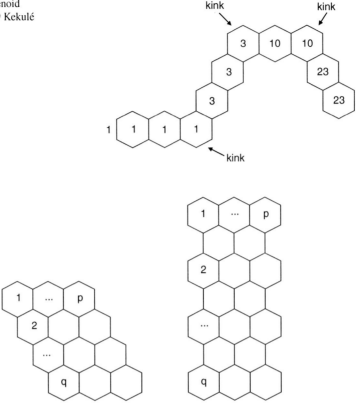

Fig. 19.2 Graphs $L(p,q)$ and $R(p,q)$

Let us illustrate this procedure in Fig. 19.1.

Benzenoid on this figure indeed has $4 \cdot 1 + 3 \cdot 3 + 2 \cdot 10 + 2 \cdot 23 = 79$ Kekulé structures. Also for some classes of benzenoid explicit formulas are known. Two such classes are presented in Fig. 19.2 [8, 10].

It has been shown [8] that $L(p,q)$ has $\binom{p+q}{q}$ Kekulé structures and that $R(p,q)$ has $(p+1)^q$ Kekulé structures. For more information about enumeration the interested reader is referred to [8, 10, 11].

19.3 Applications of the Number of Kekulé Structures in Chemistry

Let G be any graph. Denote by $\mu(G)$ the number of Kekulé structures. Also denote by $\mu_3(G)$ the number of unordered pairs of Kekulé structures M_1 and M_2 such that symmetric difference $M_1 \Delta M_2$ is the cycle of length 6 and by $\mu_5(G)$ the number of unordered pairs of Kekulé structures M_1 and M_2 such that symmetric difference $M_1 \Delta M_2$ is the cycle of length 10. We present several chemically important relations that are solely based on these three numbers.

Resonance energy RE is defined in chemistry as the difference in potential energy between the actual molecular entity and the contributing structure of lowest potential energy. The resonance energy cannot be measured, but only estimated, since contributing structures are not observable molecular entities[1] (for more information on resonance energy, see [12]). It has been shown that resonance energy can be well approximated by the following formula [8]:

$$RE \approx \frac{2}{\mu(G)} (0.841eV \cdot \mu_3(G) + 0.336eV \cdot \mu_5(G)).$$

Another, even simpler formula also gives a good approximation of resonance energy [13]:

$$RE \approx 1.185 \cdot \ln(\mu(G)) \ [eV].$$

Electrons that create Kekulé structures are called π-electrons. Total π-electron energy E can be estimated by [14, 15]:

$$E \approx 0.7578 \cdot \sqrt{2 \cdot |E| \cdot |V|} + 0.1899 \cdot |V| \cdot \mu(G)^{|V|/2}.$$

One of the most interesting results is the classical result of L. Pauling [16] that estimates bond length D_{uv} between two adjacent atoms u and v using the value of $P_{uv} = \mu(G - uv)/\mu(G)$ as input:

$$D_{uv} = D_0 - (D_0 - D_1) \cdot \frac{1.84 \cdot P_{uv}}{0.84 \cdot P_{uv} + 1},$$

where $D_0 = 150.4pm$ and $D_1 = 133.4pm$ (values of D_0 and D_1 are not random; D_0 is the length of the single bond between sp^2 hybridized carbon atoms and D_1 is the length of a double bond).

19.4 Algebraic Kekulé Structures

Let G be a benzenoid graph with at least one Kekulé structure. Let K be a Kekulé structure in G. Denote by SF the set of all congruent hexagonal faces in G. The algebraic Kekulé structure [17] $AKS(K)$ is the function $AKS(K) : SF \to \mathbb{N}_0$ that assigns to each face $F \in SF$ number of edges in K that are on its boundary in such a way that edges that are on the boundary of the infinite face and F are counted twice.

---

[1] http://www.iupac.org/goldbook/R05333.pdf.

Remark 2. The chemical motivation for such a definition is the following: each double bond consists of two π-electrons. Hence, if it belongs to a single hexagon, we may assume that both electrons belong to this hexagon. If it is on the border of two hexagons, then we may assume that each hexagon contains one of these two electrons. In this way distribution of all electrons to faces is described which gives chemically relevant data.

This concept is also defined for fullerenes. Fullerenes are allotropic modifications of carbon. In chemistry, they are defined as closed carbon-cage molecules containing only pentagonal and hexagonal rings [18]. In mathematics fullerene graphs (we call them fullerenes also for the sake of brevity) are defined as planar 3-regular 3-connected graphs in which all faces are pentagons and hexagons. Note that fullerenes don't have an irrelevant outer face. Hence for each Kekulé structure K, $AKS(K)$ assigns to each face the number of edges in K. Algebraic Kekulé structures are also analyzed for nanotubes, phenylenes, and so on, but here we restrict our attention to benzenoids and fullerenes.

Catacondensed benzenoids are benzenoids in which all vertices are on the outer face and pericondensed benzenoids are benzenoids that contain internal vertices (vertices that are not on the outer face). It can be easily seen that

Lemma 4. *Let H be a graph whose vertices are faces of catacondensed benzenoid and two vertices are adjacent if two faces are adjacent. Then, H is a tree.*

It is of interest to see if the algebraic Kekulé structures encode all information incorporated in Kekulé structures, i.e., if assignment $K \leftrightarrow AKS(K)$ is bijective for every benzenoid. It is shown in paper [19] that there is a pericondensed benzenoid in which this assignment is not bijective. Two Kekulé structures K_1 and K_2 such that $AKS(K_1) = AKS(K_2)$ are presented in Fig. 19.3.

However, it has been proved that:

Theorem 5. *Let G be a catacondensed benzenoid with at least two hexagons. Then, Kekulé structures of G are in 1-to-1 correspondence with algebraic Kekulé structures.*

Proof. We prove the claim by the induction on the number of hexagons h of G. If $h = 2$, the claim is obvious. Hence, suppose that graph G has $h > 2$ hexagons and two different Kekulé structures K_1 and K_2 such that $AKS(K_1) = AKS(K_2)$.

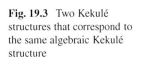

Fig. 19.3 Two Kekulé structures that correspond to the same algebraic Kekulé structure

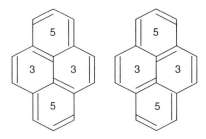

Note that G has at least one pendant hexagon H_1 (a hexagon adjacent to only one other hexagon H_2). Note that $AKS(K_1)(H_1) = AKS(K_2)(H_1)$ can only be 4, 5, or 6 and in each case arrangement of edges in K is fixed. Let G' be a benzenoid that has all faces the same as G, but has no face H_1 and let K_1' and K_2' be restrictions of K_1 and K_2 to G. It can be easily seen that $AKS(K_1') = AKS(K_2')$. From the inductive hypothesis, it follows that $K_1' = K_2'$, but then $K_1 = K_2$ which is a contradiction. $\qquad\square$

19.5 Coding of Kekulé Structures

Every Kekulé structure K of graph $G = (V, E)$ can be represented by the function $f_K : E \to \{0, 1\}$. However, this representation is not particularly efficient. Namely, the number of Kekulé structures is much smaller then $2^{|E|}$. If one is interested in a the efficient storing of Kekulé structures in a computer, then alternative strategies may be of interest. It can be shown that for the family of catacondensed benzenoids, it is possible to reduce the number of bits requireds for the storage of Kekulé structures from $|E|$ to the number of hexagons h [20]. Since, $|E| = 1 + 5h$, this is a significant reduction. Let K be any Kekulé structure and let H_0 be a pendant hexagon in G. Let g_K be the function defined by

$$g_K(H_0) = \begin{cases} 1, & AKS(K)(H_0) > 4 \\ 0, & \text{otherwise}; \end{cases}$$

and by

$$g_K(H) = AKS(K)(H) \bmod 2$$

for every hexagon $H \neq H_0$. It can be shown that [20]:

Theorem 6. *Let G be a catacondensed benzenoid with at least two Kekulé structures; then function g_K uniquely determines the Kekulé structure K.*

Another idea of compacting the information about Kekulé structures is to try to find the smallest set of edges $E' \subseteq E$ such that from restriction of f_K to E, function f_K can be completely reconstructed. More precisely, let $\mathcal{K}$ be the set of all functions f_K that correspond to some Kekulé structure of G and let $\phi : \mathcal{K} \to 2^{E'}$ be a function defined as a restriction to E'. We say that E' is the total forcing set [21] (or global forcing set [22]) if function ϕ is an injective function. The cardinality of the smallest total forcing set is called the total forcing number.

In paper [21] the total forcing number *tfn* of triangular grids has been analyzed and it has been shown that

$$\frac{5}{4}n^2 - \frac{21}{2}n + \frac{41}{4} \le tfn \le \frac{5}{4}n^2 + n - 2.$$

19 Applications of Perfect Matchings in Chemistry   473

Moreover, when n tends to infinity, then ratio $tfn/|E|$ tends to $5/12$. Let us present the results of paper [22]. Square grid R_{pq} is defined as a Cartesian product $P_p \times P_q$ of paths P_p and P_q with p and q vertices, respectively. Recall that Cartesian product $G_1 \times G_2$ of graphs G_1 and G_2 is defined by

$$V (G_1 \times G_2) = \{(u, v) : u \in V (G_1), v \in V (G_2)\} ;$$
$$E (G_1 \times G_2) = \{(uv_1)(uv_2) : u \in V (G_1), v_1 v_2 \in E (G_2)\} \cup$$
$$\{(u_1 v)(u_2 v) : u_1 u_2 \in E (G_1), v \in V (G_2)\} .$$

It is proved in [22] that the total forcing number of rectangular grid R_{pq} is $(p - 1)(q - 1) - \left\lfloor \frac{p-1}{2} \right\rfloor \left\lfloor \frac{q-1}{2} \right\rfloor$. Also total forcing numbers of benzenoids [23] and toroidal polyhexes have been analyzed [24].

19.6 Classification of Kekulé Structures

It is well known in chemistry that all Kekulé structures are not of the same significance. There are five simple purely graph-theoretical measures that can well predict the importance of Kekulé structures in benzenoids and fullerenes. These five measures are [25, 26]:

1. Number of π-electrons belonging to hexagons (applicable only for fullerenes).
2. Number of conjugated hexagons.
3. Degree of freedom.
4. Number of independent conjugated hexagons.
5. Number of independent conjugated cycles.

Let us explain these notions. In chemistry edges contained in the observed Kekulé structures are called double bonds and others are called single bonds. A conjugated cycle is a cycle in which single and double bonds alternate and a conjugated hexagon is a conjugated cycle of length 6. We say that two cycles are independent if they don't have joint vertices. The degree of freedom of a Kekulé structure is the smallest number of edges that completely determine the Kekulé structure. More precisely, the forcing set of Kekulé structure K is the set $K' \subseteq K$ such that K is the only Kekulé structure that contains K'. The degree of freedom is the cardinality of the smallest forcing set. The number of π-electrons that belong to a hexagon in a Kekulé structure K is assumed to be

$$\sum_{H \text{ is hexagon}} AKS (K)(H) .$$

Remark 3. Note that the forcing set identifies a single Kekulé structure while total forcing set identifies all Kekulé structures.

Fig. 19.4 Buckminsterfullerene

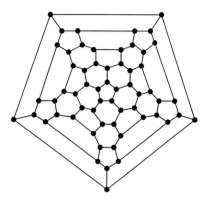

Fig. 19.5 Most significant Kekulé structure

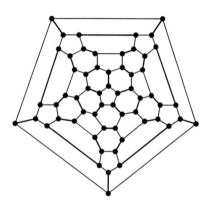

The most famous fullerene is Buckminsterfullerene whose discovery was rewarded by the Nobel prize [27]. It looks like a football. It consists of 60 carbon atoms located in the vertices of the football and bonds on the football represent chemical bonds between atoms. A planar projection of this molecule is presented in Fig. 19.4.

This molecule has a single most significant Kekulé K^* structure that is presented in Fig. 19.5.

Let us explain that this is indeed the most significant Kekulé structure using three of the five proposed measures.

Number of π-electrons belonging to hexagons. Note that each double bond is shared by two hexagons, hence the number of π-electrons belonging to hexagons is equal to 60. Also, note that all single edges are shared by the pentagon and hexagons, hence no other structure can have the same number of π-electrons that belong to hexagons.

Number of conjugated hexagons. All 20 hexagons are conjugated and in order for all of them to be conjugated each of them has to be adjacent to three double bonds. Since there are only 30 double bonds, it follows that each of them has to be shared by two hexagons. Hence, K^* is the only structure with 20 conjugated hexagons.

19 Applications of Perfect Matchings in Chemistry 475

Degree of freedom. It is calculated by computer; details of the algorithm are presented in [28]. The degree of freedom of K^* is 10, while degrees of freedom of all other structures go from 5 to 9.

Let K_1 and K_2 be Kekulé structures of graph G. We say that K_1 and K_2 are isomorphic if there is automorphism $f = (f_V, f_E)$ of G such that $f_E(K_1) = K_2$. In the last section, it was emphasized that many important measures of significance of Kekulé structures are computationally demanding. It can be easily seen that isomorphic Kekulé structures have the same values of these invariants, hence it is sufficient to calculate them for only one representative of each class. Buckminsterfullerene has 120 automorphisms. They reduce 12,500 Kekulé structures to only 158 nonisomorphic Kekulé structures. Hence, one can process just a fraction of the original number of Kekulé structures. In paper [29] all these structures are presented and some of their properties are described.

19.7 Resonance Graph

Let G be a planar graph with at least one Kekulé structure. Resonance graph $R = R(G)$ of G is a graph whose vertices are Kekulé structures and two Kekulé structures K_1 and K_2 are adjacent if $K_1 \Delta K_2$ is a face of G. Especially, if we observe fullerenes or benzenoids, then we define that K_1 and K_2 are adjacent if $K_1 \Delta K_2$ is a hexagon. The resonance graph has been independently introduced by several authors [30–32] and by Zhang et al. under the name Z-transformation [33]. These graphs have been extensively studied in mathematics and chemistry. It has been proved that resonance graphs of catacondensed benzenoids are connected, bipartite, and that they are either a path or have girth four. Also, it has been proved that the resonance graph of catacondensed benzenoid has a Hamilton path [34]. One of the most interesting results is that the resonance graphs of catacondensed benzenoid graphs are medians [35]. Let us explain the concept of the median graphs.

The interval $I(u, v)$ consists of all vertices on the shortest paths between u and v. A median of vertices u, v and w is a vertex that lies in $I(u, v) \cap I(u, w) \cap I(v, w)$. A connected graph is a median graph if every triple of its vertices has a unique median.

This result has been generalized and published in the paper [36]. These results are very important, because median graphs have been extensively studied (see references in [36]). Another interesting consequence of these results is the fact that every resonance graph can be isometrically embedded into a hypercube. These results have been a basis for the creation of software for visualization of the resonance graph of the benzenoid graph. This software can be downloaded from the Web page.[2]

---

[2] http://www-mat.pfmb.uni-mb.si/personal/vesel/visual/visualHBG.html.

Fig. 19.6 Fullerene C_{70}

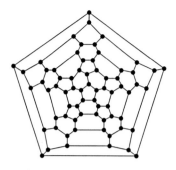

As we have mentioned, the resonance graph of every catacondensed benzenoid is connected. However, this is not the case for arbitrary graphs. Resonance graphs of fullerenes are usually not connected, but have one large component. This large component is very important in chemistry, because almost all most significant Kekulé structures are contained within this component. For instance Buckminsterfullerene has 5,828 out of 12,500 Kekulé structures in this component [28] and 5,828 structures (less then 50%) are responsible for more than 99.82% of the energy of the whole space of Kekulé structures [37].

Let us in more detail present analyses of the fullerene C_{70} (the second most stable fullerene). Its graph is presented in Fig. 19.6.

Denote by R the resonance graph of C_{70}. Let $K \in V$ the (R) be any Kekulé structure. Denote by $[K]$ the set of all Kekulé structures isomorphic to x and denote $QS = \{[K] : K \in \mathcal{K}(C_{70})\}$. Denote by RI the graph with the set of vertices $V(RI) = QS$ and set of edges

$$E(RI) = \{[K_1][K_2] : (\exists K_1' \in [K_1])\,(\exists K_2' \in [K_2])\,(K_1 K_2 \in E(R))\}.$$

By K' and K'' we denote respectively two Kekulé structures given in Fig. 19.7.

It has been shown in paper [38] that these are the only two structures that have 20 conjugate hexagons. Also, they have a maximal number of independent conjugated hexagons, independent conjugated cycles, and degree of freedom. However, there are other structures that have the same values of these three measures, but a smaller value of the number of conjugated hexagons. Therefore, we may conclude that these are two of the most important Kekulé structures. One can easily see that the distance of these the two Kekulé structures in graph R is $d_R(K', K'') = 5$. Hence, $K_1 \sim_R K_2$; i.e., they are in the same component in R. Our aim is to identify the structures in the component that contains K' and K''. Graph R has 52,168 Kekulé structures which is quite, a substantial number, but only 2,780 nonisomorphic ones. Hence, it is much faster to detect the main component in RI (i.e., the component that contains $[K'] = [K'']$) than in R. The following theorem [38] allows us just to detect the main component in RI.

Theorem 7. *Let* $K \in K(C_{70})$. *Then,* $K \sim_R K'$ *if and only if* $[K] \sim_R [K']$.

19 Applications of Perfect Matchings in Chemistry   477

Fig. 19.7 Kekulé structures K' and K''

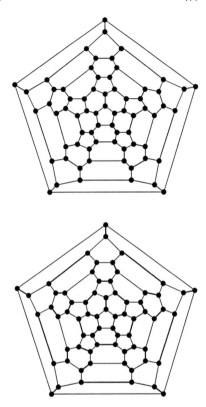

Proof. It is easy to prove that $K \sim_R K'$ implies $[K] \sim_R [K']$. Let us prove by induction on $d_{RI}([K],[K'])$ that the opposite implication holds. Assume that $d_{RI}([K],[K']) = 0$. Then, $K = K'$ or $K = K''$ and in both cases $K \sim_R K'$. Now, suppose that $d = d_{RI}([K],[K']) > 0$ and that claim holds for all Kekulé structures on the smaller distance. Let $K^* \in V(R)$ be such that $d([K'],[K^*]) = d-1$ and that $[K^*][K] \in E(RI)$. From $[K^*][K] \in E(RI)$, it follows that there are $K_0^* \in [K^*]$ and $K_0 \in [K]$ such that $K_0^* K_0 \in E(R)$. Let ϕ be an automorphism such that $\phi(K_0) = K$. Note that $\phi(K_0^*)\phi(K_0) \in E(R)$, i.e., that $\phi(K_0^*)K \in E(R)$. Since $\phi(K_0^*) \in [K^*]$, it follows that $K' \sim_R \phi(K_0^*)$. This implies that $K \sim_R K'$. □

Using the computer [38] it has been found that there are 932 out of 2,780 vertices in the main component of RI. Then, it was discovered that 17,454 out of 52,168 vertices are in the main component of R. Denote by $ch_m(n)$, $ich_m(n)$, $icc_m(n)$, and $df_m(n)$ the numbers of Kekulé structures in the main component that have n conjugated hexagons, maximally n independent conjugated hexagons, maximally n, independent conjugated cycles, and degree of freedom equal to n, respectively. Denote analogously by $ch_o(n)$, $ich_o(n)$, $icc_o(n)$, and $df_o(n)$, numbers of Kekulé structures that are not in the main component.

Tables 19.1–19.4 show that the term *main component* is indeed justified.

It can be readily seen that almost all very significant Kekulé structures are within the main component, while almost all not very significant Kekulé structures are outside the main component.

19.8 Anti-Kekulé Number

Let G be a connected graph with at least one Kekulé structure. Anti-Kekulé set $E' \subseteq E$ is the set such that $G - E'$ is connected, but has no Kekulé structures. Anti-Kekulé number $akn(G)$ of graph G is the smallest cardinality of an anti-Kekulé set. If there is no such set, then $akn(G) = \infty$. An anti-Kekulé set can be observed as the measure of the aromaticity of the graph. Anti-Kekulé numbers of fullerenes C_{20} [39] and C_{60} and C_{70} have been determined.

A very important class of fullerenes are leapfrog fullerenes. Leapfrog fullerenes are fullerenes obtained from other smaller fullerenes by leapfrog transformation. Let us explain the concept of leapfrog transformation [18, 40].

To define the leapfrog transformation of a fullerene we have to introduce stellation and dualization. Stellation, St, of a face is achieved by adding a new vertex in its center followed by connecting it with each boundary vertex. It is also called a

Table 19.1 Distribution of conjugated hexagons among Kekulé structures

n	$ch_m(n)$	$ch_o(n)$	$100 \cdot ch_m(n) / (ch_m(n) + ch_o(n))$
0	0	205	0.00
1	0	340	0.00
2	0	1,730	0.00
3	0	2,520	0.00
4	0	3,270	0.00
5	0	5, 400	0.00
6	50	6, 235	0.80
7	700	6,270	10.04
8	2,240	4,910	31.33
9	3,300	2,190	60.11
10	3,532	1,294	73.19
11	3,050	170	94.72
12	2,050	170	92.34
13	1,370	0	100.00
14	600	10	98.36
15	350	0	100.00
16	150	0	100.00
17	30	0	100.00
18	30	0	100.00
19	0	0	not defined
20	2	0	100.00

19 Applications of Perfect Matchings in Chemistry

Table 19.2 Distribution of maximal number of independent conjugated hexagons among Kekulé structures

n	$ich_m(n)$	$ich_o(n)$	$ich_m(n) / (ich_m(n) + ich_o(n))$
0	0	205	0.00
1	0	500	0.00
2	0	3,675	0.00
3	0	3,250	0.00
4	0	5,951	0.00
5	0	14,893	0.00
6	80	6,240	1.27
7	4,702	0	100.00
8	8,800	0	100.00
9	3,872	0	100.00

Table 19.3 Distribution of maximal number of independent conjugated cycles among Kekulé structures

n	$icc_m(n)$	$icc_o(n)$	$100 \cdot icc_m(n) / (icc_m(n) + icc_o(n))$
4	0	1,645	0.00
5	0	8,675	0.00
6	80	12,661	0.63
7	4,702	11,733	28.61
8	8,800	0	100.00
9	3,872	0	100.00

Table 19.4 Distribution of degrees of freedom among Kekulé structures

n	$df_m(n)$	$df_o(n)$	$100 \cdot df_m(n) / (df_m(n) + df_o(n))$
5	0	2,080	0.00
6	70	13,540	0.51
7	3,020	15,390	16.40
8	8,750	3,652	70.55
9	4,902	52	98.95
10	680	0	100.00
11	32	0	100.00

capping operation or triangulation. When all the faces of a graph are thus operated on, it is referred to as an omnicapping operation and the resulting graph is denoted by $St(G)$. Dualization, Du, of a graph is built as follows. Locate a point in the center of each face. Join two such points if their corresponding faces have a common edge. The new edge is called the edge dual. Leapfrog, Le, is a composite operation that can be written as: $Le(G) = Du(St(G))$.

In paper [18], it has been proved that the anti-Kekulé number of all fullerenes is either 3 or 4 and that for each leapfrog fullerene the anti-Kekulé number can be established by observing a finite number of cases not depending on the size of the fullerene.

19.9 Conclusion

In this chapter, several mathematical methods that provide chemically relevant data have been illustrated. There are many more methods [1] and chemical applications of perfect matchings [2]. There are whole theories about perfect matchings in mathematics, chemistry, and computer science. Also, there are a lot of problems that are still open. Probably, the hardest and the most interesting one is:

Provide an algorithm for efficient enumeration of the Kekulé structure of any graph or prove that such an algorithm does not exist.

This problem is directly related to one of the most important problems of theoretical computer science, namely: *is it true that $P = NP$?*

A simpler problem is to analyze different classes of chemically relevant nonplanar compounds (i.e., some classes on nanotubes with specified types of junctions) and to try to provide an efficient method for enumeration of Kekulé structures for these types of chemical compounds.

Classification of Kekulé structures is also based on some nonpolynomial algorithms (i.e., an algorithm for finding the degree of freedom). Hence, it is of interest to try to find as efficient algorithms as possible for some interesting classes of these structures.

Kekulé structures probably hide some still unknown information about corresponding chemical compounds. Hence, it is of interest to further study these structures.

In conclusion, we may say that Kekulé structures play an important role in chemistry and that this fact provides a series of interesting issues in mathematics. Many relevant problems are solved and many remain open. Many theoretical results have been obtained and a lot of them are open challenges.

Acknowledgments Partial support of the Ministry of Science, Education and Sports of the Republic of Croatia is gratefully acknowledged (grant no. 177–0000000-0884 and grant no. 037-0000000-2779).

References

1. Lovász L, Plummer MD (1986) Matching theory. North Holland, Amsterdam
2. Randić M (2003) Aromaticity of polycyclic conjugated hexagons. Chem Rev 103:3449–3605
3. Kuhn HW (1955) The Hungarian method for the assignment problem. Nav Res Logist Q 2:83–97
4. Veljan D (2001) Combinatorial and discrete mathematics, algoritam, zagreb (in Croatian)
5. Valiant L (1979) The complexity of computing the permanent. Theor Comput Sci 8:189–201
6. Kasteleyn PW (1967) Chapter 2. In: Harary F (ed) Graph theory and theoretical physics. Academic, New York
7. Jerrum M. Lecture Notes from a Recent Nachdiplomvorlesung at ETH-Zürich "Counting, sampling and integrating: algorithms and complexity" (draft, under construction). http://www.dcs.ed.ac.uk/home/mrj/pubs.html.
8. Gutman I, Cyvin SJ (1999) Introduction to the theory of benzenoid hydrocarbons. Springer, Berlin

19 Applications of Perfect Matchings in Chemistry

9. Cyvin SJ, Gutman I (1986) Topological properties of benzenoid systems. Part XXXVI. Algorithm for the number of Kekulé structures in some pericondensed benzenoids. MATCH Commun Math Comput Chem 19:229–242
10. Klein DJ, Babić D, Trinajstić N (2002) Enumeration in chemistry. Chem Model Appl Theory 2:56–95
11. Cyvin SJ, Gutman I (1988) Kekulé Structures in benzenoid hydrocarbons. Springer, Berlin
12. Morrison R, Boyd R (1992) Organic chemistry. Prentice-Hall, Englewood Cliffs, NJ
13. Swinborne-Sheldrakem R, Herndon WC, Gutman I (1975) Kekulé structures and resonance energies of benzenoid hydrocarbons. Tetrahedron Lett 16:755–758
14. Cioslowski J (1986) The generalized McClelland formula. MATCH Commun Math Comput Chem 20:95–101
15. Gutman I, Markovic S, Marinkovic M (1987) Investigation of the Cioslowski formula. MATCH Commun Math Comput Chem 22:277–284
16. Pauling L (1960) The nature of the chemical bond and the structure of molecules and crystals: an introduction to modern structural chemistry. Cornell University Press, Ithaca, NY
17. Randic M (2004) Algebraic Kekulé formulas for benzenoid hydrocarbons. J Chem Inform Comput Sci 44:365–372
18. Fowler PW, Manolopoulos DE (1995) An atlas of fullerenes. Clarendon Press, Oxford
19. Gutman I, Vukičević D, Graovac A, Randić M (2004) Algebraic kekulé structures of benzenoid hydrocarbons. J Chem Inform Comput Sci 44:296–299
20. Vukičević D, Žigert P (2008) Binary coding of algebraic kekulé structures of catacondensed benzenoid graphs. Appl Math Lett 21(7):712–716
21. Vukičević D, Sedlar J (2004) Total forcing number of the triangular grid, Math Commun 9:169–179
22. Vukičević D, Došlić T (2007) Global forcing number of grid graphs. Australas J Combinator 38:47–62
23. Došlić T (2007) Global forcing number of benzenoid graphs. J Math Chem 41:217–229
24. Wang H, Ye D, Zhang H, Wang H (2008) The forcing number of toroidal polyhexes. J Math Chem 43:457–475
25. Vukičević D, Randić M (2005) On kekulé structures of buckminsterfullerene. Chem Phys Lett 401:446–450
26. Vukičević D, Gutman I, Randić M (2006) On instability of fullerene C72. Croat Chem Acta 79:429–436
27. Kroto HW, Heath JR, O'Brien SC, Curl RF, Smalley RE (1985) C60: Buckminsterfullerene. Nature 318:162–163
28. Randić M, Kroto H, Vukičević D. Kekulé structures of buckminsterfullerene, Adv Quantum Chem (submitted)
29. Vukičević D, Kroto HW, Randić M (2005) Atlas of Kekulé valence structures of buckminster-fullerene. Croat Chem Acta 78:223–234
30. El-Basil S (1993) Kekulé structures as graph generators. J Math Chem 14:305–318
31. Gründler W (1982) Signinkante elektronenstrukturen fur benzenoide kohlenwasserstoffe. Wiss Z Univ Halle 31:97–116
32. Randić M (1997) Resonance in catacondensed benzenoid hydrocarbons. Int J Quantum Chem 63:585–600
33. Zhang F, Guo X, Chen R (1988) Z-transformation graphs of perfect matchings of hexagonal systems. Discrete Math 72:405–415
34. Chen R, Zhang F (1997) Hamilton paths in Z-transformation graphs of perfect matchings of hexagonal systems. Discrete Appl Math 74:191–196
35. Klavžar S, Žigert P. Resonance graphs of catacondensed benzenoid graphs are median, Manuscript
36. Klavžar S, Žigert P, Brinkmann G (2002) Resonance graphs of catacondensed even ring systems and medians. Discrete Math 253:35–43
37. Flocke N, Schmalz TG, Klein DJ (1998) Variational resonance valence bond study on the ground state of C_{60} using the Heisenberg model. J Chem Phys 109:873–880

482   D. Vukičević

38. Randić M, Vukičević D (2006) Kekulé structures of Fullerene C70. Croat Chem Acta 79:471–481
39. Vukičević D. Total forcing number and Anti-forcing number of C_{20}, preprint
40. Kutnar K, Sedlar J, Vukičević D (2009) On the Anti-Kekulé number of Leapfrog Fullerenes. J Math Chem 45:406–416

Index

$G - X$, 320
$G - e$, 320
$G - v$, 320
$\mathcal{D}(G)$, 320
$\exists \mathrm{rn}(G)$, 322
$\forall \mathrm{rn}(G)$, 321
$\mathrm{sim}(G, H)$, 320
k-connected, 221
n-cube, 56
r-regular graphs, 327
$\mathcal{G}(n, \frac{1}{2})$, 323

a.e., 322
active edge
    externally, 225, 247
    internally, 225, 247
adversary reconstruction number, 321
alien, 337, 340, 352
ally reconstruction number, 322
almost every, 322
anticircuit, 231
automorphism group, 322
avalanche, 237
average distance, 58

bad coloring polynomial, 234
    symmetric function extension, 276
β invariant, 242
boundary components, 268
bouquet graph, 269
bridge, 221

caterpillar, 329
cellular embedding, 267

centre, 50
centroid, 55
characteristic polynomial, 258
characteristic polynomials, 321
chord diagram, 269
    signed, 269
chordal graph, 55
chromatic distance, 352
chromatic metric, 352
chromatic polynomial, 275
chromatic polynomials, 321
chromatically alien, 337, 340
chromatically related, 337, 340
circle graph, 274
circuit, 221
circuit partition polynomial, 270, 282, 283
clique number, 348
cocycle, 221
λ-coloring, 232
complement of class, 338
complex
    face, 240
    facet, 240
    pure, 240
    simplicial, 240
configuration
    critical, 237
    ice, 236
    level, 237
    of a system, 237
    stable, 237
    weight, 237
convex hull, 64
convexity, 64
convolution, 248
cover polynomial, 287
critical configuration polynomial, 238
cut, 221
    -edge, *see* bridge

483

484 Index

cycle, 221
cycle-complete graph, 51
cyclic graph polynomial, 267
cyclomatic number, 238

deck, 320
degree sequence, 321, 328
diameter, 50
Dijkstra's algorithm, 67
dissipation, 237
distance, 50
distance between graphs, 320
distance hereditary, 57
distant relatedness, 349

eccentricity, 50
edge
    ordinary, 221
edge-deleted subgraph, 320
edge-difference polynomial, 258
Eulerian digraph, 270
Eulerian graph, 280
Eulerian graph state, 270
existential reconstruction number, 322

F-polynomial, 287
f-vector, 240
face enumerator, 240
fat graph, 267
flow
    H-flow, 235
    nowhere zero, 235
Floyd–Warshall algorithm, 67
forest, 221
four-thirds conjecture, 61

generalized transition polynomial, 281
geodesic, 50
GI, 320, 331
girth-endvertex property, 343
GRAFFITI, 60
graph
    bicycle space, 229
    block of, 246
    bouquet, 269
    card, 262
    contraction, 221
    cycle space, 229
    deck, 262

deletion, 221
dual, 222
dual-chordal, 244
Eulerian, 270, 280
Eulerian digraph, 270
induced, 221
invariant, 221
medial, 229
minor, 221
one-point join, 224
orientation, 230
planar, 222
plane, 222
reconstructible, 262
series–parallel, 243
spanning, 221
topological, 267
weighted, 276
graph bipartite, 341
graph entropy, 105 ff
Graph Isomorphism Problem, 320, 331
graph polynomial, 221
graph state, 259, 280

h-vector, 241
Hamiltonian, 285
Hamming graph, 56

idiosyncratic polynomial, 258
induced subgraph, 320
inexact graph matching, 320
interlace polynomial, 272, 283
isomorphic, 220
isotropic system, 272

Kauffman bracket, 282

Laplacian matrix, 228
linear subgraph, 261
local complementations, 273
loop, 221

Martin polynomial, 229, 282
i-matching, 262
matching generating polynomial, 263, 276
matching polynomial, 262, 276
matrix
    totally unimodular, 266

Index 485

median, 54
metric dimension, 66

network motifs, 332
nullity, 222

orientation
    acyclic, 230
    totally cyclic, 230
oriented graph, 63

pair weight, 281
parametrized Tutte polynomial, 278
path tree, 264
path-complete graph, 52
Penrose polynomial, 259, 282
pivot operation, 272
polychromate, 277
polynomial
    acyclic, 263
    bad coloring, 234
    Bollobás–Riordan, 267
    characteristic, 258, 260
    chromatic, 275
    circuit partition, 270, 282, 283
    cover, 287
    critical configuration, 238
    cyclic graph, 267
    edge-difference, 258
    F, 287
    generalized transition, 281
    idiosyncratic, 258
    interlace, 283
    Kauffman bracket, 282
    Martin, 229, 282
    matching, 260, 262
    matching defect, 263
    matching generating, 263
    parametrized Tutte, 278
    Penrose, 259, 282
    quasi-, 265
    reference, 263
    reliability, 238
    ribbon graph, 267
    rook, 263
    symmetric bad coloring, 276
    symmetric chromatic, 275
    topological Tutte, 267, 268
    transition, 281, 282

Tutte, 274, 276, 280, 285, 286
Tutte-Martin, 272
U-, 276
vertex-nullity, 272
W-, 276
polytope
    convex, 265
Potts model partition function, 285
property A_k, 323

radius, 50
rank, 222
RC, 321
Reconstruction Conjecture, 321
reconstruction number, 322
relatedness, 337, 352
reliability polynomial, 238
resolving set, 66
ribbon graph, 267
ribbon graph polynomial, 267
rotation scheme, 267

score vector, 230
shelling, 241
shelling polynomial, 241
signed chord diagram, 269
similarity between two graphs, 320
skein relation, 281
state model, 259
state weight, 281
status, 54
Steiner distance, 68
Subgraph Isomorphism Problem, 331
subgraph similarity, 320
sunshine graph, 329
symmetric chromatic polynomial, 275
systems biology, 332

topological graphs, 267
topological Tutte polynomial, 268
tournament, 64
trail, 221
transcription networks, 332
transition polynomial, 281, 282
tree, 221
triangle inequality, 50
Tutte polynomial, 274, 276, 280, 285, 286
    uniqueness property, 227
    universal property, 226

486 Index

Tutte-Gröthendieck invariant, 226
Tutte-Martin polynomial, 272

U-polynomial, 276
unicyclic graphs, 329
universal reconstruction number, 321

vertex state, 280
vertex state weight, 281

vertex-deleted subgraph, 320
vertex-nullity interlace polynomial, 272

W-polynomial, 276
weight system, 281

zonotope, 266
    unimodular, 266

Breinigsville, PA USA
28 October 2010

248220BV00006B/87/P